Crop Production: Methods and Technologies

Crop Production: Methods and Technologies

Editor: Stanley Hill

Callisto Reference

www.callistoreference.com

Callisto Reference,
118-35 Queens Blvd., Suite 400,
Forest Hills, NY 11375, USA

Visit us on the World Wide Web at:
www.callistoreference.com

ISBN: 978-1-64116-525-9 (Hardback)

Cataloging-in-Publication Data

Crop production : methods and technologies / edited by Stanley Hill.
p. cm.
Includes bibliographical references and index.
ISBN 978-1-64116-525-9
1. Crops. 2. Crop science. 3. Agricultural innovations. 4. Agriculture. I. Hill, Stanley.
SB91 .C76 2021
630--dc23

Table of Contents

Preface

Crops are plants that are grown and harvested for food or fodder. Non-food crops are also cultivated in floriculture or horticulture, or for industrial purposes. Crop production systems vary from farm to farm, and from region to region, depending on the available resources and limiting constraints of the climate and geography, government policies, and socioeconomic factors. Common crop production methods include monoculture, multiple cropping and intercropping. A particular cropping system may not be suitable to a particular climate, such as multiple cropping would require extensive irrigation in arid and subtropical regions, while annual farming can be productive in areas rich in grasslands or prairies. Crop yields can be increased using various practices such as tillage, nutrient management and water management. This book is a compilation of chapters that discuss the most vital concepts and emerging trends in crop production. The various advancements in crop production methods and technologies are glanced at and their applications as well as ramifications are looked at in detail. This book is meant for students who are looking for an elaborate reference text on modern crop production.

The researches compiled throughout the book are authentic and of high quality, combining several disciplines and from very diverse regions from around the world. Drawing on the contributions of many researchers from diverse countries, the book's objective is to provide the readers with the latest achievements in the area of research. This book will surely be a source of knowledge to all interested and researching the field.

In the end, I would like to express my deep sense of gratitude to all the authors for meeting the set deadlines in completing and submitting their research chapters. I would also like to thank the publisher for the support offered to us throughout the course of the book. Finally, I extend my sincere thanks to my family for being a constant source of inspiration and encouragement.

Editor

1

Culture systems in the production and quality of strawberry cultivars

Ana Paula Cecatto*, Eunice Oliveira Calvete, Alexandre Augusto Nienow, Rosiani Castoldi da Costa, Heloísa Ferro Constâncio Mendonça and Aislam Celso Pazzinato

*Faculdade de Agronomia e Medicina Veterinária, Universidade de Passo Fundo, Cx. Postal 611, 99001-070, Passo Fundo, Rio Grande do Sul, Brazil. *Author for correspondence. E-mail: ana_cecatto@yahoo.com.br*

ABSTRACT. Until recently, strawberry cultivation was exclusively performed in soil using conventional cultivation methods, which resulted in many environmental and phytosanitary problems. Currently, soilless culture is the production method advocated for environmental reasons because it greatly reduces the use of chemical pesticide and fungicide inputs. This study assessed the yield and quality of fruits from strawberry cultivars grown in two systems (soil and substrate) in a greenhouse. The experiment was performed from September 2010 to January 2011 in the University of Passo Fundo, Rio Grande do Sul State. Treatments (cultivars x culture systems) were arranged in a random block design with a 7 x 2 factorial arrangement. Evaluation included number, total and commercial fresh weight of fruits per plant, transversal diameter, total titratable acidity (TTA), total soluble solids (TSS), TSS/TTA ratio and the pH of fruits. The cultivars Camarosa, Florida Festival and Portola excelled in relation to yield when grown in the soil system. In substrate culture, all cultivars had similar performance. The yield was higher in soil culture, while the quality was higher in the substrate system.

Keywords: *Fragaria X ananassa* Duch., soil culture, substrate culture.

Introduction

Strawberry culture is highly important for economic and social reasons, mainly for aggregating family labor, as cultivation on small rural properties is predominant (PONCE et al., 2010). Properties in the states of Minas Gerais, Rio Grande do Sul and São Paulo enjoy nationwide prominence, where nearly all of the production is derived from soil culture (RADIN et al., 2011).

One of the problems related to conventional strawberry cultivation is the excessive use of pesticides. In the predominant strawberry cultivation system, high loads of pesticides are employed, making it among the four crops with the highest rates of pesticide contamination. Several solutions have been proposed for the minimization of these problems, among which greenhouse cultivation stands out because in addition to protecting the crop from wind, hail, rain, frost and low temperatures, it also reduces the prevalence of pests and diseases, providing better conditions for plant development and improving total fructification and commercial production (ANTUNES et al., 2007; CALVETE et al., 2008). However, strawberry culture in the greenhouse environment, which uses primarily soil, has productivity issues due to the susceptibility of this crop to damage by soil fungi and bacterioses. The phytosanitary limitations of soil motivated the development and use of soilless culture techniques, including hydroponics.

Hydroponic production has been used for thousands of years where suitable soil is unavailable for growth or where the soil is contaminated (JAFARNIA et al., 2010). Until 2004, this type of system occupied an area of approximately 1,140 ha in Europe (LIETEN et al., 2004), but these cultivation systems are expanding around the world.

Soilless strawberry culture using greenhouses extends the crop period, allows out of season production and increases the yield (JAFARNIA et al., 2010). This system has many other advantages as it enables the removal of disinfection products as well as reduces pesticide use, decreases the consumption of contaminated fruits and decreases environmental harm while increasing fruit quality and allowing greater ability to manage cultivation (CALVETE et al., 2007).

In the last few decades, agronomic research has prioritized the attainment of high productivity, improved resistance to diseases and pests, transportability and extended shelf life of the fruit. Moreover, programs are improving the types and quality of fruits (in term of sensory and nutritional aspects) and the production systems (CAPOCASA et al., 2008).

This study provides an evaluation of the yield and quality of fruits from different strawberry cultivars produced in a greenhouse using soil and substrate media.

Material and methods

The experiment was performed from May 12, 2010 to January 19, 2011 in the greenhouse of the Horticulture Sector of the School of Agronomy and Veterinary Science of the University of Passo Fundo, Rio Grande do Sul State, Brazil, situated at 28°15'41" S, 52°24'45" W with an altitude of 709 m.

The research was conducted in a 3.5 m high agricultural greenhouse with a semicircular roof and 510 m^2 of covered area (51 m long x 10 m wide), oriented in a northeast-southeast direction. The structure was constructed of galvanized steel and covered with a 150 micron thick low-density polyethylene (LDPE) diffused film with an anti-ultraviolet additive. A wire screen was attached to the greenhouse sides, along with mobile curtains, to prevent the entry of persons and birds while allowing access for pollinating insects.

The treatments consisted of 7 strawberry cultivars and 2 growing systems (soil and substrate media). The cultivars assessed were as follows: Camarosa, Florida Festival, Camino Real, San Andreas, Monterey, Portola and Ventana. The treatments were arranged in a randomized block design (RBD) using a 7 x 2 factorial arrangement with 3 repetitions and 10 plants per parcel; six plants per parcel was considered serviceable.

Seedlings were transplanted during the end of May 2010. In the substrate system, seedlings planted in a 30 x 30 cm spacing arrangement were grown in 150 micron white LDPE (with an anti-UV additive) filled with Tecnomax® commercial substrate; two plantation lines were used per 1.5 m x 0.79 m bag. In the soil system, seedlings were transplanted into 0.8 x 16 m beds covered with 30-micron-thick black LDPE mulch in the same plant arrangement as the substrate system.

Drip irrigation was used in both systems. In the substrate system, a single hose with 30 cm drippers (securely attached and located inside each bag) was installed using the same spacing between the plants. In the soil system, two hoses were placed beside each cultivation line, with drippers (spaced every 30 cm) positioned under the mulching. Fertigation in both cases was performed according to the formula described by Calvete et al. (2007).

Fruit was gathered monthly from September 2010 to January 2011. The maturation point used was the commercial standard (i.e., with 75% of the skin presenting a red color).

Post-harvest evaluations assessed the following parameters: yield determined by the total number and fresh weight of both fruits per plant and commercial fruits per plant, diameter of fruits, total titratable acidity (TTA), total soluble solids (TSS), TSS/TTA ratio, pH and external color. Fruits without injuries, diseases and deformities weighing more than 6 g were considered commercial (ANTUNES et al., 2007). The transversal diameter of fruits was determined using a digital pachymeter, and the result was expressed in millimeters (mm). Total titratable acidity (TTA), expressed in citric acid percentage, was determined by titrating 10 mL of juice in 100 mL distilled water with 0.1 N NaOH solution to pH 8.1. Brix degree (°Brix), as a measure of total soluble solids (TSS), was determined through refractometry. The TSS/TTA ratio was obtained through the coefficient between the two constituents; pH was obtained through juice analysis using a potentiometer (peagameter). External color was determined by taking a diffuse reflectance spectrophotometer reading (Hunter Lab) using a sphere geometry optical sensor; data correspond to the L^*, a^* and b^* analytical components (as described by Ferreira, 1981), where L^* values (luminosity) vary from light to dark, with 100 corresponding to white and 0 (zero) corresponding to black. These values were then used to calculate degrees of hue angle (H° = arctan [b★/a★]) and chroma (C★ = [a★2 + b★2]½). H° is an angle in the color wheel where 0° = red-purple, 90° = yellow, 180° = green-blue and 270° = blue. Chroma indicates intensity or saturation of color. All the evaluations were performed using 5 fruits from each parcel to determine the mean in variance analysis.

The results of the evaluations were assessed through variance analysis (with significant differences between means compared using the tukey test at 5% of error probability), and regression analysis using the CoStat statistical program (COHORT SOFTWARE, 2003).

Results and discussion

The results indicate differences between cultivars and the growing media system used, as well as the number of fruits and fresh weight (Table 1). Only the San Andreas and Monterey cultivars were similar in both systems for total and commercial fresh weight. According to Gruda (2009), the use of soilless culture technology does not imply an automatic increase in the production and quality of greens.

The cultivars Camarosa, Florida Festival and Portola grown in the soil system yielded a higher fresh weight than in the substrate medium. However, there was a general trend of a reduction in the number of fruits and fresh weight, both total and commercial, in the substrate culture.

Total fresh weight in the soil culture system was 877.51 g plant^{-1} for cv. Camarosa and 771.09 g plant^{-1} for cv. Florida Festival (Table 1), which was higher than those obtained by Antunes et al. (2010); this was also the case in the number of fruits per plant. The yields in the present study for cv. Camarosa grown in soil were also greater than those of Vignolo et al. (2011), Oliveira et al. (2008) and Watthier et al. (2011), with their recorded yields of 43.6 fruits and 813.2 g plant^{-1}, 55 fruits and 1,038 g plant^{-1} and 37.1 fruits and 600.1 g plant^{-1}, respectively. The results of this study for cv. Portola grown in soil were also higher than the 38 fruits and 600.1 g plant^{-1} recorded by Watthier et al. (2011). Only the values found by Oliveira et al. (2008) for Camino Real (44 fruits and 1,121 g plant^{-1}) were higher than the yield obtained in this study for this cultivar.

In contrast, for the substrate culture system, despite not presenting differences between the cultivars studied in relation to the number of fruits and fresh weight (both total and commercial), cv. Camarosa obtained higher fresh weight values (225 g plant^{-1}) than Radin et al. (2011) in Caxias do Sul, Rio Grande do Sul State. Caruso et al. (2011), studying *Fragaria vesca* L. in NFT system during two production cycles (summer-spring and autumn-spring), obtained a mean production of 185.2 g plant^{-1} during the cycle summer-spring and 128.2 g plant^{-1} during autumn-spring; values that were considerably lower to those obtained in the present study, where production occurred during the spring-summer cycle.

The yield performance of cultivars in relation to the total fresh weight of fruits produced during the production cycle is shown in Figure 1.

Table 1. Fruit production of 7 strawberry cultivars produced in 2 culture systems - Passo Fundo, Rio Grande do Sul State, cycle 2010/2011.

	Number of fruits per plant					
	Total			Commercial		
Cultivars	Soil	Substrate	Mean	Soil	Substrate	Mean
Camarosa	A 100 a	B 34 a	67	A 81 a	B 22 a	52
Florida Festival	A 105 a	B 37 a	71	A 91 a	B 29 a	60
San Andreas	A 31 b	A 23 a	27	A 26 b	A 18 a	22
Portola	A 55 b	B 29 a	42	A 46 b	A 23 a	3
Monterey	A 32 b	A 16 a	24	A 26 b	A 14 a	20
Ventana	A 33 b	A 23 a	28	A 29 b	A 19 a	24
Camino Real	A 30 b	A 28 a	29	A 25 b	A 24 a	25
Mean	55	27	41	46	21	34
CV (%)	18.38	27.91	55.75	18.59	30.87	60.7
	Fresh Weight (g plant^{-1})					
Camarosa	A 1.437 a	B 322 a	880	A 1.203 a	B 245 a	724
Florida Festival	A 1.430 a	B 403 a	917	A 1.292 a	B 350 a	821
San Andreas	A 498 c	A 301 a	400	A 434 c	A 266 a	350
Portola	A 1.009 ab	B 355 a	682	A 906 ab	B 310 a	608
Monterey	A 542 bc	A 218 a	380	A 463 c	A 205 a	334
Ventana	A 632 bc	B 308 a	470	A 592 bc	B 280 a	435
Camino Real	A 565 bc	B 375 a	470	A 493 bc	B 346 a	419
Mean	873	326	599	769	286	527
CV (%)	19.75	29.46	65.27	19.71	29.12	65.83

Means followed by letters in lower case in the column and preceded by the same letter capitalized in the line do not differ by Tukey test at 5% of error probability.

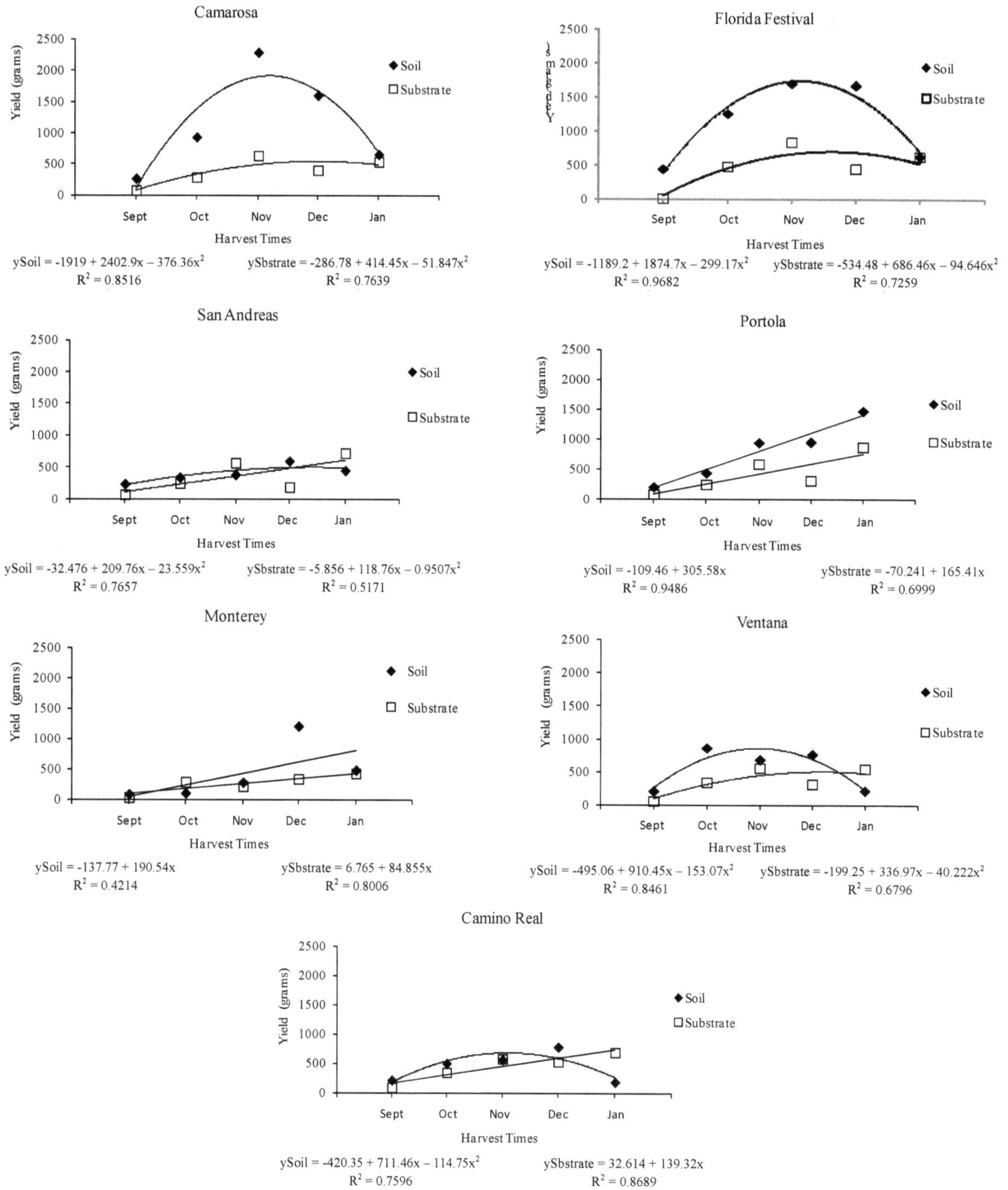

Figure 1. Yield performance during the 2010-2011 crop of 7 strawberry cultivars grown in soil and substrate, Passo Fundo, Rio Grande do Sul State, cycle 2010-2011.

Monthly yields for cultivars Camarosa, Florida Festival and Ventana resulted in a quadratic behavior for both media systems. San Andreas and Camino Real exhibited this trend only when cultivated in soil. In these two cultivars, the production peak occurred later when grown in substrate than when grown in the soil system. Using the regression equation –b $2c^{-1}$, cv. Camarosa and Ventana were observed to have a one-month delay in crop peak when grown in the substrate system. Cv. Florida Festival, on the other hand, had a 15-day delay.

For both the soil and substrate systems, the peak production period for cvs. Camarosa and Florida Festival was in November (Figure 1). San Andreas, Monterey and Camino Real, when cultivated in soil, had their peak in December. In contrast, cv. San

Andreas grown in the substrate system; Portola grown in either system and Monterey, Ventana and Camino Real grown in substrate, had their production peak in January. Only Ventana, when cultivated in the soil system, had its' production peak in October. Cvs. San Andreas and Camino Real, when grown in the substrate system, and Portola and Monterey, in either of the systems, presented an upper linear growth.

Results observed for short-day strawberry cultivars confirm findings in the literature, which note that as temperature and photoperiod increase, fruit production decreases while the development of stolons increase (BRANZANTI, 1989). Neutral-day cultivars provide greater production during the hotter periods of the year than short-day cultivars because they are less sensitive to the stimulation exerted by photoperiod and temperature for the emergence of stolons and, consequently, delay their fructification period (STRASSBURGER et al., 2010). This knowledge concerning cultivar traits is important for managing their culture as this information is beneficial to producers for staggering strawberry production.

In spite of the substrate system having displayed lower production than the soil system, this medium can increase a plants' density and correspondingly, productivity, which according to Gimenez et al. (2008) is one of the advantages of the soilless system along with eliminating the ergonomic difficulties in cultivating plants in the soil, which has hindered the recruitment of labor (GODOI et al., 2009).

The quality of the fruits is highly important because, in addition to presenting an excellent appearance, they must meet the consumers' demands in relation to flavor, sweetness, acidity and color. From the post-harvest quality parameters assessed, the pH did not present any significant differences, with a mean of 2.83 (Table 2).

Table 2. Total soluble solids (TSS) and pH in strawberry fruits produced in soil and substrate, Passo Fundo, Rio Grande do Sul State, cycle 2010/2011.

Cultivar	TSS (°Brix)	pH
Camarosa	6.90 ab	3.11^{ns}
Florida Festival	7.48 a	3.1
San Andreas	6.23 abc	2.68
Monterey	5.70 bc	2.75
Portola	5.91 bc	2.73
Ventana	6.4 abc	2.99
Camino Real	5.43 c	2.44
Mean	6.29	2.83
CV (%)	13.69	12.12
Culture System		
Soil	5.76 b	2.73^{ns}
Substrate	6.81 a	2.93
Mean	6.29	2.83
CV (%)	13.92	13.36

Means followed by the same letter in lower case in the columns do not differ by the tukey test at 5% of error probability. ^{ns} – Not significant for variance analysis.

Cekic and Yilmaz (2011) evaluation of cvs. Camarosa and Maraline in soilless culture inoculated with 2 arbuscular mycorrhizal fungi and 3 doses of phosphor, obtained a mean pH of 3.87 and 3.9, respectively. These values are somewhat higher than those found in the present study and the 3.5 pH obtained by Calvete et al. (2010) for cv. Camarosa when grown in a greenhouse using the bee species *Apis mellifera* as the pollinator.

The pH measure is very important both microbiologically and chemically because most chemical reactions occurring during the processing and storing phases (i.e., during the post-harvest stage) are profoundly altered by the variation in the environment's hydrogen concentration (GOMES; OLIVEIRA, 2011). According to Figueiredo et al. (2010) the pH reduction is not due to an increase in the citric acid content, but likely to the elevation of other acids which reduce the pH and increase the pulp's acidity.

The variable, total soluble solids, differed among the cultivars and culture systems (Table 2). Fruit from cv. Florida Festival produced the highest sugar content, whereas Camino Real, Monterey and Portola had the lowest values.

TSS values in this study were lower compared to those found by other authors in the growing conditions of Rio Grande do Sul State. Malgarim et al. (2006), in Pelotas, had 7.35 °Brix for Camarosa. However, Cekic and Yilmaz (2011) had 13.31 and 14.10% for Camarosa and Maraline, respectively, much higher than the values obtained in this study using substrate culture. Brackmann et al. (2011), assessing the quality of strawberry cultivars and clones grown in soil, had a mean 7.12 °Brix, which was also higher than the results in this study. In relation to the culture system used, the greatest sugar content was found in fruits produced in substrate (6.81 °Brix).

According to Chitarra and Chitarra (2005), total soluble solids levels vary according to species, cultivar, maturation stage and climate, with a 2-25% interval and mean values of 8-14%. Soluble solids content indicates the quantity of sugar present in the fruit, considering that acids, vitamins, amino acids and pectins are also constituent parts (albeit in a smaller proportion), in the composition of a fruit's soluble solids. However, this attribute along with the acidity, is controlled by variable levels of additive and dominant variance, i.e., they are parameters that depend on heredity (BRACKMANN et al., 2011).

In relation to a fruit's diameter, titratable total acidity and TSS/TTA ratio, an interaction between cultivars and culture systems was observed (Table 3).

It was observed that the type of media system influenced the fruit diameter of the cultivars evaluated in this study.

Table 3. Diameter (mm), total titratable acidity (TTA) and TSS/TTA ratio of the fruits in strawberry cultivars produced in 2 culture systems in greenhouse, Passo Fundo, Rio Grande do Sul State, cycle 2010-2011.

	Diameter	
Cultivar	Soil	Substrate
Camarosa	A 31.2 a	A 28.22 a
Florida Festival	A 28.64 ab	A 26.81 a
San Andreas	A 25.03 ab	A 30.73 a
Monterey	A 19.75 b	A 30.48 a
Portola	A 27.72 ab	A 28.74 a
Ventana	A 27.26 ab	A 29.30 a
Camino Real	A 23.42 ab	A 24.78 a
Mean	26.14	28.44
CV (%)	9.08	5.7
	TTA	
Cultivar	Soil	Substrate
Camarosa	A 0.81 ab	A 0.95 a
Florida Festival	A 0.72 bc	A 0.73 b
San Andreas	A 1.00 a	A 0.76 ab
Monterey	B 0.48 d	A 0.72 b
Portola	A 0.62 bcd	A 0.61 b
Ventana	A 0.69 bcd	A 0.69 b
Camino Real	A 0.52 cd	A 0.66 b
Mean	0.69	0.73
CV (%)	7.56	6.5
	Ratio (TSS/TTA)	
Cultivar	Soil	Substrate
Camarosa	A 8.19 ab	A 8.97 a
Florida Festival	A 10.77 a	A 10.07 a
San Andreas	A 5.26 b	A 9.16 a
Monterey	B 5.91 b	A 9.34 a
Portola	B 6.70 b	A 11.09 a
Ventana	A 8.08 ab	A 10.26 a
Camino Real	A 6.60 b	A 8.37 a
Mean	7.6	9.61
CV (%)	12.52	10.46

Means followed by the same letter in lower case in the column and preceded by the same letter capitalized in the line do not differ by the tukey test at 5% of error probability.

However, in both systems, the fruits were considered Class I (> 25 mm), according to the Mercosul Technical Regulation Identity and Quality for Strawberry n. 85/96. Among cultivars evaluated for this parameter, cv. Camarosa grown in soil was the only one to produce larger fruits than cv. Monterey. In contrast, both cultivars had similar size fruit when grown in substrate.

The size of the fruit depends on the interaction between flower position, the number of achenes developed, competition between fruits and the plant's vigor. In addition, larger fruits are produced during the beginning of the harvest period. Branzanti (1989) perceived a reduction in the size of fruits derived from primary flowers to those developed from secondary and tertiary flowers.

In relation to the variable total titratable acidity (Table 3), only cv. Monterey was observed to have a higher increase in TTA in substrate culture than when grown in soil. When grown in soil, cv. San Andreas had a higher concentration of citric acid and cv. Monterey had lower concentrations than the other cultivars. Cv. Camarosa grown in the substrate system had a higher citric acid content in relation to the other cultivars, which may indicate a higher concentration of other acids such as ascorbic acid.

Malgarim et al. (2006), evaluating the quality of Camarosa strawberries subjected to a modified atmosphere and resveratrol during different storage periods, had 0.60% citric acid content at harvest. Zaicovski et al. (2006) had 0.66% citric acid for cv. Camarosa subjected to different treatments with resveratrol. Both studies had lower contents than that obtained in this study for cv. Camarosa. Antunes et al. (2010), evaluating the quality of cv. Florida Festival fruits grown in soil, found a similar value to that of the present study (0.74% citric acid), but for cv. Camarosa, the content found by the authors was lower (0.76% citric acid) than that observed in this study. Brackmann et al. (2011) had a mean 0.76% citric acid content, similar to that obtained for substrate culture in this study, but higher than the soil system. Cekic and Yilmaz (2011) had 0.95% citric acid for Camarosa, the same value found in this study for soil culture.

Except for Portola and Camino Real grown in a soil culture, the citric acid contents of the other cultivars corroborate Atkinson et al. (2006) who noted that the concentration of this acid in strawberry fruits varies between 0.64-1.15%.

The relation between sugar and acidity is an important parameter in determining the fruits' maturation. With respect to this parameter, only fruits from cvs. Monterey and Portola were different when produced in soil and substrate systems; those produced in substrate had a pronouncedly sweeter flavor than an acidic sensation. In soil culture, cv. Florida Festival stood out, followed by Camarosa and Ventana (Table 3). The TSS/TTA ratio noted by Antunes et al. (2010) for cv. Florida Festival (10.83) grown in soil was similar to that found in this study. However, cv. Camarosa (9.94) had a higher ratio. The same held true in comparison to the study by Malgarim et al. (2006), which obtained a 12.25 ratio for Camarosa. Brackmann et al. (2011) found fruits with a mean 9.78 ratio, a value close to that found when a substrate system is used. Resende et al. (2008), linking the relationship of TSS/TTA with consumer acceptance, found that the highest TSS/TTA ratios were associated with a better perception of the fruit's flavor and are preferred by consumers. These authors recorded a value of 13.50 for the cultivar Camp-Dover. Thus, in this study, Florida Festival would have a greater consumer acceptance than the other cultivars.

In relation to the fruits' external color, the culture system used did not appear to have an influence;

additionally, differences were not observed between cultivars in any of the variables analyzed (L*, c* and °Hue). Regardless of the cultivar and the culture system, the fruits presented a dark external color, i.e., low luminosity (L* < 29.24) and less chromatics (c* > 36.08) (Figure 2).

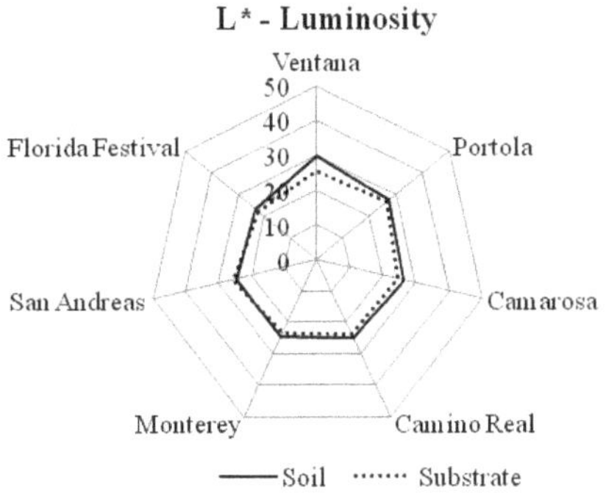

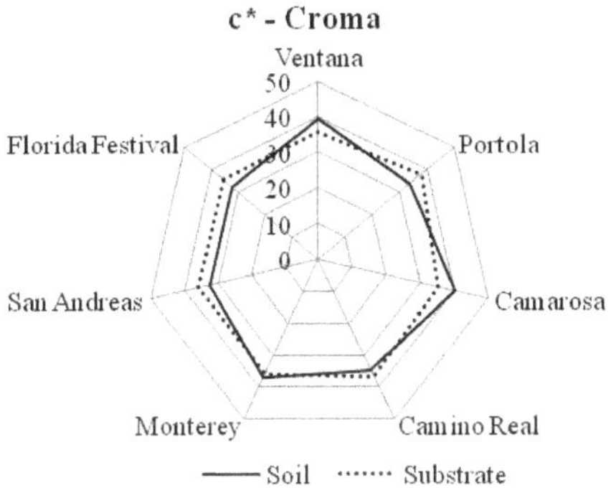

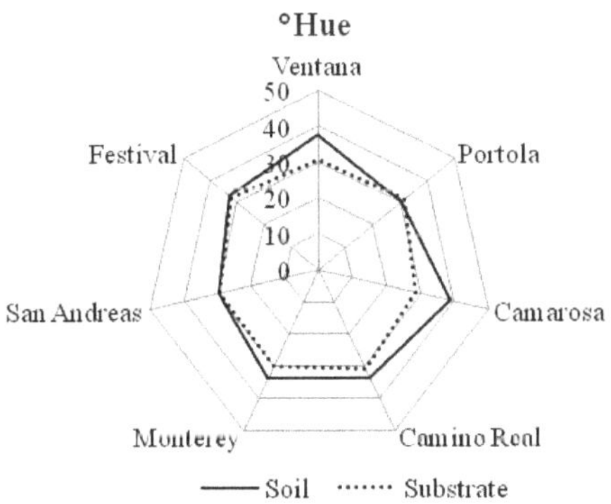

Figure 2. External color of strawberry fruits, L*, c* and °Hue values in different culture systems, Passo Fundo, Rio Grande do Sul State, cycle 2010-2011.

Color evaluation is another important attribute for strawberries because it determines the ideal conditions for harvesting and the commercialization value of the fruit. Fruit with a strong, bright red coloration are preferred, although color in most cases does not contribute an actual increase in the nutritional value or sensory quality of the product (CHITARRA; CHITARRA, 2005).

Conclusion

The cultivation of strawberries grown in two different media systems in a greenhouse generated the following conclusions:

The yield per strawberry plant was higher in soil culture, while there was better fruit quality with regard to total soluble solids, pH, diameter, acidity, titratable total acidity and TSS/TTA ratio using the substrate system.

The greatest yield in fruits per plant, when cultivated in soil in the greenhouse, was found in cvs. Camarosa and Florida Festival. The greatest yields in soil systems were cvs. Camarosa, Florida Festival and Portola.

In the substrate system, no difference was found among the cultivars in terms of production and yield of fruits per plant.

The greatest sugar contents were those of cvs. Florida Festival, followed by Camarosa, Ventana, and San Andreas. Regardless of the cultivar, when the strawberry was grown in substrate, sweeter fruits were obtained.

Fruits, regardless of cultivar or media system used, displayed a dark red color and less chromatics.

References

ANTUNES, L. E. C.; RISTOW, N. C.; KROLOW, A. C. R.; CARPENEDO, S.; REISSER JÚNIOR, C. Yield and quality of strawberry cultivars. **Horticultura Brasileira**, v. 28, n. 2, p. 222-226, 2010.

ANTUNES, O. T.; CALVETE, E. O; ROCHA, H. C.; NIENOW, A. A.; CECCHETTI, D.; RIVA, E.; MARAN, R. E. Produção de cultivares de morangueiro polinizadas pela abelha jataí em ambiente protegido. **Horticultura Brasileira**, v. 25, n. 1, p. 94-99, 2007.

ATKINSON, C. J.; DODDS, P. A. A.; FORD, Y. Y.; MIÈRE, J. L. E.; TAYLOR, J. M.; BLAKE P. S.; PAUL, N. Effects of cultivar, fruit number, and reflected photosynteticaly active radiation on *Fragaria* x *ananassa* productivity and fruit ellagic acid and ascorbic acid concentrations. **Annals of Botany**, v. 97, n. 3, p. 429-441, 2006.

BRACKMANN, A.; PAVANELLO, E. P.; BOTH, V.; JANISCH, D. I.; SCHMITT, O. J.; GIMÉNEZ, G. Avaliação de genótipos de morangueiro quanto à qualidade e potencial de armazenamento. **Revista Ceres**, v. 58, n. 5, p. 542-547, 2011.

BRANZANTI E. E. **La fresa**. Madri: Mundi-Prensa, 1989.

CALVETE, E. O.; MARIANI, F.; WESP, C. L.; NIENOW, A. A.; CASTILHOS, T.; CECCHETTI, D. Fenologia, produção e teor de antocianinas de cultivares de morangueiro em ambiente protegido. **Revista Brasileira de Fruticultura**, v. 30, n. 2, p. 396-401, 2008.

CALVETE, E. O.; NIENOW, A. A.; WESP, C. L.; CESTONARO, L.; MARIANI, F.; FIOREZE, I.; CECCHETTI, D.; CASTILHOS, T. Produção hidropônica de morangueiro em sistema de colunas verticais, sob cultivo protegido. **Revista Brasileira de Fruticultura**, v. 29, n. 3, p. 524-529, 2007.

CALVETE, E. O.; ROCHA, H. C.; TESSARO, F.; CECCHETTI, D.; NIENOW, A. A.; LOSS, J. T. Polinização de morangueiro por *Apis mellifera* em ambiente protegido. **Revista Brasileira de Fruticultura**, v. 32, n. 1, p. 181-188, 2010.

CAPOCASA, F.; SCALZO, J.; MEZZETI, B.; BATTINO, M. Combining quality and antioxidant attributes in the strawberry: The role of genotype. **Food Chemistry**, v. 111, n. 4, p. 872-878, 2008.

CARUSO, G.; VILLARI, G.; MELCHIONNA, G.; CONTI, S. Effect of cultural cycles and nutrient solutions on plant growth, yield and fruit quality of alpine strawberry (*Fragaria vesca* L.) grown in hydroponics. **Scientia Horticulturae**, v. 129, n. 3, p. 479-485, 2011.

CEKIC, C.; YILMAZ, E. Effect of arbuscular mycorrhiza and different doses of phosphor on vegetative and generative components of strawberries applied with different phosphor doses in soilless culture. **African Journal of Agricultural Research**, v. 6, n. 20, p. 4736-4739, 2011.

CHITARRA, M. I. F.; CHITARRA, A. B. **Pós-colheita de frutos e hortaliças**: fisiologia e manuseio. 2. ed. Lavras: UFLA, 2005.

COHORT SOFTWARE. **Graphics and Statistics Software for Scientists and Engineers**. Monterey, Califórnia: CoHort Software, 2003. Available from: <http://www.cohort.com>. Access on: Mar. 20, 2010.

FERREIRA, V. L. P. **Princípios e aplicações da colorimetria em alimentos**. Campinas: ITAL, 1981. (Instruções Técnicas, n. 19).

FIGUEIREDO, F. C.; BOTREL, P. P.; TEIXEIRA, C. P.; PETRAZZINI, L. L.; LOCARNO, M.; CARVALHO, J. G. de. Pulverização foliar e fertirrigação com silício nos atributos físico-químicos de qualidade e índices de coloração do morango. **Ciência e Agrotecnologia**, v. 34, n. 5, p. 1306-1311, 2010.

GIMENEZ, G.; ANDRIOLO, J.; GODOI, R. Cultivo sem solo do morangueiro. **Ciência Rural**, v. 38, n. 1, p. 273-279, 2008.

GODOI, R. S.; ANDRIOLO, J. L.; FRANQUÉZ, G. G.; JÄNISCH, D. I.; CARDOSO, F. L.; VAZ, M. A. B. Produção e qualidade do morangueiro em sistemas fechados de cultivo sem solo com emprego de substratos. **Ciência Rural**, v. 39, n. 4, p. 684-690, 2009.

GOMES, J. C.; OLIVEIRA, G. F. **Análises físico-químicas de alimentos**. Viçosa: UFV, 2011.

GRUDA, N. Do Soilless culture systems have an influence on product quality of vegetables? **Journal of Applied Botany and Food Quality**, v. 82, n. 2, p. 141-147, 2009.

JAFARNIA, S.; KHOSROWSHAHI, S.; HATAMZADEH, A.; TEHRANIFAR, A. Effect of substrate and variety on some important quality and quantity characteristics of strawberry production in vertical hydroponics system. **Advances in Environmental Biology**, v. 3, n. 4, p. 360-363, 2010.

LIETEN, F.; LONGUESSERRE, J.; BARUZZI, G.; LOPEZ-MEDINA, J.; NAVATEL, J. C.; KRUEGER, E.; MATALA, V.; PAROUSSI, G. Recent situation of strawberry substrate culture in Europe. **Acta Horticulturae**, v. 649, p.193-196, 2004.

MALGARIM, M. B.; FLORES, C. R. F.; COUTINHO, E. F. Sistemas e condições de colheita e armazenamento na qualidade de morangos cv. Camarosa. **Revista Brasileira de Fruticultura**, v. 28, n. 2, p. 185-189, 2006.

OLIVEIRA, R. P.; SCIVITTARO, W. B.; FINKENAUER, D. Produção de morangueiro da cv. 'Camino Real' em sistema de túnel. **Revista Brasileira de Fruticultura**, v. 30, n. 3, p.681-684, 2008.

PONCE, A. R.; BASTIANI, M. I. D.; MINIM, V. P.; VANETTI, M. C. D. Características físico-químicas e microbiológicas de morango minimamente processado. **Ciência e Tecnologia de Alimentos**, v. 30, n. 1, p. 113-118, 2010.

RADIN, B.; LISBOA, B. B.; WITTER, S.; BARNI, V.; REISSER JUNIOR, C.; MATZENAUER, R.; FERMINO, M. H. Desempenho de quatro cultivares de morangueiro em duas regiões ecoclimáticas do Rio Grande do Sul. **Horticultura Brasileira**, v. 29, n. 3, p. 287-291, 2011.

RESENDE, J. T. V.; CAMARGO, L. K. P.; ARGANDOÑA, E. J. S.; MARCHESE, A.; CAMARGOL, C. K. Sensory analysis and chemical characterization of strawberry fruits. **Horticultura Brasileira**, v. 26, n. 3, p. 371-374, 2008.

STRASSBURGER, A. S.; PEIL, R. M. N.; SCHWENGBER, J. E.; MEDEIROS, C. A. B.; MARTINS, D. A.; SILVA, J. B. Crescimento e produtividade de cultivares de morangueiro de "dia neutro" em diferentes densidades de plantio em sistema de cultivo orgânico. **Bragantia**, v. 69, n. 3, p. 623-630, 2010.

VIGNOLO, G. K.; ARAÚJO, V. F.; KUNDE, R. J.; SILVEIRA C. A. P.; ANTUNES, L. E. C. Produção de morangos a partir de fertilizantes alternativos em pré-plantio. **Ciência Rural**, v. 41, n. 10, p. 1755-1761, 2011.

WATTHIER, M.; SILVA, D. R.; MARTINS, D. S.; SCHWENGBER, J. E. Desempenho de cultivares de morangueiro manejadas em sistema de produção de base ecológica. **Horticultura Brasileira**, v. 29, n. 2, p. 4564-4570, 2011.

ZAICOVSKI, C. B.; TIBOLA, C. S.; MALGARIM, M. B.; FERRI, V. C.; PEGORARO, C.; CERO, J. D.; SILVA, P. R. Resveratrol na qualidade pós-colheita de morangos 'Camarosa'. **Revista Brasileira de Agrociência**, v. 12, n. 4, p. 443-446, 2006.

2

Different nitrogen and boron levels influence the grain production and oil content of a sunflower cultivar

Luiziene Soares Alves[1], Elvia Mariam Lis Martinez Stark[2], Everaldo Zonta[2], Manlio Silvestre Fernandes[2], André Marques dos Santos[3*] and Sonia Regina de Souza[3]

[1]*Programa de Pós-graduação em Agronomia, Ciência do Solo, Universidade Federal Rural do Rio de Janeiro, Seropédica, Rio de Janeiro, Brazil.* [2]*Departamento de Solos, Universidade Federal Rural do Rio de Janeiro, Seropédica, Rio de Janeiro, Brazil.* [3]*Departamento de Química, Laboratório de Bioquímica de Plantas, Universidade Federal Rural do Rio de Janeiro, BR-465, km 7, 23897-000, Seropédica, Rio de Janeiro, Brazil. *Author for correspondence. E-mail: amarques@ufrrj.br; amarques.ufrrj@outlook.com.br*

ABSTRACT. Nitrogen is essential for plant growth and development and directly influences both grain production and oil content. Boron has been studied in sunflower cultivation due to high demand of this culture for this nutrient. This study aimed to evaluate the N and B differential supply effect on the grain and oil production of sunflower plants. The experiment was conducted in a greenhouse and used two sunflower hybrids: BRS 321 and Neon. N and B were applied via nutritive solution according to the following treatments: B + 50 N (2.49 kg ha^{-1} B and 10 kg ha^{-1} N at planting and 40 kg ha^{-1} B as additional fertilization); B + 10 N (2.49 kg ha^{-1} B and 10 kg ha^{-1} N at planting); 50 N (0 B and 10 kg ha^{-1} N at planting and 40 kg ha^{-1} N as additional fertilization) and 10 N (0 B and 10 kg ha^{-1} N at planting). Analyses of total N and B in all collected plant parts were performed, as well as measurements of grain production for each chapter and oil extraction at the end of the cycle. The association between B fertilization and a greater amount of N fertilization was essential to BRS 321 oil production. Regarding grain production at lower levels of N fertilization, Neon presented the best response to B application.

Keywords: *Helianthus annuus* L., oilseed, total N, oil extraction.

Introduction

Nitrogen (N) is one of the most essential elements for plants and is a limiting factor for their growth. Because crops are demanded in such large amounts, the agricultural sector relies greatly on N and often improperly uses it, which threatens air, water and soil quality (Xu, Fan, & Miller, 2012). The search for genotypes that efficiently use N by optimizing N fertilization has been the subject of several studies on important agronomical crops (Avice & Etienne, 2014).

N is essential for sunflower grain production and oil extraction (Rasool, Hassan, & Jahangir, 2013). Abdel-Motagally and Osman (2010) observed how increased soil N through fertilization influenced several growth parameters (plant height, stem diameter, head diameter, 100-seed weight, seed yield $plant^{-1}$, seed yield ha^{-1} and oil yield ha^{-1}), causing production increases. N fertilization affects many important agronomical parameters such as leaf area, capitulum diameter, achenes weight and yield (Ali & Noorka, 2013).

According to the dosage of N used, either decreases or increases in oil yield may be reported.

In addition to N, Boron (B) is also an essential nutrient for plant growth. This micronutrient has been associated with various plant functions, such as water availability, sugar translocation, anion and cation uptake, N, phosphorus (P), carbohydrate and lipid metabolism (Al-Amery, Hamza, & Fuller, 2011). Sunflower is sensitive to nutrient shortages and can also be used as a tool for evaluating soil B availability (Oyinlola, 2007). Beneficial effects of major B doses used for sunflower productivity include: increases in achenes production (Lima et al., 2013), increases in oil extraction (Tahir, Ashraf, & Ibrahim, 2013), pollen viability (Krudnak, Wonprasaid, & Machikowa, 2013), and increases in photosynthetic activity and total fresh weight (Zahoor et al., 2011).

Taking into account the essentiality of N and B in sunflower, this study aimed to evaluate how differential supplies of nutrients might influence the grain and oil production of BRS 321 and Neon genotypes.

Material and methods

Experimental design

The experiment was carried out in a greenhouse at the Soil Sciences Department of Universidade Federal Rural do Rio de Janeiro from June to August 2013.

A completely randomized design in a factorial arrangement (2 genotypes x 2 N managements x 2 Boron levels) with eight replicates was used. Sunflower seeds from BRS 321 (EMBRAPA) and Neon (SEMBRAS) hybrids were used and were seeded on dried-air soil (DAS) trays from a sandy ULTISOL horizon surface collected at EMBRAPA – Agrobiologia. Nine days after germination in 20.7 dm^3 pots containing the same DAS, plantlets were transferred. Plants were fertilized with Hoagland solution containing urea as the N source (0.19 $g.pot^{-1}$ at planting and 0.36 $g.pot^{-1}$ as additional fertilization, corresponding to 10 kg ha^{-1} N and 40 kg ha^{-1} N, respectively) and boric acid as the B source (22.38 mg pot^{-1} B, corresponding to 2.49 kg ha^{-1} B) according to the following treatments: B + 50 N (2.49 kg ha^{-1} B and 10 kg ha^{-1} N at planting and 40 kg ha^{-1} N as additional fertilization); B + 10 N (2.49 kg ha^{-1} N and 10 kg ha^{-1} N at planting); 50 N (0 B and 10 kg ha^{-1} N at planting and 40 kg ha^{-1} N as additional fertilization); 10 N (0 B and 10 kg ha^{-1} N at planting).

Analysis of plant materials, grains and oil production

At the end of the cycle, 105 days (BRS 321) and 120 days (Neon) after germination, plants (one plant per pot) were collected and separated into capitulum, leaf blades, petiole, stem, roots and grains. Plant material was dried in a stove at 60°C until reaching a constant weight, and the dry mass was then determined. Grains were counted, weighed and divided into void and filled grains. Total N (Tedesco et al., 1995) and total B (Malavolta, Vitti, & Oliveira, 1997) analyses of the dry material from each plant part as well as filled grains were conducted.

Oil extraction using hexane by Sohxlet extractor was performed for four hours (Yaniv et al., 1998).

Statistical analysis

Statistical analysis was performed using SISVAR software for Windows (Ferreira, 2011), and the averages were compared using Tukey's test at a 5% significance level.

Results and discussion

Dry mass

The Neon hybrid presented similar a capitulum mass and number of foliar blades upon treatment with 50 N and B + 10 N, showing a possible boron benefit in conditions of low nitrogen availability for this genotype. No significant difference between treatments for the BRS 321 hybrid was reported (Table 1). No grain dry mass differences for Neon were detected (Table 1). Treatment with 10 N resulted in lower grain mass for BR 321 when compared to that of the others, suggesting that BR 321 was more sensitive to low N supply associated with B (Table 1).

Contents and N and B amounts

Neon and BRS 321 grains presented the highest grain N content compared to the other plant compartments (Tables 2 and 3).

Neon and BRS 321 grains presented the highest grain N content (Table 2), indicating an apparent influence of B on increasing N remobilization efficacy, particularly for this genotype.

When only 50 kg ha^{-1} N (50 N) was applied, leaf blades from both genotypes presented higher N contents compared to other treatments (Table 2). Therefore, we propose that the N concentration of leaf blades is a sensitive parameter for evaluating a plant's N *status* once the petiole does not present a significant difference regarding total N content (Table 2). The fourth leaf from the apex to the plant base is currently being used as an index for nutritional status evaluation of several crops, including sunflower (Silva, Aquino, & Batista, 2011). The results from this survey corroborated the use of leaf blades as a parameter in sunflower crop as well.

The capitulum of both genotypes showed high N content in grains, as it is in this compartment that the grains are inserted, consequently accepting remobilized N for grain development from other plant parts. Braz and Rossetto (2010) also reported the highest N content in sunflower grains and capitulum at the end of the cycle.

The BRS 321 genotype presented higher shoot N content than the Neon genotype with B treatment (Figure 1B). Similarly, with B and 50 kg ha^{-1} N treatments, the highest N contents in the BRS 321 grains were observed (Figure 1A). These data suggest that there is a beneficial effect of B in terms of N content in the BRS 321 genotype.

Table 1. Sunflower plants dry mass (g plant^{-1}) of BRS 321 and Neon hybrids cultivated in a greenhouse at different N and B treatments: B + 50 N (2.49 kg ha^{-1} B and 50 kg ha^{-1}); B + 10 N (2.49 kg ha^{-1} B and 10 kg ha^{-1} N); 50 N (0 B and 50 kg ha^{-1} N) and 10 N (0 B and 10 kg ha^{-1} N).

Treatments	Grains	Capitulum	Leaf blades	Petiole	Stem
			Neon		
B + 50 N	29.30 a	15.67 ab	12.36 b	2.41 c	30.19 a
B + 10 N	52.20 a	17.27 a	17.66 a	3.31 ab	28.01 a
50 N	49.62 a	16.44 a	16.97 a	3.36 a	30.01 a
10 N	39.34 a	13.89 b	14.90 ab	2.84 bc	28.96 a
CV (%)	30.36	7.90	11.49	9.55	8.34
			BRS321		
B + 50 N	25.46 a	16.03 a	12.36 b	1.52 b	18.76 a
B + 10 N	20.80 a	12.29 a	17.66 a	1.78 ab	19.82 a
50 N	23.38 a	16.31 a	16.97 a	1.89 a	19.79 a
10 N	10.38 b	15.88 a	14.90 ab	1.46 b	15.61 b
CV (%)	16.23	16.01	11.49	11.58	8.74

Averages followed by the same letters in the columns for each genotype do not differ by Tukey's test at the 5% significance level.

Table 2. Total N grains, capitulum, leaf blades, petiole, stem, roots and soil contents of BRS 321 and Neon genotypes cultivated in a greenhouse at different N and B treatments.

Treatments	N-total (dry mass mg g^{-1} N)						
	Grains	Capitulum	Leaf blades	Petiole	Stem	Roots	Soil
				Neon			
B + 50 N	14.34 Ac	7.92 Bab	6.16 Bc	2.99 Ca	1.96 Ca	1.40 Cb	1.25 Ca
B + 10 N	16.07 Ab	6.43Bc	3.20 Cd	3.01 Ca	1.36 Da	1.60 Cb	1.16 Ca
50 N	17.78 Aa	8.36 Ba	9.18 Ba	3.61 Ca	1.37 Ca	2.82 CDa	1.09 Da
10 N	15.77 Ab	7.08 Bbc	7.82 Bb	3.45 Ca	2.27 Da	1.82 CDab	1.20 Da
				BRS321			
B + 50 N	36.66 Aa	6.99 ABb	9.84 Bab	4.03 CDa	3.39 CDa	4.17 CDa	1.09 Da
B + 10 N	30.27 Ab	9.86 Bb	7.92 BCb	3.47 CDa	2.42 Da	2.13 Da	1.32 Da
50 N	35.75 Aa	14.89 Ba	11.18 Ba	4.35 Ca	3.35 Ca	1.59 Ca	1.59 Ca
10 N	33.99 Aa	13.89 Ba	9.21 Cab	3.16 Da	2.86 Da	1.53 Da	1.54 Da

Averages followed by the same capital letters in lines and small letters in columns for each genotype do not differ by Tukey's test at the 5% significance level.

Table 3. Grains, capitulum, leaf blades, petiole and stem N contents of BRS 321 and Neon genotypes cultivated in a greenhouse at different N and B treatments.

Treatments	N content (mg plant^{-1} N)					
	Grains	Capitulum	Leaf blades	Petiole	Stem	Roots
			Neon			
B + 50 N	475.81 Ac	118.69 Ba	81.11 BCab	12.54 Ca	62.29 BCa	*
B + 10 N	659.21 Ab	107.61 Ba	67.58 BCb	8.84 Ca	32.34 BCa	*
50 N	923.97 Aa	138.09 Ca	168.82 Ba	11.04 Da	59.89 CDa	*
10 N	675.81 Ab	110.98 Ba	118.08 Bab	7.04 Ca	61.44 CBa	*
			BRS321			
B + 50 N	741.73 Aa	133.49 Ba	161.59 Ba	7.38 Ca	55.16 Bca	*
B + 10 N	662.24 Aab	158.82 Ba	93.41 BCa	6.67 Ca	46.55 Ca	*
50 N	727.97 Aa	223.23 Ba	94.87 Ca	8.77 Ca	73.39 Ca	*
10 N	607.86 Ab	131.14 Ba	87.16 BCa	5.75 Ca	58.17 Bca	*

*Root N content analyses were not performed. Averages followed by the same capital letters in lines and small letters in columns for each genotype do not differ by Tukey's test at the 5% significance level.

The Neon genotype presented the highest grain N content when 50 kg ha^{-1} N, but no Boron, was applied (Figure 1A); on the other hand, shoot N content did present the same result (Figure 1B), indicating that this genotype displays promising performance regarding grain and oil production. In conditions of high N availability, the Neon genotype does not accumulate Nitrogen in the shoot, directly transmitting N uptake by grains; this results in high grain production as well as higher oil yield (Table 7).

In both genotypes, the leaf blades presented the highest B contents (Table 4). Boron is responsible for increases in efficiency of photosynthesis light energy utilization (Broadley et al., 2012). The highest B content was observed in the 151.38 mg kg^{-1} B + 50 N (Neon) and 126.14 mg kg^{-1}(BRS 321) treatments (Table 4). With no B treatment, independently of N performance, leaf blades presented medium N contents (from 48.86 to 57.98 mg kg^{-1} B). These concentrations were lower than those reported by Aquino, Silva, and Berger (2013) in their study of hybrids and sunflower varieties. They observed medium leaf blade B contents ranging from 119.25 to 140.35 mg kg^{-1} in hybrids and sunflower varieties, respectively, once plants did not receive B fertilization and were only absorbing soil nutrients.

The capitulum also presented substantial B content. Leaf blades and capitulum presented the highest B contents when plants were fertilized by these nutrients (B + 50 N and B + 10 N treatments) (Table 4).

In general, Neon presented higher B contents than those of BRS 321, even showing less responsiveness to the nutrient. The Neon genotype presented the highest leaf blade B content with B + 50 N and B + 10 N treatments (Table 5).

The capitulum also presented significant B contents in both genotypes (Table 5). This might be explained by the nutritional demand for Boron in pollen tube growth; Boron is an important nutrient for reproductive structures (Broadley et al., 2012). Small amounts of B on the pollen tube were observed when BRS 321 axillary inflorescence analysis was performed.

The genotypes presented different traits regarding B exportation to the grains. The BRS 321 and Neon genotypes grains presented similar B contents, independently of the treatment (Table 5). Lobo, Grassi Filho, and Brito (2011) reported minor micronutrient exportation with the lowest N content, including exportation of B, but did not observe any interactions between N and B doses.

Grain and oil production

The Neon genotype presented the highest oil content of grains with the 50 N treatment (Table 6). N application in a large amount (50 N) and a small N amount associated with B (B + 10 N) resulted in higher grain production for this genotype, leading to high oil content under these N and B availabilities (Table 7). In the B + 10 N treatment, filled grains production was similar to that presented with the 50 N treatment (Tables 6 and 7). This result suggests that B favors the production of filled grains, an agronomically significant trait, and might also reduce the need for N fertilizers.

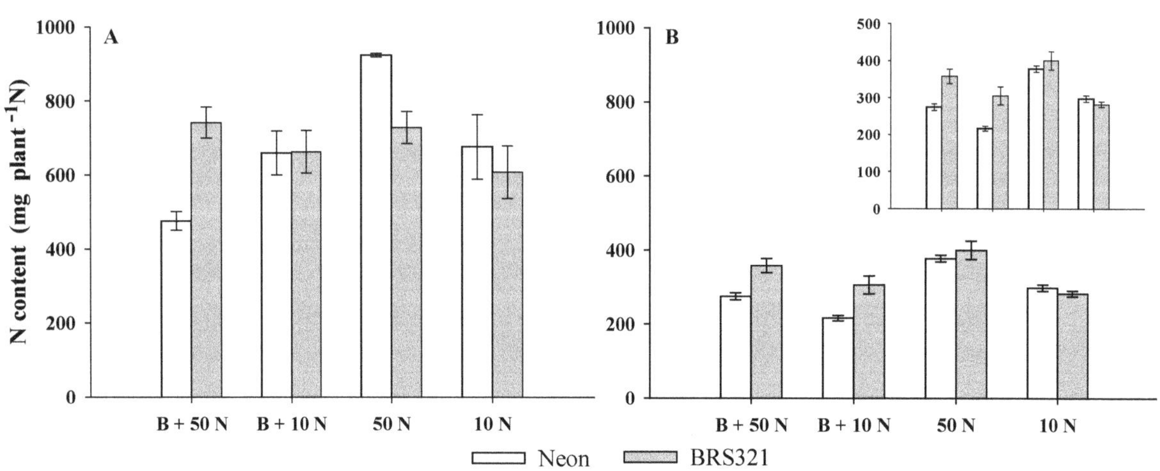

Figure 1. Grains (A) and aerial part (B) contents of BRS 321 and Neon genotypes cultivated in a greenhouse at different N and B treatments. Bars represent eight replicates, medium ± standard error.

Table 4. Grains, capitulum, leaf blades, petiole, stem, roots, axillary inflorescence and soil B contents (mg.kg^{-1}) of BRS 321 and Neon genotypes cultivated in a greenhouse at different N and B treatments.

	B content (mg kg^{-1})						
Treatments	Grain	Capitulum	Leaf blades	Petiole	Stem	Roots	Axillary inflores-cence
				Neon			
B + 50 N	0.08 Da	77.77 Ba	151.38 Aa	42.10 Ca	0.13 Da	20.43 CDab	*
B + 10 N	0.07 Ca	82.09 Aa	82.25 Ab	42.09 Ba	0.14 Ca	24.25 BCa	*
50 N	0.11 Ba	57.33 Ab	48.86 Ac	42.13 Aa	0.19 Ba	8.03 Bb	*
10 N	0.10 Ca	51.96 Ab	56.57 Ac	42.09 Aba	0.15 Ca	11.27 BCab	*
				BRS321			
B + 50 N	0.07 Da	80.35 Ba	126.14 Aa	42.14 Ca	0.15 Da	24.33 CDa	0.10 Da
B + 10 N	0.08 Ba	67.10 Bb	103.36 Ab	41.63 Bca	0.15 Da	23.59 CDab	0.30 Da
50 N	0.08 Da	48.16 Ac	52.17 Ac	42.10 Aa	0.17 Ba	14.35 Bb	0.14 Ba
10 N	0.08 Ca	53.34 Ac	57.98 Ac	42.12 Aba	0.17 Ca	16.55 BCab	0.17 Ca

*Neon genotype did not present axillary inflorescence. Averages followed by the same capital letters in lines and small letters in columns for each genotype do not differ by Tukey's test at the 5% significance level.

Table 5. Grains, capitulum, leaf blades, petiole, stem and inflorescence B contents (mg.plant^{-1}) of BRS 321 and Neon genotypes cultivated in a greenhouse at different N and B treatments.

	B content (mg plant^{-1})					
Treatments	Grain	Capitulum	Leaf blades	Petiole	Stem	Axillary inflores-cence
				Neon		
B + 50 N	3.51 Ca	1479.13 Ba	1702.58 Aa	109.62 Ca	4.69 Ca	*
B + 10 N	3.74 Ca	1456.74 Aa	1254.10 Ba	140.48 Ca	4.60 Ca	*
50 N	3.92 Ca	852.37 Ab	571.84 Bc	94.48 Ca	4.44 Ca	*
10 N	3.95 Ba	724.50 Ab	568.70 Ac	95.04 Ba	4.54 Ba	*
				BRS321		
B + 50 N	2.66 Ca	994.89 Ba	1484.39 Aa	72.54 Ca	3.66 Ca	0.44 Ca
B + 10 N	1.52 Ca	904.21 Bab	1202.53 Ab	84.78 Ca	2.77 Ca	0.40 Ca
50 N	2.06 Ca	875.26 Ab	464.79 Bc	75.87 Ca	3.14 Ca	0.64 Ca
10 N	0.80 Ca	751.97 Ac	563.29 Bc	59.28 Ca	3.54 Ca	0.60 Ca

*Neon genotype did not present axillary inflorescence. Averages followed by the same capital letters in lines and small letters in columns for each genotype do not differ by Tukey's test at the 5% significance level.

In contrast to Neon, B fertilization in the BRS 321 genotype was a determinant for the production of filled grains as 10 kg ha^{-1} N as 50 kg ha^{-1} N fertilization (Table 6). The BRS 321 genotype presented the highest oil yield with B + 50 N treatment; in other words, B fertilization might influence BRS 321 oil yield (Table 6). By testing increasing B doses (0, 0.25, 0.5 and 2 mg dm^{-3} B), Castro et al. (2006) observed the influence of B application on grain production and oil yield, mainly at 0.5 mg dm^{-3} doses and independently of water conditions.

BRS 321 genotype plants presented higher production of empty grains when they were restricted to B fertilization (Table 6). This fact might be explained by the higher demand for this nutrient in the reproductive stage; plants can present a deformed capitulum with large numbers of empty grains (Lima et al., 2013). Tahir et al. (2013) reported grain production and oil yield increases by applying B as boric acid in differentiated doses on Hysun-33 hybrid sunflower plants.

BRS 321 genotype presented the highest oil content, mainly on 50 kg ha^{-1} N and B (68.75%), apart from 50 N treatment when received the highest N doses (Table 6). According to Empresa Brasileira de Pesquisa Agropecuária (EMBRAPA, 2011) BRS 321 medium oil contents ranged from 40 to 44%, however, this survey reported 68.75%, as previously mentioned above. Probably due to a characteristic of this genotype, which proved to be responsive to B and N, especially for oil production. However, it is important confirmation of those results in a field experiment in these same conditions of N and B fertilization. Zahoor et al. (2011) also observed the highest oil contents (48%) on sunflower plants fertilized on 2 kg ha^{-1} B. According to these same authors the highest oil content might be related to greater N uptake by seed, which might subsequently be incorporated into the protein molecule. Neon genotype also presented significative oil content values, although less than BRS 321 one.

Neon genotype presented filled grains production with medium oil contents (61.54%) on 50 kg ha^{-1} N. Abdel-Motagally and Osman (2010) studying two Giza Sahka-102 and 53 genotypes obtained the highest oil yield at higher N doses, however, when analyzing seed oil percentage values observed decrease with N fertilization increase.

Table 6. Filled and void grains number (number plant^{-1}) and oil yield (percentage) of BRS 321 and Neon genotypes cultivated in greenhouse at different N and B treatments.

Treatments	Filled grains (number plant^{-1})		Void grains (number plant^{-1})		Oil yield (%)	
	Neon	BRS321	Neon	BRS321	Neon	BRS321
B + 50 N	536 Ab	235 Ba	310 Ab	271 Ab	50.65 Bb	68.75 Aa
B + 10 N	760 Aa	264 Ba	443 Aa	311 Bb	47.88 Ab	51.92 Ab
50 N	858 Aa	54 Bb	437 Aa	471 Aa	61.54 Aa	53.91 Bb
10 N	608 Ab	81 Bb	372 Aab	448 Aa	42.02 Bb	49.58 Ab
CV (%)	11.29	41.45	17.51	15.18	10.09	9.11

Averages followed by the same capital letters on line and small letters on column for each genotype do not differ by Tukey test at 5% significance level.

Significative positive correlation between plant filled grains and oil percentage on both genotypes was observed (Figure 2A and B). Curiously, BRS 321 genotype produced smaller filled grains amount than Neon one, and even so, higher or similar oil percentage when compared to Neon (Figure 2) possibly due to Neon grains dilution effect.

The production of filled grains of the BRS 321 genotype was the highest with N treatment and B application, and consequently the oil yield was also the highest (Table 7). Rasool, Hassan, and Jahangir (2013) observed that as N doses increased, grains production and oil yield also increased; additionally, grains oil content decreased when high N doses were applied.

The production of capitulum filled grains in the Neon genotype was higher with lower N doses and N supplementation, as well as with N treatment once these treatments presented the highest capitulum oil yield (Table 7). Both genotypes presented beneficial effects of B, mainly for the Neon genotype, apart from the lower N doses applied with B. This resulted in the production of filled grains and oil yield similar to that of the BRS 321 genotype. Foloni et al. (2010) also demonstrated the beneficial effects of B in sunflower crop. These authors reported that B doses applied only on the planting furrow, regardless of leaf B, provided the highest aerial part total phytomass production; this positively influenced capitulum grains total mass. Leaf application increased the total number of grains only on plots receiving no B in the soil (Foloni et al., 2010).

When the BRS 321 genotype received lower N doses with no B application, it led to the lower production of filled grains compared to that of other treatments. The beneficial effect of B was also demonstrated in this study (Table 7).

It can be concluded from results of this analysis that both the Neon and BRS 321 genotypes presented different traits regarding nutritional demand, as well as in the production of grains and oil yield

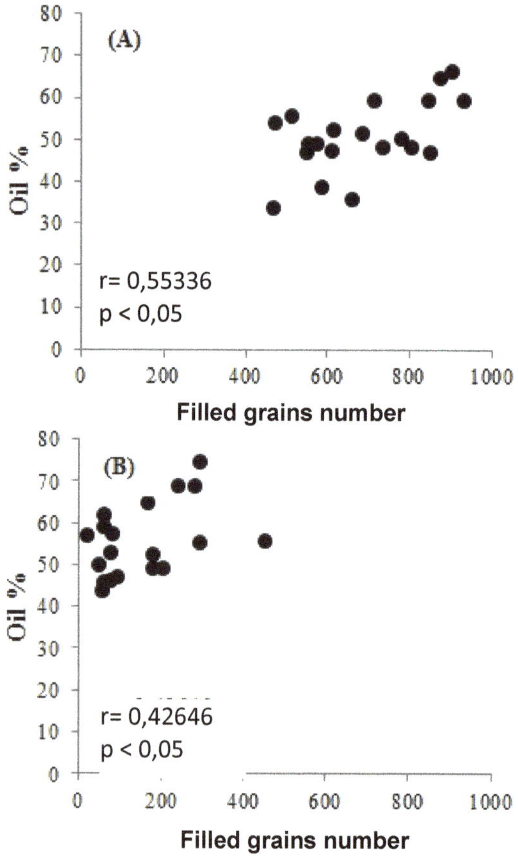

Figure 2. Correlation between the number of filled grains (number plant^{-1}) and oil yield of Neon (A) and BRS 321 (B) genotypes.

Table 7. Production of filled grains (g capitulum^{-1}) and oil yield (g capitulum^{-1}) of BRS 321 and Neon genotypes cultivated in a greenhouse at different N and B treatments.

Treatments	Filled grain production (g capitulum^{-1})		Oil yield (g capitulum^{-1})	
	Neon	BRS321	Neon	BRS321
B + 50 N	28.23 Ac	21.03 Ba	14.28 Ab	13.07 Aa
B + 10 N	50.21 Aa	18.32 Ba	24.06 Aa	9.63 Bab
50 N	47.35 Aab	17.42 Bab	29.21 Aa	9.54 Bab
10 N	38.37 Ab	8.58 Bb	16.18 Ab	4.24 Bb
CV (%)	19.13		22.85	

Averages followed by the same capital letters in lines and small letters in columns for each genotype do not differ by Tukey's test at the 5% significance level.

Conclusion

Sunflower hybrids presented different traits regarding B fertilization. The combination of a larger supply of N associated with B is essential for BRS 321 grain production and oil yield. The Neon genotype is more responsive in terms of grain production, exhibiting less input when associated with B fertilization. Results indicate leaf blades as the most sensitive part for evaluating sunflower plant N *status*.

Acknowledgements

The authors are grateful to Petrobras Biocombustível (Petro Bio Girassol) and the National Council for the Improvement of Higher Education (CAPES) for the scholarship and grants that supported this work. This work is part of the MSc. dissertation of the first author in Programa de Pós-Graduação em Agronomia – Ciência do Solo (CPGA-CS) at Universidade Federal Rural do Rio de Janeiro (UFRRJ).

References

Abdel-Motagally, F. M. F., & Osman, E. A. (2010). Effect of Nitrogen and Potassium Fertilization Combinations on Productivity of Two Sunflower Cultivars under East of El-ewinate Conditions. *American-Eurasian Journal of Agricultural & Environmental Sciences*, *8*(4), 397-401.

Al-Amery, M. M., Hamza, J. H., & Fuller, M. P. (2011). Effect of Boron Foliar Application on Reproductive Growth of Sunflower (*Helianthus annuus* L.). *International Journal of Agronomy*, *2011*(2011), ID 230712. doi: 10.1155/2011/230712

Ali, A., & Noorka, I. R. (2013). Nitrogen and phosphorus management strategy for better growth and yield of sunflower (*Helianthus annuus* L.) hybrid. *Soil Environment*, *32*(1), 44-48.

Aquino, L. A., Silva, F. D. B., & Berger, P. G. (2013). Características agronômicas e o estado nutricional de cultivares de girassol irrigado. *Revista Brasileira de Engenharia Agrícola e Ambiental*, *17*(5), 551-557.

Avice, J.-C., & Etienne, P. (2014). Leaf senescence and nitrogen remobilization efficiency in oilseed rape (*Brassica napus* L.). *Journal of Experimental Botany*, *65*(14), 3813-3824.

Braz, M. R. S., & Rossetto, C. A. V. (2010). Acúmulo de nutrientes e rendimento de óleo em plantas de girassol influenciados pelo vigor dos aquênios e pela densidade de semeadura. *Semina: Ciências Agrárias*, *31*(suppl. 1), 1193-1204.

Broadley, M., Brown, P., Cakmak, I., Rengel, Z., & Zhao, F. (2012). Function of nutrients: micronutrients. In *Marschner's Mineral Nutrition of Higher Plants* (3nd ed.). Londres, UK: Academic Press is an imprint of Elsevier.

Castro, C., Moreira, A., Oliveira, R. F., & Dechen, A. R. (2006). Boro e estresse hídrico na produção do girassol. *Ciência e Agrotecnologia*, *30*(2), 214-220.

Empresa Brasileira de Pesquisa Agropecuária [EMBRAPA]. (2011). *Novas cultivares de girassol da Embrapa chegam ao mercado*. Retrieved on May 15, 2014 from www.spm.embrapa.br/noticias/noticia_completa/ 115/

Ferreira, D. F. (2011). SISVAR: A Computer statistical analysis system. *Ciência e Agrotecnologia*, *35*(6), 1039-1042.

Foloni, J. S. S., Garcia, R. A., Cardoso, C. L., Teixeira, J. P., & Grassi Filho, H. (2010). Desenvolvimento de grãos e produção de fitomassa do girassol em função de adubações boratadas. *Bioscience Journal*, *26*(2), 273-280.

Krudnak, A., Wonprasaid, S., & Machikowa, T. (2013). Boron affects pollen viability and seed set in sunflowers. *African Journal of Agricultural Research*, *8*(2), 162-166.

Lima, A. D., Viana, T. V. A., Azevedo, B. M., Marinho, A. B., & Duarte, J. M. (2013). Adubação borácica na cultura do girassol. *Revista Agro@ambiente On-Line*, 7(3), 269-276.

Lobo, T. F., Grassi Filho, H., & Brito, I. C. A. (2011). Efeito do nitrogênio na nutrição do girassol. *Bioscience Journal*, *27*(3), 380-391.

Malavolta, E., Vitti, G. C., Oliveira, S. A. (1997). *Avaliação do estado nutricional de plantas: princípios e aplicações*. Piracicaba, SP: Potafós.

Oyinlola, E. Y. (2007). Effect of Boron Fertilizer on Yield and Oil Content of Three Sunflower Cultivars in the Nigerian Savanna. *Journal of Agronomy*, *6*(3), 421-426.

Rasool, F. U., Hassan, B., & Jahangir, I. A. (2013). Growth and yield of sunflower (*Helianthus annus* L.) as influenced by nitrogen, sulphur and farmyard manure under temperate conditions. *SAARC Journal of Agriculture*, *11*(1), 81-89. doi.org/10.3329/sja.v11i1.18386

Silva, H. R. F., Aquino, L. A., & Batista, C. H. (2011). Residual effect of phosphate fertilizer on sunflower productivity in succession to cotton. *Bioscience Journal*, *27*(5), 786-793.

Tahir, M., Ashraf, S., & Ibrahim, M. (2013). Effect of Foliar Application of Boron on Yield and Quality of Sunflower (*Helianthus annuus* L.). *Crop & Environment*, *4*(1), 23-27.

Tedesco, J. M., Gianello, C., Bissani, C. A., Bohnen, H., & Volkweiss, S. J. (1995). *Análise de solo, plantas e outros materiais* (2a ed.). Porto Alegre, RS: UFRGS.

Xu, G., Fan, X., & Miller, A. J. (2012). Plant nitrogen assimilation and use efficiency. *Annual Review of Plant Biology*, *63*, 153-82.

Yaniv, Z., Shabelsky, E., Schafferman, D., Granot, I., & Kipnis, T. (1998). Oil and fatty acid changes in *Sinapsis* and *Crambe* seeds during germination and early development. *Industrial Crops and Products*, *9*(1), 1-8.

Zahoor, R., Ahmad Basra, S. M., Munir, H., Nadeem, M. A., Yousof, S., & Ousaf, S. H. Y. (2011). Role of Boron in Improving Assimilate Partitioning and Achene Yield in Sunflower. *Journal of Agricultural & Social Sciences*, 7(2), 49-55.

3

The spatial and temporal independence of Italian Zucchini production

Vilson Benz*, Alessandro Dal'Col Lúcio and Sidinei José Lopes

*Programa de Pós-graduação em Agronomia, Departamento de Fitotecnia, Universidade Federal de Santa Maria, Av. Roraima, 1000, 97105-900, Camobi, Santa Maria, Rio Grande do Sul, Brazil. *Author for correspondence. E-mail: vibenz@gmail.com*

ABSTRACT. It is necessary to know the production variability of experimental units in a protected environment, both in terms of space and throughout different harvests. This knowledge helps reduce experimental error and increases result reliability in combination with the adoption of appropriate experimental techniques and cultural management. The objective of this study was to characterize the spatial and temporal independence of Italian Zucchini fruit production as cultivated in a protected environment. The study used production data from exploratory experiments under protected cultivation in the Department of Plant Sciences of the Federal University of Santa Maria. Different units sizes were simulated on the basis of known production values for the number of plants per row. To verify the independence or randomness of fruit production, a test run was applied to the units within the rows of individual and grouped harvests and in the same unit between individual and grouped harvests. There was a greater randomness of production in units with a higher number of plants. Using a lower number of harvests randomizes the production of units as the harvests are carried out. Groupings of harvests and units formed by four and five basic units are the most efficient combination for reducing non-randomness in the production of Italian Zucchinis.

Keywords: randomness, protected cultivation, experimental accuracy, experimental planning, plot size, test of uniformity.

Introduction

Horticulture is a highly intensive activity. It requires sophisticated technology and is constantly changing, demanding a higher quality of technical information than other fields. The cultivation of vegetable crops is characterized by a need for less physical space; however, it is widely used both in space and time (FILGUEIRA, 2003).

No matter how well-controlled an experiment is, there are always random variations within the same treatment, and these variations are known as the experimental error (STEEL et al., 1997). The higher the experimental error, the lower the probability of having significant differences between the treatment's mean. According to Storck et al. (2011), if the causes of the error are known, the error can be controlled and maintained at an acceptable level.

One way to reduce experimental error arising from unit heterogeneity is good experimental planning with an appropriate experimental design, number of repetitions, size and form of units and sample sizes. In protected cultivation experiments, the following factors become a source of variability and contribute to the growth of heterogeneity among units: the presence or absence of harvested fruits, the difference in air temperature due to row positions, damage caused to plants by fruit harvesting, small differences in the quantity of irrigation water, drainage differences, trampling between rows, presence of insect pests, diseases and weeds (COUTO et al., 2009; LOPES et al., 1998; LORENTZ et al., 2005; LÚCIO et al., 2008; ZANARDO et al., 2010).

One method for evaluating experimental area heterogeneity is the use of uniformity testing or exploratory experiments. In these cases, the individual plant production data are collected to identify the existing area variation (STORCK et al., 2011). Based on the data generated in exploratory experiments, the productivity behavior within the units or harvests can be identified. It is necessary to know the production variability of experimental units within the protected environment, both in terms of space and different harvest times because the dependence of experimental units is very often disregarded or ignored. There are not many studies that would identify the production behavior while observing the tendencies within the environment, with the objective of reducing experimental error and ensuring higher reliability of results and which would adopt appropriate experimental techniques and cultural management.

The objective of this study was to characterize the spatial and temporal dependence of Italian Zucchini fruit production within a protected environment.

Material and methods

This study employed Italian Zucchini fruit production data from two exploratory experiments that were performed in plastic greenhouses at the Department of Plant Sciences of the Federal University of Santa Maria. The experiments were carried out at coordinates 29° 43' 23" S and 53° 43' 15" W and an altitude of 95 m. The climate of this region is classified as Cfa humid subtropical with no dry season and a hot summer according to the Köppen classification (MORENO, 1961).

The first experiment was performed during the summer/autumn and the second in the winter/spring of 2001. Both experiments were composed of eight rows of 20 plants, and each plant constituted a basic unit (BU). The first experiment included 12 harvests and the second one had 30 harvests.

The plants were identified by row number and by their position inside the row. The values of fresh fruit biomass (FFB, in g) per individual BU and at sizes of 1, 2, 4 and 5 BU per unit were simulated for the rows.

To verify the randomness of data distribution, a run test was applied to characterize the spatial and temporal dependence of production (BEAVER et al., 1974; COSTA NETO, 2002) among units within the row in individual and grouped harvests and to the same unit between individual and grouped harvests.

For each harvest unit and grouping size, a median value was calculated for the formation of classes (1 and 2). Class 1 was formed by elements below the median (n) and class 2 was formed by the elements above the median (m) where the signs "-" and "+" were adopted, respectively. The elements equal to the median were included in class 1.

The n elements of class 1 and the m elements of class 2 were numbered on the basis of the order in which they occurred. The number of equal symbols were preceded and followed by different symbols or by no symbol to determine the number of sequences, or in other words, the "c" value.

When the number of units per row was lower than 20, the test was carried out with a determination of the least significant difference (LSD), which is equal to double the lowest value between α1 p ($c \leq c_1$) and α2 p ($c \geq c_2$) (tabulated by the run test). The sequence was considered random when the LSD was higher than 5% and non-random when it was lower than 5%. When the number of units per row was higher than 20, the test was completed by using an approximation of the normal distribution based on the equation $Z = c - E(c)/ \sqrt{V}\ (c)$, where c is the number of observed sequences, $V(c) = 2{\star}m{\star}n(2{\star}m{\star}n - m - n) / (m + n)^2 \star (m + n - 1)$ and $E(c) = 1 + (2{\star}m{\star}n / m + n)$. The assumption of randomness was rejected if $z > |z\ \alpha/2|$ at a 5% probability level.

Results and discussion

There were cases of non-random fresh fruit biomass (FFB) distribution within the rows during the first cultivations in summer-autumn and the second cultivation during winter-spring. These cases were composed of units of one or two plants, and they included individual or grouped harvests (Table

1). These results may have occurred as a result of environmental influences and plant rotation, in which the production is concentrated in rows, contrary to what happened in units of more than two plants and with the harvest groupings in which the variability was lower in space and time. While evaluating the air temperature distribution in a protected environment, Furlan and Folegatti (2002) verified the distribution to be non-homogeneous and noted that opening the side curtains provided isotherms in which higher air temperatures occurred in the center of the protected environment. In the evaluation of Italian Zucchini growth dynamics, Strassburger et al. (2011) found that the availability of solar radiation during the cultivation period strongly affected the production and distribution of dry matter, as well as all the culture growth indexes of Italian Zucchini. The growth rates of zucchini plants and their fruits are higher when the available solar radiation is higher. Considering the influence of these factors on the development of the plant culture, it is possible that the lack of randomness observed in smaller row units was caused by environmental variations.

While evaluating the harvest totals in the first experiment, there was randomness in the production regardless of the unit size. However, there were rows without randomness in the second experiment that were formed by units of one BU, the difference of which could be explained by a higher number of harvests.

In the experimental units composed of one or two BU, a higher number of percentage of zero values was observed. In other words, there were units with no row production (Tables 2 and 3). The lack of production per plant had a direct influence on the results of the sequence test, giving rise to rows with lack of randomness. In other words, within the units composed of two plants, a lack of production by one plant influenced the neighboring plant. The same pattern occurred in the units that yielded no harvest, where in some cases the groupings had a lower randomness than the individual harvests. According to Couto et al. (2009), smaller unit sizes are often used in vegetable crops experiments because of area limitations, which reflect data dispersion problems by the emergence of zero values in the variables indicating fruit production.

Table 1. Percentage of rows with spatial and temporal independence for different sizes of units and groupings of Italian Zucchini fruits (*Cucurbita pepo*) harvests in two plastic greenhouse experiments. Santa Maria, Rio Grande do Sul State, 2012.

		no. of basic units per plot			
Experiment	Harvests	1	2	4	5
	Spatial Grouping				
1st	Individual	98.96	98.66	100	100
	2 in 2	97.92	95.83	100	100
	3 in 3	100	90.62	100	100
	4 in 4	95.83	95.83	100	100
	6 in 6	100	100	100	100
	Total	100	100	100	100
2nd	Individual	96.25	96.66	100	100
	2 in 2	96.66	96.66	100	100
	3 in 3	98.75	98.75	100	100
	5 in 5	100	93.75	100	100
	6 in 6	97.50	97.50	100	100
	10 in 10	100	95.83	100	100
	15 in 15	93.75	87.50	100	100
	Total	87.50	100	100	100
	Temporal Grouping				
1st	Individual	98.75	96.25	100	100
	2 in 2	100	100	100	100
	3 in 3	100	100	100	100
	4 in 4	100	100	100	100
2nd	Individual	94.38	95.00	82.50	93.75
	2 in 2	100	98.75	100	100
	3 in 3	100	100	100	100
	5 in 5	100	100	100	100
	6 in 6	100	100	100	100
	10 in 10	100	100	100	100

Different simulations showed that the growth of a unit's size and/or harvest groupings led to a reduction in the number of zeros in rows, with an increased average, median and standard deviation and reduction of the coefficient of variation (CV%). Spatial evaluation indicated that the sizes of four and five BU in the unit provided values of CV%, which was similar to the harvest groupings of the units with one and two plants. However, these results did not represent lack of randomness behavior within the rows (Tables 2 and 3).

To overcome the problem of lack of randomness in the production of FFB at certain points in the row, the best option is to use units formed by four and five BU. However, because there is often a lack of space for experiments, units formed by two plants and groupings of the total harvests can be a way to reduce experimental error within the rows. Carpes et al. (2010) found similar results in which the variances became homogeneous, starting with the harvest grouping with the presence of units with no fruits ready to be harvested. Thus, they reduced interference, preventing the disruption by zero values of variance estimations between the harvests. Moreover, Feijó et al. (2005) suggest using units with six plants and groupings of the initial half of the productive period to estimate experimental error with a greater precision for Italian Zucchini experiments.

Table 2. Medium, median and standard deviation (S.D.) in grams, percentage of zero and coefficient of variation (CV%) of the spatial evaluation of fresh Italian Zucchini (*Cucurbita pepo*) fruit biomass in rows of different unit sizes and harvest groupings in summer-autumn. Santa Maria, Rio Grande do Sul State, 2012. 1st Experiment.

Basic Units (BU) per plot	Grouping	Medium	Median	S.D.	% Zero	CV%
	Individual	92.09	51.96	122.38	61.88	132.88
	2 in 2	184.18	148.39	179.93	38.44	97.69
1 BU	3 in 3	276.28	241.59	232.22	26.09	84.05
	4 in 4	368.37	360.85	262.68	17.92	71.31
	6 in 6	552.55	550.53	353.11	11.56	63.90
	Total	1105.11	1154.00	613.17	9.38	55.49
	Individual	184.18	151.31	167.84	40.83	91.13
	2 in 2	368.37	333.90	238.74	16.04	64.81
2 BU	3 in 3	552.55	503.91	318.80	5.63	57.70
	4 in 4	736.74	717.67	352.62	2.92	47.86
	6 in 6	1105.11	1072.31	467.06	0.00	42.26
	Total	2210.21	2216.25	783.51	0.00	35.45
	Individual	368.37	339.58	228.61	22.08	62.06
	2 in 2	736.74	731.21	315.92	3.75	42.88
4 BU	3 in 3	1105.11	1094.59	448.26	0.63	40.56
	4 in 4	1473.48	1554.17	464.22	0.00	31.51
	6 in 6	2210.21	2303.31	662.49	0.00	29.97
	Total	4420.43	4381.75	1094.03	0.00	24.75
	Individual	460.46	437.17	270.82	18.23	58.82
	2 in 2	920.92	920.28	374.85	2.08	40.70
5 BU	3 in 3	1381.38	1410.34	536.92	0.78	38.87
	4 in 4	1841.84	1880.38	604.26	0.00	32.81
	6 in 6	2762.77	2800.31	850.29	0.00	30.78
	Total	5525.53	5547.06	1503.60	0.00	27.21

Table 3. Medium, median and standard deviation (S.D.) in grams, percentage of zero and coefficient of variation (CV%) in the spatial evaluation of fresh Italian Zucchini (*Cucurbita pepo*) fruit biomass from rows of different unit sizes and harvest groupings in winter-spring. Santa Maria, Rio Grande do Sul State, 2012. 2nd Experiment.

Basic Units (BU) per plot	Grouping	Medium	Median	S.D.	% Zero	CV%
	Individual	188.77	159.85	167.99	43.54	88.99
	2 in 2	377.55	357.92	245.38	22.08	64.99
	3 in 3	566.32	552.53	307.78	13.13	54.35
1 BU	5 in 5	943.87	956.70	395.61	4.27	41.91
	6 in 6	1132.65	1153.79	446.43	2.50	39.42
	10 in 10	1887.75	1909.02	632.74	0.83	33.52
	15 in 15	2831.62	2948.72	852.29	0.31	30.10
	Total	5663.24	5826.50	1454.35	0.00	25.68
	Individual	377.55	349.10	235.49	23.33	62.37
	2 in 2	755.10	744.56	344.07	7.75	45.57
	3 in 3	1132.65	1128.98	434.62	2.63	38.37
2 BU	5 in 5	1887.75	1903.57	559.73	0.63	29.65
	6 in 6	2265.30	2298.05	654.44	0.25	28.89
	10 in 10	3775.49	3801.27	920.66	0.00	24.39
	15 in 15	5663.24	5749.03	1212.02	0.00	21.40
	Total	11326.48	11450.75	1967.80	0.00	17.37
	Individual	755.10	730.24	331.93	10.25	43.96
	2 in 2	1510.20	1492.43	482.03	1.83	31.92
	3 in 3	2265.30	2274.74	606.10	0.00	26.76
4 BU	5 in 5	3775.49	3733.94	762.30	0.00	20.19
	6 in 6	4530.59	4556.60	966.54	0.00	21.33
	10 in 10	7550.98	7482.25	1256.33	0.00	16.64
	15 in 15	11326.48	11292.44	1741.67	0.00	15.38
	Total	22652.95	22865.13	2616.99	0.00	11.55
	Individual	943.87	926.94	368.38	7.40	39.03
	2 in 2	1887.75	1861.55	547.52	0.83	29.00
	3 in 3	2831.62	2825.24	701.66	0.00	24.78
5 BU	5 in 5	4719.36	4640.19	883.36	0.00	18.72
	6 in 6	5663.24	5602.25	1085.57	0.00	19.17
	10 in 10	9438.73	9293.94	1483.06	0.00	15.71
	15 in 15	14158.09	14016.47	2052.83	0.00	14.50
	Total	28316.19	28100.69	3205.52	0.00	11.32

During the evaluation of randomness in harvests of the units in the time the influence of the total number of harvests was verified (Table 1). With a smaller number of harvests, there was a greater tendency to observe random production values, which can be verified by the frequency of high randomness percentage values in the first experiment (12 harvests) in comparison to the second one (30 harvests).

In the first experiment, an absence of randomness was observed in the temporal production of individual harvests, regardless of the size of the BU. A rotation occurred among low and high fruit production values, which concentrated in lower production values during the first harvests, intermediate and final ones.

A total randomness in the time of production was observed in the first experiment in all of the harvest groupings. However, there were some units formed by two BUs during the second experiment that exhibited a lack of randomness among harvests grouped two by two, which indicated that fruit production was concentrated mainly in the first half of the productive period.

While grouping the harvests every three in three, compliance with randomness occurred. However, this grouping in different sizes of unit was very close to non-randomness based on the high number of formed sequences, showing a rotation among groups of three harvests with greater and smaller fruit production and also a concentration of production between the tenth and fifteenth harvests.

The low number of harvests and, consequently, the sequences formed in the first experiment may have influenced the randomness of the grouping production. This situation was not verified in the second experiment, even with a higher number of harvests and sequences (30). This result is based on the work of Lorentz et al. (2005), who suggested a low number of harvests instead of total production to achieve homogeneous production and faster and more reliable experiments.

The first experiment showed a lack of randomness in individual harvests with units of one and two BU. The second experiment resulted in a lack of randomness at all unit sizes when simulated in the individual harvests. Both experiments were predominated by a high percentage of units with no production and high coefficients of variation for the individual harvests. However, the unit sizes from four and five BUs in the second experiment showed coefficient of variation and percentage of zero values lower than those in the first experiment. These results did not present the problems of high randomness, indicating that the number of harvests may have been more influenced than the variability of the data (Tables 4 and 5).

To minimize the observed lack of randomness in the temporal production of FFB within the units, it is possible to reduce the number of harvests and avoid using the data from individual units. Using harvest groupings and experimental units of more than two plants were the most efficient combinations for reducing non-randomness in the cultivation of Italian Zucchini. This result is attributed to tendencies within the rows, in space or throughout the harvests over time.

Regardless of the cultivation period or the form of randomness evaluation (temporal or spatial), the factors that lacked randomness did not depend on the variability of production, but on how this production was distributed within the rows or in the harvests, forming the sequences above and below the median.

Generally, for different cultivation periods and situations in which some rows were lacking randomness, this result was not associated with the use of smaller unit sizes. In these cases, the class 1 and 2 sequences did not happen by chance; they was caused by the low number of formed sequences. This result shows that the use of smaller unit sizes can help concentrate production in specific regions within the rows. This trend can be attributed to temperature variations, humidity and damage caused during harvests and/or crop handling, among other factors.

The higher the size of the experimental unit in BU numbers, the smaller the number of units within the rows. Because the sequence test was based on the number of observed sequences, the bigger the size the simulated unit is, the smaller the number of sequences within the rows are and, therefore, the higher the probability of obtaining the results for rows with random production.

Table 4. Medium, median and standard deviation (S.D.) in grams, percentage of zero and coefficient of variation (CV%) for the temporal evaluation of fresh Italian Zucchini (*Cucurbita pepo*) fruit biomass in rows with different unit sizes and harvest groupings during the summer-autumn season. Santa Maria, Rio Grande do Sul State, 2012. 1st Experiment.

Basic Units (BU) per plot	Grouping	Medium	Median	S.D.	% Zero	CV%
	Individual	92.09	36.77	131.47	61.88	142.76
1 BU	2 in 2	184.18	135.55	188.89	38.44	102.56
	3 in 3	276.28	252.41	202.65	26.09	73.35
	4 in 4	368.37	316.72	229.50	17.92	62.30
	Individual	184.18	121.48	213.16	40.83	115.73
2 BU	2 in 2	368.37	297.94	320.04	16.04	86.88
	3 in 3	552.55	519.28	332.61	5.63	60.19
	4 in 4	736.74	618.16	412.13	2.92	55.94
	Individual	368.37	303.24	349.89	22.08	94.98
4 BU	2 in 2	736.74	609.53	553.97	3.75	75.19
	3 in 3	1105.11	1063.70	548.71	0.63	49.65
	4 in 4	1473.48	1242.88	741.40	0.00	50.32
	Individual	460.46	378.20	421.90	18.23	91.63
5 BU	2 in 2	920.92	777.88	668.73	2.08	72.62
	3 in 3	1381.38	1355.08	646.80	0.78	46.82
	4 in 4	1841.84	1586.47	895.79	0.00	48.64

Table 5. Medium, median and standard deviation (S.D.) in grams, percentage of zero and coefficient of variation (CV%) for the temporal evaluation of fresh Italian Zucchini (*Cucurbita pepo*) fruit biomass in rows of different unit sizes and harvest groupings during the winter-spring season. Santa Maria, Rio Grande do Sul State, 2012. 2nd Experiment.

Basic Units (BU)per plot	Grouping	Medium	Median	S.D.	% Zero	CV%
	Individual	188.77	130.70	225.26	43.54	119.33
	2 in 2	377.55	314.73	343.28	22.08	90.92
1 BU	3 in 3	566.32	492.44	449.53	13.13	79.38
	5 in 5	943.87	853.68	543.62	4.27	57.59
	6 in 6	1132.65	1053.09	519.32	2.50	45.85
	10 in 10	1887.75	1877.31	536.05	0.83	28.40
	Individual	377.55	289.18	380.06	23.33	100.66
	2 in 2	755.10	636.36	604.47	7.75	80.05
2 BU	3 in 3	1132.65	982.26	814.14	2.63	71.88
	5 in 5	1887.75	1697.74	995.99	0.63	52.76
	6 in 6	2265.30	2040.61	940.69	0.25	41.53
	10 in 10	3775.49	3690.04	936.22	0.00	24.80
	Individual	755.10	577.93	677.63	10.25	89.74
	2 in 2	1510.20	1272.33	1106.92	1.83	73.30
4 BU	3 in 3	2265.30	1984.65	1526.94	0.00	67.41
	5 in 5	3775.49	3399.90	1878.49	0.00	49.75
	6 in 6	4530.59	4040.18	1743.81	0.00	38.49
	10 in 10	7550.98	7474.70	1612.09	0.00	21.35
	Individual	943.87	739.30	823.91	7.40	87.29
	2 in 2	1887.75	1590.41	1356.37	0.83	71.85
5 BU	3 in 3	2831.62	2434.41	1887.82	0.00	66.67
	5 in 5	4719.36	4266.09	2324.08	0.00	49.25
	6 in 6	5663.24	5127.28	2148.13	0.00	37.93
	10 in 10	9438.73	9257.09	1967.29	0.00	20.84

Conclusion

There is greater randomness of production in units with a higher number of plants.

Units composed of two basic units grouping harvest totals or units of four or five BU, in any grouping of harvests, generate rows with randomness in the production of FFB in the units.

A lower number of harvests randomizes the production of units while the harvests are carried out.

Harvest groupings and units formed by four and five basic units are the most efficient combination for reducing non-randomness for Italian Zucchini production. This efficiency is attributed to tendencies within the rows and associated with the space or time of the harvests.

Acknowledgements

To National Council of Technological and Scientific Development (Conselho Nacional de Desenvolvimento Científico e Tecnológico – CNPq) for granting the scholarship.

References

BEAVER, R.; MENDENHALL, W.; REINMUTH, J. **Statistics for management and economics**. 2nd ed. North Scituate: Duxbury, 1974.

CARPES, R. H.; LÚCIO, A. D.; LOPES, S. J.; BENZ, V.; HAESBAERT, F. M.; SANTOS, D. Variabilidade produtiva e agrupamentos de colheitas de abobrinha italiana cultivada em ambiente protegido. **Ciência Rural**, v. 40, n. 2, p. 294-301, 2010.

COSTA NETO, P. L. O. **Estatística**. 2. ed. São Paulo: Edgard Blücher, 2002.

COUTO, M. R. M.; LÚCIO, A. D.; LOPES, S. J.; CARPES, R. H. Transformação de dados em experimentos com abobrinha italiana em ambiente protegido. **Ciência Rural**, v. 39, n. 6, p. 1701-1707, 2009.

FEIJÓ, S.; OLIVEIRA, S. J. S.; STORCK, L.; LÚCIO, A. D.; DAMO, H. P.; MARTINI, L. F. D. Repetibilidade da produção de frutos de abobrinha italiana. **Revista Brasileira de Agrociência**, v. 11, n. 1, p. 39-43, 2005.

FILGUEIRA, F. A. R. Agrotecnologia moderna na produção e comercialização de hortaliças. **Novo manual de olericultura**. 2. ed. Viçosa: UFV, 2003.

FURLAN, R. A.; FOLEGATTI, M. V. Distribuição vertical e horizontal de temperaturas do ar em ambientes protegidos. **Revista Brasileira de Engenharia Agrícola e Ambiental**, v. 6, n. 1, p. 93-100, 2002.

LOPES, S. J.; STORCK, L.; HELDWEIN, A. B.; FEIJÓ, S.; ROS, C. A. Técnicas experimentais para tomateiro tipo salada sob estufas plásticas. **Ciência Rural**, v. 28, n. 2, p. 193-197, 1998.

LORENTZ, L. H.; LUCIO, A. D.; BOLIGON, A. A.; LOPES, S. J.; STORCK, L. Variabilidade da produção de frutos de pimentão em estufa plástica. **Ciência Rural**, v. 35, n. 2, p. 316-323, 2005.

LÚCIO, A. D.; CARPES, R. H.; STORCK ,L.; LOPES, S. J.; LORENTZ, L. H.; PALUDO, A. L. Variância e média da massa de frutos de abobrinha-italiana em múltiplas colheitas. **Horticultura Brasileira**, v. 26, n. 3, p. 335-341, 2008.

MORENO, J. A. **Clima no Rio Grande do Sul**. Porto Alegre: Secretaria da Agricultura, 1961.

STEEL, R. G. D.; TORRIE, J. H.; DICKEY, D. A. **Principles and procedures of statistics**: a biometrical approach. New York: McGraw- Hill, 1997.

STORCK, L.; GARCIA, D. C.; LOPES, S. J.;

ESTEFANEL, V. **Experimentação vegetal**. 3. ed. Santa Maria: UFSM, 2011.

STRASSBURGER, A. S.; PEIL, R. M. N.; FONSECA, L. A.; AUMONDE, T. Z.; MAUCH, C. R. Dinâmica de crescimento da abobrinha italiana em duas estações de cultivo. **Acta Scientiarum. Agronomy**, v. 33, n. 2, p. 283-289, 2011.

ZANARDO, B.; LÚCIO, A. D.; PALUDO, A. L.; SANTOS, D.; BENZ, V. Posições das mudas de alface nas bandejas de poliestireno e efeitos na normalidade e homogeneidade dos erros na produção de plantas. **Revista Ciência Agronômica**, v. 41, n. 2, p. 285-293, 2010.

4

The production of *Physalis* spp. seedlings grown under different-colored shade nets

Daniel Fernandes da Silva[1*], Rafael Pio[2], Joyce Dória Rodrigues Soares[2], Paulyene Vieira Nogueira[3], Pedro Maranha Peche[3] and Fabíola Villa[4]

*[1]Programa de Pós-graduação em Botânica Aplicada, Departamento de Biologia, Universidade Federal de Lavras, 37200-000, Cx. Postal 3037, Lavras, Minas Gerais, Brazil. [2]Departamento de Agricultura, Universidade Federal de Lavras, Lavras, Minas Gerais, Brazil. [3]Programa de Pós-graduação em Fitotecnia, Departamento de Agricultura, Universidade Federal de Lavras, Lavras, Minas Gerais, Brazil. [4]Centro de Ciências Agrárias, Universidade Estadual do Oeste do Paraná, Marechal Cândido Rondon, Paraná, Brazil. *Author for correspondence. E-mail: daniel_eafi@yahoo.com.br*

ABSTRACT. The objective of this study was to evaluate the production of seedlings of *Physalis* L. species under different-colored shade nets. Four shade nets individually stained white, blue, red and black, all with 50% shading, were used in this study, and an additional treatment (control) was used in which seedlings were grown in full sun. The study examined four species of *Physalis*, namely, *P. peruviana*, *P. pubescens*, *P. minima* and *P. ixocarpa*. The experiment followed a randomized block design with three blocks and 25 seeds per plot. The species were sown in styrofoam trays. Germination was monitored daily to calculate the Emergency Velocity Index (EVI) and stabilize the overall percentage of emergence. Height, stem diameter, number of leaves, leaf area index and dry mass of seedlings were assessed at 50 days after sowing. The study found that these species react differently to changes in the light spectrum. Seedlings of *P. peruviana* should be grown under a white or red shade net; of *P. pubescens* under a white or black shade net; of *P. minima* under a white, red or black shade net; and of *P. ixocarpa* under a white shade net. For all species, 50% shade should be used.

Keywords: *Physalis* L., sexual propagation, seedling morphology, light spectrum.

Introduction

Physalis, commonly known as fisális in Brazil and as Physalis in English, is a genus comprising more than over one hundred species. Belonging to the family Solanaceae these species are characterized by a permanent and inflated calyx that surrounds and protects the fruit against herbivores and weather. Among the many species of the genus, some have potential for agronomic exploitation due to their nutritional and medicinal compounds, known for some time in popular culture and, more recently, through studies by researchers from several institutions (Silva et al., Villa, Barp, Rotili, & Stumm, 2013; Muniz et al., 2014).

Among the *Physalis* species whose fruits, which are rich in compounds beneficial to human health, have potential as food are *Physalis peruviana* L., *Physalis pubescens* L., *Physalis minima* L. and *Physalis ixocarpa* Brot.

Seedling growth is one of the most important stages in orchard production. The seedling stage directly influences performance and production, the return on invested capital and the quality of the fruit produced (Campanharo, Rodrigues, Lira Junior, Espindula, & Costa, 2006).

The attenuation of solar radiation is one of the most important factors in the production of seedlings, as this factor acts directly to affect the energy balance and, consequently, the environmental conditions (Hernandes, Pedro-Junior, & Bardin, 2004).

One method of reducing the radiation intensity and also modulating the quality of the incident radiation is to use colored shade nets.

These nets may differ in their transmission efficiency or diffusion of scattered light and also in their ability to distribute the light passing directly through them. These characteristics and differences depend on the physical properties of the nets (Oren-Shamir et al., 2001). The use of colored shade nets increases the growth capacity of the plants and, therefore, influences the quality of the product.

The quality of seedlings can be specified in terms of appropriate morphological and physiological parameters. The use of morphological evaluation by researchers is increasing because it is relatively easy to understand and intuitive for nursery professionals, whereas physiological assessment requires a knowledge of parameters that are not always simple and easy to measure (Gomes, Couto, Leite, Xavier, & Garcia, 2002).

In view of the importance of quality seedlings and the ability to enhance seedling quality by manipulating the light spectrum with shade nets, this study evaluated the seedling production of four species of Physalis grown under shade nets of different colors.

Material and methods

The shade materials used for the plants consisted of Cromatinet® shade nets, Polysack brand, with 50% shading, in four different colors: white, blue, red and black. The nets were deployed on a wooden frame from which they could be removed. The frame measured 3.0 x 3.0 x 1.5 m in width, length and height, respectively. In a control treatment, plants were grown in full sunlight.

Seeds of four species of Physalis were used, namely, *P. peruviana*, *P. pubescens*, *P. minima* and *P. ixocarpa*. The seeds were collected from mature fruits, which were dried in the shade, and then placed into polypropylene trays with 128 cells, using one seed per cell. The substrate used was a mix of soil (dystrophic red clay latosol characteristic of the region) plus commercial substrate, 1:1 v v^{-1}. After sowing, the trays were watered and placed under shade nets, and the seeds were allowed to germinate.

We used a DBC statistical design, a 4 x 5 factorial scheme with four species of Physalis and four color shade nets, in addition to the full-sun control treatment. Each treatment consisted of three replications of 25 seeds each. From the time of sowing, plants were monitored daily.

We evaluated daily the number of seedlings to calculate the Emergence Speed Index according to the equation proposed by Maguire (1962). At the end of this evaluation, 30 days after sowing, we calculated the total percentage of emergence.

Seedlings grown under shade conditions were taken into the field after 50 days. New evaluations were then performed, namely, height (cm), measured with a ruler; stem diameter (mm), measured with calipers; number of leaves; leaf area, measured with the LICOR® LAI-2200 apparatus; and dry mass, measured by drying of the seedlings in the greenhouse for three days and then weighing.

Five seedlings were evaluated as replicates, and an overall average was calculated for each parameter evaluated. We analyzed the data obtained in the experiment with Sisvar software (Ferreira, 2011). An analysis of variance was used. A Scott-Knott cluster test was used. The statistical significance level applied was 5%.

Results and discussion

The results showed significant interaction for the variables emergence speed index (ESI), height, stem diameter, number of leaves and total dry mass. The treatments were also significant for the species, for the color of the shade net, for the leaf area index, and, only among species, for the percentage of emergence.

There was no significant interaction between the color of the shade net and the species studied in relationship to the percentage of emergence. The results were significant only among species, where *P. ixocarpa* had a lower percentage of emergence than the other species (Table 1).

These results demonstrate that the four species of Physalis studied did not have their germination inhibited by the presence of light, demonstrating that positive or neutral photoblastic behavior was operative.

Environmental variations have the ability to facilitate the expression of certain genetic characteristics that may or may not be suitable for adaptive responses for a given plant at a given site

(Botezelli, Davide, & Malavasi, 2000). Thus, seeds of different origins show genetic variability and show relationships between this variability and the environment. This observation sheds light on the lower percentage of emergence in *P. ixocarpa*.

Table 1. Emergency percentage and leaf area index (LAI) of seedlings of four *Physalis* species grown under different-colored shade nets.

Species	Emergency percentage (%)*
P. peruviana	80.53a
P. pubescens	77.86a
P. minima	78.66a
P. ixocarpa	68.53b
CV (%)	1.54
Species	**Leaf Area Index***
P. peruviana	1.89a
P. pubescens	1.72a
P. minima	2.16a
P. ixocarpa	1.30b
CV (%)	5.97
Shade net color	**Leaf Area Index***
Pleno sol	2.58a
Branca	1.51b
Azul	1.71b
Vermelha	1.53b
Preta	1.43b
CV (%)	5.97

*Means followed by the same lower case letter in the column do not differ by Skott-Knott test at 5% probability.

The Emergence Speed Index (ESI) of *Physalis* seeds showed a significant interaction between species and the color of the shade net. However, *P. peruviana* was the only species to show a significant difference in ESI when grown under different-colored shade nets (Table 2). For this species, the red and white nets accelerated the germination of the seeds.

These results corroborate those obtained by Lessa, Ferreira, Araújo Neto, and Souza (2013) and Ferraresi, Yamashita, and Carvalho (2009), who also found superiority of red and white light for the germination of *Emilia coccinea* (Sims) G. Don. and *Murdannia nudiflora* (L.) Brenans seeds, respectively. The high ESI for the seeds germinated under the red and white nets is associated with phytochrome activity. To absorb light in the red range, phytochrome switches to its active form, promoting germination.

The white light treatment includes similar amounts of red and far-red light. However, under these light conditions, the predominant active form of phytochrome facilitates the outcome that the germination percentage and the speed thereof are magnified, accelerating the emergence of seedlings.

The lower value of ESI for *P. peruviana* seed occurred in full sun. This can be explained by the greater loss of substrate water when compared with the treatments under shade nets, regardless of color, due to the direct incident solar radiation on the plants. The presence of shade nets helps in maintaining the relative humidity by hindering ventilation (Oliveira, Leitão, & Rocha, 2012). Thus, the presence of the substrate screens can reduce water loss to the environment, allowing higher rates of imbibition of the seed by accelerating seed germination.

Table 2. Emergency Speed Index (ESI), length (cm), diameter (mm), number of leaves and dry mass (g) of seedlings of four Physalis species grown under different-colored shade nets.

Shade net color	*P. peruviana*	*P. pubescens*	*P. minima*	*P. ixocarpa*
	Emergency Speed Index (ESI)*			
Full sun	6.17Cc*	6.06Ac	18.27Aa	12.96Ab
White	11.83Ab	8.30Ac	18.24Aa	11.26Ab
Blue	9.37Bc	7.23Ac	16.94Aa	12.92Ab
Red	11.03Ac	6.59Ad	17.48Aa	13.85Ab
Black	8.60Bc	8.14Ac	15.04Aa	11.04Ab
CV (%)	13.37			
	Stem length (cm)*			
Full sun	1.95Ac*	1.67Ac	3.51Bb	10.15Ca
White	2.57Ac	2.09Ac	6.41Ab	17.00Aa
Blue	2.22Ac	2.27Ac	4.58Bb	8.40Da
Red	2.63Ac	2.20Ac	5.71Ab	12.85Ba
Black	2.21Ac	1.89Ac	5.38Ab	12.68Ba
CV (%)	13.18			
	Stem diameter (mm)*			
Full sun	1.54Ab*	1.33Ab	2.62Aa	2.74Ba
White	1.63Ac	1.08Ad	2.69Ab	3.93Aa
Blue	1.41Ac	1.20Ac	2.46Aa	1.99Cb
Red	1.54Ab	1.36Ab	2.98Aa	3.02Ba
Black	1.53Ab	1.03Ac	2.81Aa	2.97Ba
CV (%)	11.97			
	Number of leaves*			
Full sun	4.87Ac*	4.80Ac	7.20Bb	10.47Ba
White	5.67Ac	5.07Ac	10.13Ab	12.13Aa
Blue	5.40Ab	5.20Ab	8.20Ba	6.53Cb
Red	5.80Ab	5.20Ab	10.27Aa	10.53Ba
Black	4.33Ab	4.67Ab	9.73Aa	10.73Ba
CV (%)	12.12			
	Dry mass (g)*			
Full sun	0.51Aa	0.51Aa	0.48Aa	0.36Ab
White	0.05Cc	0.08Cc	0.20Bb	0.42Aa
Blue	0.06Cb	0.12Cb	0.26Ba	0.19Ba
Red	0.07Cb	0.06Cb	0.33Ba	0.36Aa
Black	0.19Bb	0.36Ba	0.51Aa	0.44Aa
CV (%)	24.35			

*Means followed by the same lowercase letter in uppercase line and column do not differ by Skott-Knott test at 5% probability.

The effects of the white net on the light spectrum are neutral, not interfering with the transmitted light spectrum. However, differences between the ESI seeds germinated under greenhouse white and full sun were observed (Table 2). This result is consistent with the environmental conditions provided by the white net.

Although this shade net does not cause changes in light quality, it can promote increases of ambient temperature relative to the external environment by hindering the air circulation due to the resistance to airflow, as mentioned above (Oliveira et al., 2012), and this factor is decisive for the germination of *P. peruviana* seed. However, it was quite close to the results found by the same authors when testing a second batch of seeds. For the species in all treatments, the result for *P. minima* was higher, which may be related to genetic factors linked to the colonizing ability of the environments by the various species.

There was a significant interaction between species and color of shade net for the length of the Physalis seedlings. The results showed that *P. minima* and *P. ixocarpa* were sensitive to the color of the shade net, displaying significantly different results. For *P. minima*, greater lengths were observed in seedlings grown under greenhouse white, red and black, and *P. ixocarpa* presented greater length seedlings when grown only under greenhouse white (Table 2, Figure 1).

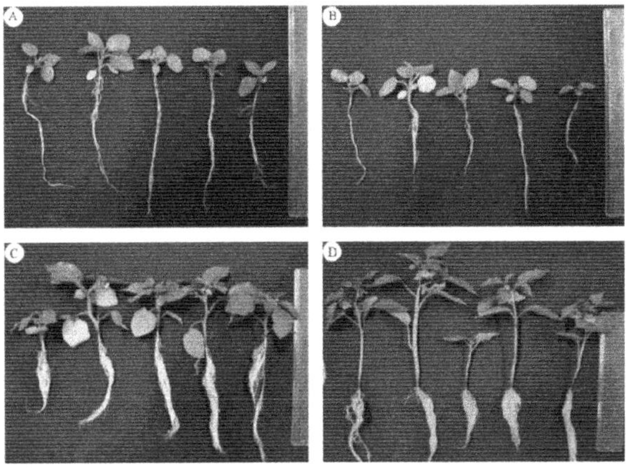

Figure 1. Seedlings of *P. peruviana* (A), *P. pubescens* (B), *P. minimum* (C) and *P. ixocarpa* (D) with 50 days of age, growing under full sun, white, blue, red and black shading net (left to right).

The longer length of the seedlings when grown under red and white shade nets is related to the greater balance in the relationship R:FR that they offer. Variations in the ratios R:FR stimulate responses to stem elongation. This stretching is connected to the phytochrome, which regulates the transport of plant growth regulators, among them, auxin. These regulators are observed in plants grown in bright cultures with red light and are degraded in crops cultivated under blue light. Thus, the shoot elongation and apical dominance are common under red light. However, the blue light acts to inhibit the growth of the plant. Thus, it is common to observe of side shoots in plants grown under blue light, depending on the breaking of the apical dominance due to the degradation of auxins (Oren-Shamir et al., 2001).

Thus, the effect of the increased intensity of the spectrum in the red range caused by manipulating with shade nets in stem elongation, together with the adverse effect of the blue shade net, was previously reported, e.g., by Oren-Shamir et al. (2001) and Ovadia, Dori, Nissim-Levi, Shahak, and Oren-Shamir (2009) in five ornamental species.

The plants may exhibit greater length when exposed to favorable environmental conditions, which allow them a high photosynthetic performance, resulting in greater amount of photosynthate, or when exposed to light deficiency, elongating their strategy to achieve organs such as those induced by light in higher strata. The reduction of light intensity without improvement in the quality of the light offered for black shade net can justify the growth of *P. minima* seedlings under the greenhouse conditions in this same growth intensity of seedlings grown under the red and white shade nets, which contain a more favorable light spectrum; however, in seedlings grown under black shade nets, rapid growth can lead to the formation of poor etiolated seedlings (Santos, Morais, Borsoi, Secco, & Moreira, 2010).

The occurrence of low rates in the growth of plants grown in full sun, relative to plants under shade nets, is a widespread result in the literature (Oliveira, Castro, Costa, & Oliveira, 2009; Holcman & Sentelhas, 2013; Souza, Oliveira, Silva, & Lima, 2013). This effect is a result of excessive light intensity, capable of generating damage to photosynthetic tissues (Araújo & Deminicis, 2009). However, the most different results are observed across species, although some have greater height in unshaded conditions. Examples include rosemary (*Rosmarinus officinalis* L.) and guanandi (*Calophyllum brasiliensis* Cambess.); under blue screens, raffia palm (*Rhapis excelsa* (Thunberg) Henry ex Rehder.) and basil (*Ocimum gratissimum* L.); under black shade nets, the *Anthurium andraeanum* Lindl, or exhibit similar growth in different color screens as the case of coffee seedlings (*Coffea arabica* L.) (Meirelles, Paiva, Oliveira, & Tavares, 2007; Martins, Alvarenga, Castro, Pinto, & Silva, 2008; Nomura et al., 2009; Henrique, Alves, Deuner, Goulart, & Livramento, 2011; Souza, Silva, Oliveira, Santos Neto, & Santos, 2014).

In increasing the thickness of the stems of Physalis seedlings in different colors, there was also a significant interaction between the factors. *P. ixocarpa* was the only species to show a significant difference between the treatments, showing diameter of the thicker stem when grown under a white shade net, similar to the pattern observed for seedling height (Table 2).

Corroborating Souza, Silva, Oliveira, Santos Neto, and Santos (2014), increased seedling stem diameter is a characteristic of the great phenotypic plasticity of some species. This occurs with the function of supporting the plant canopy. Thus, similarly to the results that the above authors found in rosemary plants, the treatment in which plants had higher length was also presented the greatest thickness of the stem of the seedlings of *P. ixocarpa*.

Although *P. minima* also showed differences in the length of the plants, the smaller amplitude between the results in relation to the seedlings of *P. ixocarpa*, made this species did not present significant difference for stem diameter. The lengthening of the changes, not accompanied by an increase in the thickness of the stem, can be unbalanced evidence provided in the light and may lead the seedling to shading and loss of quality.

The increased thickness of the stem is a desirable feature, as well as giving support to the mechanical room (Souza et al., 2013), a higher stem thickening can meet the increased need for transport of sap that feeds this room (Freitas, Oliveira, Carvalho, Santos, & Santos, 2007).

For the number of leaves in Physalis seedlings, *P. minima* showed a greater number of leaves under greenhouse white, red and black, *P. ixocarpa* only under the white greenhouse (Table 2). Both species showed higher number of leaves on the same treatment that was also observed higher stem length. These changes observed in the seedlings are results of photomorphogenesis, where low intensity and/ or light quality makes for greater growth and increase in the number of leaves as a strategy to meet the verified deficiency.

The increase in the number of leaves directly reflects the plant leaf area because this factor is dependent on the number, size and residence time leaf in plant (Monteiro et al., 2005). Among the species, decreasing values of the LAI were presented by *P. minima*, *P. peruviana*, *P. pubescens* and *P. ixocarpa*, in that order, that being only the last was significantly different from the others (Table 1).

Increased leaf area in plants, in turn, allows more light interception, with total potential favoring photosynthesis and, consequently, increased production of photosynthate that can be made available for plant growth and fruit production (Reis, Azevedo, Albuquerque, & Silva Junior, 2013).

The leaf area index (LAI) is of paramount importance for us to model the growth and development of plants, and therefore the productivity of the crop (Reis et al., 2013). In IAF analyses of four species of Physalis grown under shade nets of different colorations, each factor alone was significant (Table 1).

The order found for LAI in four species of *Physalis* studied is attributed to the morphology of each species where *P. peruviana* and *P. pubescens* have wider and long leaves for *P. minima* and *P. ixocarpa*. *P. minima*, although with relatively small leaves, also exhibits greater proximity between these, by having short internodes due to the small size of the species, which causes the projection to the ground to be greater than those in the other species, although this presents the only significant difference from *P. ixocarpa*.

The full sun environment elicited an increase in LAI fisális seedlings compared with the other treatments (Table 1). Unlike most results obtained in the literature, where the leaf expansion for light disability compensation is observed, fisális species had higher LAI in full sun.

Most of the projection area in the shadow of the leaves on the ground (LAI) in full sun were driven by the growth of these species in this treatment, so that the presentation of changes in leaf disposition in angularity, size, shape and curvature of the leaves can be expressed better as horizontal distribution.

Similar results were reported by Fagundes et al. (1999) employed in forage grasses, that the largest observed LAI was attributed to better architecture and arrangement of leaves. In addition, Lima, Silva, Moraes, Dantas, and Almeida (2008) found higher leaf area in *Caesalpinia ferrea* Mart ex Tul. seedlings when grown in full sun.

Another hypothesis supported by Bezerra Neto et al. (2005) based on studies with lettuce is that when subjected to excessive radiation, leaves of certain species can acquire particular morphology as an adaptation strategy and may thus alter its leaf area.

The dry mass of the analysis of the seedlings showed a significant interaction between species and shade net color (Table 2). In the dry mass of seedlings, full sun was more significant, providing greater accumulation in all species, but in *P. minima* and *P. ixocarpa*, other shade net colors also provided the same dry mass accumulation, which was not different from the full-sun treatment; they are the black shade net for *P. minima* and the white, red and black shade nets for *P. ixocarpa*.

The results showed that seedlings of *P. peruviana* and *P. pubescens* grown in the climatic conditions of Lavras, are insensitive to the intensity and quality of light, with no significant difference in height growth, stem diameter, number of leaves, which resulted in seedlings with equal dry weight, in all treatments for these species.

In addition to the full-sun treatment, under a black shade net, *P. minima* seedlings also had better outcomes in relation to dry biomass accumulation. This mass accumulation is one plant response to the reduction in the intensity of quality without implementation of light quality because the black shade net only reduces the light, without converting light to the red and blue spectrum, which is theoretically more favorable for photosynthesis.

High dry mass accumulation with a black shade net was also reported by Henrique, Alves, Deuner, Goulart, and Livramento (2011) in coffee seedlings, which had lower dry mass compared to the seedlings grown under a red shade net, from five luminous qualities evaluated.

The quality of light offered after passing through the blue shade net inhibited the growth of *P. ixocarpa* seedlings. Due to the low performance achieved in the evaluation of the length, diameter and number of leaves, the seedlings also had a lower dry mass accumulation when grown under this condition. The low solar radiation in the red band and the high incidence of blue light can affect auxin synthesis, also affecting the growth of plants, and consequently its biomass accumulation.

Lower dry mass results in plants under blue shade nets compared with other colors were also reported by Corrêa, Pinto, Reis, and Moreira (2012) and Henrique, Alves, Deuner, Goulart, and Livramento (2011), in *Origanum vulgare* L. and coffee, respectively. However, the accumulation of dry biomass is related to the response of each species to the light spectrum (Brant et al., 2009) and is not standardized to the lowest and highest accumulation in a single color.

After evaluating all of the parameters, you can see that the modulation of light quality during the training period of *Physalis* seedlings is able to influence its quality. Therefore, the analysis of aspects performed in this study, combined with the physiological and anatomical reviews context, become necessary for a final definition of the best light spectrum for the formation of seedlings of each species to reduce the time, cost and increase the quality of the seedlings to the producer.

Conclusion

Physalis peruviana, *Physalis pubescens*, *Physalis minima* and *Physalis ixocarpa* react differently to the light spectrum used for training quality seedlings. Seedlings of *P. peruviana* should be formed under white or red shade nets; *P. pubescens* under white or black shade nets; *P. minima* under white, red or black shade nets; and *Physalis ixocarpa* under white shade nets.

Acknowledgements

The authors would like to thank the Capes and Fapemig for the financial support.

References

Araújo, S. A. C., & Deminicis, B. B. (2009). Fotoinibição da fotossíntese. *Revista Brasileira de Biociências, 7*(4), 463-472.

Bezerra Neto, F., Rocha, R. H. C., Rocha, R. C. C., Negreiros, M. Z., Leitão, M. M. V. B. R., Nunes, G. H. S., ... Queiroga, R. C. L. F. (2005). Sombreamento para produção de mudas de alface em alta temperatura e ampla luminosidade. *Horticultura Brasileira, 23*(1), 133-137.

Botezelli, L., Davide, A. C., & Malavasi, M. M. (2000). Características dos frutos e sementes de quatro procedências de *Dipteryx alata* Vogel (Baru). *Cerne, 6*(1), 9-18.

Brant, R. S., Pinto, J. E. B. P., Rosa, L. F., Albuquerque, C. J. B., Ferri, P. H., & Corrêa, R. M. (2009). Crescimento, teor e composição do óleo essencial de melissa cultivada sob malhas fotoconversoras. *Ciência Rural, 39*(5), 1401-1407.

Campanharo, M., Rodrigues, J. J. V., Lira Junior, M. A., Espindula, M. C., & Costa, J. V. T. (2006). Características físicas de diferentes substratos para produção de mudas de tomateiro. *Caatinga, 19*(2), 140-145.

Corrêa, R. M., Pinto, J. E. B. P., Reis, E. S., & Moreira, C. M. (2012). Crescimento de plantas, teor e qualidade de óleo essencial de folhas de orégano sob malhas coloridas. *Global Science and Technology, 5*(1), 11-22.

Fagundes, J. L., Silva, S. C., Pedreira, C. J. S., Sbrissia, A. F., Carnevalli, R. A., Carvalho, C. A. B., & Pinto, L. F. M. (1999). Índice de área foliar, interceptação luminosa e acúmulo de forragem em pastagens de *Cynodon* spp. sob diferentes intensidades de pastejo. *Scientia Agricola, 56*(4), 1141-1150.

Ferraresi, D. A., Yamashita, O. M., & Carvalho, M. A. C. (2009). Superação da dormência e qualidade de luz na germinação de sementes de *Murdannia nudiflora* (L.) Brenans. *Revista Brasileira de Sementes, 31*(4), 126-132.

Ferreira, D. F. (2011). SISVAR: A computer statistical analysis system. *Ciência e Agrotecnologia, 35*(6), 1039-1042.

Freitas, Z. M. T. S., Oliveira, F. J., Carvalho, S. P., Santos, V. F., & Santos, J. P. O. (2007). Avaliação de caracteres quantitativos relacionados com o crescimento vegetativo entre cultivares de café arábica de porte baixo. *Bragantia, 66*(2), 267-275.

Gomes, J. M., Couto, L., Leite, H. G., Xavier, A., & Garcia, S. L. R. (2002). Parâmetros morfológicos na avaliação da qualidade de mudas de *Eucalyptus grandis*. *Revista Árvore, 26*(6), 655-664.

Henrique, P. C., Alves, J. D., Deuner, S., Goulart, P. F. P., & Livramento, D. E. (2011). Aspectos fisiológicos do desenvolvimento de mudas de café cultivadas sob telas de diferentes colorações. *Pesquisa Agropecuária Brasileira, 46*(5), 458-465.

Hernandes, J. L., Pedro-Junior, M. J., & Bardin, L. (2004). Variação estacional da radiação solar em ambiente externo e no interior de floresta semidecídua. *Revista Árvore, 28*(2), 167-172.

Holcman, E., & Sentelhas, P. C. (2013). Crescimento e desenvolvimento de bromélias em ambiente protegido, cobertos com PEBD e diferentes malhas de sombreamento. *Horticultura Brasileira, 31*(3), 386-391.

Lessa, B. F. T., Ferreira, V. M., Araújo Neto, J. C., & Souza, R. C. (2013). Germinação de sementes de *Emilia coccinea* (Sims) G. DON em função da luminosidade, temperatura, armazenamento e profundidade de semeadura. *Semina: Ciências Agrárias, 34*(6), 3193-3204.

Lima, J. D., Silva, B. M. S., Moraes, W. S., Dantas, V. A. V., & Almeida, C. C. (2008). Efeitos da luminosidade no crescimento de mudas de *Caesalpinia ferrea* Mart. ex Tul. (Leguminosae, Caesalpinoideae). *Acta Amazônica, 38*(1), 5-10.

Maguire, J. D. (1962). Speeds of germination-aid selection and evaluation for seedling emergence and vigor. *Crop Science, 2*(2), 176-177.

Martins, J. R., Alvarenga, A. A., Castro, E. M., Pinto, J. E. B. P., & Silva, A. P. O. (2008). Avaliação do crescimento e do teor de óleo essencial em plantas de *Ocimum gratissimum* L. cultivadas sob malhas coloridas. *Revista Brasileira de Plantas Medicinais, 10*(4), 102-107.

Meirelles, A. J. A., Paiva, P. D. O., Oliveira, M. I. O., & Tavares, T. S. (2007). Influência de diferentes sombreamentos e nutrição foliar no desenvolvimento de mudas de Palmeiras Ráfia *Rhapis excelsa* (Thunberg) Henry ex. Rehder. *Ciência e Agrotecnologia, 31*(6), 1884-1887.

Monteiro, J. E. B. A., Sentelhas, P. C., Chiavegato, E. J., Guiselini, C., Santiago, A. V., & Prela, A. (2005). Estimação da área foliar do algodoeiro por meio de dimensões e massa das folhas. *Bragantia, 64*(1), 15-24.

Muniz, J., Kretzschmar, A. A., Rufato, L., Pelizza, T. R., Rufato, A. R., & Macedo, T. A. (2014). General aspects of *physalis* cultivation. *Ciência Rural, 44*(6), 964-970.

Nomura, E. S., Lima, J. D., Rodrigues, D. S., Garcia, V. A., Fuzitani, E. J., & Silva, S. H. M. (2009). Crescimento e produção de antúrio cultivado sob diferentes malhas de sombreamento. *Ciência Rural, 39*(5), 1394-1400.

Oliveira, G. M., Leitão, M. M. V. B. R., & Rocha, R. C. (2012). Temperatura do ar no interior e exterior de ambientes protegidos. *Verde de Agroecologia e Desenvolvimento Sustentável,* 7(2), 250-257.

Oliveira, M. I., Castro, E. M., Costa, L. C. B., & Oliveira, C. (2009). Características biométricas, anatômicas e fisiológicas de *Artemisia vulgaris* L. cultivada sob telas coloridas. *Revista Brasileira de Plantas Medicinais, 11*(1), 56-62.

Oren-Shamir, M., Gussakovsky, E. E., Shpiegel, E., Nissim-Levi, A., Ratner, K., Ovadia, R., ... Shahak, Y. (2001). Coloured shade nets can improve the yield and quality of green decorative branches of *Pittosporum variegatum*. *Journal of Horticultural Science and Biotechnology, 76*(3), 353-361.

Ovadia, R., Dori, I., Nissim-Levi, A., Shahak, Y., & Oren-Shamir, M. (2009). Coloured shade-nets influence stem length, time to flower, flower number and inflorescence diameter in four ornamental cut-flower crops. *Journal of Horticultural Science and Biotechnology, 84*(2), 161-166.

Reis, L. S., Azevedo, C. A. V., Albuquerque, A. W., & Silva Junior, J. F. (2013). Índice de área foliar e produtividade do tomate sob condições de ambiente protegido. *Revista Brasileira de Engenharia Agrícola e Ambiental, 17*(4), 386-391.

Santos, R. F., Morais, L., Borsoi, A., Secco, D., & Moreira, D. C. (2010). Níveis de sombreamento na produção e desenvolvimento de mudas *Eucalyptus grandis* Hill ex Maiden. *Pesquisa Aplicada e Agrotecnologia, 3*(3), 201-206.

Silva, D. F., Villa, F., Barp, F. K., Rotili, M. C. C., & Stumm, D. R. (2013). Conservação pós-colheita de fisális e desempenho produtivo em condições edafoclimáticas de Minas Gerais. *Revista Ceres, 60*(6), 826-832.

Souza, G. S., Oliveira, U. C., Silva, J. S., & Lima, J. C. (2013). Crescimento, produção de biomassa e aspectos fisiológicos de plantas de *Mentha piperita* L. cultivadas sob diferentes doses de fósforo e malhas coloridas. *Global Science and Technology, 6*(3), 35-44.

Souza, G. S., Silva, J. S., Oliveira, U. C., Santos Neto, R. B., & Santos, A. R. (2014). Crescimento vegetativo e produção de óleo essencial de plantas de alecrim cultivadas sob telas coloridas. *Bioscience Journal, 30*(1), 232-239.

5

Rhizobia inoculation and liming increase cowpea productivity in Maranhão State

Thiago Palhares Farias[1], André Trochmann[1], Bruno Lima Soares[2] and Fatima Maria Souza Moreira[1*]

*[1]Departamento de Ciências do Solo, Universidade Federal de Lavras, Cx. Postal 3037, 37200-000, Lavras, Minas Gerais, Brazil. [2]Departamento de Agricultura, Universidade Federal de Lavras, Lavras, Minas Gerais, Brazil. *Author for correspondence. E-mail: fmoreira@dcs.ufla.br*

ABSTRACT. The objetives of the study were to evaluate the agronomic efficiency of strains approved as inoculants for cowpea and of three new strains under selection, as well as to evaluate the influence of liming on the symbiosis and productivity. Two field experiments were conducted using cowpea (BRS Guariba) from June to September 2013. In the first experiment, a completely randomized block experimental design was used, with four replicates and seven treatments (INPA03-11B, UFLA03-84, UFLA03-153, UFLA03-154, UFLA03-164 and two controls). In the second experiment, a 7 x 2 factorial design was used, with the above described treatments being the first factor and liming as the second factor. Parameters evaluated were: nodule dry matter, shoot dry matter, shoot N content and accumulation, number of pods per plant, number of seeds per pod, weight of 100 seeds, grain yield and grain N content and accumulation. Strains UFLA03-153 and UFLA03-164 were more efficient for cowpea inoculation than the strains approved as inoculants, UFLA03-84 and INPA03-11B. Although the tested rhizobia strains and cultivar BRS Guariba are tolerant to soil acidity, productivity washigher when soil was limed. Yields obtained with fertilized and inoculated treatments were well above Maranhão state average.

Keywords: *Vigna unguiculata*, *Bradyrhizobium*, acidity tolerance, efficiency, production system, biology nitrogen fixation.

Introduction

The expansion of cowpea [*Vigna unguiculata* (L.) Walp.] throughout Brazil has consolidated the importance of this crop at national level. Cowpea, in addition to being a traditional crop in northern and northeastern Brazil, is being integrated into the strategic plan for grain production in the Brazilian Cerrado (Freire Filho et al., 2011). Cowpea is grown throughout the State of Maranhão, in different production systems. This state is the fourth biggest producer of cowpea in Northeast Brazil, but it has low productivity indexes, estimated as 510 kg ha^{-1} (Instituto Brasileiro de Geografia e Estatística [IBGE], 2015). The crop is traditional in family-based agricultural systems but has been introduced to the production systems of the Maranhão "Cerrado" over the past few years, and the southern part of this region is becoming the largest cowpea producer in the State.

This crop establishes symbioses with nitrogen-fixing bacteria, which allow plants to satisfy their nutritional nitrogen needs, resulting in low production costs and increased productivity. However, the efficiency of the symbiosis depends on both the environment and genetic factors of both symbionts (Moreira & Siqueira, 2006).

Highly weathered soils predominate in Brazil, which can present chemical restrictions to agricultural production. Acidity is one of the most limiting features, and it can restrict the productivity of leguminous plants and their biological nitrogen fixation (BNF) (Kopittke, Blamey, Wang, & Menzies, 2011).

Many plants have developed morpho-physiological mechanisms to adapt to environmental adversities, making the effects dependent on the species and even on the genotype (Bian, Zhou, Sun, & LI, 2013). Due to its tropical origin, certain genotypes of cowpea developed tolerance to soil acidity (Jemo, Abaidoo, Nolte, & Horst, 2007). Because BNF evolved under the same conditions, various species and strains of N_2-fixing bacteria have also adapted to the same constraints (Soares et al., 2014).

However, depending on the level of tolerance, leguminous plants and N_2-fixing bacteria present varied responses to acidity (Zahran, 1999). Thus, if limiting factors such as soil pH, aluminum and manganese toxicity, and nutrient deficiency restrict the development of the macrosymbiont, the diazotrophs will not be able to efficiently fix N_2. Chemical soil restrictions connected to acidity therefore need to be corrected, and liming is often used because it acts directly on chemical attributes linked to acidity, thereby improving soil fertility. However, the recommendation of liming for production of cowpea inoculated with N_2-fixing bacteria must be based on the different symbiont responses because their magnitude will define the need for soil correction and its economic viability.

Selection of new bacterial strains more efficient in nitrogen fixation in cowpea is important for the optimization of crop nutrition techniques. N_2-fixing efficiency is the main criterion to recommend a new rhizobia strain, and the process of selecting new strains should be rigorous to differentiate the levels of efficiency between different strains (Moreira & Siqueira 2006).

Despite the importance of this crop, few studies have been performed for the selection of new rhizobia strains that are efficient at nitrogen fixation in symbiosis with cowpea at Maranhão, and this practice is little used in this State, especially in family-based agriculture systems. Besides, even large producers who adopt inoculation for cowpea production mostly use strains approved for soybean. Furthermore, no studies on the influence of liming on this symbiosis and on cowpea productivity have been reported for Maranhão. Therefore, the objective of the study was to evaluate the agronomic efficiency of two strains approved as inoculants for cowpea, and three new strains under selection, as well as the influence of liming on the symbiosis and cowpea productivity in Maranhão State, Brazil.

Material and methods

Two experiments were conducted in the State of Maranhão (MA), one in the Center Region, in an area of family-based agriculture at Dom Pedro (MA), elevation 171 m, latitude 5° 05'S, longitude 44° 21' W, with typic Dystrophic Red-Yellow Argisol, and one in the North Region, at the Federal Institute of Education, Science and Technology of Maranhão (Instituto Federal de Educação, Ciência and Tecnologia do Maranhão - IFMA), São Luís Campus, Maracanã, at São Luís, Maranhão State, elevation 41 m, latitude 2° 36'S, longitude 44° 16'W, with typic Dystrophic Yellow Latosol, between June and September 2013. The climate of the first region is sub-humid ($C_2W_2A'a'$), and that of the second region is humid ($B_1WA'a'$), according to the Thornthwaite classification. Minimal and maximal temperature, and rainfall data for the duration of the experiments are presented in Figure 1. The chemical characteristics of the soil at 0-20 cm are shown in Table 1.

A completely randomized experimental blocks design was used at Dom Pedro, with four replicates and seven treatments [five inoculant strains: INPA03-11B, UFLA03-84, UFLA03-153, UFLA03-154 and UFLA03-164 + two non-inoculated controls – one with application of 80 kg ha^{-1} urea-N (W/N) and another without mineral N (N/N)]. A randomized experimental design blocks was used at São Luís, with four replicates and a 7 x 2 factorial scheme [two liming treatments (with and without)].

Three new nitrogen-fixing bacterial strains were tested, based on previous selection in experiments under axenic conditions and using pots containing soil (Soares et al., 2014). The tested strains were UFLA03-153, UFLA03-154 and UFLA03-164, which were isolated from areas of bauxite mining in the Poços de Caldas mountain range (Serra de Poços de Caldas), Minas Gerais state, and identified as *Bradyrhizobium* sp., *Burkholderia fungorum* and *B. elkanii*, respectively (Oliveira-Longatti et al., 2014; Soares et al., 2014). In addition, two strains previously approved as inoculants for cowpea, UFLA03-84 (*Bradyrhizobium* sp.) and INPA03-11B (*B. elkanii*), were also tested (Soares et al., 2006). Seeds were treated with 250 g peat-based inoculants per 10 kg seeds. The inoculants consisted of cultures of the various tested strains at the log phase, grown on a growth medium: peat mixture (7:4, w:v) with 10^{-9} CFU g^{-1}. Experimental plots consisted of six rows, 4 m long, sown with cowpea cv. BRS Guariba, with 0.50 m spacing between lines and eight plants per linear meter.

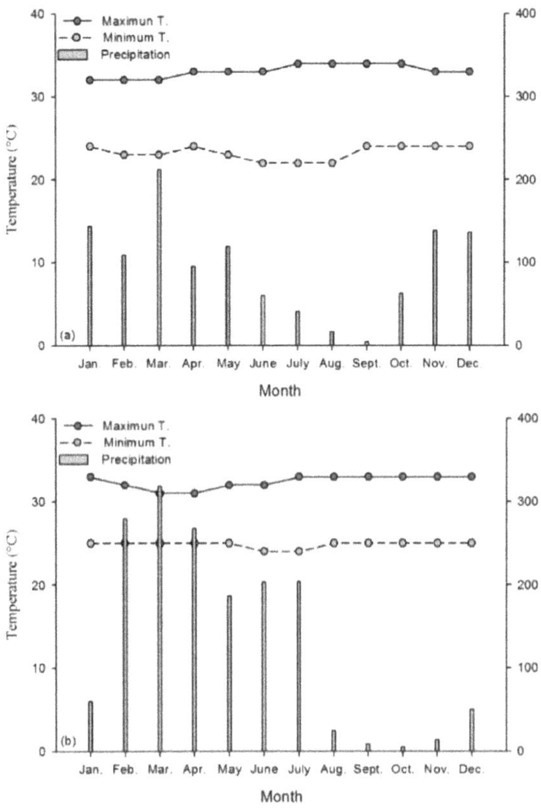

Figure 1. Minimum and maximun temperatures and mean precipitation by month in 2013 at Dom Pedro, Maranhão State (a) and São Luís, Maranhão State (b); Source: Instituto Nacional de Meteorologia – INMET.

Before planting, soils were sampled for assessment of the native rhizobia population density (approximate number g $soil^{-1}$, CFU) in a greenhouse experiment. Cowpea cv. BRS Guariba was grown as the trap plant, for 25 days, in wide-mouthed glass bottles containing Hoagland and Arnon (1950) nutrient solution inoculated with six soil dilutions (10^{-1} to 10^{-6}).

At Dom Pedro, there was no need for correction of soil acidity. Plantation fertilization was performed with application of 90 kg ha^{-1} P_2O_5 as triple superphosphate (45% P_2O_5). Potassium was not applied because the soil analysis revealed high potassium content. At São Luís, 2.53 t ha^{-1} dolomitic limestone, 40 kg ha^{-1} P_2O_5 as triple superphosphate (45% P_2O_5), and 40 kg ha^{-1} K_2O as potassium chloride (60% K_2O) were applied. The need for liming for neutralization of soil acidity was determined by increasing the base saturation to 70%.

Table 1. Chemical composition of soil samples (layer 0-20 cm)[1].

Soil Property	Unity	Municipality	
		Dom Pedro	São Luís
pH (H_2O)		6,2	5,2
P (Mehlich I)	mg dm^{-3}	3,84	24,62
K (Mehlich I)	mg dm^{-3}	288,0	22,0
Ca	$cmol_c$ dm^{-3}	2,60	0,50
Mg	$cmol_c$ dm^{-3}	1,30	0,20
Al	$cmol_c$ dm^{-3}	0,00	0,60
H+Al	$cmol_c$ dm^{-3}	2,08	3,62
SB	$cmol_c$ dm^{-3}	4,64	0,76
t	$cmol_c$ dm^{-3}	4, 64	1,36
T	$cmol_c$ dm^{-3}	6,72	4,38
m	%	0,00	44,12
V	%	69,02	17,27
OM	dag kg^{-}	3,84	1,99
Zn	mg dm^{-3}	4,05	1,04
Fe	mg dm^{-3}	182,02	212,56
Mn	mg dm^{-3}	116,40	1,60
Cu	mg dm^{-3}	0,89	0,24
B	mg dm^{-3}	0,35	0,26
S	mg dm^{-3}	4,38	3,32

(1)Analysis performed by the Soil Fertility Laboratory, Soil Science Department - UFLA. Ca and Mg were extracted with 1 mol L^{-1} KCl, S was extracted with monocalcium phosphate in acetic acid, H+Al was extracted with SMP, (OM) organic matter was extracted through Na2Cr2O7 4N + H2SO4 10 N oxidation, micronutrients were extracted with Mehlich 1, and B was extracted in hot water.

At full bloom, 35 days after emergence (DAE), ten plants were collected from the third and fourth rows of each plot. The following variables were quantified: nodule number (NN) and dry matter (NDM), shoot dry matter (SDM), relative efficiency (Efr) and shoot N content (SNC) and accumulation (SNA). At full maturity, 75 DAE, all plants were collected from the fourth and fifth rows of each plot. The variables quantified were number of pods per plant (NPP), number of seeds per pod (NSP), weight of 100 grains (p100), grain yield (GY), nitrogen grain content (GNC) and accumulation (GNA).

Shoot and nodule dry matters were determined following drying at 60°C in a forced-air circulation oven until constant weight was reached. N levels were determined according to the micro Kjeldahl method. N accumulation was calculated using the shoot and grain N contents. The former was calculated by multiplying shoot nitrogen content by the shoot dry matter and the latter by multiplying the grain nitrogen content by the grain yield with humidity corrected to 13%. Relative efficiency was calculated by dividing the shoot dry matter of inoculated plants by the shoot dry matter of plants with 80 kg ha^{-1} N and multiplying by 100.

The data were subjected to a variance analysis, at $p < 0.05$. When significant differences were observed, averages were compared using the Scott-Knott test, at $p < 0.05$, using the R software (R Development Core Team, 2013).

Results and discussion

Dom Pedro experimental area

Table 2 lists the mean values of the variables analyzed at the flowering stage (35 DAE) and maturation stage (75 DAE) of the experiment performed at Dom Pedro. The treatments had a significant effect on all the variables.

At the first assessment (35 DAE), there were no differences ($p > 0.05$) in the number of nodules between the treatments inoculated with the strains UFLA03-84, UFLA03-153, UFLA03-154 and UFLA03-164 and the uninoculated control without mineral nitrogen (N) supplementation (Table 2). Nodulation of the control without N revealed the nodulating ability of native rhizobia populations, estimated to have a density of 140 CFU g^{-1} soil, that are adapted to the local environmental and soil conditions.

The nodule dry matter was not significantly different ($p > 0.05$) among the treatments inoculated with the three new strains (UFLA03-153, UFLA03-154 and UFLA03-164) and the strain approved as an inoculant (UFLA03-84) (Table 2). Studies performed by Soares et al. (2014) using potted soil in Minas Gerais and by Ferreira et al. (2013) and Costa, Nóbrega, Martins, Amaral, and Moreira (2011) using field experiments in Piauí showed similar results. The Scott-Knott test indicated that the uninoculated treatments without mineral N supplementation and the treatments inoculated with INPA03-11B (another strain approved as an inoculant strain) had intermediate nodule dry matter values (Table 2).

The greatest shoot dry matter was observed in the treatments inoculated with UFLA03-164, which did not significantly differ ($p < 0.05$) from the treatments with application of 80 kg ha^{-1} mineral N. The remaining treatments exhibited shoot dry matter values that did not differ from each other (Table 2).

There were no significant differences in the relative efficiencies between the treatment inoculated with UFLA03-164 and the control with N fertilization (Table 2). According to Moreira and Siqueira (2006), BNF is dependent on the interaction between rhizobia, the legume host and the environment, further adding to the effects of genetic variation among symbionts. This interaction can explain the low efficiency of the strains UFLA03-84 and INPA03-11B, which did not provide different results ($p > 0.05$) from the uninoculated treatment without mineral N supplementation.

Table 2. Average number of nodules per plant (NN), nodule dry matter (NDM, mg $plant^{-1}$), shoot dry matter (SDM, g $plant^{-1}$), relative efficiency (Efr, %), shoot nitrogen content (SNC, %), and shoot nitrogen accumulation (SNA, mg $plant^{-1}$) at 35 days after emergence (DAE), number of pods per plant (NPP) (Unit), number of seeds per pod (NSP) (Unit), weight of 100 seeds (w100seeds, g), grain yield (GY, kg ha^{-1}), grain nitrogen content (GNC, %), and grain nitrogen accumulation (GNA, mg $plant^{-1}$) 75 DAE, of BRS Guariba cultivar at Dom Pedro municipality, Maranhão(1).

Treatment	NN Unid.	NDM mg $planta^{-1}$	SDM g $planta^{-1}$	Efr %	SNC %	SNA mg $planta^{-1}$
UFLA03-153	24,50a	137,45a	4,40b	53,02b	4,18b	184,91b
UFLA03-154	30,50a	158,30a	5,75b	68,93b	3,96c	232,48b
UFLA03-164	38,50a	185,84a	8,55a	104,01a	4,37b	375,38a
UFLA03-84	28,75a	160,42a	6,75b	80,87b	4,07c	274,56b
INPA03-11B	18,25b	97,85b	6,05b	74,40b	3,85c	232,98b
Control N/N[2]	34,25a	110,95b	4,90b	60,40b	4,26b	209,28b
Control W/N[2]	8,00c	13,57c	8,28a	100,00a	5,47a	454,47a
CV (%)	23,27	24,78	23,40	23,57	4,34	22,31

Treatment	NPP Unid.	NSP Unid.	p.100 g	GY kg ha^{-1}	GNC %	GNA mg $planta^{-1}$
UFLA03-153	4,09b	8,91a	21,38b	1.245,07b	4,22b	328,44b
UFLA03-154	3,43c	8,80a	20,90b	1.021,54c	4,04b	259,21c
UFLA03-164	3,93b	9,07a	21,28b	1.227,92b	4,27b	328,81b
UFLA03-84	3,35c	6,92b	21,66b	796,56c	4,57a	226,61c
INPA03-11B	3,04c	9,39a	21,11b	959,85c	4,54a	270,83c
Control N/N	2,93c	9,57a	21,74b	959,76c	4,35b	259,81c
Control W/N	4,83a	8,95a	23,22a	1.621,38a	4,87a	497,13a
CV (%)	12,46	9,93	4,38	15,82	2,40	16,75

(1)Means followed by the same lowercase letter within the same column or by the same uppercase letter within the same row belong to the same group according to the Scott-Knott test, at $p < 0.05$.
2N/N - uninoculated and with no mineral Nitrogen, W/N- uninoculated and with 80 kg ha^{-1} mineral N.

The N content in the shoot dry matter was higher for the treatment with N fertilization (80 kg ha^{-1}) compared to the other treatments (Table 2). There was no difference ($p > 0.05$) between the treatments inoculated with UFLA03-164 and UFLA03-153 and the uninoculated control without N supplementation, which exhibited intermediate levels. The treatments with UFLA03-154 had N levels in the shoot dry matter that were similar to the treatments inoculated with UFLA03-84 and INPA03-11B, forming a group that had the lowest mean values for this variable (Table 2). N levels in all treatments can be considered adequate for cowpea growth as all were above 3% (Anele et al., 2010; Costa et al., 2011).N accumulation in the shoot dry matter for the treatment inoculated with strain UFLA03-164 was not significantly different from the uninoculated treatment without mineral N supplementation (Table 2). Soares et al. (2014) in experiments with potted soil in Minas Gerais and Ferreira et al. (2013) in field experiments in Piauí have both reported similar results.

At the second evaluation time-point (75 DAE), the treatments showed significant effects for the variables analyzed during the maturation stage (Table 2). The treatment with the greatest number of pods per plant treatment was the uninoculated control supplemented with 80 kg ha^{-1} mineral N, followed by the treatments inoculated with strains UFLA03-153 and UFLA03-164. The Scott-Knott test ($p > 0.05$) indicated that there were no differences in the number of seeds per pod except for in the treatment inoculated with strain UFLA03-84, a result that may be related to the genotype of the cultivar BRS Guariba. The control treatment with N supplementation yielded the highest weight of 100 seeds (Table 2). The other treatments were not significantly different from each other. In the overall comparison for the three variables described above, the strain UFLA03-84 had the lowest values.

The highest grain yields for the inoculated treatments were observed for UFLA03-153 and UFLA03-164, which were not significantly different from each other (Table 2). The treatment with N supplementation (80 kg ha^{-1}) had the highest grain yield of all the treatments. Greater grain production as a result of N fertilization and increased yield of cowpeas inoculated with rhizobia were also reported by Santos, Aguiar, and Moura, (2011). These authors studied the effects of soil management on the efficiency of rhizobia-inoculated cowpeas in Maranhão and reported yields of approximately 1,700 kg ha^{-1}. This high yield was also obtained with 98 kg ha^{-1} of single superphosphate. Gualter, Boddey, Rumjanek, Freitas, and Xavier (2011) also examined the efficiency of rhizobial strains inoculated onto cowpeas in Maranhão and reported a maximum yield of 893 kg ha^{-1} on family farms. The treatments that these authors performed also included liming and phosphorus and potassium fertilization. The same authors observed a maximum yield of 679 kg ha^{-1} for the treatments inoculated with strains approved as inoculants, such as INPA03-11B, which was well below the yield observed for the same treatment at Dom Pedro (Table 2).

The different treatments could be divided into two different groups ($p < 0.05$) based on the grain N content (Table 2). The first group had higher values and consisted of the uninoculated treatment with mineral N supplementation, UFLA03-84 and INPA03-11B. The second group consisted of the remaining treatments. The highest grain N accumulation was observed for the uninoculated treatment with 80 kg ha^{-1} mineral N supplementation. The second group, in which the treatments had intermediate values of grain N accumulation, consisted of strains UFLA03-164 and UFLA03-153. The last group consisted of the treatments inoculated with strains UFLA03-154, UFLA03-84 and INPA03-11B and the uninoculated control without mineral N supplementation (Table 2). Considering all parameters the BNF efficiency for strains UFLA03-164 and UFLA03-153 can be considered satisfactory in the municipality of Dom Pedro.

São Luís experimental area

Table 3 lists the mean values of the variables analyzed at the flowering stage (35 DAE) and maturation stage (75 DAE) of the experiment performed at São Luís.

The interaction between liming and inoculants did not have a significant effect ($p > 0.05$) on the variables analyzed at flowering (Table 3). The N source had a significant effect on all the variables analyzed, and the Scott-Knott test ($p < 0.05$) revealed the agronomic efficiency of the strain UFLA03-153.

The average acidity of the soil, the high level of Al^{3+} saturation and Ca^{2+} and Mg^{2+} deficiency did not limit the nodulation of the strains evaluated (Table 3). According to Correa and Barneix (1997), rhizobia decrease their growth rate under acidic conditions as an adaptive response to their environment. However, high concentrations of H^+ and Al^{3+} can lead to a significant reduction in population density and survival of rhizobia in the soil. An analysis of the number and dry matter of nodules showed that liming did not have a

Table 3. Average number of nodules per plant (NN), nodule dry matter (NDM, mg plant^{-1}), shoot dry matter (SDM, g plant^{-1}), relative efficiency (Efr, %), shoot nitrogen content (SNC, %), and shoot nitrogen accumulation (SNA, mg plant^{-1}) at 35 days after emergence (DAE), number of pods per plant (NPP), number of seeds per pod (NSP), weight of 100 seeds (w100seeds, g), grain yield (GY, kg ha^{-1}), grain nitrogen content (GNC, %), and grain nitrogen accumulation (GNA, mg plant^{-1}) at 75 DAE, of BRS Guariba cultivar at São Luís municipality, Maranhão State[(1)].

Treatment	NN Unid.	NDM mg planta^{-1}	SDM g planta^{-1}	Efr %	SNC %	SNA mg planta^{-1}
UFLA03-153	23,38a	165,32a	4,75b	63,08b	4,25b	201,19b
UFLA03-154	28,25a	206,03a	4,83b	61,54b	4,41b	213,70b
UFLA03-164	22,50a	177,57a	5,00b	64,50b	4,22b	213,73b
UFLA03-84	22,50a	164,24a	4,83b	61,05b	4,25b	205,84b
INPA03-11B	27,25a	174,18a	5,50b	69,01b	3,97b	219,37b
Control N/N[2]	24,88a	149,84a	4,15b	52,03b	3,88b	160,71b
Control W/N	7,25b	33,77b	8,03a	100,00a	4,80a	386,27a
Liming						
Without	21,32a	151,37a	5,07a	66,75a	4,16b	214,57a
With	23,25a	154,61a	5,52a	67,60a	4,35a	242,80a
CV (%)	23,25	16,34	22,52	24,13	8,24	25,95

Treatment	NPP Unid.	NSP Unid.	p.100 g	GY kg ha^{-1}	GNC %	GNA mg planta^{-1} Limed	GNA mg planta^{-1} Without Liming
UFLA03-153	3,64b	9,34a	21,35b	1.153,41b	4,62a	367,41bA	300,91aB
UFLA03-154	3,56b	8,28b	20,92b	978,27c	4,22b	276,88cA	240,15bA
UFLA03-164	3,38b	8,61b	21,00b	970,20c	4,07c	274,11cA	220,28bA
UFLA03-84	3,68b	7,64b	20,83b	923,04c	3,94c	234,55cA	217,28bA
INPA03-11B	3,14b	9,59a	20,66b	980,36c	4,27b	274,94cA	248,18bA
Control N/N	3,59b	8,04b	21,12b	982,84c	4,14b	298,35cA	210,86bB
Control W/N	4,79a	8,00b	23,26a	1.430,08a	4,78a	504,38aA	354,90aB
Without liming	3,59a	8,15b	20,90b	959,31b	4,23b		
Limed	3,78a	8,85a	21,72a	1.160,18a	4,35a		
CV (%)	13,01	11,73	4,03	12,67	2,20	14,06	

[(1)]Means followed by the same lowercase letter within the same column or the same uppercase letter within the same row belong to the same group according to the Scott-Knott test, at $p < 0.05$. [2]N/N - uninoculated and with no mineral Nitrogen, W/N- uninoculated and with 80 kg ha^{-1} mineral N.

significant effect ($p > 0.05$) on the treatments (Table 3), indicating that the evaluated strains as well as the native population of rhizobia are tolerant to the soil pH and aluminum levels. These results are consistent with those reported by Soares et al. (2014), who showed "*in vitro*" tolerance of strains UFLA03-153, UFLA03-154, UFLA03-164, UFLA03-84 and INPA03-11B to a pH of 5.0 and to aluminum.

The development of adaptation mechanisms to adverse soil conditions in slow-growing strains has been reported by several authors (Keyser & Munns 1979; Correa & Barneix 1997; Lima et al., 2009; Ferreira, Bomfeti, Soares, & Moreira, 2012).

The number and dry matter of nodules were not significantly different ($p > 0.05$) between treatments except for the control with N supplementation, which had the fewest and lowest dry matter of nodules (Table 3). The values for these two variables were similar to those observed by other authors in field experiments conducted in Maranhão (Gualter et al., 2011; Santos et al., 2011).

The evaluation of the shoot dry matter, relative efficiency and shoot N content and accumulation indicated that the interaction between liming and the inoculants did not significantly affect the treatments ($p > 0.05$) (Table 3). The cultivar BRS Guariba showed acid tolerance given that the low pH of 5.2 and low natural fertility of the soil did not limit nodulation. In addition, root growth inhibition or symptoms of nutrient deficiency, especially for N, were not visually observed during the experiment, confirming the adaptation of cowpea cultivars to the acidity and low fertility of the soil observed by other authors (Jemo et al., 2007). However, the liming factor had a significant effect on the N content of the shoots (Table 3). The shoot N content was relatively high compared to levels reported by other authors (Costa et al., 2011; Anele et al., 2010), indicating that N was not a limiting factor for plant growth (Table 3). The N source had a significant effect ($p < 0.05$), and the uninoculated treatment with 80 kg ha^{-1} mineral N supplementation exhibited the highest shoot dry matter (Table 3). The other treatments showed intermediate shoot dry matter values and did not significantly differ from each other according to the Scott-Knott test.

There was no significant difference in the relative efficiency

between the inoculated treatments and the uninoculated control treatment without mineral N supplementation (Table 3). This similarity demonstrates the ability of the native population to establish effective symbioses with cowpeas, a legume that is considered promiscuous (Guimarães et al., 2012; Jaramillo et al., 2013). Similar results were reported by Rufini et al. (2014), who studied the symbiotic efficiency of rhizobia inoculants on cowpeas grown in eutrophic Red Latosol soils in Minas Gerais.

The control with N supplementation (80 kg ha^{-1}) had significantly higher N content and accumulation in its shoot dry matter ($p < 0.05$) compared to the remaining treatments. There were no differences between the inoculated treatments and the control treatments without N supplementation, demonstrating the ability of the native populations to establish effective symbioses with cowpea in São Luís (Table 3). All the treatments had satisfactory N content, with levels greater than 3%.

In the second evaluation (75 DAE), except for grain N accumulation, the interaction between liming and the inoculants did not have a statistically significant effect ($p > 0.05$) on the variables analyzed in the maturation stage. This finding is consistent with the tolerance of the strains evaluated and the native rhizobia population to the pH and aluminum levels in the soil (Table 3). However, liming did have a significant effect on several variables, including the number of seeds per pod, seed weight, grain yield and grain N content, indicating that although tolerant, BRS Guariba is still responsive to increased nutrient availability in the soil, especially regarding yield gain. Some authors have found that liming has no effect on N_2 fixation or cowpea production in the northeast (Costa & Stamford 1991), but Cravo, Smyth, and Brasil, (2012) reported that cowpea yields increased significantly in response to liming in the northeast and Amazon regions, respectively.

Regarding the effect of the N sources, a significant effect ($p < 0.05$) of the treatments was observed for all the other variables analyzed at this stage.

The Scott-Knott tests indicated that the treatment with 80 kg ha^{-1} mineral N supplementation yielded the greatest number of pods per plant (Table 3). The inoculated treatments were not significantly different from the control treatment without mineral N supplementation.

The number of seeds per pod significantly differed ($p < 0.05$) among treatments, with the strains UFLA03-153 and INPA03-11B showing the most seeds per pod. The remaining inoculated treatments did not differ from the uninoculated controls with and without mineral N supplementation regarding this variable (80 kg ha^{-1}).

The uninoculated treatment with 80 kg ha^{-1} N supplementation had the highest seed weights (Table 3). The inoculated treatments were grouped into a second group and did not significantly differ ($p > 0.05$) from the uninoculated treatment without mineral N supplementation.

Grain yield varied significantly ($p < 0.05$) among the treatments, and increased cowpea yields were observed for the uninoculated control with 80 kg ha^{-1} mineral N supplementation (Table 3). The strain UFLA03-153 was superior to the strains approved as inoculants (UFLA03-84 and INPA03-11B) and resulted in the highest grain yield among the inoculated treatments. Other authors observed similar grain yields between an uninoculated control without mineral N supplementation and the strains UFLA03-84 and INPA03-11B, with values close to the ones reported in São Luís (Rufini et al., 2014). The lowest mean yield was observed for the treatment inoculated with strain UFLA03-84.

The treatments could be divided into three significantly different groups ($p < 0.05$) based on the grain N content (Table 3). The first group with the highest N content consisted of the uninoculated treatment with mineral N supplementation and UFLA03-153. The second group consisted of the treatments UFLA03-154, INPA03-11B and the control without mineral N supplementation. The last group consisted of the treatments inoculated with strains UFLA03-164 and UFLA03-84 (Table 3).

Grain N accumulation was the one variable that was significantly influenced ($p < 0.05$) by the interaction of the liming and inoculants (Table 3). When soil correction was performed, the Scott-Knott test indicated three treatment groups, with the uninoculated control with mineral N supplementation showing the highest levels of grain N accumulation, followed by the treatment inoculated with UFLA03-153, with intermediate levels, and with the remaining treatments forming the last group. In the absence of soil correction, the treatment inoculated with strain UFLA03-153 was similar to the control with N supplementation, showing the highest levels of grain N accumulation, followed by the remaining treatments (Table 3). This pattern shows that the efficiency of chemical N fertilizer is more dependent on improvements in chemical properties of the soil than on the efficiency of the inoculants, confirming the tolerance of the evaluated strains to acidic soils. Similar results have been reported by other authors, who found no difference in the grain N content compared to treatments inoculated with the strains UFLA03-84 and INPA03-11B (Soares et al., 2006). Both with and without the liming, the non-inoculated treatment without mineral N supplementation had similar levels of grain N accumulation as the treatments inoculated with strains UFLA03-154, UFLA03-164, UFLA03-84 and INPA03-11B (Table 3). Other authors have observed similar results (Rufini et al., 2014).

It is noteworthy that the yields obtained in these experiments (Dom Pedro: 797-1,621 kg ha^{-1}; São Luís: 923-1,430 kg ha^{-1}) were well above the state average (510 kg ha^{-1}). This increase is not only due to the rhizobial inoculants but is also due to the phosphate fertilization, which supplies one of the primary limiting factors in tropical soils, especially in Dom Pedro. Although the yields obtained are higher with N fertilization than with rhizobial inoculants, the economic and ecological benefits of the latter outweigh the benefits of the former (Sousa & Moreira, 2011).

Notably, interaction between liming and inoculants only had a significant effect on one variable, and liming is therefore not recommended for use in combination with rhizobial inoculants. However, increased cowpea production was observed when the effects of liming were isolated.

Conclusion

Inoculation of cowpea with strains UFLA03-153 and UFLA03-164 is recommended for family-based agriculture systems and for intensive agriculture systems.

The strains UFLA03-153 and UFLA03-164 are more efficient for atmospheric nitrogen fixation than the strains approved as inoculants, UFLA03-84 and INPA03-11B, resulting in increased productivity.

Strain UFLA03-164 is recommended for the production of forage with high protein content in Dom Pedro.

Liming increases the grain yield of cultivar BRS Guariba in production systems.

The use of liming together with inoculation is not recommended for the production of cowpea in Dom Pedro and São Luís.

Acknowledgements

The authors wish to thank the Federal Institute for Education, Science and Technology of the state of Maranhão (Instituto Federal de Educação, Ciência and Tecnologia do Maranhão- IFMA), The Federal University of Lavras (Universidade Federal de Lavras- UFLA), National Council for Scientific and Technological Development (Conselho Nacional de Desenvolvimento Científico e Tecnológico - CNPq) and the Brazilian Federal Agency for the Support and Evaluation of Graduate Education (Coordenação de Aperfeiçoamento de Pessoal de Nível Superior- CAPES) for financing the present study and for the research scholarships granted to the students and researcher.

References

Anele, U. Y., Arigbede, O. M., Sudekum, K. H., Ike, K. A., Oni, A. O., Olanite, J. A., ... Jolaosho, A. O. (2010). Effects of processed cowpea (*Vigna unguiculata* L. Walp) haulms as a feed supplement on voluntary intake, utilization and blood profile of West African dwarf sheep fed a basal diet of Pennisetum purpureum in the dry season. *Animal Feed Science and Technology, 159*(1-2), 10-17.

Bian, M., Zhou, M., Sun, D., & LI, C. (2013). Molecular approaches unravel the mechanism of acid soil tolerance in plants. *The Crop Journal, 1*(2), 91-104.

Correa, O. S., & Barneix, A. J. (1997). Cellular mechanisms of pH tolerance in *Rhizobium loti*. *World Journal of Microbiology and Biotechnology, 13*(2), 153-157.

Costa, E. M., Nóbrega, R. S. A., Martins, L. V., Amaral, F. H. C., & Moreira, F. M. S. (2011). Nodulação e produtividade de *Vigna unguiculata* (L.) Walp por cepas de rizóbio no Pólo de produção Bom Jesus, PI. *Revista Ciência Agronômica, 42*(1), 1-7.

Costa, J. P. V., & Stamford, N. P. (1991). Efeito de fontes, doses, e granulometrias de calcários na fixação do N_2 e no estado nutricional de caupi. *Ciência Agrícola, 1*(1), 37-48.

Cravo, M. S., Smyth, T. J., & Brasil, E. C. (2012). Calagem em Latossolos Amarelo Distrófico da Amazônia e sua influência em atributos químicos do solo e na produtividade de culturas anuais. *Revista Brasileira de Ciência do Solo, 36*(3), 895-908.

Ferreira, L. V. M., Nóbrega, R. S. A., Nóbrega, J. C. A., Aguiar, F. L. de, Moreira, F. M. S., & Pacheco, L. P. (2013). Biological Nitrogen Fixation in Production of *Vigna unguiculata* (L.) Walp, Family Farming in Piauí, Brazil. *Journal of Agricultural Science, 5*(4), 153-160.

Ferreira, P. A. A., Bomfeti, C. A., Soares, B. L., & Moreira, F. M. S. (2012). Efficient nitrogen-fixing *Rhizobium* strains isolated from amazonian soils are highly tolerant to acidity and aluminium. *World Journal Microbiology and Biotechnology, 28*(5), 1947-1959.

Freire Filho, F. R., Ribeiro, V. Q., Rocha, M. M., Silva, K. J. D., Nogueira, M. S. R., & Rodrigues, E. V. (2011). *Feijão-caupi no Brasil: produção, melhoramento genético, avanços e desafios* (p. 84) Teresina, PI: Embrapa Meio-Norte.

Gualter, R. M. R., Boddey, R. M., Rumjanek, N. G., Freitas, A. C. R., & Xavier, G. R. (2011). Eficiência agronômica de estirpes de rizóbio em feijão-caupi cultivado na região da Pré-Amazônia maranhense. *Pesquisa Agropecuária Brasileira, 46*(3), 303-308.

Guimarães, A. A., Jaramillo, P. M., Nóbrega, R. S. A., Florentino, L. A., Silva, K. B., & Moreira, F. M. S. (2012). Genetic and Symbiotic Diversity of Nitrogen-Fixing Bacteria Isolated from Agricultural Soils in the Western Amazon by Using Cowpea as the Trap Plant. *Applied and Environmental Microbiology, 78*(18), 6726-6733.

Hoagland, D. R., & Arnon, D. I. (1950). The water-culture method for growing plants without soil (2a ed.) Berkeley, California: University of California, College of Agriculture, Agricultural Experiment Station.

Instituto Brasileiro de Geografia e Estatística [IBGE]. (2015). Pesquisa mensal de previsão e acompanhamento das safras agrícolas no ano civil. *Levantamento Sistemático da Produção Agrícola, 29*(12), 1-88.

Jaramillo, P. M. D., Guimarães, A. A., Florentino, L. A., Silva, K. B., Nóbrega, R. S. A., & Moreira, F. M. S. (2013). Symbiotic nitrogen-fixing bacterial populations trapped from soils under agroforestry systems. *Scientia Agricola, 70*(6), 397-404.

Jemo, M., Abaidoo, R. C., Nolte, C., & Horst, W. J. (2007). Aluminum resistance of cowpea as affected by phosphorus-deficiency stress. *Journal of Plant Physiology, 164*(4), 442-451.

Keyser, H. H., & Munns, D. N. (1979). Tolerance of rhizobia to acidity, aluminum, and phosphate. *Soil Science Society of America Journal, 43*(3), 519-523.

Kopittke, P. M., Blamey, F. P. C., Wang, P., & Menzies, N. W. (2011). Calculated activity of Mn^{2+} at the outer surface of the root cell plasma membrane governs Mn nutrition of cowpea seedlings. *Journal of Experimental Botany, 62*(11), 3993-4001.

Lima, A. S., Nóbrega, R. S. A., Barberi, A., Silva, K. Ferreira, D. F., & Moreira, F. M. S. (2009). Nitrogen-fixing bacteria communities occurring in soils under different uses in the Western Amazon Region as indicated by nodulation of siratro (*Macroptilium atropurpureum*). *Plant and Soil, 320*(1-2), 127–145.

Moreira, F. M. S., & Siqueira, J. O. (2006). *Microbiologia e bioquímica do solo* (2a ed.). Lavras, MG: UFLA.

Oliveira-Longatti, S. M., Marra, L. M., Soares, B. L., Bomfeti, C. A., Silvia, K., Ferreira, P. A. A., & Moreira, F. M. S. (2014). Bacteria isolated from soils of the western Amazon and from rehabilitated bauxite-mining areas have potential as plant growth promoters. *World Journal of Microbiology and Biotechnology, 30*(4), 1239-1250.

R Development Core Team (2013). R: *A language and environment for statistical computing.* R Foundation for Statistical Computing, Vienna, Austria.

Rufini, M., Silva, M. A., Ferreira, P. A. A., Souza, C. A., Soares, B. L., Andrade, M. J., & Moreira, F. M. S. (2014). Symbiotic efficiency and identification of rhizobia that nodulate cowpea in a Rhodic Eutrudox. *Biology and Fertility of Soils, 50*(1), 115-122.

Santos, J. G. D., Aguiar, A. C. F., & Moura, E. G. (2011). Soil management and efficiency of rhizobia strains of cowpea *Vigna unguiculata* (L.) Walp. in the tropics. *Chilean Journal of Agricultural Research, 71*(4), 594-600.

Soares, B. L., Ferreira, P. A. A., Oliveira-Longatti, S. M., Marra, L. M., Rufini, M., Andrade, M. J. B., & Moreira, F. M. S. (2014). Cowpea sybiotic eficiency, pH and aluminium tolerance in nitrogen-fixing bacteria. *Scientia Agricola, 71*(3), 171-180.

Soares, A. L. L., Pereira, J. P. A. R., Ferreira, P. A. A., Vale, H. M. M., Lima, A. S., Andrade, M. J. B., & Moreira, F. M. S. (2006). Eficiência agronômica de rizóbios selecionados e diversidade de populações nativas nodulíferas em perdões (MG): I - caupi. *Revista Brasileira de Ciência do Solo, 30*(5), 795-802.

Sousa, P. M., & Moreira, F. M. S. (2011). Potencial econômico da inoculação de rizóbios em feijão-caupi na agricultura familiar: um estudo de caso. *Revista em Extensão, 10*(2), 37-54.

Zahran, H. H. (1999). *Rhizobium*-legume symbiosis and nitrogen fixation under severe conditions and in an arid climate. *Microbiology and Molecular Biology Reviews, 63*(4), 968-989.

6

Optimizing row spacing and plant population arrangement for a new short-height castor genotype in fall-winter

Genivaldo David de Souza-Schlick[1], Rogério Peres Soratto[2*] and Mauricio Dutra Zanotto[2]

*[1]Programa de Pós-graduação em Agronomia-Agricultura, Faculdade de Ciências Agronômicas, Universidade Estadual Paulista, Botucatu, São Paulo, Brazil. [2]Departamento de Produção e Melhoramento Vegetal, Faculdade de Ciências Agronômicas, Universidade Estadual Paulista, Rua José Barbosa de Barros, 1780, 18610-307, Botucatu, São Paulo, Brazil. *Author for correspondence. E-mail: soratto@fca.unesp.br*

ABSTRACT. Plant population and arrangement should be properly designed to maximize the seed yield of short-height genotypes of castor (*Ricinus communis* L.) in each cropping season. Experiments were performed in the in the fall-winter cropping season of 2008 and 2009 in Botucatu, State of São Paulo, Brazil, to evaluate the effects of row spacing (0.45, 0.60, 0.75, and 0.90 m) and plant population (25,000, 40,000, 55,000, and 70,000 plants ha^{-1}) on agronomical traits, yield components, seed yield, and oil yield of the genotype FCA-PB. A high plant density caused a reduction in plant survival and increased the first raceme insertion height, particularly in the year with the higher rainfall. The basal stem diameter, number of racemes per plant, and seeds per raceme were reduced by increasing the plant density. However, the seed and oil yields were minimally influenced by plant population density and row spacing because the castor plant compensated for a low plant population with a high number of racemes and seeds. The maximum seed yield was achieved with an estimated 0.64 m of row spacing and 50,107 plants ha^{-1} (i.e., 3.2 plants m^{-1}).

Keywords: *Ricinus communis*, spatial plant distribution, plant density, grain yield, oil content.

Introduction

Castor (*Ricinus communis* L.) oil is distinct from other vegetable oils, mainly because it consists of up to 90% of a hydroxylated fatty acid called ricinoleic acid (SEVERINO et al., 2012), and it has many applications in the chemical industry (BARNES et al., 2009), including biodiesel production (BALDWIN; COSSAR, 2009).

The castor plant is tolerant to drought and adapted to many cropping conditions (BABITA et al., 2010; CARVALHO et al., 2010; ZHOU et al., 2010). In Brazil, new castor genotypes have been developed with a high yield potential, a short growing cycle, even maturation, indehiscent fruit, and a short stature and are suitable for mechanical harvesting techniques. Furthermore, castor is an attractive option for growth as a second crop (out-of-season harvest) in the fall-winter cropping season, particularly in succession to soybean [*Glycine max* (L.) Merr.]. In fall-winter, traditional crops, such as corn (*Zea mays* L.) and sunflower (*Helianthus annuus* L.), are not grown because rain is scarce and erratic (SAVY FILHO, 2005; SILVA et al., 2010).

The optimization of row spacing and in-row plant density is a simple procedure with a low cost but has a significant influence on yield (SEVERINO et al., 2006a and b; SEVERINO et al., 2012; SORATTO et al., 2012) and is essential to maximize seed production (COX; CHERNEY, 2011). A high plant density may result in overgrown plants (CARVALHO et al., 2010) and subsequent lodging, whereas a low plant population may favor weed infestation, late flowering, long lateral branches, and wide stems, which impair mechanical harvesting (LOPES et al., 2008; SEVERINO et al., 2006b, 2012). Light interception by plants strongly influences the crop yield when other environmental factors are favorable, and it is modified by the plant spatial distribution in a given area (SEVERINO et al., 2006a and b; 2012).

Nevertheless, the optimum plant arrangement depends on the characteristics of each genotype, such as the growth habit, plant stature and architecture (SEVERINO et al., 2006a and b), and environmental conditions (SEVERINO et al., 2006a; SORATTO et al., 2012). Kittock and Williams (1970) observed that, for short-height castor in a 1.02 m row spacing, the best plant population was 58,000 plants ha^{-1} under irrigation and 30,000 plants ha^{-1} under rainfed conditions. For the IAC 2028 cultivar, plant populations greater than 25,000 plants ha^{-1} did not increase the seed yield, regardless of the row spacing used, in the fall-winter cropping season (SOUZA-SCHLICK et al., 2011). Soratto et al. (2012) reported that the seed yield of castor genotypes in fall-winter was greater for more equidistant plant arrangements and that an increase in the plant population did not affect the seed yield when the same row spacing (0.90 m in this case) was used. However, there are no conclusive studies and recommendations to date for the optimum row spacing and plant population arrangement for the growth of new short-height castor genotypes as a second crop in fall-winter under a mechanized cropping system.

This study aimed to evaluate the effects of row spacing and plant population arrangement on agronomical traits, yield components, seed yield, and oil yield of the short-height castor genotype FCA-PB grown in the fall-winter season.

Material and methods

The experiment was performed in Botucatu, São Paulo State, Brazil (48° 23' W; 22° 51' S; 740 m above sea level), during the fall-winter cropping season 2008 and 2009. The soil was a sandy clay loam, kaolinitic, thermic Typic Haplorthox (distroferric Red Latosol). The chemical characteristics of the topsoil (0-0.20 m) were determined at the beginning of each experiment. In 2008, the soil characteristics were as follows: pH ($CaCl_2$) of 5.0; 38 g dm^{-3} of total organic matter; 39.0 mg dm^{-3} of p (resin); 4.1, 31.0, 9.0, and 51.0 $mmol_c$ dm^{-3} of exchangeable K, Ca, Mg, and H+Al, respectively; and base saturation of 46%. The values in 2009 were as follows: pH ($CaCl_2$) of 5.6; 40 g dm^{-3} of total organic matter; 26.0 mg dm^{-3} of p (resin); 2.6, 43.0, 22.0, and 29.0 $mmol_c$ dm^{-3} of exchangeable K, Ca, Mg, and H+Al, respectively; and base saturation of 70%.

The climate of the region is Cwa according to the Köppen classification, which is characterized as a tropical, with a dry winter and a hot rainy summer. The data for the rainfall and temperature registered during the experiments are presented in Figure 1.

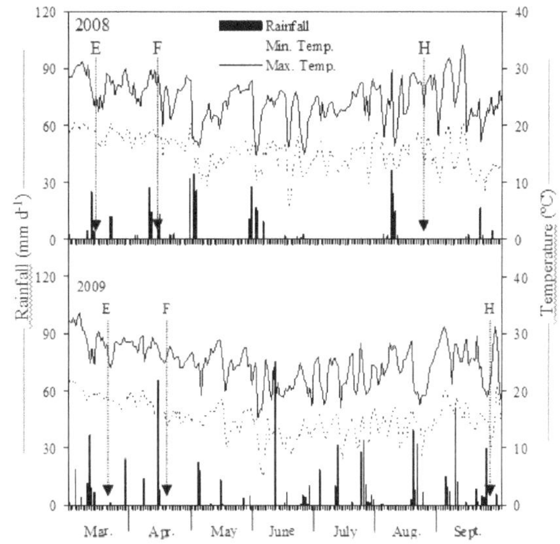

Figure 1. Rainfall, maximum and mnimum air temperature in the experimental area in Botucatu, São Paulo State, Brazil, from March to September 2008 and 2009. E: plant emergence; F: flowering of the first raceme; H: harvesting.

The experimental design was a completely randomized block in a split-plot arrangement, which was replicated four times, with the row spacings assigned to the main plots and the plant populations assigned to the subplots. The treatments were a factorial distribution of four row spacings (0.45, 0.60, 0.75, and 0.90 m) and four plant populations (25,000, 40,000, 55,000, and 70,000 plants ha^{-1}) as the subplots. The main plots were 20 m long and consisted of ten rows when the spacing was 0.45 m and six rows when the spacing was 0.60, 0.75, or 0.90 m. The subplots were 5 m long. Data were not obtained for the plants at 0.5 m at the end of each row and in the lateral rows of each subplot.

The FCA-PB genotype was developed in the breeding program of the College of Agricultural Sciences - São Paulo State University (UNESP). Its characteristics are a short, indehiscent fruits, susceptibility to gray mold [*Botryotinia ricini* (G.H. Godfrey) Whetzel], an oil content higher than 47%, and a short cycle (130-150 days).

A no-tillage system was adopted. Two weeks prior to sowing, the area was sprayed with 1.44 kg a.i. ha^{-1} of glyphosate. Furrow opening and fertilizer distribution was performed using a sowing-fertilizer machine; mineral fertilizer was applied at a dose of 150 kg ha^{-1} of NPK 8-28-16. Sowing was manually performed in the first half of March 2008 and 2009 using an excessive number of seeds. The seeds were placed at a 4 cm depth and covered with soil. Seedling emergence occurred 10 days after sowing, and thinning was performed at 10 days after emergence (DAE) to achieve the assigned plant population. Topdressing fertilization using 60 kg ha^{-1} of N (ammonium sulfate) was performed at 30 DAE.

Weeding was performed by hand. Flowering of the first raceme occurred at 42 DAE in 2008 and 59 DAE in 2009. After the beginning of flowering, fungicides were sprayed to prevent gray mold infestation. In 2008, procymidone at 0.5 kg a.i. ha^{-1} was sprayed three times, and iprodione at 0.5 g i.a. ha^{-1} was sprayed once. The plants were sprayed with 0.5 kg ha^{-1} of procymidone four times in 2009.

The plants were harvested at 153 DAE in 2008 and 185 DAE in 2009. The plant survival rate was calculated for all the plants in two 4-m-long rows. The basal stem diameter (at the first internode), plant height (distance from the soil to the highest point of the plant), and first raceme insertion height (distance from the soil to the point of raceme insertion) were measured for ten plants in each subplot. All the racemes in two 4-m-long rows were hand-harvested. The fruits were counted and hand-threshed, and the seeds were counted and weighed; the seed yield was adjusted to 130 g kg^{-1} of moisture content.

The seeds were oven dried at 70°C for 24 hours to maximize the sample homogeneity. A subsample (12 to 15 g) from each plot was weighed, and the seed oil content was determined using a bench-top NMR spectrometer model SLK-100 (SpinLock, Cordoba, Argentina), following the method described by Colnago et al. (2007). The oil content (%) was expressed on a seed dry matter basis.

The Lilliefors test was used to verify the normality of the data. The homogeneity of the variances was evaluated using the Cochran-Bartlett test. The data were subjected to an ANOVA using a split-plot scheme in the statistical program SISVAR. The blocks and all the block interactions were considered random effects; the year, row spacing, and plant population were considered fixed effects. The row spacing, plant population, year, and their interaction means were subjected to a regression analysis using the software SigmaPlot 10.0. The criterion adopted to select a model was the magnitude of the regression coefficients ($p \leq 0.05$).

Results and discussion

The growing conditions, particularly the rainfall, were different between 2008 and 2009 (Figure 1). The rainfall during the castor cycle totaled 359 mm in 2008 and 655 mm in 2009; the 38-yr average is 555 mm. Because of the greater amount and better distribution of rainfall in 2009, the castor cycle was 36 days longer than in 2008. According to Souza et al. (2007), the excessive availability of water can favor the vegetative growth of castor plants and delay flowering and fruit ripening.

The row spacing, plant population, and their interaction were significant to the plant survival rate. Increasing the plant population density, particularly in a wide row spacing, decreased plant survival (Figure 2A), most likely because this combination led to higher plant densities within a row. The widest row spacing (0.90 m) combined with the highest plant population (70,000 plants ha^{-1}) reduced the plant survival rate by approximately 11.4% compared to the combination of the narrowest row spacing (0.45 m) and sparsest plant population (25,000 plants ha^{-1}) (Figure 2A). The plant population density had a greater effect on plant survival rate than the row spacing. Although weaker plants died before harvest because they were less competitive under a higher plant density, a lower plant density provided better conditions for proper growth. Soratto et al. (2012) also observed decreasing survival of castor plants due to an increasing number of plants per area (i.e., higher plant densities within a row). Similarly, Cox and Cherney (2011) found that higher plant densities in a row reduced the survival of soybean plants. In these situations, there is strong competition among the plants for water, light, and nutrients, and the weaker plants may not complete their growth cycle.

The basal stem diameter was more strongly affected by the plant population than by the row spacing (Figure 2B). Increasing the plant population reduced the stem diameter in all row spacings, but the stem diameter decreased more in the 0.90 m row spacing compared to the narrower row spacings. Distinct plant arrangements may affect the quality and quantity of the sunlight intercepted by plants. Kittock and Williams (1970) also found that an increased number of plants within a row reduced the diameter of castor stems. Bizinoto et al. (2010) found that the castor stem diameter decreased in response to an increased plant density. According to these authors, the reduction in stem diameter was possibly due to the etiolation caused by a high competition for light at a high plant density within a row or a dense plant population. Because castor plants with thin stems are favorable for mechanical harvesting (LOPES et al., 2008), our results indicate that the implementation of a high plant population can promotes plants with thinner stems, thus facilitating mechanical harvesting techniques (Figure 2B).

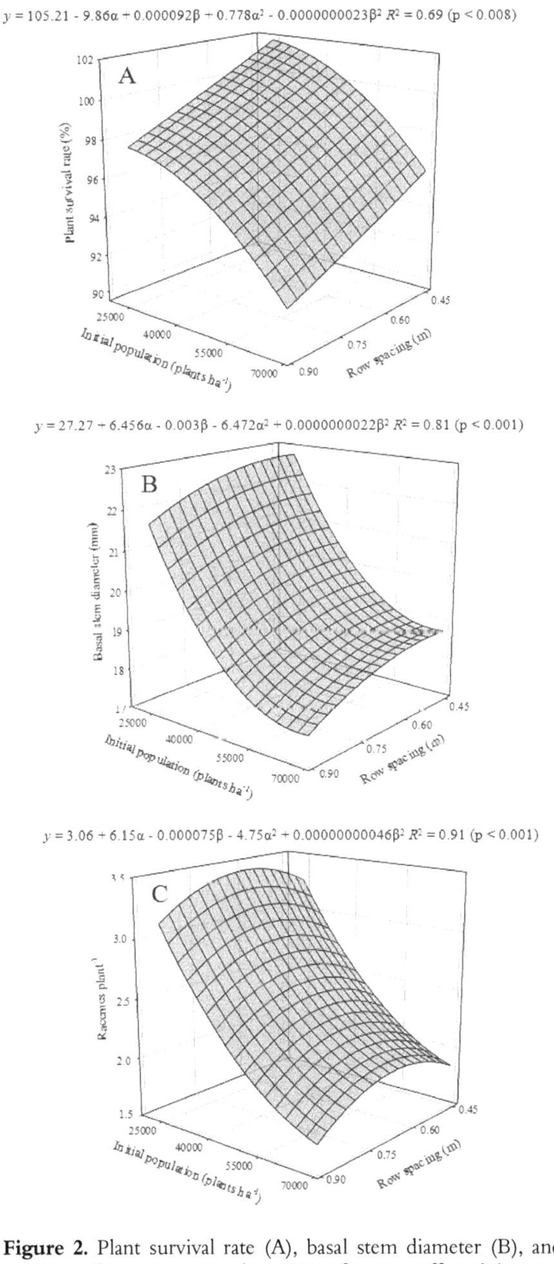

Figure 2. Plant survival rate (A), basal stem diameter (B), and number of racemes per plant (C) of castor affected by row spacing (α) and plant population (β). Data are averaged across year and replication.

Plant height was not significantly affected by the row spacing and plant population, and the average plant height was 1.15 m. Kittock and Williams (1970) and Bizinoto et al. (2010) also found no effect of plant density on the height of castor plants. The effects of plant density on plant height is very complex and depends on both environmental factors and plant architecture. Plant growth is slow under the low water availability and temperature during the fall-winter cropping season, and the effect of row spacing and plant population on plant height appears to be lower than that observed in the spring-summer season, a time when the conditions are more favorable to plant growth (SEVERINO et al., 2006a and b; SORATTO et al., 2012).

The height of insertion of the first raceme was linearly increased by increments in the plant population in both years, regardless of the row spacing (Figure 3A). In addition, the insertion height differences between the lower and higher plant populations were greater in the second year compared to the previous year, and more available water (Figure 1) may have caused stem etiolation. Kittock and Williams (1970), Bizinoto et al. (2010), and Soratto et al. (2012) reported that higher plant densities increased the first raceme insertion height of castor plants, without affecting plant height.

An increasing plant population decreased the number of racemes per plant at any row spacing (Figure 2C); similar results were found by Kittock and Williams (1970) and Bizinoto et al. (2010). The number of racemes per plant increased with increasing row spacing up to 0.65 m and then decreased with wider row spacings. The greatest number of racemes per plant was estimated for the sparsest plant population (25,000 plants ha^{-1}) and the 0.65 m row spacing, i.e., plants more equidistantly distributed within (0.62 m) and between rows (0.65 m), with 1.62 plants m^{-1} within row. Whenever plants are more widely spaced, there is a larger area for light interception and lower competition or interference among them. Thus, photoassimilate production is higher, and plants display the appropriate architecture, which allows them to express their potential by forming more racemes (BIZINOTO et al., 2010; KITTOCK; WILLIAMS, 1970).

The number of seeds per raceme decreased linearly as the plant population increased in both years, regardless of row spacing (Figure 3B). It was notable that the number of seeds per raceme obtained in 2008 was higher than in 2009, particularly for the lower plant populations, possibly because of the higher rainfall in the second year (Figure 1), which occurred at the fruit formation stage (June and July). A high relative humidity increases the incidence of gray mold in racemes and, thus, may reduce the number of seeds per raceme (SUSSEL et al., 2009).

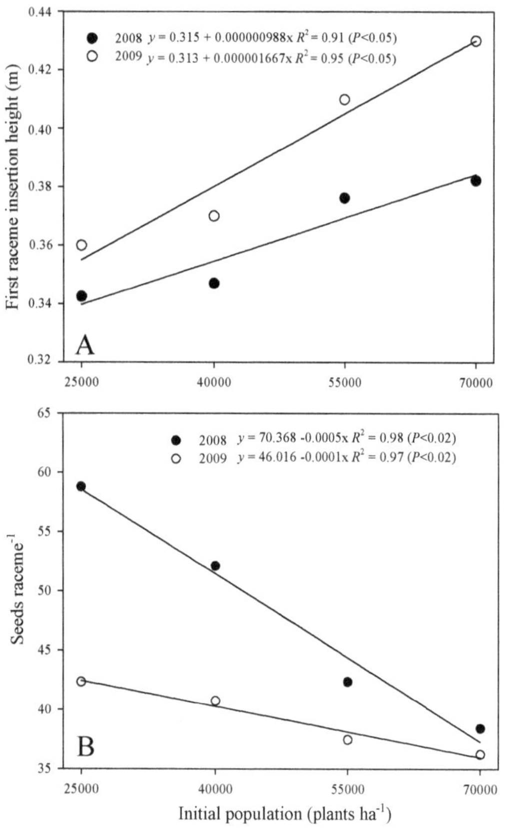

Figure 3. First raceme insertion height (A) and number of seeds per raceme (B) of castor affected by plant population in two years. Data are averaged across row spacing and replication.

The seed weight was not significantly affected by the factors studied. The seed weight average was higher in 2009 (463 mg seed^{-1}) than in 2008 (376 mg seed^{-1}), possibly because of the increased water availability during the seed formation and filling stages (Figure 1) and the lower number of seeds per raceme (Figure 3B).

Seed yield was affected by the row spacing × plant population interaction, and both row spacing and plant population had quadratic effects on the grain yield (Figure 4A). Considering the average for the two years, the highest seed yield (2,248 kg ha^{-1}) was estimated for the combination of 0.64 m row spacing and 50,107 plants ha^{-1} (i.e., 3.2 plants m^{-1}). However, the effects of plant population and row spacing on the seed yield were not as strong as observed for the number of racemes per plant and seeds per raceme (Figure 2C and 3B). No correlation between this yield components and the seed yield was observed, which indicates that the castor plants compensated at the lower population densities while increasing the number of racemes per plant and seeds per raceme, without an equivalent variation in the seed yield. Severino et al. (2006a and b) and Carvalho et al. (2010) found higher castor seed yields by increasing the plant population, but those studies did not include very high plant populations, such as those used in the present research. Soratto et al. (2012) reported that, for a high plant population (55,000 plants ha^{-1}), the seed yield of two castor genotypes in fall-winter was greater at more equidistant plant arrangements because the plant density decreased within a row, promoting a better plant distribution in the area and lower competition among the plants.

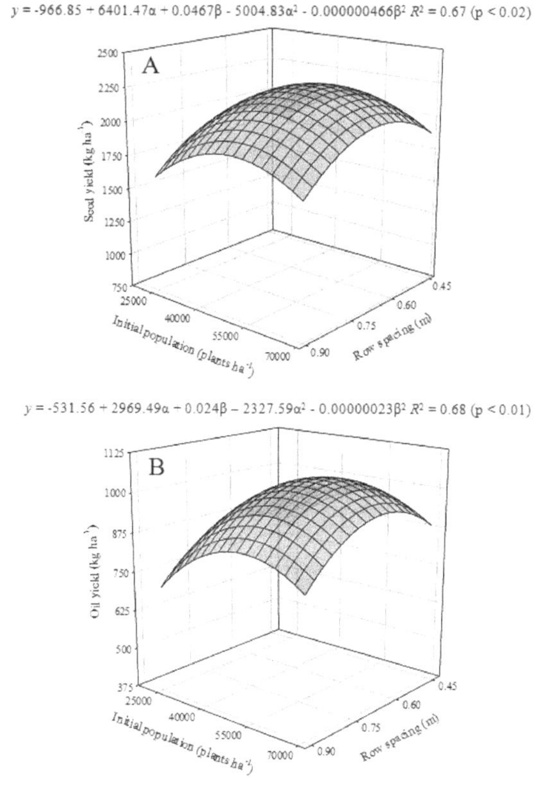

Figure 4. Seed yield (A) and oil yield (B) of castor affected by row spacing (α) and plant population (β). Data are averaged across year and replication.

The oil content in the castor seeds was not significantly affected by the treatments. Averaging the two years, the seed oil content of the subplot samples varied within the range of 43.4% to 48.5%, values that are similar to those reported in the literature (SEVERINO et al., 2006a and b). According to Severino et al. (2006b), the oil content in castor seeds is a characteristic of high hereditability. Kittock and Williams (1970) also observed no influence of the plant population on oil content. According to Koutroubas et al. (1999), higher seed yields are usually obtained at the expense of the oil content, and there is a negative correlation between these two variables; however, this correlation was not found in the present study.

Because the seed oil content was not influenced by the treatments, the seed yield and oil yield were similarly influenced by the row spacing and plant population (Figure 4A and B). The seed yield was positively and strongly correlated to the oil yield ($r = 0.97$, $p < 0.001$) (i.e., the oil yield was mainly determined by the seed yield).

The results of this study showed that the castor plant responses to changes in the plant population and arrangement are very complex and may be dependent on the weather conditions. At a plant population that is higher than generally recommended and used in Brazil (CARVALHO et al., 2010; SEVERINO et al., 2006a and b), the castor plants compensated for a lower plant population with a high number of racemes per plant and seeds per raceme in such a way that confined the seed and oil yields to a narrow range. Thus, for castor genotype FCA-PB grown in the fall-winter season, higher plant populations (~50,000 plants ha^{-1}) and narrower row spacing (~0.65 m) can be useful for providing greater seed and oil yields (Figure 4A and B) and plants with thinner stems (Figure 2B), thus facilitating mechanical harvesting.

Conclusion

A high plant density within a row caused a reduction in the survival rate of castor plants. The first raceme insertion height was increased by increasing the plant population, particularly in the year with the higher rainfall. The castor plants compensated for a low plant population with a high number of racemes per plant and seeds per raceme such that the seed and oil yields varied only within a narrow range. The maximum seed yield of castor genotype FCA-PB in the fall-winter season was achieved with an estimated 0.64 m row spacing and 50,107 plants ha^{-1} (i.e., 3.2 plants m^{-1}).

Acknowledgements

To FAPESP (State of São Paulo Research Foundation) for supporting this research and providing scholarship to the first author (Proc. 2007/05851-8). To CNPq (The National Council for Scientific and Technological Development) for providing research award to the second author. To Dr. Luiz Alberto Colnago, from Embrapa Agricultural Instrumentation, for assistance in determining seed oil content.

References

BABITA, M.; MAHESWARI, M.; RAO, L. M.; SHANKER, A. K.; RAO, D. G. Osmotic adjustment, drought tolerance and yield of castor (*Ricinus communis* L.) hybrids. **Environmental and Experimental Botany**, v. 69, n. 3, p. 243-249, 2010.

BALDWIN, B. S.; COSSAR, R. D. Castor yield in response to planting date at four locations in the south-central United States. **Industrial Crops and Products**, v. 29, n. 2-3, p. 316-319, 2009.

BARNES, D. J.; BALDWIN, B. S.; BRAASCH, D. A. Degradation of ricin in castor seed meal by temperature and chemical treatment. **Industrial Crops and Products**, v. 29, n. 2-3, p. 509-515, 2009.

BIZINOTO, T. K. M. C.; OLIVEIRA; E. G.; MARTINS; S. B.; SOUZA, S. A.; GOTARDO, M. Cultivo da mamoneira influenciada por diferentes populações de plantas. **Bragantia**, v. 69, n. 2, p. 367-370, 2010.

CARVALHO, E. V.; SÁ, C. H. A. C.; COSTA, J. L.; AFFÉRRI, F. S.; SIEBENEICHLER, S. C. Densidade de plantio em duas cultivares de mamona no Sul do Tocantins. **Revista Ciência Agronômica**, v. 41, n. 3, p. 387-392, 2010.

COLNAGO, L. A.; ENGELSBERG, M.; SOUZA, A. A.; BARBOSA, L. L. High-throughput, non-destructive determination of oil content in intact seeds by continuous wave-free precession NMR. **Analytical Chemistry**, v. 79, n. 3, p. 1271-1274, 2007.

COX, W. J.; CHERNEY, J. H. Growth and yield responses of soybean to row spacing and seeding rate. **Agronomy Journal**, v. 3, n. 1, p. 123-128, 2011.

KITTOCK, D. L.; WILLIAMS, J. H. Effects of plant population on castorbean yield. **Agronomy Journal**, v. 62, n. 4, p. 527-529, 1970.

KOUTROUBAS, S. D.; PAPAKOSTA, D. K.; DOITSINIS, A. Adaptation and yielding ability of castor plant (*Ricinus communis* L.) genotypes in a Mediterranean climate. **European Journal of Agronomy**, v. 11, n. 3-4, p. 227-237, 1999.

LOPES, F. F. M.; BELTRÃO, N. E. M.; LOPES NETO, J. P.; PEDROZA, J. P. Crescimento inicial de genótipos de mamoneira com sementes submetidas ao envelhecimento acelerado. **Revista Brasileira de Oleaginosas e Fibrosas**, v. 12, n. 1, p. 69-79, 2008.

SAVY FILHO, A. **Mamona**: tecnologia agrícola. Campinas: Emopi, 2005.

SEVERINO, L. S.; AULD, D. L.; BALDANZI, M.; CÂNDIDO, M. J. D.; CHEN, G.; CROSBY, W.; TAN, D.; HE, X.; LAKSHMAMMA, P.; LAVANYA, C.; MACHADO, O. L. T.; MIELKE, T.; MILLER, T. D.; MORRIS, J. B.; MORSE, S. A.; NAVAS, A. A.; SOARES, D. J.; SOFIATTI, V.; WANG, M. L.; ZANOTTO, M. D.; ZIELER, H. A Review on the challenges for increased production of castor. **Agronomy Journal**, v. 104, n. 4, p. 853-880, 2012.

SEVERINO, L. S.; COELHO, D. K.; MORAES, C. R. A.; GONDIM, T. M. S.; VALE, L. S. Otimização do espaçamento de plantio para a mamoneira cultivar BRS Nordestina. **Revista Brasileira de Oleaginosas e Fibrosas**, v. 10, n. 1-2, p. 993-999, 2006a.

SEVERINO, L. S.; MORAES, C. R. A.; GONDIM, T. M. S.; CARDOSO, G. D.; BELTRÃO, N. E. M. Crescimento e produtividade da mamoneira influenciada por plantio em diferentes espaçamentos entre linhas. **Revista Ciência Agronômica**, v. 37, n. 1, p. 50-54, 2006b.

SILVA, A. G.; CRUSCIOL, C. A. C.; SORATTO, R. P.; COSTA, C. H. M.; FERRARI NETO, J. Produção de fitomassa e acúmulo de nutrientes por plantas de cobertura e cultivo da mamona em sucessão no sistema plantio direto. **Ciência Rural**, v. 40, n. 10, p. 2092-2098, 2010.

SORATTO, R. P.; SOUZA-SCHLICK, G. D.; FERNANDES, A. M.; ZANOTTO, M. D.; CRUSCIOL, C. A. C. Narrow row spacing and high plant population to short height castor genotypes in two cropping seasons. **Industrial Crops and Products**, v. 35, n. 1, p. 244-249, 2012.

SOUZA, A. S.; TÁVORA, F. J. A. F.; PITOMBEIRA, J. B.; BEZERRA, F. M. L. Épocas de plantio e manejo da irrigação para a mamoneira. II – crescimento e produtividade. **Revista Ciência Agronômica**, v. 38, n. 4, p. 422-429, 2007.

SOUZA-SCHLICK, G. D.; SORATTO, R. P.; PASQUALI, C. B.; FERNANDES, A. M. Desempenho da mamoneira IAC 2028 em função do espaçamento entre fileiras e população de plantas na safrinha. **Bragantia**, v. 70, n. 3, p. 519-528, 2011.

SUSSEL, A. A. B.; POZZA, E. A.; CASTRO, H. A. Elaboração e validação de escala diagramática para avaliação da severidade do mofo cinzento em mamoneira. **Tropical Plant Pathology**, v. 34, n. 3, p. 186-191, 2009.

ZHOU, G.; MA, B. L.; LI, J.: FENG, C.; LU, J.; QIN, P. Determining salinity threshold level for castor bean emergence and stand establishment. **Crop Science**, v. 50, n. 5, p. 2030-2036, 2010.

7

Growth and production of tomato fertilized with ash and castor cake and under varying water depths, cultivated in organic potponics

Daniela Pinto Gomes[1*], Daniel Fonseca de Carvalho[1], Marinaldo Ferreira Pinto[1], David da Cunha Valença[1] and Leonardo Oliveira Medici[2]

*[1]Departamento de Engenharia, Universidade Federal Rural do Rio de Janeiro, Br-465, km 7, 23897-000, Seropédica, Rio de Janeiro, Brazil, [2]Departamento de Ciências Fisiológicas, Universidade Federal Rural do Rio de Janeiro, Seropédica, Rio de Janeiro, Brazil. *Author for correspondence. Email: danielagomesagro@hotmail.com*

ABSTRACT. Two experiments were performed to evaluate the effect of ash (40, 80, and 120 g per plant), castor cake (140 and 280 g per plant) and water depth (135, 165, 191, and 213 mm) on the growth and production of organic tomato cultivated in pots in a greenhouse. The experimental design was randomized blocks, and the irrigation was managed using an automatic irrigation device. The following variables were evaluated: plant heights, numbers of leaves, bunches, flowers and fruits, total mass of fruits, mass of marketable fruits, mass of fruits with blossom-end rot, total diameter of fruits, and diameter of marketable fruits. Most of the growth variables showed gains with the application of 140 g of ash and 280 g of cake. The dose of 280 g of castor cake was responsible for the greatest mass of marketable fruits (1.78 kg per plant), regardless of the ash dose. The water deficit reduced values of most of the variables of growth and production. The irrigation depth of 213 mm was responsible for the greatest mass of marketable fruits (4.04 kg per plant). The highest water use efficiencies, 37.00 and 37.93 kg m^{-3}, were observed at irrigation depths of 191 and 213 mm, respectively.

Keywords: *Solanum lycopersicum* L., organic agriculture, water management.

Introduction

Organic agriculture is a production system based on sustainability and has less harmful effects on the environment than conventional agriculture (Lee, Choe, & Park, 2015), which relies on larger amounts of external inputs (Gomiero, Paoletti, & Pimentel, 2008). Organic potponics is a new cultivation system within organic agriculture, using residual materials as a source of nutrients associated with adequate irrigation management, performed by a low-cost automatic device. This new system aims to apply water and nutrients independently, as opposed to hydroponics and fertigation systems, which require, respectively, the disposal of part of the nutrient solution and the application of a leaching fraction to avoid an imbalance of nutrients or salinization of the substrate (Choi, Lee, & Park, 2015).

Although many studies present different residual materials as a source of nutrients to plants, the use of wood ash and castor cake for tomato crops (*Solanum lycopersicum* L.) is still novel. Ash is a residue from burning wood (Nabeela et al., 2015), and depending on its origin, it may have high contents of K, P, Ca, and Mg (Oburger et al., 2016), which allows its use as a fertilizer or to correct soil acidity (Kuba, Tscholl, Partl, Meyer, & Insam, 2008). Castor cake is a byproduct of the extraction of castor oil (Magriotis et al., 2014). It contains 4.15, 0.61, and 0.96% of N, P, and K, respectively, and a C/N ratio of approximately 11.6 (Zapata, Vargas, Reyes, & Belmar, 2012). Castor cake is considered a high-quality fertilizer (Lima et al., 2011).

Adequate irrigation management (Wang, Kang, Du, Li, & Qiu, 2011) is vital to avoid the leaching of nutrients, especially in tomato production systems, which demand substantial amounts of N fertilizers (Zotarelli, Dukes, Sholberg, Muñoz-Carpena, & Icerman, 2009).

Higher values of water use efficiency can be obtained by irrigation management with water deficit (Cantore et al., 2016), especially in the cultivation of tomatoes, which are sensitive or moderately tolerant to water deficit (Zheng et al., 2013). However, irrigation is one of the most complex techniques in agriculture because many factors are involved in its management (Escarabajal-Henarejos, Molina-Martínez, Fernández-Pacheco, & García-Mateos, 2015). Irrigation automation systems can be used to apply water deficit to the crops and help farmers to perform adequate irrigation management. Automatic device for irrigation (Gomes, Carvalho, Almeida, Medici, & Guerra, 2014; Medici, Rocha, Carvalho, Pimentel, & Azevedo, 2010) are available that control irrigation based on soil water tension (Medici et al., 2010) using low-cost materials, compared to other controllers available in the market (Gonçalves et al., 2014).

This study aimed to evaluate the effects of wood ash, castor cake, and varying irrigation depths on the growth and productivity of tomato cultivated in an organic potponic system.

Material and methods

Two experiments were conducted in a greenhouse in the Campus of the Federal Rural University of Rio de Janeiro (UFRRJ), Seropédica, Rio de Janeiro State, Brazil (22°48'S; 43°41'W; 33 m), in 2014 (experiment I) and 2015 (experiment II). Tomato seedlings, cv. 'Dominador', were transplanted to 8-L pots filled with soil collected from the A horizon of a Planosol, whose chemical analysis showed the following results: Experiment I – $pH_{(H2O)}$: 5.5, Al: 0 $mmol_c$ dm^{-3}, Ca: 18 $mmol_c$ dm^{-3}, Mg: 5 $mmol_c$ dm^{-3}, Na: 1 $mmol_c$ dm^{-3}, P: 18.0 mg dm^{-3}, and K: 24.0 mg dm^{-3}; Experiment II – $pH_{(H2O)}$: 6.6, Al: 0 $mmol_c$ dm^{-3}, Ca: 10 $mmol_c$ dm^{-3}, Mg: 2 $mmol_c$ dm^{-3}, Na: 0 $mmol_c$ dm^{-3}, P: 33.0 mg dm^{-3}, and K: 50.0 mg dm^{-3}.

The temperature and relative air humidity inside the greenhouse were monitored using a thermo-hygrograph (IMPAC/IP-747RH). The means of the maximum daily temperatures were 39.4°C (2014) and 40.4°C (2015), which are considered above the recommended level for the development of tomato plants (Filgueira, 2008). However, no damage was observed in fruiting and fruit set. The average daily minimum and mean temperatures were 17.7 and 26.0°C (2014) and 18.2 and 25.5°C (2015), respectively. The relative air humidities were 65.9% (2014) and 71.4% (2015), which are considered adequate for tomato cultivation (Guimarães, Caliman, Silva, Flores, & Elsayed, 2007).

Tomatoes were planted in on polystyrene trays of 128 cells, filled with organic substrate, and transplanted at 32 days after sowing, on April 17, 2014, and on June 22, 2015. Weed control and disbudding were performed manually, starting at 15 days after transplanting (DAT). Staking started at 20 DAT, and tomato plants were trained using one single stake with a plastic ribbon. After three leaves were formed from the 6th bunch, the tomato plants were pruned (the main stem was cut). At 95 (experiment I) and 30 (experiment II) DAT, the control of tomato diseases was initiated, using weekly applications of Bordeaux or lime-sulfur mixtures. Fruit harvest occurred from 78 (2014) and 71 (2015) DAT to 127 (2014) and 112 (2015) DAT.

The experimental design included randomized blocks totaling 240 pots spaced 1.0 x 0.5 m apart. Each experimental plot consisted of 10 plants that were arranged in a row.

Experiment I used a factorial scheme with 3 doses of ash and 2 doses of castor cake, with 4 replications, using the following treatments: T1 – 40 g of ash and 140 g of castor cake per plant; T2 – 80 g of ash and 140 g of castor cake per plant; T3 – 120 g of ash and 140 g of castor cake per plant; T4 – 40 g of ash and 280 g of castor cake per plant; T5 – 80 g of ash and 280 g of castor cake per plant; and T6 – 120 g of ash and 280 g of castor cake per plant. The doses of ash are equivalent to 0.8 t ha^{-1}(30 kg ha^{-1} of K), 1.6 t ha^{-1}(60 kg ha^{-1} of K), and 2.4 t ha^{-1} (90 kg ha^{-1} of K), respectively, whereas the doses of castor cake are equivalent to 2.8 t ha^{-1} (140 kg ha^{-1} of N) and 5.6 t ha^{-1} (280 kg ha^{-1} of N), respectively. In addition, 50 g of reactive natural phosphate were added once to each pot at the time of planting. The castor cake was applied at planting and at 30 DAT.

The ash used was obtained from the burning of Eucalyptus wood, and the castor cake was obtained at the market. The chemical analyses of these materials showed the following results: ash – pH $_{(H2O)}$: 10.4, P: 17.0 g kg^{-1}, K: 37.8 g kg^{-1}, Ca: 2.1 cmolc dm^{-3}, and Mg: 1.3 cmolc dm^{-3}; and castor cake – N: 49.8 g kg^{-1}, P: 4.7 g kg^{-1}, K: 1.7 g kg^{-1}, Ca: 8.5 g kg^{-1}, and Mg: 14.5 g kg^{-1}.

In experiment II, 4 irrigation depths were tested, with 6 replications: L4 – the volume applied with irrigation management performed using the automatic device for irrigation, promoted a soil water tension of approximately 12 kPa; and L3, L2, and L1 – fractions corresponding to 90, 77, and 63% of the volume applied in L4, respectively. Based on the results obtained in experiment I, fertilization for experiment II contained 280 g of castor cake, 25 g of potassium sulfate, 50 g of reactive natural phosphate, and 8 g of calcitic limestone (RNV 88 and 45% of Ca). The doses of potassium sulfate, reactive natural phosphate, and limestone were applied at planting, while the castor cake was applied as in experiment I.

A drip irrigation system was used in both experiments. Experiment I used drippers (Supertif/John Deere Water), and experiment II used spaghetti tubing (nominal diameter of 0.7 mm, Plasnova). Micro tubes of different lengths were used to allow the application of the different irrigation depths: L1 – 2 micro tubes with a length of 36 cm (2.5 L h^{-1}); L2 – 2 micro tubes with a length of 42 cm (3.0 L h^{-1}); L3 – 2 micro tubes with a length of 52 cm (3.5 L h^{-1}); and L4 – 2 micro tubes with a length of 65 cm (3.9 L h^{-1}), at a pressure of 24.5 kPa. In both cultivation years, tests were performed in each experimental plot to quantify the water distribution uniformity, and a Christiansen's uniformity coefficient of 97% was obtained in both experiments.

For irrigation management, two automatic devices for irrigation were installed in one of the four plots (replicates) representing each treatment (one in each pot), totaling 12 devices for experiment I. In experiment II, the automatic device for irrigation was installed only in one plot (replicate) representing treatment T4, one in each pot. The applied volume was recorded daily through the readings of hydrometers installed at the beginning of each irrigation line (block). Thus, it was possible to calculate the water depth applied by emitters (localized form) in each treatment from the ratio between the applied water volume and the area of the pot (from 1 to 30 DAT) or the area shaded by the plant (from 31 to 126 [2014] or to 112 [2015] DAT).

Additionally, in experiment II, the soil water tension was monitored using tensiometers connected to pressure transducers, which were linked through cables to 3 data acquisition systems (data loggers), referring to each evaluated block. The data loggers were programed to record soil water tension every hour when the irrigation system was off and every 10 seconds when the irrigation system was on.

Non-destructive growth analysis was performed by measuring plant height and the numbers of leaves, bunches, flowers, and fruits. The four most representative plants of each plot were evaluated, and the measurements were performed biweekly.

The following data were collected: total mass of fruits, mass of marketable fruits, mass of fruits with blossom-end rot, total diameter of fruits, and diameter of marketable fruits. Unmarketable fruits were considered those with equatorial diameter smaller than 40 mm and/or defects such as rot, overripe fruits, blossom-end rot, sunburned fruits, empty fruits, fruits with open locule, yellowish fruits, cracked fruits, and fruits with deep damages.

Water use efficiency (WUE) was obtained by the ratio between the produced mass of marketable fruits and the applied irrigation depth (Lovelli, Perniola, Ferrara, & Tommaso, 2007).

Analysis of variance was performed for all the data, and when significance was observed using an F test, the data were subjected to regression analysis.

Results and discussion

Experiment I (2014) – Effects of ash and castor cake

There was no significant interaction between the factors ash and castor cake on any of the growth variables. The dose of 280 g of castor cake promoted increments in plant height and in the number of leaves, bunches, flowers, and fruits (Table 1) in relation to the dose of 140 g. This residue has been studied to meet the nutritional needs of other crops, such as castor, and increments in plant height and number of inflorescences have been observed with increases in its dosage (Oliveira Filho, Oliveira, Medeiros, Mesquita, & Zonta, 2010). Similar results were found by Vignolo, Araújo, Kunde, Silveira, and Antunes (2011) in the number of strawberry fruits and by Martins, Suguino, Dias, and Perdoná (2011) in the number of leaves of banana seedlings.

The ash doses promoted effects on all variables, except on the number of flowers. The dose of 120 g led to increments in plant height (18 and 12% at 60 and 75 DAT, respectively) (Figure 1a) and in the numbers of leaves (13 and 12% at 45 and 60 DAT, respectively) (Figure 1b), bunches (14 and 19% at 45 and 60 DAT, respectively) (Figure 1c), and fruits (27% at 75 DAT) (Figure 1d), in relation to the dose of 40 g. In a study conducted by Nabeela et al. (2015), there was a positive increase in the height of canola (*Brassica napus* L.) seedlings in response to increasing doses of ash. For the grape crop, Piva, Botelho, Ortolan, Müller, and Kawakami (2013) obtained a positive increment in leaf area index with the increase in ash doses associated with bovine manure.

Table 1. Growth variables for the doses of castor cake, evaluated at 1 and at 15, 30, 45, 60, 75, 90, 105, and 120 days after transplanting (DAT).

Doses	Days after transplanting (DAT)								
	1	15	30	45	60	75	90	105	120
(g)					Plant height (cm)				
140	17.8a	37.3a	74.3a	99.8b	123.5b	151.1b	171.6a	174.4a	173.4a
280	17.8a	37.2a	77.0a	111.7a	139.0a	162.5a	176.1a	181.4a	179.7a
					Number of leaves				
140	5.38a	10.0a	10.9b	12.0b	16.9b	20.0b	22.2b	21.9a	21.6a
280	5.33a	10.2a	12.0a	14.2a	19.6a	22.3a	23.5a	22.2a	22.0a
					Number of bunches				
140	-	-	1.5b	2.9b	4.3b	5.2b	6.1a	6.0a	-
280	-	-	2.3a	3.7a	5.5a	5.9a	6.6a	6.2a	-
					Number of flowers				
140	-	-	3.5a	6.4a	0.7b	3.8b	4.2a	1.5a	-
280	-	-	3.7a	6.7a	2.0a	5.5a	4.8a	1.8a	-
					Number of fruits				
140	-	-	0.3b	3.1b	7.9b	10.7b	13.1b	11.7b	-
280	-	-	0.7a	5.0a	12.8a	15.9a	17.5a	16.0a	-

The – symbol indicates that the variable was not evaluated. Means followed by the same letter in the columns do not differ significantly by an F test at the 0.05 probability level.

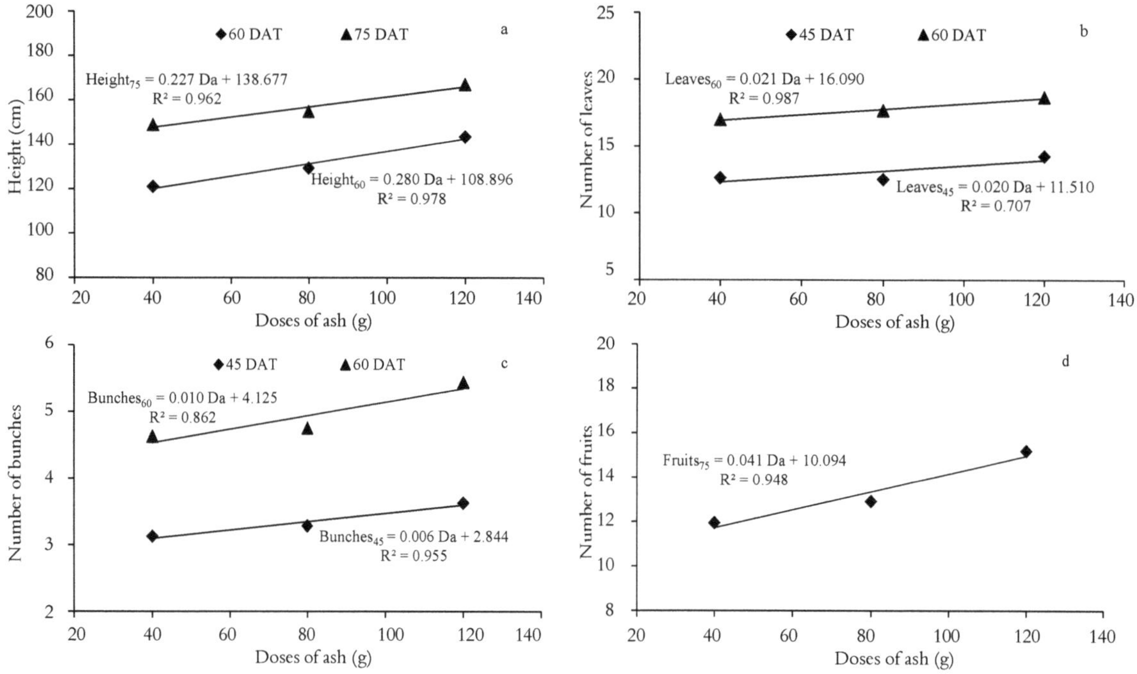

Figure 1. Plant height (a) and number of leaves (b), bunches (c), and fruits (d) at different doses of ash.

The increase in the dose of castor cake resulted in positive increments in all the tested growth variables, and this effect was observed in most of the evaluations performed. To a lesser degree, the increase in ash doses caused significant effects only on the variables plant height (observed at 60 and 75 DAT) and numbers of leaves and fruits, observed only at 45, 60, and 75 DAT. The increase in the doses of castor cake on the growth variables was more pronounced than the effect of the ash doses and resulted in mean increments of 85.3 and 79.2%, respectively, in the total mass of fruits and mass of marketable fruits (Table 2). Additionally, the increase in the total diameter of fruits and diameter of marketable fruits was 5.9%. The effect of ash doses on the growth variables was not sufficient to promote increments in the total mass fruits and mass of marketable fruits or in the total diameter of fruits and diameter of marketable fruits, which can be partially explained by the fact that the K doses (30, 60, and 90 kg ha^{-1}), applied in the form of ash, are not sufficient to promote increments in the tomato fruit production.

The beneficial effects on growth variables, and consequently, on the production variables obtained in the present study, are due to the increase in N provided to the plants through the castor cake, corroborating Ferreira, Ferreira, and Fontes (2010), who reported that an increment in the level of N generated increased plant height, number of leaves, flowering, and fruiting, resulting in greater production of tomato fruits. Mueller, Wamser, Suzuki, and Becker (2013) obtained positive increments in the total yield of tomato with the increase in the doses of N in the form of poultry litter.

Table 2. Total mass of fruits (TMF), mass of marketable fruits (MMF), total diameter of fruits (TDF), and diameter of marketable fruits (DMF) for the doses of castor cake.

Dose	TMF	MNF	TDF	DMF
(g)	(kg per plant)		(mm)	
140	1.02b	0.99b	52.80b	53.49b
280	1.89a	1.78a	55.93a	56.59a

Means followed by the same letter in the columns do not differ significantly based on an F test at 0.05 probability level.

In response to crop development in the different treatments, the irrigation depths applied through the automatic device for irrigation management were equal to 132, 177, 177, 175, 195, and 191 mm for T1, T2, T3, T4, T5, and T6, respectively, which shows that as the lowest doses of ash and castor cake (40 and 140 g, respectively) were applied, the lowest irrigation depth was obtained in response to the lower growth of the plants. When higher doses of ash (80 and 120 g) and castor cake (280 g) were applied, the greatest irrigation depths (195 and 191 mm) were obtained in response to the greater growth of the plants, in agreement with the results found in the present study.

Experiment II (2015) – Effect of water deficit

The applied irrigation depths were equal to 135, 165, 191, and 213 mm, respectively, for L1, L2, L3, and L4. Aiming for a greater yield of tomato plants in a protected environment, Santana, Vieira, Barreto, and Cruz (2010) and Macêdo and Alvarenga (2005) applied irrigation depths varying from 372 to 802 mm and from 158.1 to 399.2 mm, respectively. Also in a protected environment and producing tomatoes in pots, Silva et al. (2013) applied irrigation depths ranging from 180 to 828 mm. In the present study, the lower values of applied irrigation depth are essentially due to the cultivation system in pots and in a protected environment and due to the irrigation method, in which water is locally applied, associated with the adequate management of the irrigation system with an automatic device for irrigation. All these factors contributed to the lower water consumption during the productive process.

With a reduction of 37% in irrigation depth (213 to 135 mm) or increase in soil water tension for the activation of irrigation (11.6 to 47.2 kPa), there was a reduction in plant height (11, 12, 7, 8, 7, and 7% at 45, 60, 75, 90, 105, and 111 DAT, respectively) and in the numbers of leaves (11% at 45 DAT) and fruits (32 and 20% at 60 and 75 DAT, respectively) (Figure 2). Similar results were found for plant height by Soares, Lima, Brito, Sá, and Araújo (2011), Macêdo and Alvarenga (2005) and Santana et al. (2010), for number of leaves by Soares et al. (2011), and for fruits by Silva et al. (2013) and Macêdo and Alvarenga (2005). The numbers of bunches and flowers were not influenced by the irrigation depths, which is consistent with the result found by Cantore et al. (2016) in a study of water deficit in tomato crops.

The irrigation depth of 213 mm (11.6 kPa) promoted the highest values of the evaluated growth variables, resulting in greater total mass of fruits, mass of marketable fruits, and total diameter of fruits (Figures 3a and 4, respectively). This result demonstrates that yield was directly proportional to the irrigation depth, corroborating the results found by Mukherjee, Kundu, and Sarkar (2010) and Zheng et al. (2013). The diameter of marketable fruits was not influenced by the irrigation depths.

The mass of fruits with blossom-end rot (Figure 3b) varied from 1.34 to 0.83 kg per plant with the increment in irrigation depth, showing that even when plants were not subjected to water deficit (100% of automatic device for irrigation; 213 mm and 11.6 kPa), the incidence of fruits with blossom-end rot was high. That pattern can be explained by the high K:Ca ratio of 2.7. According to Jones Júnior (2005), excess K fertilizers in the nutrient solution reduce the absorption of Ca because K is preferentially absorbed and transported in the plant. Fanasca et al. (2006) observed increases in the mass of fruits with blossom-end rot from 0.34 to 0.98 kg per plant with variation of 0.23 to 4.4 in the K:Ca ratio.

The mass of marketable fruits declined by 45% with the variation in irrigation depth from 213 to 135 mm as a result of the effect of water deficit that caused losses in the production through fruits with blossom-end rot. As in the present study, Cantuário, Luz, Pereira, Salomão, and Rebouças (2014) observed increases in the incidence of fruits with blossom-end rot as the soil water tension increased for bell peppers. It is possible that there was a reduction in the transport of water in the xylem to the shoots, leading to a lower transport of Ca, especially to the apex of the fruit, due to the imposition of water deficit (Adams & Ho, 1993; Ho, Belda, Brown, Andrews, & Adams, 1993), and this Ca limitation could promote the formation of small fruits and a larger amount of fruits with rot (Madrid, Barba, Sánchez, & Garcia, 2009; Wang et al., 2011).

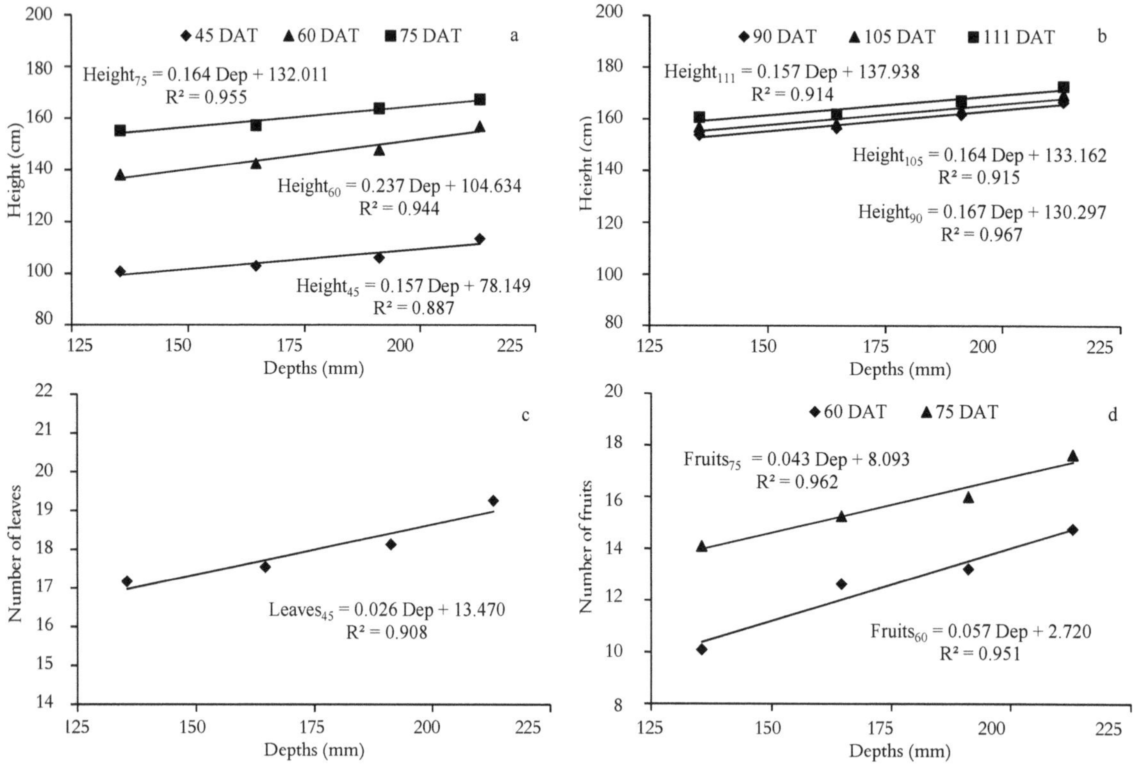

Figure 2. Plant height from 45 to 75 days after transplanting (DAT) (a) and from 90 to 111 DAT (b), number of leaves at 45 DAT (c), and number of fruits at 60 and 75 DAT (d) at different irrigation depths.

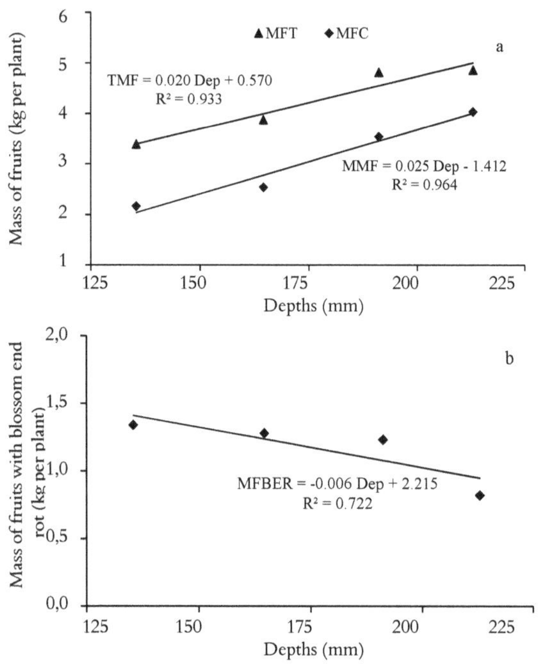

Figure 3. Total mass of fruits (TMF), mass of marketable fruits (MMF) (a), and mass of fruits with blossom-end rot (MFBER) (b) at different irrigation depths.

Marouelli and Silva (2007) observed maximum marketable yields for a tomato crop using soil water tensions of 12 kPa for the stage of fruit development and 15 kPa for the maturation stage. These values of soil water tension are similar to the mean tension for the treatment with the greatest irrigation depth used in the present study.

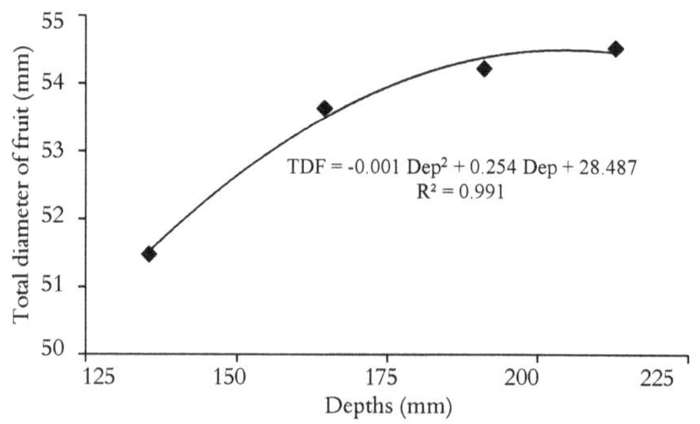

Figure 4. Total diameter of fruits (TDF) at different irrigation depths.

The reduction in irrigation depth promoted a reduction in most of the variables evaluated in the present study, indicating that any degree of water stress can harm the growth and yield of tomatoes (Saif, Maqsood, Farooq, Hussain, & Habib, 2003). Water deficit results in numerous physiological alterations, such as reductions in root matter, leaf area, and number of leaves and, consequently, reductions in plant growth (Mao et al., 2003; Patanè,

Tringali, & Sortino, 2011). When the plant is subjected to water stress, almost all of its aspects of growth and development are affected, which can modify its anatomy and morphology as well as interfere with many metabolic reactions (Soares et al., 2011). The lack of water reduces cell turgor pressure and consequently decreases cell elongation and, therefore, plant growth and development (Taiz & Zeiger, 2009).

The highest values of water use efficiency (WUE) were observed for the irrigation depths of 191 mm (37.0 kg m^{-3}) and 213 mm (37.9 kg m^{-3}) (Figure 5) and are above the mean value presented by Doorenbos and Kassam (1994) for a tomato crop (11.0 kg m^{-3}). This pattern occurred because the yield of the crop was close to the mean yield, while the applied irrigation depth was much lower than the range of irrigation depths cited by the previously mentioned authors.

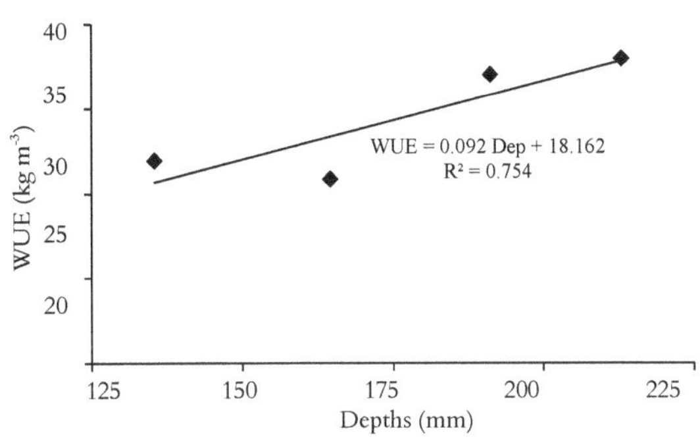

Figure 5. Water use efficiency (WUE) at different irrigation depths.

This results indicate that the cultivation system and the irrigation management were sufficient, leading to a WUE approximately 3.5 times higher than the mean value reported in the literature. Furthermore, the WUE presented by Doorenbos and Kassam (1994) derives from mean values of conditions that are different from those in the present study, only serving as a reference. In more recent studies and under more similar conditions, i.e., tomato cultivation in pots, WUE values higher than those presented by the above-mentioned authors have also been observed, such as in the study of Reina-Sánchez, Romero-Aranda and Cuartero (2005), who obtained a WUE of 25.0 kg m^{-3}.

In general, the irrigation depth that promotes the greatest WUE is lower than the irrigation depth that promotes maximum yield (Nangare, Singh, Kumar, & Minhas, 2016), which justifies the use of deficit irrigation aiming at better WUE. However, this pattern was not observed in the present study because WUE was calculated based on the marketable yield, which was linearly influenced by the irrigation depths due to the losses caused by the incidence of fruits with blossom-end rot. Under the conditions of the present study, water deficit would not be recommended because any level of deficit would lead to a reduction in WUE.

Conclusion

Tomato plants grew larger and produced more when fertilized with 140 g of ash or 280 g of castor cake. With the dose of 280 g of castor cake, the ash did not promote further plant growth.

Water deficit caused reductions in the growth and yield of tomatoes.

Higher values of water use efficiency, 37.0 and 37.9 kg m^{-3}, were observed at the irrigation depths of 191 and 213 mm, respectively.

References

Adams, P., & Ho, L. C. (1993). Effects of environment on the uptake and distribution of calcium in tomato and on the incidence of blossom-end rot. *Plant and Soil, 154*(1), 127-132.

Cantore, V., Lechkar, O., Karabulut, E., Sellami, M. H., Albrizio, R., Boari, F., ... Todorovic, M. (2016). Combined effect of deficit irrigation and strobilurin application on yield, fruit quality and water use efficiency of "cherry" tomato (*Solanum lycopersicum* L.). *Agricultural Water Management, 167*, 53-61.

Cantuário, F. S., Luz, J. M. Q., Pereira, A. I. A., Salomão, L. C., & Rebouças, T. N. H. (2014). Blossom-end rot and scald in fruits of sweet pepper submitted to water stress and silicon rates. *Horticultura Brasileira, 32*(2), 215-219.

Choi, J. M., Lee, C. W., & Park, J. S. (2015). Performance of seedling grafts of tomato as influenced by root substrate formulations, fertigation leaching fractions, and N concentrations in fertilizer solution. *Horticulture, Environment, and Biotechnology, 56*(1), 17-21.

Doorenbos, J., & Kassam, A. H. (1979). *Yield response to water.* Rome, IT: FAO.

Escarabajal-Henarejos, D., Molina-Martínez, J. M., Fernández-Pacheco, D. G., & García-Mateos, G. (2015). Methodology for obtaining prediction models of the root depth of lettuce for its application in irrigation automation. *Agricultural Water Management, 151*, 167-173, 2015.

Fanasca, S., Colla, G., Maiani, G., Venneria, E., Rouphael, Y., Azzini, E., & Saccardo, F. (2006). Changes in antioxidant content of tomato fruits in response to cultivar and nutrient solution composition. *Journal Agricultural and Food Chemistry, 54*(12), 4319-4325.

Ferreira, M. M. M., Ferreira, G. B., & Fontes, P. V. R. (2010). Nitrogen fertilization efficiency in tomato at two sowing times. *Revista Ceres, 57*(2), 263-273.

Filgueira, F. A. R. (2008). *Novo Manual de Olericultura: Agrotecnologia moderna na produção e comercialização de hortaliças*. Viçosa, MG: UFV.

Gomes, D. P., Carvalho, D. F., Almeida, W. S., Medici, L. O., & Guerra, J. G. M. (2014). Organic carrot-lettuce intercropping using mulch and different irrigation levels. *Journal of Food, Agriculture & Environment*, *12*(1), 323-328, 2014.

Gomiero, T., Paoletti, M. G., & Pimentel, D. (2008). Energy and environmental issues in organic and conventional agriculture. *Critical Reviews in Plant Sciences*, *27*(4), 239-254.

Gonçalves, F. V., Medici, L. O., Almeida, W. S., Carvalho, D. F., Santos, H. T., & Gomes, D. P. (2014). Irrigation with Irrigás, Class A pan and an low cost controller in the organic cultivation of lettuce. *Ciência Rural*, *44*(11), 1950-1955.

Guimarães, M. A., Caliman, F. R. B., Silva, D. J. H., Flores, M. P., & Elsayed, A. Y. A. M. (2007). Exigências climáticas da cultura do tomateiro. In: D. J. H. Silva, & F. X. R. VALE, (Eds). *Tomate: Tecnologia de produção* (p. 85-99). Visconde do Rio Branco, MG: Suprema Gráfica.

Ho, L. C., Belda, R., Brown, M., Andrews, J., & Adams, P. (1993). Uptake and transport of calcium and the possible causes of blossom-end rot in tomato. *Journal of Experimental Botany*, *44*(259), 509-518.

Jones Júnior, J. R. (2005). *Hidroponics: a practical guide for soilless grower*. Boca Raton, FL: CRC press.

Kuba, T., Tscholl, A., Partl, C., Meyer, K., & Insam, H. (2008). Wood ash admixture to organic wastes improves compost and its performance. *Agriculture, Ecosystems and Environment*, *127*(1-2), 43-49.

Lee, K. S., Choe, Y. C., & Park, S. H. P. (2015). Measuring the environmental effects of organic farming: A meta-analysis of structural variables in empirical research. *Journal of Environmental Management*, *162*, 263-274. doi: 10.1016/j.jenvman.2015.07.021

Lima, R. L. S., Severino, L. S., Sampaio, L. R., Sofiatti, V., Gomes, J. A., & Beltrão, N. E. M. (2011). Blends of castor meal and castor husks for optimized use as organic fertilizer. *Industrial Crops and Products*, *33*(2), 364-368. doi: 10.1016/j.jenvman.2015.07.021

Lovelli, S., Perniola, M., Ferrara, A., & Tommaso, T. D. (2007). Yield response factor to water (ky) and water use efficiency of *Carthamus tinctorius* L. and *Solanum melongena* L. *Agricultural Water Management*, *92*(1-2), 73-80.

Macêdo, L. S., & Alvarenga, M. A. R. (2005). Effects of water nivels and potassium fertirrigation on growth, production and quality of tomato fruits in greenhouse. *Ciência e Agrotecnologia*, *29*(2), 296-304.

Madrid, R., Barba, E. M., Sánchez, A., & Garcia, A. L. (2009). Effects of organic fertilisers and irrigation level on physical and chemical quality of industrial tomato fruit (cv. Nautilus). *Journal of the Science of Food and Agriculture*, *89*(15), 2608-2615.

Magriotis, Z. M., Carvalho, M. Z., Sales, P. F., Alves, F. C., Resende, R. F., & Saczk, A. A. (2014). Castor bean (*Ricinus communis* L.) presscake from biodiesel production: An efficient low cost adsorbent for removal of textile dyes. *Journal of Environmental Chemical Engineering*, *2*(3), 1731-1740.

Mao, X., Liu, M., Wang, X., Liu, C., Hou, Z., & Shi, J. (2003). Effects of deficit irrigation on yield and water use of greenhouse grown cucumber in the North China Plain. *Agricultural Water Management*, *61*(3), 219-228.

Marouelli, W. A., & Silva, W. L. C. (2007). Water tension thresholds for processing tomatoes under drip irrigation in Central Brazil. *Irrigation Science*, *25*(4), 411-418.

Martins, A. N., Suguino, E., Dias, N. M. S., & Perdoná, M. J. (2011). Effect of addition of castor bean pie in substrates in acclimatization of micropropagated banana plantlets. *Revista Brasileira de Fruticultura*, *33*(1), 198-207.

Medici, L. O., Rocha, H. S., Carvalho, D. F., Pimentel, C., & Azevedo, R. A. (2010). Automatic controller to water plants. *Scientia Agricola*, *67*(6), 727-730.

Mueller, S., Wamser, A. F., Suzuki, A., & Becker, W. F. (2013). Tomato yield under organic fertilization and supplementation with mineral fertilizers. *Horticultura Brasileira*, *31*(1), 86-92.

Mukherjee, A., Kundu, M., & Sarkar, S. (2010). Role of irrigation and mulch on yield, evapotranspiration rate and water use pattern of tomato (*Lycopersicon esculentum* L.). *Agricultural Water Management*, *98*(1), 182-189.

Nabeela, F., Murad, W., Khan, I., Mian, I. A., Rehman, H., Adnan, M., & Azizullah, A. (2015). Effect of wood ash application on the morphological, physiological and biochemical parameters of *Brassica napus* L. *Plant Physiology and Biochemistry*, *95*, 15-25. doi: 10.1016/j.plaphy.2015.06.017

Nangare, D. D., Singh, Y., Kumar, P. S., & Minhas, P. S. (2016). Growth, fruit yield and quality of tomato (*Lycopersicon esculentum* Mill.) as affected by deficit irrigation regulated on phenological basis. *Agricultural Water Management*, *171*, 73-79. doi: 10.1016/j.agwat.2016.03.016

Oburger, E., Jäger, A., Pasch, A., Dellantonio, A., Stampfer, K., & Wenzel, W. W. (2016). Environmental impact assessment of wood ash utilization in forest road construction and maintenance - A field study. *Science of the Total Environment*, *544*, 711-721. doi: 10.1016/j.scitotenv.2015.11.123

Oliveira Filho, A. F., Oliveira, F. A., Medeiros, J. F., Mesquita, T. O., & Zonta, E. (2010). Growth of castor cultivars under castor presscake doses. *Revista Verde*, *5*(5), 18-24.

Patanè, C., Tringali, S., & Sortino, O. (2011). Effects of deficit irrigation on biomass, yield, water productivity and fruit quality of processing tomato under semi-arid Mediterranean climate conditions. *Scientia Horticulturae*, *129*(4), 590-596.

Piva, R., Botelho, R. V., Ortolan, C., Müller, M. M. L., & Kawakami, J. (2013). Fertilization in organic vineyard cv. Isabel using wood ash and cattle manure. *Revista Brasileira de Fruticultura*, *35*(2), 608-615.

Reina-Sánchez, A., Romero-Aranda, R., & Cuartero, J. (2005). Plant water uptake and water use efficiency of greenhouse tomato cultivars irrigated with saline water. *Agricultural Water management*, *78*(1-2), 54-66.

Saif, U., Maqsood, M., Farooq, M., Hussain, S., & Habib, A. (2003). Effect of planting patterns and different irrigation levels on yield and yield component of maize (*Zea mays* L.). *International Journal of Agriculture and Biology*, *5*(1), 64-66.

Santana, M. J., Vieira, T. A., Barreto, A. C., & Cruz, O. C. (2010). Tomato response irrigated with different levels of soil water replacement. *Irriga*, *15*(4), 443-454.

Silva, J. M., Ferreira, R. S., Melo, A. S., Suassuna, J. F., Dutra, A. F., & Gomes, J. P. (2013). Cultivation of tomato in greenhouse under different replenishment rates of evapotranspiration. *Revista Brasileira de Engenharia Agrícola e Ambiental*, *17*(1), 40-46.

Soares, L. A. A., Lima, G. S., Brito, M. E. B., Sá, F. V. S., & Araújo, T. T. (2011). Growth of tomato and physical quality of fruits under water stress in greenhouse. *Revista Verde*, *6*(3), 203-212.

Taiz, L., & Zeiger, E. (2009). *Fisiologia vegetal*. Porto Alegre, RS: Artmed.

Vignolo, G. K., Araújo, V. F., Kunde, R. J., Silveira, C. A. P., & Antunes, L. E. C. (2011). Pre-planting fertilization effects on strawberry fruit yield. *Ciência Rural*, *41*(10), 1755-1761.

Wang, F., Kang, S., Du, T., Li, F., & Qiu, R. (2011). Determination of comprehensive quality index for tomato and its response to different irrigation treatments. *Agricultural Water Management*, *98*(8), 1228-1238.

Zapata, N., Vargas, M., Reyes, J. F., & Belmar, G. (2012). Quality of biodiesel and press cake obtained from *Euphorbia lathyris*, *Brassica napus* and *Ricinus communis*. *Industrial Crops and Products*, *38*, 1-5. doi: 10.1016/j.indcrop.2012.01.004

Zheng, J., Huang, G., Jia, D., Wang, J., Mota, M., Pereira, L.S., ... Liu, H. (2013). Responses of drip irrigated tomato (*Solanum lycopersicum* L.) yield, quality and water productivity to various soil matric potential thresholds in an arid region of Northwest China. *Agricultural Water Management*, *129*, 181-193. doi: 10.1016/j.agwat.2013.08.001

Zotarelli, L., Dukes, M. D., Sholberg, J. M. S., Muñoz-Carpena, R., & Icerman, J. (2009). Tomato nitrogen accumulation and fertilizer use efficiency on a sandy soil, as affected by nitrogen rate irrigation scheduling. *Agricultural Water Management*, *96*(8), 1247-1258.

8

Productivity and chemical composition of food-type soybeans sown on different dates

Gustavo Henrique Freiria[1], Wilmar Ferreira Lima[2], Rodrigo Santos Leite[3], José Marcos Gontijo Mandarino[3], Josemeyre Bonifácio da Silva[1*] and Cássio Egídio Cavenaghi Prete[1]

*[1]Programa de Pós-graduação em Agronomia, Centro de Ciências Agrárias, Universidade Estadual de Londrina, Rod. Celso Garcia Cid, Pr-445, Km 380, Campus Universitário, Cx. Postal 10011, 86057-970, Londrina, Paraná, Brazil. [2]Instituto Agronômico do Paraná, Londrina, Paraná, Brazil. [3]Empresa Brasileira de Pesquisa Agropecuária, Centro Nacional de Pesquisa de Soja, Embrapa/Soja, Londrina, Paraná, Brazil. *Author for correspondence. *E-mail: josibonifacio@uel.br*

ABSTRACT. The objective of this study was to investigate the effect of different sowing dates on the yield, 100-grain weight, oil, protein and isoflavone contents of food-type soybean breeding lines UEL 110, UEL 115 and UEL 123 and a cultivar BRS 257. The materials were seeded on four sowing dates, and the experiment was conducted in a randomized block design with four replications. The productivity and 100-grain weight varied with the sowing dates and soybean genotypes. The protein content ranged from 36.40 to 42.44%, and the oil content ranged from 18.29 to 22.71%. No significant interaction was found between the genotype and sowing dates for the protein content. The isoflavone content also varied with the sowing dates and soybean genotypes. The cultivar BRS 257 had the highest isoflavone content, including the β-glucoside, malonyl glucoside and aglycones forms. The different sowing dates influenced the productivity, 100-grain weight, oil and protein content and isoflavone levels. Higher temperatures and lower rainfall during the grain filling decreased the productivity and isoflavone content and increased the protein content. For all sowing dates, the BRS 257 soybean food-type cultivar showed the highest isoflavone content, indicating that the effect of genotype is more important.

Keywords: soybean, oil, protein, isoflavones.

Introduction

Soybean is a versatile grain due to its high protein content (40%), high lipid content (20%), high mineral content, especially of K, Na, Ca, Mg, S and P, and vitamin content, including thiamine, riboflavin, niacin, pantothenic acid, biotin, folic acid, inositol and choline (Liu, 1997).

In addition to this rich chemical composition, soybean grains have isoflavones. Isoflavones are secondary plant metabolites, a group of natural bioflavonoids synthesized exclusively by the Leguminosae (Fritsche & Steinhart, 1999), and they are associated with the risk reduction or prevention of various diseases, such as breast cancer and prostate cancer (Liggins et al., 2000), cardiovascular disease (Rimbach et al., 2008), and estrogenic and antioxidant activity (Liu, Kanjo, & Mizutani, 2010; Ma et al., 2010). Twelve isoflavone forms have been reported in soybeans, including three aglycones:

genistein, daidzein and glycitein and their respective 7-O-β-D-glucosides (genistin, daidzin and glycitin), 6"-O-malonyl-7-O-β-D-glucosides (malonylgenistin, malonyldaidzin and malonylglycitin) and 6"-O-acetyl-7-O-β-D-glucosides (acetylgenistin, acetyldaidzin and acetylglycitin (Shao et al., 2009).

The yield and the isoflavone, protein and lipid contents can be influenced by the genetic and climatic conditions. Air temperature during the grain formation stage is the most critical environment factor affecting the final content of oil, protein and isoflavones. However, other factors also affect these characteristics, such as humidity, light intensity and rainfall (Carrão-Panizzi, Berhow, Mandarino, & Oliveira, 2009). Regarding the protein and oil level, Marques et al. (2011) observed that the oil content of the Brazilian soybean cultivars ranged from 14 to 24%, and the protein content ranged from 33 to 49%; they concluded that the sowing dates affected only the protein content of the cultivars. Carrão-Panizzi et al. (2009) observed that the total isoflavone content ranged from 12 mg 100 g^{-1} (cv. Embrapa 48) to 461 mg 100 g^{-1} (cv. CS 305) and concluded that the differences among the soybean cultivars were due to genetic effects. Ávila et al. (2007) observed that the oil and isoflavone content varied in range among the Brazilian soybean cultivars, but the protein content was not changed. Carrão-Panizzi, Simão and Kikuchi (2003) studied the effects of the genotype, environment and interactions on the isoflavone contents and reported significant genotype x environment interactions. The average of the total isoflavone content of the Brazilian soybean cultivars is approximately 150 mg 100 g^{-1} (Carrão-Panizzi & Kitamura, 1995).

Lipoxygenases (LOXs) (EC 1.13.11.12) are a class of enzymes that produce the volatile compounds associated with the rancid or beany soybean flavor (Fukushige, Wang, Simpson, Garden, & Hildebrand, 2005). Genetic elimination of lipoxygenase enzymes can improve the flavor of the soybean by reducing the formation of the volatile compounds associated with the characteristic flavor. The lipoxygenase-free soybean cultivars are classified as food-type soybeans because they show this special characteristic (Silva et al., 2012; Liu, 1999).

The selection of genotypes with high grain yield is the main goal of soybean breeding programs. In addition to high grain yield, currently, high-quality soybean grains must have a mild flavor and nutritive and functional properties. Therefore, the objective of this study was to investigate the effect of different sowing dates on the yield, oil, protein content and the isoflavone contents of food-type breeding lines and soybean cultivars and their interactions.

Material and methods

Soybean food types

Three lines (UEL 110, UEL 115, and UEL 123) of the breeding program of The State University of Londrina (UEL), located at Parana state, Brazil and the BRS 257 soybean cultivar supplied by Embrapa Soybean, which is located in the city of Londrina, Parana State, Brazil, were evaluated. All seeds of the breeding lines and the cultivar are lipoxygenase-free and classified as food-type soybeans (Silva et al., 2012). The materials differed in maturity, height, seed weight, growth habit, and other characteristics (Table 1).

Table 1. Agronomic characteristics of the genotypes studied.

Agronomic	Lines/Cultivar			
Characteristics	UEL 110	UEL 115	UEL 123	BRS 257
Florewing dates (days)	44	41	44	43
Maurity group	Early	Middle	Middle	Early
Stem lenght (cm)	97.08	109.39	106.56	67.97
Growth Type	Determined	Indeterminate	Indeterminate	Determined
Seed coat color	Yellow	Yellow	Yellow	Yellow
Hilum color	Yellow	Brown	Brown	Yellow
Lipoxygenases	Null	Null	Null	Null
Lipoxygenases	Null	Null	Null	Null

Grain yield (kg ha^{-1}) and 100-grain weight (g).

The experiment was laid out in a randomized block design with four replications and plot areas of 5 x 0.45 x 0.07 m. The sowing dates were October 7 (sowing date 1), October 14 (sowing date 2), October 29 (sowing date 3) and November 9 (sowing date 4) in the 2013/2014 crop season. The seeds were sown in the farm of The UEL's Agronomy School (23° 20' 23" S, 51° 12' 32" W, 535 m altitude), Londrina, Paraná State, Brazil. According to Instituto Agronômico do Paraná (IAPAR, 1987), the climate classification is Köppen Cfa, with an annual precipitation average ranging from 1400 to 1600 mm. The average rainfall and daily temperatures during the experimental period are shown in Figure 1.

After being manually harvested, the grain yield was calculated as productivity in kg ha^{-1}, with correction for 13% moisture. For the calculation of the 100-grain weight (g), four replicates of 100 grains were weighed.

Oil and protein content

Oil and protein contents were determined in the whole soybean grains by near-infrared spectroscopy (NIR) according to the Heil (2012) method using an Antaris™ II FT-NIR analyzer from Thermo Scientific (Thermo Electron Co., USA). The determinations were performed in triplicate, with a resolution of 2.0 cm^{-1} and an average of 32 scans per determination. The predictions were made by mathematical models for oil (R = 0.9487) and crude protein (R = 0.9675).

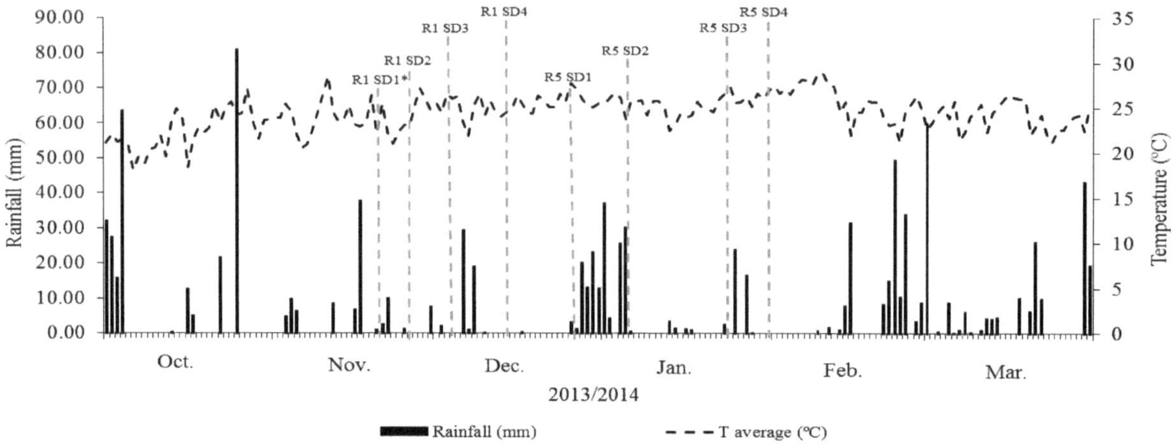

Figure 1. Rainfall and daily temperatures averages observed during the experiment and soybean stages of development.
*Reproductive stage R1 and R5 (Fehr & Caviness, 1977) in the first, second, third and fourth sowing date (SD).

Isoflavone analysis

For the isoflavone content determination, the soybean grains were ground in a TE 631 knife grinder mill from Tecnal, Brazil and defatted with hexane in a 1:10 ratio (w/v) for 1 hour at room temperature by continuous and rotary agitation, followed by vacuum filtration. The extraction was performed in duplicate with 0.5 g of each sample ground with 6 mL of organic solvent containing 70% ethanol acidified with 0.001% glacial acetic acid and then stirred once every 15 min. for 1 hour at room temperature. Then, the mixture was placed in an ultrasonic bath for 15 min. at room temperature, centrifuged at 21,000 xg at 4°C for 15 min. in an Eppendorf 5804 R Centrifuge (Hamburg, GE) and filtered in a 0.22 μm Millex filter. The separation and quantification of isoflavones were performed in duplicate according to Berhow (2002) using an Ultra-Performance Liquid Chromatography (UPLC) system. The column was a reversed-phase type (model ACQUITY UPLC BEH C18, Waters, United States) with dimensions of 2.1 mm (ID) x 50 mm length and a particle size of 1.7 μm. Elution was performed in a linear binary gradient system using mobile phase A, with methanol with 0.025% trifluoroacetic acid, and mobile phase B with ultrapure deionized water with 0.025% trifluoroacetic acid. The flow rate applied was 0.35 mL min.$^{-1}$ at 24°C. The gradient begins with 25% eluent A; at the 7 min gradient, the elution ratio ultimately reached 80% A, and then, the initial conditions had returned by 8 min., with a total run time of 10 min. The detector was a diode array (Waters) with a 254 nm wavelength. Standard solutions for the calibration curve construction (peak area of isoflavone contents) were made for each of the twelve isoflavone forms: daidzin, genistin, glycitin, malonyl daidzin, malonyl genistin, glycitin malonyl, acetyl daidzin, acetyl genistin, acetyl glycitin, glycitein and genistein, which were purchased from Sigma and Fluka (0.5, 1, 2, 4, 6, 8, and 10 μg mL^{-1}). The standards were injected in triplicate to yield the corresponding chromatograms for the isoflavone forms, each with its corresponding retention time. The peaks for each isoflavone in each sample were identified for the comparison of the retention times and UV spectrum of the respective reference standard regions. The application was coupled in the chromatograph-generated calibration curves, and the isoflavone concentrations were calculated and expressed in mg of each isoflavone per 100 g of sample (mg isoflavone 100 g^{-1}).

Statistical analysis

The data were submitted to analysis of variance (ANOVA), and the means were compared by Tukey's test at 5% probability using the SISVAR program (Ferreira, 2011). The Pearson correlation using the GENES program was applied to the contents of the protein and oil (Cruz, 2013).

Results and discussion

According to the analysis of variance (ANOVA), the genotype x sowing date interaction showed a significant difference ($p < 0.05$) in grain yield, 100-grain weight, oil content and for the content of all isoflavone forms, suggesting different responses of the genotypes with regard to the sowing date. However, the protein content showed no significant interaction.

The breeding lines and the BRS 257 soybean cultivar did not differ in productivity for the first and second sowing date (Table 2). In the third sowing date, a higher grain yield was observed for the BRS 257 soybean cultivar, but this cultivar showed lower grain productivity from the fourth sowing date. The highest productivity was obtained for all breeding lines and for the cultivar BRS 257 from sowing date 1 (SD 1) and 3

(SD 3), and this productivity did not differ among the soybeans. The lowest grain yields were observed from sowing date 4 for the breeding line UEL 110 and for the cultivar BRS 257. The higher 100-seed weight was observed for the UEL 110 and UEL 123 breeding lines in SD 1 and for UEL 115 and UEL 110 in SD 3 and SD 4. Within the genotypes, UEL 110, BRS 257 showed the higher 100-seed weight for SD 1 and SD 4. During the experiment, a prolonged drought period was associated with high temperatures (Figure 1), which coincided with the grain filling in SD 2 and SD 4 and the flowering in the SD 4. The productivity and 100-grain weight are well known to be influenced by genetic and climatic factors. The interaction between genotype and climatic conditions results in significant differences in the performance of each genotype according to climatic conditions such as temperature and rainfall (Gauch & Zobel, 1997). Hydric stress in the flowering stage can cause flower abortion and anthesis, and during the seed filling, it can affect the 100-seed weight, reducing the productivity in both cases (Avila et al., 2007; Barbosa et al., 2013). The lower availability of water promotes a decrease in the photosynthesis and grain filling (Marcos Filho, 2005). High temperatures, especially when associated with periods of low rainfall during ripening, can cause early maturation under these conditions of low vigor, as these conditions do not support the natural deposition of carbohydrates, lipids and proteins (Marques et al., 2011; Zhe, Lauer, Borges, & Leon., 2010; França Neto, Krzyzanowski, Henning, West, & Miranda, 1993) because there is a reduction of photosynthate translocation to the grains.

No significant difference at $p < 0.05$ was observed in the genotype x sowing date interaction for protein content, indicating that the temperature variation among the sowing dates was not enough to cause differences in protein. However, some differences were observed when each factor was considered for genotype and sowing date (Table 3). Protein content ranged from 36.40 to 42.44%, with cultivar BRS 257 showing the highest content (40.77%). Sowing dates (SD) 2 and 4 were more appropriate for an increase in protein content, and the lowest content was observed when the grains were cultivated in the SD 3 (37.05%). High temperature reduced the protein content (Pípolo, Sinclair, & Camara, 2004). The higher temperatures occurred in the SD 2 and SD 4 (grain filling); however, these temperatures were probably not high enough to reduce the protein content. Similar results were shown by Bellaloui, Reddy, Gillen, Fischer, and Mengistu (2011). Some works showed a relation between higher temperatures and lower rainfall on the increase in protein content (Rangel, Minuzzi, Braccini, Scapim, & Cardoso, 2007). Within each genotype, UEL 110, UEL 115 and BRS 257 showed the lowest oil content for SD 2. Pearson's correlation between the protein and oil contents was negative and significant (-0.79), proving that when the protein content increased, the oil content decreased. These results are in accordance with those reported by Pípolo et al. (2004) and Kandil, Sharief, Morsy, and El-Sayed (2013).

Concerning the isoflavones from all sowing dates (Table 4), BRS 257 showed the highest β-glucosides content, suggesting the effect of genotype on these isoflavone forms. The breeding line UEL 110 did not differ from the cultivar BRS 257 in the second sowing date. Among the breeding lines, the observed β-glucosides content was greater in the third sowing date. The lower β-glucosides content were found in the SD 2. Regarding the malonyl glucosides, the same results were observed with BRS 257, which exhibited the highest content of malonyl glucosides in all sowing dates. All genotypes showed the highest malonyl content for SD 3 and the lowest for SD 2. Aglycone contents varied from 8.92 to 0.72 mg 100 g^{-1} of isoflavones. BRS 257 showed the higher aglycone content for SD 1, and no difference between the breeding lines UEL 110 and UEL 115 was observed for SD 2 and SD 3. The highest content of aglycones was observed for SD1, and the lowest contents were observed for all genotypes in SD 2.

Table 2. Grain yield (kg ha^{-1}) and 100 grains weight (g) of the studied genotypes in different sowing dates.

Genotype	Sowing date[1] SD 1	SD 2	SD 3	SD4	Mean
	Grain yield (Kg ha^{-1})				
UEL 110	3649.08 Aa	2745.82 Ba	3526.02 Aab	1854.05 Cab	2943.74
UEL 115	3794.17 Aa	2563.26 Ba	3305.63 Ab	2136.65 Bab	2949.93
UEL 123	3876.00 Aa	2240.85 Ba	3642.03 Aab	2214.77 Ba	2993.42
BRS 257	3443.04 Aa	2412.48 Ba	4051.92 Aa	1450.03 Cb	2839.36
Mean	3690.57	2490.60	3631.40	1913.87	
	100 grains weight (g)				
UEL 110	16.46 Aa	13.68 Bab	13.77 Bab	17.39 Aa	15.33
UEL 115	15.21 Ab	13.08 Bb	14.63 Aa	13.44 Bc	14.09
UEL 123	16.40 Aa	12.89 Db	14.04 Cab	15.34 Bb	14.67
BRS 257	16.11 Aab	14.18 Ba	13.49 Bb	15.65 Ab	14.86
Mean	16.05	13.46	13.98	15.45	

[1]Sowing date at October 7, 2013 (SD 1), October 14, 2013 (SD 2), October 29, 2013 (SD 3) and November 9, 2013 (SD 4). Means followed by the same letter (capital in the line and small in the column) do not differ by Tukey at 5% probability. C.V (%) = 13.39 (Grain yield (kg ha^{-1})) and = 3.73 (100 grains weight (g)).

Table 3. Protein and Oil Content (%) of the studied genotypes sowing dates.

Genotype	Sowing dates[1] SD 1	SD 2	SD 3	SD 4	
	Protein				Mean
UEL 110	39.85	40.49	36.65	39.43	39.11 b
UEL 115	38.24	39.92	36.41	39.16	38.43 b
UEL 123	39.89	39.32	39.60	36.40	38.80 b
BRS 257	40.33	42.44	38.72	41.60	40.77 a
Mean	39.58 B	40.54 A	37.05 C	39.95 AB	
	Oil				
UEL 110	21.39 Bab	20.40 Ca	22.66 Ab	20.86 Bcb	21.33
UEL 115	22.27 Aa	20.52 Ba	22.71 Aa	22.19 Aa	21.92
UEL 123	21.11 Bbc	20.99 Ba	22.62 Aa	21.82 Aba	21.63
BRS 257	20.31 Ac	18.29 Bb	20.38 Aa	20.33 Ab	19.83
Mean	21.27	20.05	22.09	21.30	

[1]Sowing date at October 7, 2013 (SD 1), October 14, 2013 (SD 2), October 29, 2013 (SD 3) and November 9, 2013 (SD 4). Means followed by the same letter (capital in the line and small in the column) do not differ by Tukey at 5% probability. C.V (%) = 2.58 (protein) and = 2.84 (oil).

Table 4. β-glucosides, malonylglucosides, aglycones and total isoflavones contents (mg 100 g^{-1}) of the studied genotypes.

Genotypes	Sowing dates[1] SD 1	SD 2	SD 3	SD 4	
	β-Glucosides				Mean
UEL 110	76.70 Bb	34.16 Da	84.87 Ab	42.92 Cb	59.66
UEL 115	57.86 Ac	22.47 Cb	57.05 Ad	30.63 Bc	42.00
UEL 123	53.47 Bd	20.53 Db	74.00 Ac	29.16 Cc	44.29
BRS 257	92.84 Aa	35.26 Da	88.35 Ba	49.06 Ca	66.38
Mean	70.21	28.10	76.07	37.94	
	Malonylglucosides				
UEL 110	113.76 Bb	75.89 Da	155.67 Ab	87.40 Cb	108.63
UEL 115	88.89 Bc	45.35 Db	125.65 Ad	64.06 Cd	80.98
UEL 123	81.90 Bd	37.67 Dc	150.71 Ac	68.58 Cc	84.71
BRS 257	165.12 Ba	72.28 Da	195.22 Aa	97.91 Ca	132.63
Average	112.42	57.79	156.81	79.48	
	Aglycones				
UEL 110	7.80 Ab	0.99 Db	5.90 Bb	2.32 Ca	4.25
UEL 115	4.96 Ac	1.26 Da	3.38 Bd	1.78 Cb	2.84
UEL 123	5.13 Ac	0.72 Dc	4.05 Bc	1.70 Cb	2.90
BRS 257	8.92 Aa	1.47 Da	6.77 Ba	2.49 Ca	4.91
Average	6.70	1.11	5.02	2.07	

[1]Sowing date at October 7, 2013 (SD 1), October 14, 2013 (SD 2), October 29, 2013 (SD 3) and November 9, 2013 (SD 4). Means followed by the same letter (capital in the line and small in the column) do not differ by Tukey at 5% probability. CV (%) = 2.18 (Isoflavones total); 3.09 (Aglycones); 2.11 (β-Glucosides); 2.27 (Malonylglucosides).

These results indicate that the temperature affects the contents of these isoflavones. The average temperature during the months from October to March was 24.4°C. February had the highest temperature average with 25.82°C and the highest thermal variation 3.98°C, which coincided with the beginning of grain filling. The period of the early grain filling (R5 stage) to physiological maturity (R8 stage) for genotypes sown on October 14, 2013 (SD 2), had the highest temperature average with 26.04°C. There have been many studies proving the influence of environmental conditions and genetic factors on the isoflavone contents (Tsukamoto et al., 1995; Carrão-Panizzi et al., 2003; Lee et al., 2003; Riedl et al., 2007). Tsukamoto et al. (1995) and Kim, Kim, Kim, and Chung (2012) observed that the isoflavone content was lower when the temperature was high during the grain filling stage. Carrão-Panizzi et al. (2009) observed that total isoflavone content ranged among the genotypes studied and concluded that the differences in the isoflavone content were due to genetic effects because all cultivars were grown at the same location and in the same growing season.

Conclusion

The different sowing dates influence the productivity, 100-grain weight, oil and protein content and isoflavone levels.

Higher temperatures and lower rainfall during grain filling decreased the productivity and isoflavone content and increased the protein content.

For all sowing dates, the BRS 257 food-type soybean cultivar showed high isoflavone content, indicating that the effect of genotype is more important.

References

Ávila, M. R., Braccini, A. L., Scapim, C. A., Mandarino, J. M. G., Albrecht, L. P., & Vidigal Filho, P. S. (2007). Componentes do rendimento, teores de isoflavonas, proteínas, óleo e qualidade de sementes de soja. *Revista Brasileira de Sementes, 29*(3), 111-127.

Barbosa, M. C., Braccini, A. L., Scapim, C. A., Albrecht, L. P., Piccinin, G. G., & Zucareli, C. (2013). Desempenho agronômico e componentes da produção de cultivares de soja em duas épocas de semeadura no arenito caiuá. *Semina: Ciências Agrárias, 34*(3), 945-960.

Bellaloui, N., Reddy, K. N., Gillen, A. M., Fischer, D. K., & Mengistu, A. (2011). Influence of planting date on seed protein, oil, sugars, minerals, and nitrogen metabolism in soybean under irrigated and non-irrigated environments. *American Journal of Plant Sciences, 2*(5), 702-715.

Berhow, M. A. (2002). Modern analytical techniques for flavonoid determination. In: B. S. Buslig, & J. A. Manthey (Ed.), *Flavonoids in the living cell* (p. 61-76). New York, NY: Klusher Academic.

Carrão-Panizzi, M. C., Simão, A. S., & Kikuchi, A. (2003). Efeitos de genótipos, ambientes e de tratamentos hidrotérmicos na concentração de isoflavonas agliconas em grãos de soja. *Pesquisa Agropecuária Brasileira, 28*(8), 897-902.

Carrão-Panizzi, M. C., Berhow, M., Mandarino, J. M. G., & Oliveira, M. C. N. (2009). Environmental and genetic variation of isoflavone content of soybean seeds grown in Brazil. *Pesquisa Agropecuária Brasileira, 44*(11), 1444-1451.

Carrão-Panizzi, M. C., & Kitamura, K. (1995). Isoflavone content in Brazilian soybean cultivars. *Breeding Science, 45*(1), 295-300.

Cruz, C. D. (2013). Genes – a software package for analysis in experimental statistics and quantitative genetics. *Acta Scientiarum. Agronomy, 35*(3), 271-276.

Fehr, W. R., & Caviness, C. E. (1977). *Stages of soybean development (Special report, 80)*. Ames, IA: Iowa State University of Science and Technology.

Ferreira, D. F. (2011). Sisvar: a computer statistical analysis system. *Ciência e Agrotecnologia, 35*(6), 1039-1042.

França Neto, J. B., Krzyzanowski, F. C., Henning, A. A., West, S. H., & Miranda, L. C. (1993). Soybean seed quality as affected by shiveling due to heat and drought stress during seed filling. *Seed Sciene and Technology, 21*(1), 107-116.

Fritsche, S., & Steinhart, H. (1999). Occurrence of hormonally active compounds in food: a review. *European Food Research and Technology, 209*(1), 153-179.

Fukushige, H., Wang, C., Simpson, T. D., Gardner, H. W., & Hildebrand, D. F. (2005). Purification and identification of linoleic acid hydroperoxides generated by soybean seed lipoxygenases 2 and 3. *Journal of agricultural and food chemistry, 53*(14), 5691-5694. Gauch, H. G., & Zobel, R. W. (1997). Identifying mega-environments and targeting genotypes. *Crop Science, 37*(2), 311-326.

Heil, C. (2012). *Rapid, multi-component analysis of soybeans by FT-NIR Spectroscopy*. Madison, WI: Thermo Fisher Scientific.

Instituto Agronômico do Paraná [IAPAR]. (1987). *Cartas climáticas básicas do Estado do Paraná*. Londrina, PR: Instituto Agronômico do Paraná.

Kandil, A. A., Sharief, A. E., Morsy, A. R., & El-Sayed, A. I. M. (2013). Influence of planting date on some genotypes of soybean growth, yield and seen quality. *Journal of Biological Sciences, 13*(3), 146-151.

Kim, E. H., Kim, S. L., Kim, S. H., & Chung, I. M. (2012). Comparison of isoflavones and anthocyanins in soybean [Glycine max (L.) Merril] seeds of different planting dates. *Journal of Agricultural and Food Chemistry, 60*(41), 10196-10202.

Lee, S. J., Ahn, J. K., Kim, S. H., Kim, J. T., Han, S. J., Jung, M. Y., & Chung, I. M. (2003). Variation in isoflavone of soybean cultivars with location and storage duration. *Journal of Agricultural and Food Chemistry, 51*(11), 3382-3389.

Liggins, J., Bluck, L. J. C., Runswick, S., Atkinson, C., Coward, W. A., & Bingham, S. A. (2000). Daidzein and genistein content of fruits and nuts. *Journal of Nutrition Biochemistry, 11*(6), 326-331.

Liu, K. S. (1997). Chemistry and Nutritional Value of Soybean Components. In: K. L. Liu (Ed.), *Soybean: chemistry, technology, and utilization* (p. 25-113). New York, NY: Chapman & Hall.

Liu, K. (1999). *Soybeans: chemistry, technology and utilization*. New York, NY: Chapman & Hall.

Liu, Z., Kanjo, Y., & Mizutani, S. A. (2010). Review of phytoestrogens: Their occurrence and fate in the environment. *Water Research, 44*(2), 567-577.

Ma, W., Yuan, L., Yu, H., Ding, B., Xi, Y., Feng, J., & Xiao, R. (2010). Genistein as a neuroprotective antioxidant attenuates redox imbalance induced by β-amyloid peptides 25–35 in PC12 cells. *International Journal of Developmental Neuroscience, 28*(4), 289-295.

Marcos Filho, J. (2005). *Fisiologia de sementes de plantas cultivadas*. Piracicaba, SP: Fealq.

Marques, M. C., Hamawaki, O. T., Sediyama, T., Bueno, M. R., Reis, M. S., Cruz, C. D., & Nogueira, A. P. O. (2011). Adaptabilidade e estabilidade de genótipos de soja em diferentes épocas de semeadura. *Bioscience Journal, 27*(1), 59-69.

Pípolo, A., Sinclair, T. R., & Camara, G. M. S. (2004). Effects of temperature on oil and protein concentration in soybean seeds cultured *in vitro*. *Annals of Applied Biology, 144*(1), 71-76.

Rangel, M. A. S., Minuzzi, A., Braccini, A. L., Scapim, C. A., & Cardoso, P. C. (2007). Efeitos da interação genótipos x ambientes no rendimento de grãos e nos teores de proteínas de cultivares de soja. *Acta Scientiarum. Agronomy, 29*(3), 351-354.

Riedl, K. M., Lee, J. H., Renita, M., St Martin, S. K., Schwartz, S. J., & Vodovotz, Y. (2007). Isoflavone profiles, phenol content, and antioxidant activity of soybean seeds as influence by cultivar and growing location in Ohio. *Journal of the Science of Food and Agriculture, 87*, 1997-1206.

Rimbach, G., Boesch-Saadatmandi, C., Frank, J., Fuchs, D., Wenzel, U., Daniel, H., ... Weinberg, P. D. (2008). Dietary isoflavones in the prevention of cardiovascular disease – A molecular perspective. *Food Chemical Toxicology, 46*(4), 1308-1319.

Silva, J., Prudencio, S., Carrão-Panizzi, M., Gregorut, C., Fonseca, F., & Mattoso, L. (2012). Study on the flavor of soybean cultivars by sensory analysis and electronic tongue. *International Journal of Science and Technology, 47*(1), 1630-1638.

Shao, S., Duncan, A. M., Yang, R., Marcone, M. F., Rajcan, I., & Tsao, R. (2009). Tracking isoflavones: from soybean to soy flour, soy protein isolates to functional soy bread. *Journal of Functional Foods, 1*(1), 119-127.

Tsukamoto, C., Shimada, S., Igita, K., Kudou, S., Kokubun, M., Okubo, K., & Kitamura, K. (1995). Factors affecting isoflavone contente in soybean seeds: changes in isoflavones, saponins, and composition of fatty acids at diferente temperatures during seed development. *Journal of Agricultural and Food Chemistry, 43*(5), 1184-1192.

Zhe, Y., Lauer, J. G., Borges, R., & Leon, N. (2010). Effects of genotype x environment interaction on agronomic traits in soybean. *Crop Science, 50*(1), 696-702.

9

Lettuce production in a greenhouse under fertigation with nitrogen and potassium silicate

Renan Soares de Souza, Roberto Rezende, Tiago Luan Hachmann*, Cláudia Salim Lozano, André Felipe Barion Alves Andrian and Paulo Sérgio Lourenço de Freitas

Departamento de Agronomia, Universidade Estadual de Maringá, Av. Colombo, 5790, 87020-900, Maringá, Paraná, Brazil.
**Author for correspondence. Email: tiagohach@gmail.com*

ABSTRACT. The objective of this study was to evaluate the effect of nitrogen and potassium silicate on the production and commercial aspects of curly lettuce, Vera cultivar. The experimental design was completely randomized (CRD), with ten treatments and three replications. The treatments, arranged in a factorial design according to the Plan Puebla III matrix (Turrent & Laird, 1975), consisted of the combination of five doses of nitrogen (9; 54; 90; 126, and 171 kg ha^{-1}) and five doses of potassium silicate (1.15; 6.90; 11.50; 16.10, and 21.85 kg ha^{-1}). A control treatment without application of nitrogen and potassium silicate was also inserted. The crop was grown in a greenhouse, and the doses were applied as sidedressing using a drip micro-irrigation system. Total fresh matter, commercial fresh matter, non-commercial fresh matter, number of leaves and commercial trade index were evaluated. The commercial fresh matter and the number of commercial leaves per plant were affected only by nitrogen fertigation and increased linearly with an increase in the dose of nitrogen, with the best responses observed at the highest dose of this element (171 kg ha^{-1}). Potassium silicate only had an effect on non-commercial fresh matter, with no influence on the other characteristics.

Keywords: *Lactuca sativa* L., protected cultivation, micro drip irrigation, silicon.

Introduction

Lettuce (*Lactuca sativa* L.) is a herbaceous plant that belongs to the Asteraceaes family. It is the foremost leafy vegetable consumed in Brazil and around the world, standing out for being a good source of vitamins and minerals (Stagnari, Galiene, & Pisante, 2015). In addition to the nutritional aspect, it is a crop of great importance from a socio-economic point of view, being a source of income for farmers.

Some factors are of fundamental importance in vegetable production for the proper development and growth of the crop. Plant nutrition is one of the most important factors and has a direct impact on the production and quality of vegetables. Fertigation in vegetable production is a tool that enables plants to be provided with the amount of nutrients demanded in each phenological stage, resulting in greater efficiency and economy in the use of fertilizers (Frizzone, Freitas, Rezende, & Faria, 2012).

According to Kano et al. (2010), potassium and nitrogen are the nutrients most accumulated by shoots in conventional vegetable farming systems. The importance of the functions these nutrients perform in plants has been recognized, such as in the formation of cellular components (Taiz & Zeiger, 2013) and in the regulation of osmotic potential and enzymatic activation (Stagnari et al., 2015).

While silicon is characterized as being a beneficial element for plants, it is not considered essential. This element has a direct influence on the intensification of photosynthesis and resistance to fungal and bacterial diseases, insects, drought, low temperatures and salinity (Epstein & Bloom, 2006). The use of silicon in lettuce production can promote greater physical performance and better quality (fewer senescent, insect damaged or diseased leaves) because of the positive effects of Si, Ca, Mg, and K absorption (Ferreira et al., 2010).

The adequate supply of nutrients to lettuce plants is a practice that needs to be researched to determine the best way to manage crop nutrition. In this context, given that potassium and nitrogen supply is important for lettuce, and considering the possible benefits that fertilization with silicon can offer to its culture, this study aimed to evaluate the effects of varying doses of nitrogen and potassium silicate when applied by fertigation on curly lettuce plants, cultivar Vera.

Material and methods

The experiment was conducted in a protected environment, at the Irrigation Technical Center (CTI) of the State University of Maringá (UEM) in the city of Maringá – Paraná State (23°23'56,50"S, 51°57'7,53"W). The experimental period took place between April 5, 2013 and June 18, 2013. The crop was installed in a protected environment, with an arch-type cover, measuring 30 m long, 6.9 m wide and 3.5 m in height. The regional climate is characterized as Cfa, subtropical, according to the classification of Koppen.

The soil where the experiment was installed is from the Dystroferric Red Nitosol class with clay texture (Empresa Brasileira de Pesquisa Agropecuária [EMBRAPA], 2013). For determining the chemical characterization of the soil, a sample was collected at the depth of 0.10 m, and the following results were obtained: pH ($CaCl_2$) = 7.00; pH (H_2O) = 7.80; organic matter and C = 26.43 and 15.33 g dm^{-3}, respectively; P = 105.19 mg dm^{-3}; K^+ = 0.31 $cmol_c$ dm^{-3}; Ca^{+2} = 6.20 $cmol_c$ dm^{-3}; Mg^{+2} = 1.45 $cmol_c$ dm^{-3}; Al^{+3} = 0.0 $cmol_c$ dm^{-3}; H^++Al^{+3} = 2.19 $cmol_c$ dm^{-3}; sum of bases = 7.95 $cmol_c$ dm^{-3}; CTC 10.14 $cmol_c$ dm^{-3}; base saturation = 78.41%; and S = 10.23 mg dm^{-3}.

The experimental design was completely randomized (CRD), with ten treatments and three replications. The treatments were arranged in a factorial design according to the Plan Puebla III matrix (Turrent & Laird, 1975) and consisted of the combination of five doses of nitrogen (9, 54, 90, 126 and 171 kg ha^{-1}) and five doses of potassium silicate (1.15, 6.90, 11.50, 16.10 and 21.85 kg ha^{-1}). A control treatment without the application of nitrogen and potassium silicate was also inserted.

Urea (45% N) was used as the nitrogen source (N) and Fertisilício® was used as the potassium silicate source. This product is classified as a leaf fertilizer, potassium silicate based, and contains 12% water-soluble Si (165.6 g L^{-1}) and 12% water-soluble K_2O (165.6 g L^{-1}). The application of nitrogen via irrigation water was carried out through a pre-prepared urea solution at 10% N (m v^{-1}). The doses of nutrients were divided into four applications during the cycle, using a spray volume of 15 L.

The preparation of the beds was carried out by soil tillage with rotary hoe. Subsequently, 0.5 kg of poultry waste and 40 kg ha^{-1} (urea, 45% N) per square meter were applied and incorporated into the soil.

Seedlings of curly lettuce (Cultivar Vera) were produced in polystyrene trays of 128 cells containing a commercial substrate. The seedlings were kept in a greenhouse until the time of transplanting, when they had four leaves. The plants were placed in plots with the spacing of 0.20 x 0.20 m between each seedling. Thirty-six plants, distributed in three rows of 12 plants, were allocated in each plot.

Irrigation was carried out by micro-irrigation through a drip system. The system was composed of a 0.5 hp centrifugal pump; a 500 L tank; a 32 mm diameter PVC main line; five 32 mm diameter PVC branch lines; a 32 mm diameter PVC main return line; six valves in each branch line; and two side lines for each plot, each consisting of 16 mm diameter polyethylene driplines and 12 pressure compensating emitters on each sideline.

The operating pressure used for the drip irrigation and fertigation was 10 m.w.c., with an average flow value of 0.84 L h^{-1} for each emitter and a CUC (Christiansen uniformity coefficient) of 95.47%. The irrigation was monitored using tensiometers installed at the depth of 0.10 m, and the value used as a reference was 15 kPa (Vilas Boas et al., 2007).

Evaluations were carried out on the eight central plants of each plot. Total fresh matter (MFT), commercial fresh matter (MFC), non-commercial fresh matter of shoots (MFNC), number of commercial leaves (NFC) and commercial trade index (IC) were evaluated.

The commercial fresh matter was determined, after obtaining the total fresh weight, by removing senescent and/or damaged leaves not suitable for trade from the outer region of the plant. To determine the NFC, the leaves suitable for trade and consumption were counted. The IC was determined by the ratio between the MFC and MFT (Santos, Mendonça, Silva, Espínola, & Souza, 2011).

After being tabulated, data were submitted to analysis of variance using the statistical program SISVAR (Ferreira, 2011). Then, multiple linear regression analyses were performed, selecting statistical models based on the F-test for the regression ($p < 0.05$), F-test for lack of adjustment ($p > 0.05$), *t*-test for regression coefficients ($p < 0.05$) and the adjusted determination coefficient (R^2_{aj}). In the selection of statistical models, twelve possible linear models were studied, one being considered complete, that is: $\hat{Y}=\beta_0+\beta_1(N)+\beta_2(Si \text{ and } K_2O)+\beta_3(N)^2+\beta_4(Si \text{ and } K_2O)^2+\beta_5(N)(Si \text{ and } K_2O)$, where $\hat{Y}$ is the estimated value of the characteristic in question; β_0, β_1, β_2, β_3, β_4, β_5 are regression coefficients; N is the dose of N (kg ha^{-1}); and Si and K_2O is the dose of Si and K_2O (kg ha^{-1}).

Results and discussion

The results of the descriptive statistics for total fresh matter of shoots, commercial fresh matter of shoots, non-commercial fresh matter of shoots, number of commercial leaves per plant and commercial trade index are shown in Table 1.

The fertilization with nitrogen had effects on total fresh matter, commercial fresh matter, non-commercial fresh matter and number of commercial leaves. An effect of potassium silicate was observed only in non-commercial fresh matter.

The results of the regression analysis for the total fresh matter and commercial fresh matter were similar. In these cases, there was a positive linear correlation of these characteristics due to the increase in nitrogen levels (Figures 1 and 2).

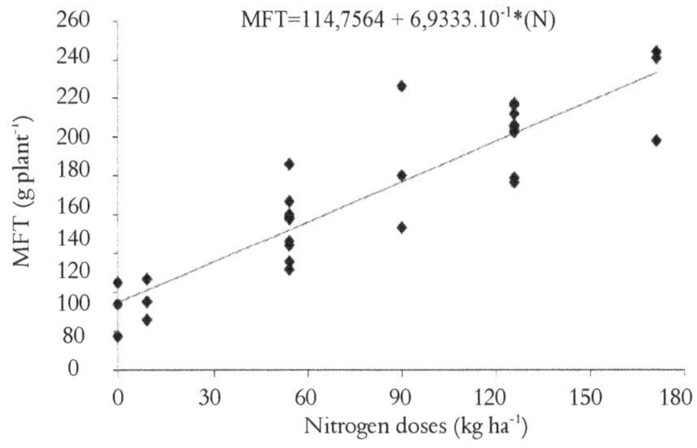

Figure 1. Total fresh matter (MFT) of curly lettuce, Vera cultivar, according to the nitrogen doses. *Significant by t-test ($p < 0.05$); [a] adjusted determination coefficient.

For total fresh matter, an increase of 25 kg ha^{-1} in the nitrogen dose was responsible for the addition of 17.33 g per plant, and the highest result (233.32 g per plant) was obtained with the maximum nitrogen dose (171 kg ha^{-1}) (Figure 1).

According to Taiz and Zeiger (2013), nitrogen participates in the formation of many cell components, including aminoacids and nucleic acids. An increase in vegetative growth, expansion of the photosynthetically active area and higher yield potential are possible results of a situation where there is an adequate supply of nitrogen (Filgueira, 2008).

Studying the application of four nitrogen sidedressing fertilization doses (0, 60, 120, and 180 kg ha^{-1}) and five doses of molybdenum via leaf fertilization (0, 35.1, 70.2, 105.3, and 140.4 g ha^{-1}) on lettuce, Raider cultivar, Resende, Alvarenga, Yuri, and Souza (2012) found that in the absence of molybdenum, the largest commercial fresh matter occurred at the dose of 66.4 kg ha^{-1} of nitrogen. In the present study, similar results were found; a productive benefit of lettuce was verified with nitrogen fertilization.

Table 1. Descriptive statstics for the characteristics total fresh matter (MFT), commercial fresh matter (MFC), non-commercial fresh matter (MFNC), number of commercial leaves (NFC) and commercial trade index (IC) of Vera lettuce.

Statistic	Characteristics				
	MFT (g)	MFC (g)	MFNC (g)	NFC	IC
Minimum	97.15	94.29	2.65	13.00	0.90
Maximum	243.84	235.83	19.71	21.63	0.98
Standard deviation	41.43	39.57	4.23	2.01	0.02
General average	170.92	161.12	9.79	16.68	0.94
C.V. 1 (%)	24.24	24.56	43.19	12.05	2.39
C.V. 2 (%)*	11.75	13.18	35.52	8.53	2.40

* Variation coefficient of the experiment, in which s = (QMR) $^{(1/2)}$, "s" is the standard deviation and "QMR" is the residual mean square.

In the present study, the productive aspect of lettuce was favored by nitrogen application, and the maximum production (219.66 g per plant) was verified at the highest dose (171 kg ha^{-1}). In a study by Silva, Pereira, Reis, Lima, & Taveira (2008) with four water depths (91.99, 142.79, 186.34, and 237.14 mm), combined with four doses of nitrogen (0, 80, 160, and 320 kg ha^{-1}), in lettuce, cultivar Raider, and grown under greenhouse conditions, it was observed that the estimated maximum productivity occurred with 208.03 mm and 290.5 kg ha^{-1} of nitrogen.

In a study that evaluated nitrogen doses (0, 141.5, 283, 566, and 1,132 mg per pot) and lettuce cultivars (Lucy Brown, Tainá, Vera, Verônica and Elisa), Mantovani, Ferreira, and Cruz (2005) observed that the maximum production of lettuce, cultivar Vera, was obtained with the application of 770 mg of nitrogen per pot, which corresponds, according to the authors, to approximately 163 kg ha^{-1} of nitrogen.

The commercial fresh matter was increased by 16.26 g per plant with an increase of 25 kg ha^{-1} in the nitrogen dose, and the best response (219.66 g per plant) occurred with the highest nitrogen dose, which was 171 kg per ha (Figure 2).

There was significant influence of nitrogen and potassium silicate application via fertigation on non-commercial fresh matter, which increased in a linear manner according to the increase in nitrogen (Figure 3) and potassium silicate doses (Figure 4). The maximum non-commercial fresh matter (13.65 g per plant) was observed at the highest dose of nitrogen (171 kg ha^{-1}).

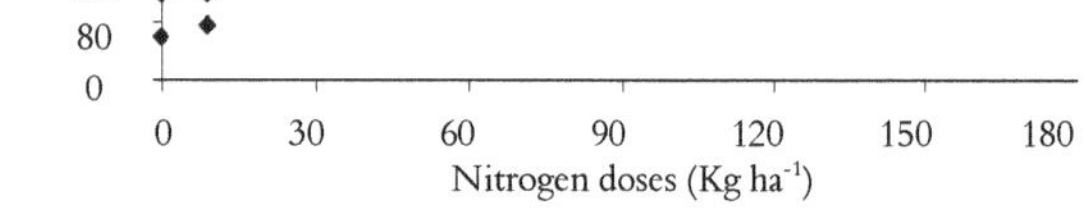

Figure 2. Commercial fresh matter (MFC) of curly lettuce, Vera cultivar, in relation to nitrogen doses. *Significant by t-test ($p < 0.05$); [(a)] adjusted determination coefficient.

In regards to the potassium silicate doses, a similar behavior was verified. The maximum non-commercial fresh matter (13.19 g per plant) was observed at the maximum dose of potassium silicate (21.85 kg ha^{-1}) (Figure 4).

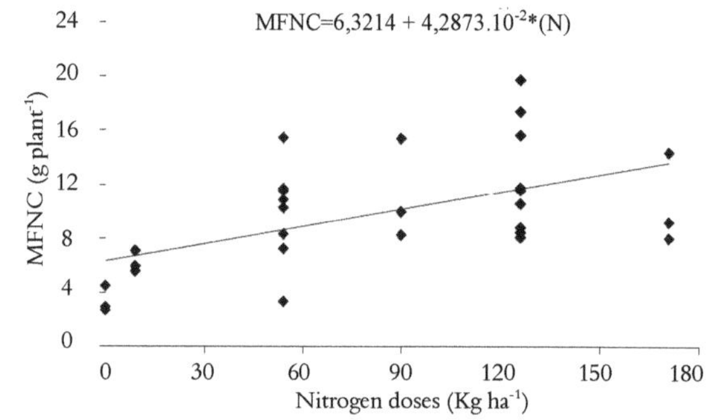

Figure 3. Non-commercial fresh matter (MFNC) of curly lettuce, Vera cultivar, in relation to nitrogen doses. *Significant by t-test ($p < 0.05$); [(a)] adjusted determination coefficient.

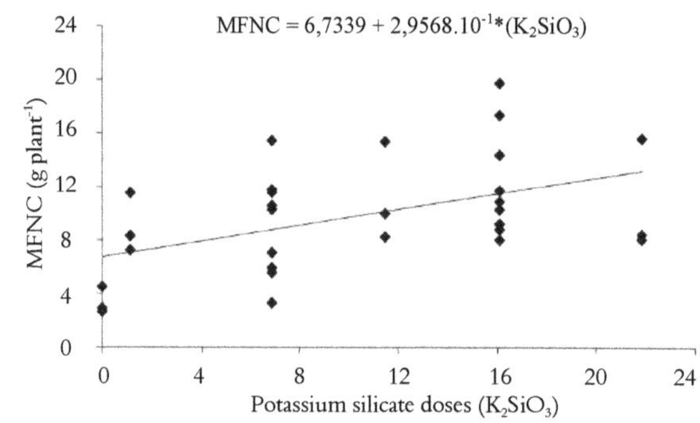

Figure 4. Non-commercial fresh matter (MFNC) of curly lettuce, Vera cultivar, in relation to potassium silicate doses (K_2SiO_3). *Significant by t-test ($p < 0.05$); [(a)] adjusted determination coefficient.

It is likely that the increase in the non-commercial fresh matter of shoots, which occurred due to the increase in both nitrogen and potassium silicate doses, followed the increase in the total fresh matter of shoots as a function of the increase in the doses of N. Still, there was no loss to the commercial quality of lettuce, as there were benefits to the commercial fresh matter and number of commercial leaves per plant characteristics as the dose of N was increased. The absence of fertigation effects with nitrogen and with potassium silicate on the commercial trade index also corroborates this finding.

Koetz, Coelho, Costa, Lima, and Souza (2006), in evaluating different irrigation frequencies associated with doses of potassium chloride (100, 150, 200, and 250 kg ha^{-1}) applied via fertigation on lettuce (Raider cultivar) in a greenhouse, observed that the maximum yield was obtained with the dose of 119.36 kg ha^{-1} of K_2O. This effect of the potassium doses was not observed in the present experiment.

With the supply of silicon and potassium to lettuce, the present study aimed to evaluate its effect on the reduction of non-commercial fresh matter of shoots (senescent or damaged leaves) and increase in the commercial trade index. However, an increase in non-commercial fresh matter of shoots and the absence of fertigation effects with potassium silicate on the commercial trade index were observed. Thus, with regard to the result obtained for non-commercial fresh matter in the present study, there was disagreement with the result obtained by Ferreira et al. (2010), where an evaluation of the application of increasing doses of calcium silicate on three lettuce cultivars (Raider, Regina and Vera) concluded that with an increase in calcium silicate doses, there was a reduction in fresh matter percentage of senescent and diseased leaves.

Only nitrogen fertigation significantly influenced the number of commercial leaves per plant, which correlated linearly with the increase in N doses (Figure 5). There was an increase of 0.72 in the number of commercial leaves per plant, conditioned by an increase of 25 kg ha^{-1} in the dose of nitrogen; in this case, the highest value (19.27 leaves per plant) was obtained with the maximum dose of N (171 kg ha^{-1}).

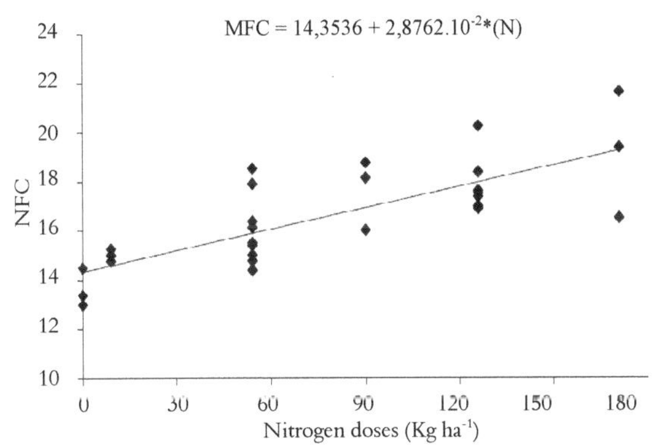

Figure 5. Number of commercial leaves (NFC) of curly lettuce, Vera cultivar, in relation to nitrogen doses. *Significant by t-test ($p < 0.05$); [a] adjusted determination coefficient.

There was no significant influence of nitrogen fertigation and potassium silicate fertigation on the commercial trade index, considering that the regression analysis selected no statistical model for this response variable, and its overall average was 0.94, as shown in Table 1.

Santos et al. (2011) studied eight lettuce cultivars and found no significant difference between cultivars, with regard to the commercial index, and obtained an average of 0.70. This value was numerically lower than the overall average for commercial trade index found in the present study, which was 0.94. It is probable that the value of 0.94, as shown in this study, is related to a good commercial quality lettuce due to the use of a protected environment and drip micro irrigation. With the use of these techniques, there is the possibility, for example, that less water accumulates on the leaves of the plants, resulting in a reduced proliferation of diseases or injuries.

Conclusion

Nitrogen had a positive effect on the production variables of lettuce.

The highest total fresh matter, commercial fresh matter and number of commercial leaves per plant were obtained at the dose of 171 kg N ha^{-1}.

There was no effect of potassium silicate on the productive characteristics of lettuce plants, except for non-commercial fresh matter.

References

Empresa Brasileira de Pesquisa Agropecuária. [EMBRAPA]. (2013). *Sistema brasileiro de classificação de solos* (3a. ed.). Rio de Janeiro, RJ: Embrapa Solos.

Epstein, E., & Bloom, A. J. (2006). *Nutrição mineral de plantas: princípios e perspectivas*. Londrina, PR: Planta.

Ferreira, R. L. F., Souza, R. J., Carvalho, J. G., Araújo Neto, S. E., Mendonça, V., & Wadt, P. G. S. (2010). Avaliação de cultivares de alface adubadas com silicato de cálcio em casa-de-vegetação. *Ciência e Agrotecnologia, 34*(5), 1093-1101.

Ferreira, D. F. (2011). Sisvar: a computer statistical analysis system. *Ciência e Agrotecnologia, 35*(6), 1039-1042.

Filgueira, F. A. R. (2008). *Novo manual de olericultura: agrotecnologia moderna na produção e comercialização de hortaliças*. Viçosa, MG: UFV.

Frizzone, J. A., Freitas, P. S. L., Rezende, R., & Faria, M. A. (2012). *Microirrigação: gotejamento e microaspersão*. Maringá, PR: Eduem.

Kano C., Cardoso A. I. I., & Villas Bôas, R. L. (2010). Influência de doses de potássio nos teores de macronutrientes em plantas e sementes de alface. *Horticultura Brasileira, 28*(3), 287-291.

Koetz, M., Coelho, G., Costa, C. C., Lima, E. P., & Souza, R. J. (2006). Efeito de doses de potássio e da freqüência de irrigação na produção da alface-americana em ambiente protegido. *Engenharia Agrícola, 26*(3), 730-737.

Mantovani, J. R., Ferreira, M. E., & Cruz, M. C. P. (2005). Produção de alface e acúmulo de nitrato em função da adubação nitrogenada. *Horticultura Brasileira, 23*(3), 758-762.

Resende, G. M., Alvarenga, M. A. R., Yuri, J. E., & Souza, R. J. (2012). Rendimento e teores de macronutrientes em alface americana em função de doses de nitrogênio e molibdênio. *Horticultura Brasileira, 30*(3), 373-378.

Santos, D., Mendonça, R. M. N., Silva, S. M., Espínola, J. E. F., & Souza, A. P. (2011). Produção comercial de cultivares de alface em Bananeiras. *Horticultura Brasileira*, *29*(4), 609-612.

Silva, P. A. M., Pereira, G. M., Reis, R. P., Lima, L. A., & Taveira, J. H. S. (2008). Função de resposta da alface americana aos níveis de água e adubação nitrogenada. *Ciência e Agrotecnologia*, *32*(4), 1266-1271.

Stagnari, F., Galiene, A., & Pisante, M. (2015). Shading and nitrogen management affect quality, safety and yield of greenhouse-grown leaf lettuce. *Scientia Horticulturae*, *192*(31), 70-79.

Taiz, L., & Zeiger, E. (2013). *Fisiologia Vegetal* (3a. ed.). Porto Alegre, RS: Artmed.

Turrent, A., & Laird, R. J. (1975). La matriz experimental Plan Puebla, para ensayos sobre prácticas de producción de cultivos. *Agrociencia*, *9*(1), 117-143.

Vilas Boas, R. C., Carvalho, J. A., Gomes, L. A. A., Souza, K. J., Rodrigues, R. C., & Sousa, A. M. G. (2007). Efeito da irrigação no desenvolvimento da alface crespa, em ambiente protegido, em Lavras, MG. *Revista Brasileira de Engenharia Agrícola e Ambiental*, *11*(4), 393-397.

10

Handling and postharvest shelf life of ora-pro-nobis leaves

Camila Karen Reis Barbosa*, Fernando Luiz Finger and Vicente Wagner Dias Casali

Departamento de Fitotecnia, Universidade Federal de Viçosa, Av. Peter Henry Rolfs, s/n, 36570-900, Viçosa, Minas Gerais, Brazil.
**Author for correspondence. E-mail: camila.barbosa@ufv.br*

ABSTRACT. The goal of this study was to assess the effects of hydrocooling and plastic bag use on the postharvest quality ora-pro-nobis (*Pereskia aculeata* Mill.) leaves stored at 5 and 25°C. The experiments were conducted in a split plot scheme, with treatments in plots and subplots storage time in a randomized block design. The leaf chlorophyll content, mass loss (ML), relative water content (RWC), soluble sugars levels, reducings (RED), non-reducings and starch were determined. In leaves stored at 25°C, the ML was higher when leaves were hydrocooled. The highest RWC resulted from the use of plastic bag, which prevented the leaves from wilting for a longer period of time. The soluble sugars levels varied according to the increase or decrease in RWC. The contents of RED decreased with the time of storage at 25°C. The cold storage increased the shelf life of ora-pro-nobis by 168 hours. The hydrocooling increased the RWC of leaves, although it did not influence the shelf life. The perforated plastic bag was effective in increasing the shelf life, and when associated with hydrocooling, the plastic bags provide a lower rate of loss of fresh mass.

Keywords: *Pereskia aculeata* Mill., hydrocooling, packing, cold storage.

Introduction

The cultivation and consumption of fresh vegetables has decreased in several regions of Brazil, as a result of the increasing use of processed foods. However, vegetables are extremely important for a well-balanced diet. This fact has led some world-recognized agencies for hunger prevention to organize programs to promote the construction of residential gardens as a way of fighting diseases related to vitamin deficiency and obesity. Ora-pro-nobis (*Pereskia aculeata* Mill.) is an unconventional rustic vegetable found in many Brazilian backyards and is used in folk medicine. The vegetables begin an accelerated deterioration process once harvested, and the shelf life is determined by a combination of internal and external factors. Some simple postharvest management techniques can extend the vegetable's shelf life without resulting in higher costs to the farmers. Plastic packages work as physical protection and reduce deterioration by diminishing water loss. Hydrocooling, which is an essential technique for removing field heat from freshly harvested products, may be beneficial in the postharvest of many perishable products (WILLS et al., 2004). The objective of this study was to evaluate the effects of hydrocooling, plastic bag use and cold storage on the postharvest quality of ora-pro-nobis leaves.

Material and methods

The experiment was conducted in Viçosa, Minas Gerais State, Brazil (20°45' S, 651 m a.s.l.), from

July to September 2011. Entire branches of ora-pro-nobis were randomly collected in the morning between 7:00 and 8:00 a.m., and they were immediately taken to the laboratory where the leaves were separated using pruning shears. The leaves with yellowing color, those showing symptoms of disease and those that were wilted were discarded. Two independent experiments were conducted simultaneously by storing the leaves at 25°C and 50% RH or at 5°C and 90% UR. Approximately 40 g of leaves was subjected to the following treatments: 1) Control without hydrocooling and packing; 2) Hydrocooling + perforated plastic bag; 3) Hydrocooling; and 4) Perforated plastic packaging without hydrocooling. Initially, tests were performed to determine the length of time for hydrocooling by submersing the leaves in a mixture of ice and water at 5°C. The leaves temperature was measured using a digital infrared thermometer. Polyethylene plastic bags with a capacity of 1 liter, containing 40 holes that were 1.0 cm in diameter were used to wrap the leaves. Shelf life of the leaves as well as their, loss of fresh weight, chlorophyll content, relative water content, total soluble sugars, reducing sugars, non-reducing sugars and starch were determined. The chlorophyll content and loss of fresh weight were measured successively in the same experimental unit from the beginning to the end of the evaluations. Destructive analyses were performed using different samples, so the experimental unit, which initially established 40 g of leaves, would not be compromised. The end of the shelf life of the ora-pro-nobis was characterized by a loss of water and wilting or blackening of the leaves. The amount of leaf chlorophyll was estimated using the portable meter SPAD-502 [Soil-Plant Analysis Development (SPAD) Section, Minolta] and readings were taken at 0, 6, 12, 18, 24, 30, 36, 42, 48, 72, 96 and 120 hours after harvesting.These readings were taken at six points that had been marked on each of two leaves prior to the procedure. The loss of fresh weight in the experimental unit (40 g) and the content of carbohydrate content present in the leaves were determined at the same frequency. The relative water content was evaluated before and every 24 hours after treatment, according to Álvares et al. (2010). In this case, eight sections (for each repetition) of leaf blade were cut and were approximately 1 cm^2 in size. To quantify the total soluble sugars, the phenol-sulfuric method described by Dubois et al. (1956) was used. The quantification of reducing sugars was determined using the Somogyi-Nelson method (NELSON, 1944). The non-reducing sugars content was obtained by subtracting the values of reducing sugar values from the total soluble sugars. The experiment was conducted with four replication in a split-plot scheme design, with treatments in plots and subplots storage time in a randomized block design. Data were submitted to variance analysis, and the means were separated using Tukey's test at 5% probability. Quantitative data were analyzed by regression analysis, with the model chosen according to the coefficient of determination and biological phenomenon. The statistical program was the Software Analysis and Experimentation Group (SAEG / UFV).

Results and discussion

Length of hydrocooling – To determine the ideal hydrocooling time, the leaves were immersed in water at 5 ± 0.5°C. A sharp decrease in temperature happened in the first 5 minutes, but the then atabilized at 5°C, which lasted more than 20 minutes (Figure 1).

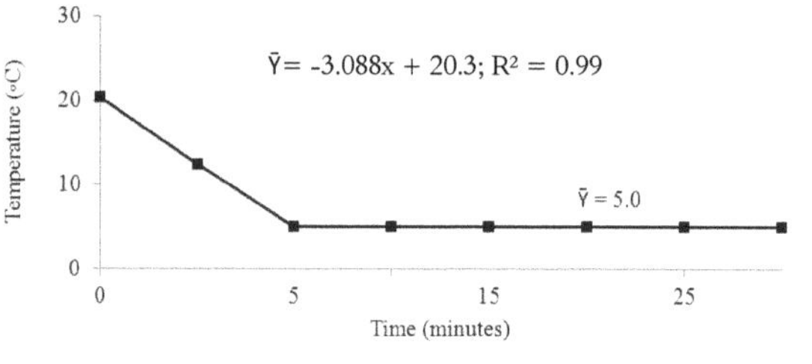

Figure 1. Temperature of ora-pro-nobis during hydrocooling.

During the cooling period, there was a reduction of 15.3°C, which occurred at a rate of 3.6°C per minute, thus amounting to a 75.4% removal of the initial heat. Such behavior was also observed by Álvares et al. (2007), who observed that parsley branches required 15 minutes for temperature stabilization.

Shelf life - Leaves stored at 25°C, regardless of the treatment applied developed darkening areas after 48 hours of harvest. Hydrocooling without the subsequent cold storage did not increase the shelf life of leaves because, despite showing the effect of relative water content, the leaves darkened, similar to leaves that were not hydrocooled. Blackening was not observed in leaves that were stored at 5°C for the entire storage period. The end of the shelf life of the ora-pro-nobis under refrigeration was characterized by loss of water and subsequent wilting of the leaves, which occurred after 120 hours in control and hydrocooled only leaves and after 216 hours for the leaves wrapped in plastic bags, with or without hydrocooling. Hydrocooling with subsequent cold storage without plastic packaging did not increase the shelf life of the leaves because, although the initial water content in the leaves increased, that water was rapidly lost to the surrounding atmosphere. At the end of 120 hours, the hydrocooled leaves that were packaged had the same water content and appearance as the leaves that were hydrocooled and not packaged. There was an increase of 168 hours (7 days) in the shelf life of the leaves that were packed and were stored under refrigeration compared to the shelf life of that were not packed and were stored at room temperature. As in present study, a previous studied noted that the longevity of parsley leaves was extended by refrigeration and hydrocooling under the same conditions of temperature (ÁLVARES et al., 2010).

Weight loss of fresh leaves - A significant correlation was found between time and the treatments for the leave's accumulated weight loss. In storage at 25°C for 12 hours, the leaves stored in perforated plastic bags presented lower accumulated weight loss, thus differing from the other treatments. At the end of the shelf life, greater accumulated weight loss was present in all treatments that had been submitted to hydrocooling, with or without perforated packaging, compared to the other treatments. Nevertheless, greater weight loss was present in the control compared to the plastic packed leaves (Table 1).

Table 1. Average accumulated weight loss values for ora-pro-nobis leaves stored at room temperature (25°C) for 48 hours over the storage time; control (C), hydrocooling (HC), packaging (P).

Time (hours)	Control (%)		HC + P (%)		P (%)		HC (%)	
0	0.00 ± 0.00	a	0.00 ± 0.00	a	0.00 ± 0.00	a	0.00 ± 0.00	a
6	1.91 ± 0.19	a	1.75 ± 0.17	a	0.71 ± 0.01	a	1.92 ± 0.19	a
12	2.96 ± 0.56	a	2.84 ± 0.56	a	1.27 ± 0.56	b	3.04 ± 1.17	a
18	5.14 ± 0.90	a	5.49 ± 0.76	a	2.78 ± 0.25	b	6.03 ± 0.35	a
24	6.37 ± 0.97	a	7.00 ± 0.95	a	3.62 ± 0.42	b	7.52 ± 0.37	a
30	7.50 ± 1.14	a	8.46 ± 1.11	a	4.45 ± 0.52	b	8.82 ± 0.42	a
36	8.81 ± 1.48	b	10.06 ± 1.18	ab	5.49 ± 0.67	c	10.40 ± 0.45	a
42	10.07 ± 1.48	b	11.56 ± 1.31	a	6.59 ± 0.82	c	12.03 ± 0.42	a
48	11.07 ± 1.45	b	12.73 ± 1.41	a	7.38 ± 0.99	c	13.10 ± 0.50	a

Means (n=4) followed by the same lowercase letter in the row do not differ according to Tukey's test at 5% probability.

The highest weight loss in the leaves subjected to hydrocooling is due to the evaporation of water adhered to the leaf surface and/or absorbed during the treatment. This behavior was also observed by Gillies and Toivonen (1995), in whose study hydrocooling added 5% to the weight of broccoli. After 36 hours of storage at 5°C, the leaves packed in perforated plastic don't hydrocooled had the lowest weight loss values and the leaves hydrocooled without plastic bags had the highest accumulated weight loss. This behavior continued until the end of the shelf life (Table 2).

Table 2. Average accumulated weight loss values for ora-pro-nobis leaves stored under refrigeration (5°C) for 120 hours over the storage time; control (C), hydrocooling (HC), packaging (P).

Time (hours)	Control (%)		HC + P (%)		P (%)		HC (%)	
0	0.00 ± 0.00	a	0.00 ± 0.00	a	0.00 ± 0.00	a	0.00 ± 0.00	a
6	1.05 ± 0.31	a	1.22± 0.32	a	0.98± 0.27	a	1.58± 0.22	a
12	1.66 ± 0.70	a	2.02± 0.71	a	1.40± 0.28	a	2.57± 0.84	a
18	3.01 ± 1.08	a	2.75± 0.76	a	2.37± 0.27	a	4.45± 0.74	a
24	3.87 ± 0.92	a	3.16± 0.84	a	2.94± 0.32	a	5.71± 0.90	a
30	4.81 ± 1.02	a	3.77± 0.93	a	3.47± 0.20	a	7.09± 1.28	a
36	5.98 ± 1.15	ab	4.62± 1.38	b	4.08± 0.21	b	8.64± 1.50	a
42	6.98 ± 1.33	ab	4.94± 1.20	b	4.61± 0.20	b	9.84± 2.09	a
48	7.80 ± 1.52	ab	5.55± 1.34	b	5.11± 0.28	b	10.98± 2.49	a
72	11.52 ± 2.39	b	7.56± 1.88	c	7.73± 0.42	c	15.67± 3.59	a
96	14.53 ± 2.35	b	9.12± 2.83	c	10.02± 0.71	c	19.71± 5.27	a
120	17.37 ± 2.47	b	11.40± 2.47	c	12.23± 0.94	c	22.82± 6.12	a

Means (n=4) followed by the same lowercase letter in the row do not differ according to Tukey's test at 5% probability.

The highest weight loss of the hydrocooled leaves, compared to that observed in the control, was due to the evaporative loss of water adhered to the surface and/or absorbed during hydrocooling. However, the appearance of the leaves was the same: both types of leaves were very wilted after 120 hours. A perforated plastic bag was used to reduce the rate of weight loss in both hydrocooling and non-hydrocooling treatments. The use of a plastic package was also effective in reducing the cumulative weight loss of broccoli (TOIVONEN, 1997). It is believed that the perforated plastic, in addition to reducing the vapor pressure deficit between air and the produce, also maintained the temperature inside the container lower than the temperature of the cooling room, thus reducing the effect of random variations in temperature in the room (GILLIES; TOIVONEN, 1995). The weight loss of leaves accumulated over time during storage at 25°C (Figure 2) and 5°C (Figure 3) increased linearly.

During storage at 25°C, the rate of mass loss varied among treatments. The rate of mass loss observed was 5.85% per day in the control, 3.65% per day in packaged leaves, 6.89% per day in hydrocooled leaves and 6.61% per day in hydrocooled and packaged leaves. During storage at 5°C, the rates of mass loss per hour also varied among treatments. The rate of mass loss observed was 3.66% per day in the control, 2.53% per day in packaged leaves, 4.96% per day in hydrocooled leaves and 2.45% per day in hydrocooled and packaged leaves.

Chlorophyll – There was no significant effect of the treatments or storage on the chlorophyll content of leaves stored at 25 or 5°C (data not show).

Relative water content - Only the treatment affected the relative water content for the leaves stored at 25°C. Higher relative water content was observed in hydrocooled and packed leaves compared to leaves that were hydrocooled without plastic bag. This effect demonstrates the importance of using perforated plastic packaging to maintain the water content. The unpacked hydrocooled leaves did not differ from the control in their relative water content. In ora-pro-nobis leaves stored at 5°C, the effect of the interaction between treatment and storage time was observed. In this case, the use of a perforated plastic package resulted in a higher relative water content in leaves with or without prior hydrocooling (Table 3).

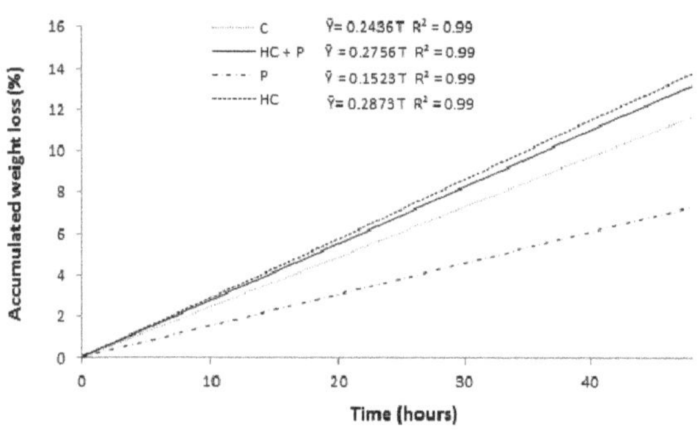

Figure 2. Estimated accumulated weight loss in ora-pro-nobis leaves during storage at room temperature (25°C) for 48 hours per treatment; control (C), hydrocooling (HC), packaging (P).

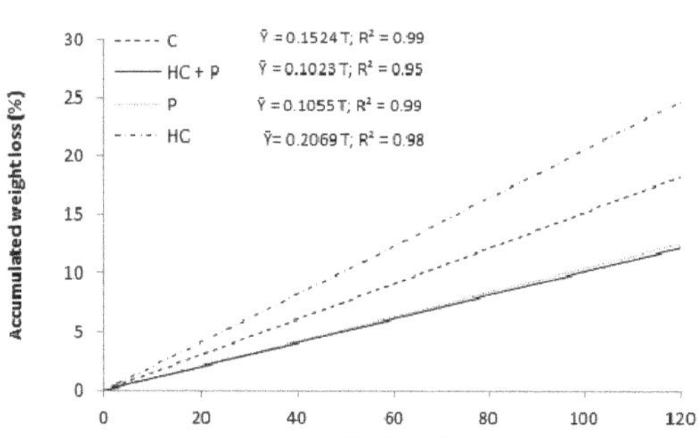

Figure 3. Estimated accumulated weight loss in ora-pro-nobis leaves during storage under refrigeration (5°C) for 120 hours per treatment; control (C), hydrocooling (HC), packaging (P).

Table 3. Average values of relative water content in ora-pro-nobis leaves stored under refrigeration (5°C) for 120 hours at room temperature (25°C).

Temp.	Time (hours)	Control (%)	HC + P (%)	P (%)	HC (%)
	0	83.74 a AB	83.74 a B	83.74 a AB	83.74 a AB
	24	77.34 b ABC	94.03 a AB	83.00 ab AB	85.69 ab AB
5°C	48	70.77 b BCD	90.03 a AB	81.13 a AB	84.43 a AB
	72	69.33 c BCD	89.37 a AB	82.14 ab AB	76.18 bc B
	96	64.14 b CD	85.90 a AB	80.67 a AB	65.25 bc C
	120	60.22 c D	84.86 a B	73.44 b B	61.85 c C
25°C		80.00 a	86.60 a	82.95 ab	82.30 b

Means (n = 4 at 5°C and n = 12 at 25°C) followed by the same lowercase letter in the row and the same uppercase letter in the column do not differ according to Tukey's test at 5% probability.

The relative water content at 24 hours was increased in hydrocooled leaves due to the absorption of water during hydrocooling process. Higher relative water content was observed in mustard leaves (*Brassica juncea*) when they were hydrocooled and subsequently packaged, regardless of the storage temperature (LAZAN et al.,1987) and in parsley leaves that were hydrocooled and stored at 5°C, compared to the control without hydrocooling (ÁLVARES et al., 2010).

Levels of carbohydrates - The treatments and length of storage affected the levels of reducing sugars, non-reducing sugars and starch leaves that were stored at 25 and 5°C. An effect on concentration was observed in all cases, except that of the reducing sugars in leaves that were stored at 25°C. With a lower relative water content, there was concentration of carbohydrate per gram of fresh leaf that lost more water, as detected in the control. In the hydrocooled and packaged leaves, there was lower content of soluble sugars due to the higher relative water content. The reducing sugar content over time in leaves that were stored at 25°C for 48 hours exhibited a linear decreasing behavior in the hydrocooled and packed leaves and a quadratic polynomial behavior in the other treatments and control. There were no changes in the levels of reducing sugars as a function of the applied treatments in parsley leaves (ÁLVARES et al., 2010).

Conclusion

The cold storage increased the shelf life of ora-pro-nobis by 168 hours (seven days). The hydrocooling increased the relative water content of leaves, although it did not influence the shelf life. The perforated plastic bag was effective in increasing the shelf life by reducing the rate of weight loss accumulated regardless the storage temperature. When associated with hydrocooling, the plastic bags provided a lower rate of loss of fresh mass.

References

ÁLVARES, V. S.; FINGER, F. L.; SANTOS, R. C. A.; NEGREIROS, J. R. S.; CASALI, V. W. Effect of pre-cooling on the postharvest of parsley leaves. **Journal of Food, Agriculture and Environment**, v. 5, n. 2, p. 31-34, 2007.

ÁLVARES, V. S.; RAMOS, P. A. S.; MAPELI, A. M.; FINGER, F. L. Pré-resfriamento e embalagem na conservação de folhas de salsa. **Brazilian Journal of Food Technology**, v. 13, n. 2, p. 107-111. 2010.

DUBOIS, M.; GILLES, K. A.; HAMILTON, J. K.; REBERS, P. A.; SMITH, F. Colorimetric method for determination of sugars and related substances. **Analytic Chemistry**, v. 28, n. 3, p. 350-356, 1956.

GILLIES, S. L.; TOIVONEN, P. M. A. Cooling method influences the postharvest quality of broccoli. **HortScience**, v. 30, n. 2, p. 313-315, 1995.

LAZAN, H.; MOHD ALI, Z.; MOHD, A.; NAHAR, F. Water stress and quality decline during storage of tropical leafy vegetables. **Journal of Food Science**, v. 52, n. 5, p. 1286-1292, 1987.

NELSON, N. A photometric adaptation of Somogyi method for determination of glucose. **Journal of Biological Chemistry**, v. 153, n. 2, p. 375-380, 1944.

TOIVONEN, P. M. A. The effects of storage temperature, storage duration, hydro-cooling, and micro-perforated wrap on shelf life of broccoli (*Brassica oleracea* L., Italica Group). **Postharvest Biology and Technology**, v. 10, n. 1, p. 59-65, 1997.

WILLS, R.; MCGLASSON, B.; GRAHAM, D.; JOYCE, D. **Postharvest**: an introduction to the physiology and handling of fruit, vegetables and ornamentals. Wallingford: New South Wales University Press, 2004.

11

Intercropping corn and kudzu in a rotation system with soybean

Núbia Maria Correia[1*], Bruno Daniel[2] and Marcela Basile Leite[2]

*[1]Departamento de Fitossanidade, Universidade Estadual Paulista "Julio de Mesquita Filho", Via de Acesso Prof. Paulo Donato Castellane, s/n., 14884-900, Jaboticabal, São Paulo, Brazil. [2]Curso de Agronomia, Universidade Estadual Paulista "Julio de Mesquita Filho", Jaboticabal, São Paulo, Brazil. *Author for correspondence. E-mail: correianm@fcav.unesp.br*

ABSTRACT. The objective of this study was to evaluate the effects of sowing procedures and the seed sowing rate of kudzu (*Pueraria phaseoloides* cv. 'Comum') in corn on corn plant development, forage dry matter production, weed infestation and soybean plant development and yield. The kudzu seeds were either broadcasted over the terrain or sown in furrows (in the main plots) at rates, in both cases, of 400, 800, 1200 and 1600 points of cultural value (on the subplots). Singly cultivated corn was considered the control treatment. Intercropping did not cause corn plant development to show any significant difference compared with that of the singly cultivated corn plants. When sown in furrows, the number and dry matter contents of the kudzu plants were higher than sowing by broadcasting. The number of plants and the dry matter of the plants increased with the number of seeds. Prior to sowing soybean, the incidence of weeds decreased as the number of kudzu seeds at sowing increased. The intercropping of corn with kudzu benefited soybean plant population, plant height, and grain yield in comparison with the single cultivating of corn in the previous year.

Keywords: *Glycine max*, *Pueraria phaseoloides*, *Zea mays*, integrated crop-livestock, no-tillage system.

Introduction

Among the main obstacles to the success of the no-tillage system in most parts of the State of São Paulo and Central Brazil is the low production of straw during the Fall/Winter and Winter/Spring periods. This low production of straw is essentially due to prolonged dry spells, which usually take place during these periods in Brazil. As a consequence, several of these areas remain in a fallow condition for up to seven months, which poses serious difficulties for the success of the no-tillage system (BARDUCCI et al., 2009). A solution for this problem would be the intercropping of plant species such as cereals and pastures, which allow the establishment of pastures after harvest of cereal crops and, simultaneously, additional feed for livestock in the dry season. At the same time, the process of cultivation improves the physical and chemical properties of the soil and also increases the amount of straw needed for the no-tillage system of cultivating intercropping species. The corn plant, because of its height, with ears at high positions, is appropriate for intercropping because these characteristics make the harvesting of the intercropped forage plants easier (ALVARENGA et al., 2006).

Plants of the genus *Urochloa* are the most used for intercropping with corn, but other forage species, such as those of tropical kudzu (*Pueraria phaseoloides*), may be used for the same reason. Kudzu is advantageous compared with other grasses because it is capable of fixing atmospheric N in the plant, thus improving the nutritional value of the feed consumed by livestock. Kudzu, a member of the Fabaceae family, is a perennial plant that originated from Asia. Kudzu has thin, flexible, and hairy stems that are easily rooted when touching the soil (SEPROTEC, 2010). Kudzu is well adapted to poor and acidic soils, showing a good competition capacity when growing close to weeds. It is also well adapted to grow in shaded areas, making it a good alternative to soil cover plants in association with rubber trees and American oil palm (EMBRAPA, 1999; MONTEIRO et al., 2009).

Cereal species and forage plant intercropping systems have consistently been found to be highly beneficial both for soil characteristics and plant yields. One of these desirable consortia is that between corn and kudzu plants. Nonetheless, the intercropping arrangement of plants of these species need to be more clearly understood with respect to, for example, sowing procedures. The broadcasting of seeds or sowing in furrows with nitrogenous fertilizer being side dressed is a simple technique that can easily be put into effect by the average farmer. However, the amount of seeds to be sown is to be calculated in consideration of each sowing modality and the seeds' cultural value. To facilitate the diffusion of information, the values should be presented in points of cultural value (PCV = quantity of seeds x cultural value). This calculation is particularly important in regions such as that of the Northeast of the State of São Paulo, where the common water deficit during the Fall/Winter period makes it difficult to establish field crops.

The present study was thus undertaken based on the assumptions that kudzu and corn intercropping is 1) not harmful to the development of the corn plant, 2) capable of reducing weed infestation, 3) capable of providing the adequate amount of straw for the subsequent sowing of soybean, and 4) capable of improving the productive performance of soybean plants. The objective of this study was to evaluate the effects of sowing methods and the amount of kudzu seeds in corn intercropping on maize development, kudzu biomass and weed infestation, and the development and production of the soybean crop in rotation.

Material and Methods

The experiment was carried out in the field from December 2008 to April 2010 at Paulista State University (UNESP) at its campus in Jaboticabal, State of São Paulo, Brazil. The local altitude is 605 m above sea level, the latitude is 21° 15'17" South, and the longitude is 48° 19'20" West. According to the Köeppen classification, the climate of the region is the Aw type and is characterized by a rainy Summer and a dry Winter.

Intercropping and its effects on a corn crop and weeds were evaluated from December 2008 to October 2009. The soybean crop in rotation with the intercropping system was evaluated from December 2009 to April 2010. Table 1 shows the data concerning the maximum and minimum temperatures, air relative humidity and precipitation during the period from December 2008 to April 2010.

Table 1. Mean monthly values of maximum and minimum air temperatures, air relative humidity, and total monthly rain during the months from December 2008 to April 2010.

Year	Month	Air temperature (°C)		Air relative humidity (%)	Rain (mm)
		Maximum	Minimum		
2008	December	31.0	19.1	77.0	278.9
2009	January	29.7	19.8	80.4	238.0
	January	31.2	20.6	80.9	190.6
	February	31.0	20.2	80.4	217.9
	March	29.5	17.2	74.9	70.8
	May	28.4	15.5	75.9	26.6
	June	25.0	12.2	76.5	51.9
	July	27.6	14.4	74.6	25.5
	August	28.0	14.6	66.3	133.1
	September	29.7	17.8	74.0	132.4
	October	30.8	18.1	72.8	101.9
	November	32.1	21.0	74.8	163.3
	December	29.8	20.5	81.8	383.7
2010	January	30.4	20.8	82.2	240.7
	January	32.2	20.4	76.6	150.7
	February	31.4	20.0	77.6	183.0
	March	29.2	17.1	74.6	95.5

Source: Agroclimatological Station of the Department of Exact Sciences, FCAV/UNESP, 1.0 km distant from the experimental area.

The experiment was set up in the field according to a randomized complete block design with four repetitions in a split-plot arrangement. The kudzu seeds were either broadcasted over the terrain or sown in furrows (in the main plots) at rates, in both cases, of 400, 800, 1200 and 1600 points of cultural value (on the subplots). The check treatment for these sowing procedures consisted of plots in which corn was singly cultivated.

The kudzu seeds used to sow the experiment had a germination of 60% and a purity of 98%, so that its cultural value was 58.8%.

The plot size was 3.6 m wide by 32 m long, and the subplots were 3.6 m wide and 8 m long. The area located in each plot from which the data were collected was formed by two 6-m-long lines, totaling 10.8 m^2.

DKB 350 YG corn hybrid seeds were sown on December 18, 2008, under conventional soil tillage. The distance between rows was 0.9 m, with 6 seeds per meter sown at a depth of 5 cm. A formulated fertilizer (8-24-12) was applied to the soil at a rate of 450 kg ha^{-1} in the sowing furrows.

Before sowing, the seeds were treated with insecticides (deltamethrin and pirimifos-metil, 0.001 mg plus 0.02 mg kg^{-1} of seeds) and fungicides (fludioxonil and metalaxyl, 0.038 mg plus 0.015 mg kg^{-1} of seeds).

For weed control, the post-emergence herbicides nicosulforon (40 g ha^{-1}) and 2.4-D (806 g ha^{-1}) were applied 18 days after the sowing of corn (DAS_C).

At 22 DAS_C, the corn plants received a side-dressed dose of 75 kg ha^{-1} of N provided by ammonium sulfate fertilizer. Kudzu seeds were sown simultaneously with the fertilizer application according to the procedures and rates specified in this study.

The fertilizer was distributed without incorporation along a straight line with the assistance of manual equipment. To have the fertilizer incorporated into the soil as thoroughly as when distributed in the field by mechanical equipment, a superficial (up to 3 cm deep) row was opened in the center between two neighboring planting lines, and the fertilizer was placed in that row. When the seeds were sown in furrows, the kudzu seeds were manually sown over the fertilizer. When broadcasted over the terrain, the kudzu seeds were manually spread between the corn rows in the corresponding plots before the fertilizer was side dressed.

The height and the dry matter of the corn plants were determined 31 and 33 days after the sowing of kudzu (DAS_K), respectively. Grain yield, the number of plants per plot, and the number of ears per plant and by plot were evaluated at the end of the crop cycle (April 10, 2009).

The number of kudzu plants was evaluated at 32 DAS_K and again before the corn harvest (86 DAS_K). This counting was done in two 0.45-m^2 areas randomly picked in each of the sub-plots. At grain harvest and at 315 DAS_K (November 20, 2009), 26 days before soybean sowing, the shoots of the plants were harvested from two 0.45-m^2 areas taken randomly from each sub-plot to determine the shoot dry matter.

At the end of the corn crop, weeds from two 0.45-m^2 areas within each sub-plot had their shoots harvested and their dry matter content evaluated. At 317 DAS_K (November 22, 2009), that is, 28 days before soybean sowing, the weeds were visually evaluated, and the various degrees of infestation were expressed as a percentage, in which zero indicated the absence of plant cover and 100 indicated that the area was completely covered by plants.

Fifteen days prior to and the day of soybean sowing, the weeds and fodder plants of the experimental area were desiccated using 1.30 kg ha^{-1} of glyphosate acid equivalent in the first application and 0.975 kg ha^{-1} in the second application.

Based on the results of the soil chemical analysis, the soil received a dose of 245 kg ha^{-1} of the formula 0-20-20 placed on the bottom of the furrows.

Soybean seeds of the cultivar NK 7059 RR (V-Max RR) were sown on December 12, 2009, according to a no-tillage system at a depth of 5 cm and 45 cm between rows at a rate of 21.4 seeds per meter of row. Before being sown, the seeds were submitted to treatment with carbendazim and imidacloprid (0.5 mg plus 1.2 g kg^{-1} of seeds) for the protection of the plants' initial growth and as inoculants (*Bradyrhizobium elkanii* at a rate of 3.5 mL kg^{-1} of soybean seeds with 5 x 10^9 viable cells mL^{-1}).

The experiment was set up with the same plot (sowing procedure) and sub-plot (seed quantities) arrangement that had been used in the experiment of the previous year. The plot size was 3.6 m wide and 32 m long, and the sub-plots were 3.6 m of wide and 8 m long with 3 central 6-m-long lines as the area in which the data were to be collected. The five soybean rows were sown in the central part of the plot, occupying a width of 2.25 m.

The herbicides glyphosate (0.975 kg ha^{-1}) and chlorimuron-ethyl (20 g ha^{-1}) were applied 31 days after soybean sowing (DAS_S), when the soybean plants had their fourth trifoliated leaves fully expanded and when the weed plants had 2 to 10 definitive leaves (the dicotyledonous) and up to 4 tillers (the grasses).

At 31 DASs, the number of weeds in two randomly chosen 0.225-m^2 areas in the sub-plot area was counted. The shoot of each weed was cut off and dried in an oven to a constant weight.

At 39 DAS_S, the number of soybean plants in each sub-plot was counted, and this value was used to calculate the number of plants per hectare. At 75 DAS_S, the plant height and dry matter were determined. Grain yield and the weight of 100 grains were determined at the end of the crop cycle.

The data obtained in both phases of the experiment were submitted to an analysis of variance and the F test. The seed amounts were submitted to regression analysis when the chosen data adjustment model was that of the largest determination coefficient and, more importantly, that of its significance ($p < 0.01$ and $p < 0.05$) in the F test. To evaluate the performance of the single cultivation in comparison with sowing procedures and sowing rates, the degrees of freedom of the treatments were determined in accordance with an orthogonal contrast of interest. The check treatment was compared with the convenient intercropping treatments. Each contrast was analyzed by means of the F test, accepting or rejecting the null hypothesis H_0 (Y = 0).

Results and discussion

The interaction between the sowing procedure and sowing rate was not significant for any evaluated characteristics.

The sowing procedure had an effect on corn grain yield: the highest production was verified when the kudzu seeds had been broadcasted (Table 2). However, the analyses of the orthogonal contrasts showed that the corn yield when kudzu had been sown in furrows (8.979.95 kg ha^{-1}) was statistically similar to that of the singly cultivated corn (9.490.33 kg ha^{-1}). The higher yield observed when the kudzu seeds were broadcasted might be due to the lower kudzu plant number, which represented a lesser degree of competition for the natural resources needed by the corn plants to grow and produce.

Table 2. Grain yield of corn intercropped with kudzu sown either in furrows or broadcasted.

Sowing procedure	Grain yield (kg ha^{-1})
Broadcast	9710.36 a[(1)]
In furrows	8979.95 b
DMS	263.47

[(1)]Means followed by the same letter do not differ at the 5% level of probability according to F test.

The further evaluated characteristics of the corn (plant height, shoot dry matter, plant population and number of ears per plant and per plot) were not affected by the isolated factors. Based on the contrast analyses, the treatments of intercropping with broadcast sowing and sowing in furrows did not differ from the control monoculture of corn.

Freitas et al. (2005a and b) report that *Urochloa brizantha* intercropped with silage corn did not cause any significant modification in the amount of both fresh and dry weight of the corn plants independently of the method of sowing and the weed plant management. Borghi and Crusciol (2007), studying in corn plant rows 0.45 or 0.90 m apart one from another, intercropped *U. brizantha* by sowing the seeds on the corn rows, between the rows, or simultaneously on and between the rows. They observed that the corn yield was significantly affected only when the spacing between corn rows was 0.45 m and when the *B. brizantha* seeds had been sown both on and between the corn rows.

Heinrichs et al. (2005), intercropping corn with jack bean (*Canavalia ensiformis* L.), pigeon pea (*Cajanus cajan*, L.), dwarf mucuna (*Mucuna deeringiana* (Bort.) Merr), and crotalaria (*Crotalaria spectabilis*, Roth), observed no significant effect on corn yield in the first year, but in the second year, the corn yield was significantly higher when the intercropping had been with jack bean. This result was probably due to the higher green matter produced by jack beans in the first year, which resulted in higher levels of available nutrients in the soil for the second-year corn plants.

The sowing procedures had a significant effect on the numbers of plants at 32 and 86 days after kudzu sowing (DAS_K) and on the kudzu shoot dry matter yield at 86 DAS_K (Table 3). At 32 and 86 DAS_K, there was greater number of kudzu plants on the furrows of the sowing treatment; consequently, at 86 DAS_K, the plants of this treatment accumulated more shoot dry matter.

Table 3. Kudzu plant numbers 32 and 86 days after in furrow or broadcasted sowing of kudzu (DAS_K) seeds in an intercropping with corn and kudzu shoot dry matter 86 DAS_K.

Sowing procedure	Number of plants m^{-2}		Dry matter
	32 DAS_K	86 DAS_K	(g m^{-2}) - 86 DAS_K
Broadcast	23.01 b[(1)]	40.42 b	7.82 b
In furrows	47.57 a	104.31 a	15.55 a
DMS	12.63	27.94	4.52

[(1)]Means followed by the same letter do not differ at the 5% level of probability according to F test.

The seed sowing rate significantly influenced the shoot dry matter at 86 and 315 DAS_K (Figure 1A and B) and the number of plants at 32 and 86 DAS_K (Figure 2). The number of plants (at 32 and 86 DAS_K) and the shoot dry matter (at 86 and 315

DAS$_K$) increased linearly with an increase in the quantity of kudzu seeds sown in the plots (Figures 1 and 2). These findings showed that sowing 1600 points of cultural value (PCV) of kudzu seeds led to an increase in the density and biomass accumulation.

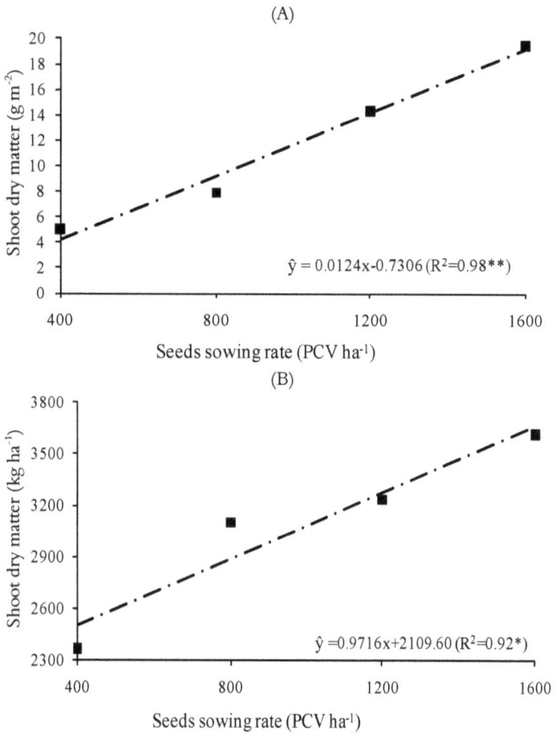

Figure 1. Kudzu shoot dry matter 86 (A) and 315 (B) days after kudzu seeds were sown at different sowing rates in an intercropping with corn.

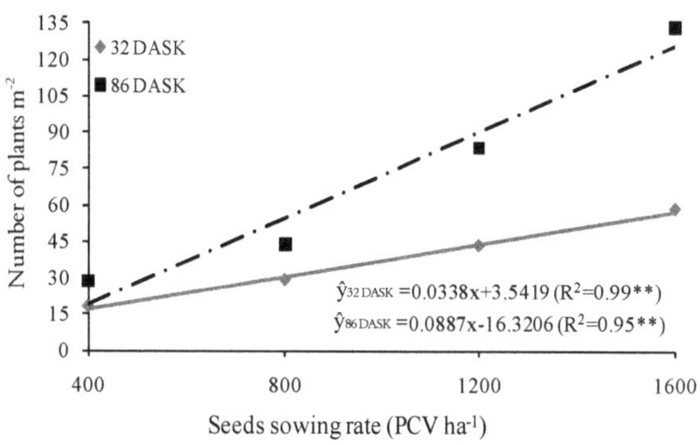

Figure 2. Number of kudzu plants intercropped with corn 32 and 86 days after sowing kudzu seeds at different sowing rates.

Freitas et al. (2005a e b), working with corn intercropped with *U. brizantha*, verified that the highest production of *U. brizantha* dry matter took place when this species' seeds had been sown in lines between the corn plant rows and that the lowest production was verified when the forage seeds had been broadcasted 30 days after corn sowing. In both sowing procedures, the amount of sown seeds was 380 PCV. The procedure by which the *U. brizantha* seeds are broadcasted among corn plant rows 30 days after corn sowing is the most adopted in Brazil (FREITAS et al., 2005a e b) because farmers consider this procedure the easiest and most practical for the intercropping of these species.

The dry matter of weeds was not significantly affected by the studied treatments. However, the degree of weed infestation 317 DAS$_K$ was significantly influenced by the sowing rate (Figure 3). The orthogonal contrasting of the treatments showed that the intercropping treatments did not differ from the treatment in which corn was singly cultivated.

The observations made at 317 DAS$_K$ showed a linear reduction in weed infestation as the kudzu seed sowing rate increased (Figure 3). This result is probably a consequence of the fact that the higher the sowing rate is, the faster and better the soil is covered by kudzu plants, which are thus rendered more capable of suppressing the growth of the weeds.

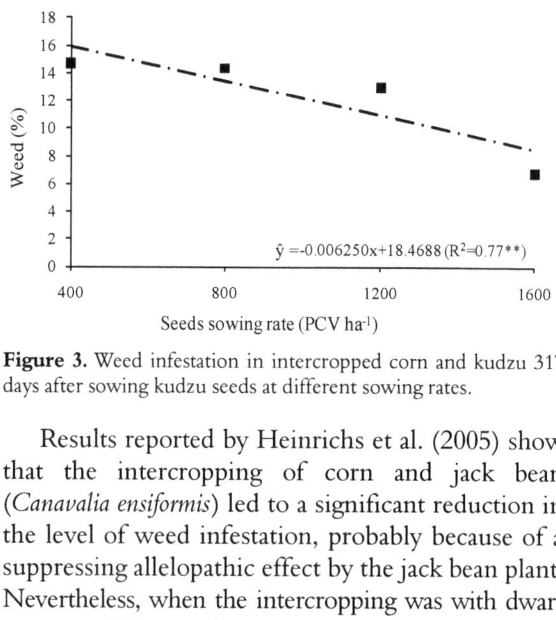

Figure 3. Weed infestation in intercropped corn and kudzu 317 days after sowing kudzu seeds at different sowing rates.

Results reported by Heinrichs et al. (2005) show that the intercropping of corn and jack bean (*Canavalia ensiformis*) led to a significant reduction in the level of weed infestation, probably because of a suppressing allelopathic effect by the jack bean plant. Nevertheless, when the intercropping was with dwarf mucuna (*Mucuna deeringiana*), pigeon pea (*Cajanus cajan*) and crotalaria (*Crotalaria spectabilis*), the authors observed no such effect. But, Linhares et al. (2009) reported that the intercropping of corn with gliciridia (*Gliciridia sepium)*, a perennial species of the Fabaceae family that displays known allelopathic properties, had no effect on weed plant population.

In the soybean crop, the kudzu sowing procedure in rotation with the corn-kudzu intercropping had a significant effect on the number of weeds (Table 4). When the kudzu was broadcasted, the number of weeds was 425.55

plants m^{-2}, whereas when the kudzu was sown in furrows, the number was 401.29 plants m^{-2}.

The soybean plants, while in rotation with the corn-kudzu intercropping, had their population, height, and grain yield significantly affected by the kudzu seed sowing rate (Figure 4). The soybean plant population and grain yield increased linearly with the increase of the sowing rate. For plant height, the best fitting was verified to be the polynomial, with the lowest mean resulting from the use of kudzu sown at a rate of 400 points of cultural value.

The analysis by contrast showed that the soybean plants growing in plots in which the previous crop was singly cultivated corn had a lower dry matter content than those grown in plots in which the corn had previously been intercropped with kudzu sown by broadcasting. The corn and kudzu intercropping, independently of the sowing procedure and of the sowing rate, benefitted the plant population and the height and grain yield of soybean in the first year of the experiment compared with the maize monoculture control plot.

Table 4. Results of F test by way of orthogonal contrast analyses of the sowing procedure of kudzu (in furrows or broadcasted), seeds sowing rate of kudzu (400, 800, 1200 and 1600 points of cultural value - PCV) and corn monoculture, beyond of the average values of the characteristics evaluated in the weed (number of plants, shoot dry matter and dry matter plant^{-1}) and in the soybean crop (plant population ha^{-1}, shoot dry matter, height, 100 grain weight and e grain yield) grown in rotation to kudzu and corn intercropping.

Variation causes		Weed			Soybean				
		Number of plants m^{-2}	Dry matter (g m^{-2})	Dry matter plant^{-1}	Plant population ha^{-1}	Dry matter (g planta^{-1})	Height (cm)	100 grain weight (g)	Grain yield (kg ha^{-1})
Corn monoc. x broadcast		0.05	0.43	0.03	6.43*	5.01*	7.20*	1.82	6.22*
Corn monoc.x in furrows		0.03	0.30	0.01	13.92**	3.32	4.88*	0.87	7.04*
Corn monoc. x 400 PCV		0.04	1.33	0.11	4.86*	2.81	5.60*	0.24	3.96*
Corn monoc. x 800 PCV		1.44	0.05	0.15	5.17*	3.75	6.17*	0.08	4.54*
Corn monoc. x 1200 PCV		1.69	0.33	0.11	19.03**	5.29	9.06**	0.06	9.80**
Corn monoc. x 1600 PCV		0.03	0.44	0.13	14.82**	3.06	5.69*	0.44	8.96**
Treatments					Average values				
Sowing	Broadcast	425.56	49.40	0.12	349382.72	24.36	49.01	14.18	3002.50
	In furrows	401.30	50.40	0.13	389506.17	22.54	47.88	13.81	3059.76
Rates	400 PVC	422.96	45.68	0.11	332098.77	23.34	44.82	13.19	2532.38
	800 PVC	344.07	53.50	0.15	340740.74	23.54	49.69	13.66	2950.57
	1200 PVC	485.19	50.57	0.11	411111.11	24.03	51.19	13.64	3347.96
	1600 PVC	401.48	49,86	0.13	393827.16	22.88	48.08	13.89	3293.62
Corn monoculture		411.85	55.47	0.14	274197.53	14.57	42.60	13.48	2303.60

*Significant at 5% probability by the F test of the analyses of variance.

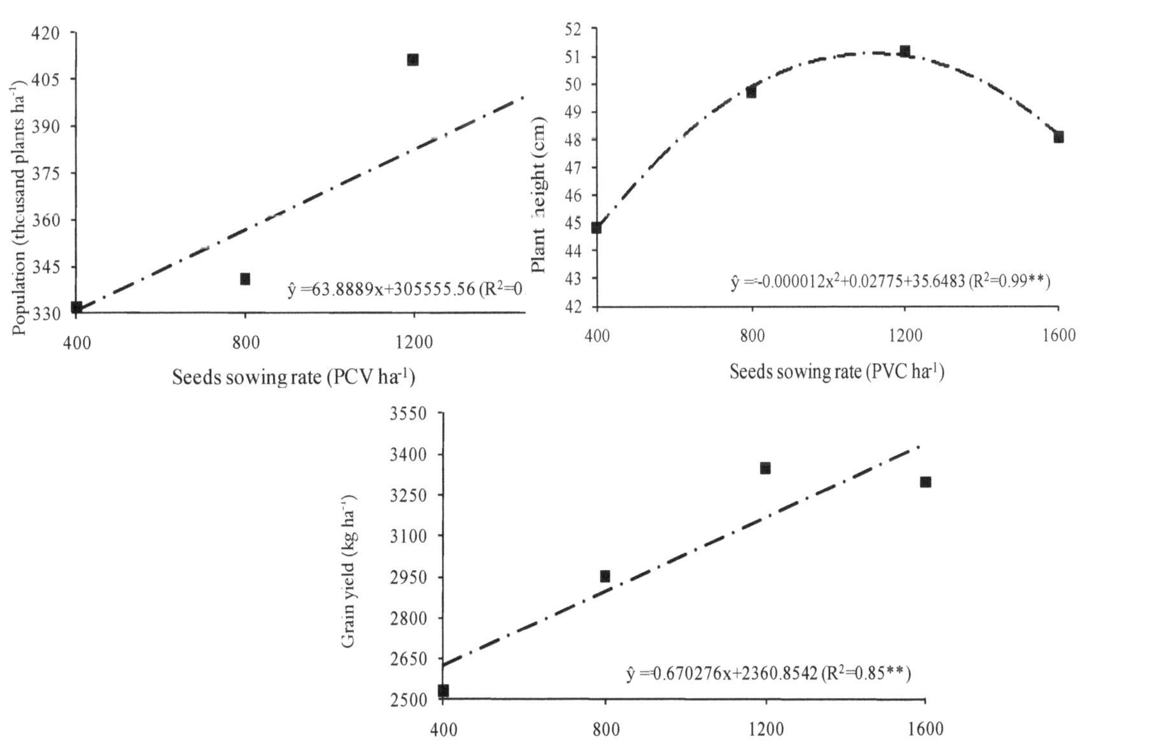

Figure 4. Soybean plant population, plant height and grain yield rotating with intercropped corn and kudzu sown at different sowing rate.

Pacheco et al. (2009) reported no significant difference in plant population, the number of pods with 2 or 3 grains or the mass of 1,000 grains in soybean plants rotated with several plant species. However, when grain yield of soybean plants growing in rotation with *U. ruziziensis* was compared with that of soybean plants growing in soil previously submitted to fallow, the results showed the grain yield of the former to be significantly higher than that of the latter. Correia and Durigan (2006) reported that the rotation of soybean with *U. brizantha* (in the two years of the experiment) and with *Eleusine coracana* (only in the second year of the experiment) resulted in higher grain yield, higher dry weight of aerial parts, and taller plants. Nunes et al. (2006) verified that rotating beans with *Panicum maximum* cv. Mombaça, *U. decumbens*, and *Panicum maximum* cv. Tanzania resulted in heavier grains (determined by the weight of 100 grains) and higher grain yield.

Conclusion

The intercropping of corn with kudzu sown either in furrows or by broadcasting has no negative effect on corn plant performance.

The seed sowing rate of 1600 points of cultural value, either in furrows or broadcasted, permits the kudzu plants to reach the maximum straw production.

The incidence of weeds is reduced by increasing the kudzu sowing rate, particularly when the kudzu seeds are sown in furrows.

The intercropping of corn with kudzu benefited soybean plant population, plant height, and grain yield in comparison with the single cultivating of corn in the previous year.

Acknowledgements

To CNPQ for their financial support.

References

ALVARENGA, R. C.; COBUCCI, T.; KLUTHCOUSKI, J.; WRUCK, F. J.; CRUZ, J. C.; NETO, M. M. G. **A cultura do milho na integração lavoura-pecuária**. Sete Lagoas: Embrapa Milho e Sorgo, 2006.

BARDUCCI, R. S.; COSTA, C.; CRUSCIOL, C. A. C.; BORGHI, E.; PUTAROV, T. C.; SARTI, L. M. N. Produção de *Brachiaria brizantha* e *Panicum maximum* com milho e adubação nitrogenada. **Revista Archivos de Zootecnia**, v. 58, n. 222, p. 211-222, 2009.

BORGHI, E.; CRUSCIOL, C. A. C. Produtividade de milho, espaçamento e modalidade de consorciação com *Brachiaria brizantha* em sistema de plantio direto. **Pesquisa Agropecuária Brasileira**, v. 42, n. 2, p. 163-171, 2007.

CORREIA, N. M.; DURIGAN, J. C. Influência do tipo e quantidade de resíduos vegetais associados a herbicidas residuais no desenvolvimento da cultura da soja. **Bragantia**, v. 65, n. 3 p. 421-432, 2006.

EMBRAPA-Empresa Brasileira de Pesquisa Agropecuária. **Redução dos impactos ambientais da pecuária de corte no Acre**. Rio Branco: Centro de Pesquisa Agroflorestal do Acre, 1999.

FREITAS, F. C. L.; FERREIRA, L. R.; FERREIRA, F. A.; SANTOS, M. V.; AGNES, E. L.; CARDOSO, A. A.; JAKELAITIS, A. Formação de pastagem via consórcio de *Brachiaria brizantha* com o milho para silagem no sistema de plantio direto. **Planta Daninha**, v. 23, n. 1, p. 49-58, 2005a.

FREITAS, F. C. L.; FERREIRA, L. R.; FERREIRA, F. A.; SANTOS, M. V.; AGNES, E. L. A. Cultivo consorciado de milho para silagem com *Brachiaria brizantha* no sistema de plantio convencional. **Planta Daninha**, v. 23, n. 4, p. 635-644, 2005b.

HEINRICHS, R.; VITTI, G. C.; MOREIRA, A.; FIGUEIREDO, P. A. M.; FANCELLI, A. L.; CORAZZA, E. J. Características químicas de solo e rendimento de fitomassa de adubos verdes e de grãos de milho, decorrente do cultivo consorciado. **Revista Brasileira de Ciência do Solo**, v. 29, n. 1, p. 71-79, 2005.

LINHARES, E. L. R.; SILVA, P. S. L.; OLIVEIRA, P. F.; OLIVEIRA, F. H. T.; TORRES, S. B. Planting density of gliricidia when intercropped with maize for weed control. **Planta Daninha**, v. 27, n. special, p. 967-975, 2009.

MONTEIRO, E. M. M.; LOURENÇO JUNIOR, J. B.; SANTOS, N. F. A.; AVIZ, M. A. B. Valor nutritivo da leguminosa *Pueraria phaseoloides* como alternativa na suplementação alimentar de ruminantes na Amazônia Oriental. **Ciência Rural**, v. 39, n. 2, p. 613-618, 2009.

NUNES, U. R.; ANDRADE JÚNIOR, V. C.; SILVA, E. B.; SANTOS, N. F.; COSTA, H. A. O.; FERREIRA, C. A. Produção de palhada de plantas de cobertura e rendimento do feijão em plantio direto. **Pesquisa Agropecuária Brasileira**, v. 41, n. 6, p. 943-948, 2006.

PACHECO, L. P.; PIRES, F. R.; MONTEIRO, F. P.; PROCÓPIO, S. O.; ASSIS, R. L.; CARGNELUTTI FILHO, A.; CARMO, M. L.; PETTER, F. A. Sobressemeadura da soja como técnica para supressão da emergência de plantas daninhas. **Planta Daninha**, v. 27, n. 3, p. 455-463, 2009.

SEPROTEC. **Forrageiras perenes - puerária**. Available from: <http://www.seprotec.com.br/produtos_forrageiras_pueraria.asp>. Access on: Jun. 16, 2010.

12

Fertilization with nitrogen and potassium in banana cultivars 'Grand Naine', 'FHIA 17' and 'Nanicão IAC 2001' cultivated in Ribeira Valley, São Paulo State, Brazil

Edson Shigueaki Nomura[1*], Francine Lorena Cuquel[2], Erval Rafael Damatto Junior[1], Eduardo Jun Fuzitani[1] and Ana Lúcia Borges[3]

*[1]Agência Paulista de Tecnologia dos Agronegócios, Polo Regional Vale do Ribeira, Cx. Postal 122, 11900-000, Registro, São Paulo, Brazil. [2]Universidade Federal do Paraná, Curitiba, Paraná, Brazil. [3]Empresa Brasileira de Pesquisa Agropecuária, Embrapa Mandioca e Fruticultura, Cruz das Almas, Bahia, Brazil. *Author for correspondence. E-mail: edsonnomura@apta.sp.gov.br*

ABSTRACT. The 'Grand Naine' banana cultivar is grown around the world, but it is susceptible to black Sigatoka fungus. The 'FHIA 17' and 'Nanicão IAC 2001' cultivars have the potential to replace susceptible cultivars, but there is little information about their nitrogen and potassium nutritional requirements. This study aimed to identify the best fertilizer recommendation with N and K for banana cultivars 'FHIA 17', 'Nanicão IAC 2001', and 'Grand Naine'. The applied fertilizer levels with N and K_2O in two production cycles were as follows: NK0, without fertilization; NK1, 175 N and 285 kg K_2O ha^{-1} $year^{-1}$; NK2, 350 N and 570 kg K_2O ha^{-1} $year^{-1}$; and NK 3, 525 N and 855 kg K_2O ha^{-1} $year^{-1}$. It was concluded that the N and K fertilization levels influenced most of the phenological and production parameters of 'Grand Naine' and 'Nanicão IAC 2001'. 'Nanicão IAC 2001' and 'Grand Naine' achieved maximum yield with an application of 525 kg N ha^{-1} $year^{-1}$ and 855 kg K_2O ha^{-1} $year^{-1}$. The 'FHIA 17' cultivar showed a low response to nitrogen and potassium fertilization. Regardless of cultivars, to achieve maximum productivity (44.2 and 43.3 ton ha^{-1} $year^{-1}$), an application at 150% and 124% of the fertilizer recommendation is required in the 1st and 2nd production cycles, respectively.

Keywords: nutrition, genetic improvement, *Musa* spp.

Introduction

The Ribeira Valley is the main banana-producing region of São Paulo State with an area of 34,500 hectares and production of approximately 952,700 tons (Nomura et al., 2013b). The region has favourable climatic conditions for banana growth during most of the year and is characterized by cultivation on small farms (10-20 ha) focusing on the production of banana type Nanica.

Banana type Nanica includes the Cavendish and Gros Michel subgroups and forms the basis of the domestic and world market (Soto-Ballestero, 2008). The commercially exploited cultivars in São Paulo are 'Grand Naine', 'Williams', and 'Nanicão', but all of them have at least one undesirable trait for commercial exploitation, such as plant height, productivity, tolerance to pests and diseases, or drought or cold resistance (Nomura, Damatto Jr., Fuzitani, Amorim, & Silva, 2013a).

The Honduran Agricultural Research Foundation (FHIA) breeding programme has developed several banana cultivars, such as 'FHIA 17' (AAAA), a hybrid of the Gros Michel cultivar that stands out among the banana type Nanica. Another cultivar that stands out among the banana type Nanica is the 'Nanicão IAC 2001' (AAA - Cavendish), which was selected by the Agronomic Institute (APTA/IAC). Both cultivars have a great potential for introduction into the Brazilian production system, especially in São Paulo, because they have fruit features similar to those of the 'Grand Naine' cultivar and because they are tolerant to black Sigatoka fungus (*Mycosphaerella fijiensis* Morelet) and Panamá disease (*Fusarium oxysporum* f. sp. *cubense*).

The final step of breeding is to evaluate plants in different regions to determine its soil and climate adaptability, as well as its resistance to pests and diseases (Roque, Amorim, Ferreira, Ledo, & Amorim, 2014; Mendonça, Duarte, Costa, Matos, & Seleguini, 2013; Nomura et al., 2013a; 2013b; Oliveira, Lessa, Silva, & Oliveira, 2008). Fruit postharvest characteristics are also an important parameter for evaluating the final quality of produced fruits (Godoy et al., 2016; Reis, Viana, Jesus, Santos, & Oliveira, 2016; Castricini, Santos, Deliza, Coelho, & Rodrigues, 2015). Few studies have looked at new banana cultivars and their nutritional requirements in different soil and climatic conditions; this is necessary because the nutritional requirements of these materials are not always the same as traditionally cultivated bananas.

Banana plants require large amounts of nutrients to maintain high yields over time in commercial crops. These nutrients may be supplied by the soil (areas with natural high fertility) or by fertilization in amounts and proportions that are at least equivalent to the nutrients extracted by harvest (Soto-Ballestero, 2008). Most of the soils cultivated with banana in the Ribeira Valley region have low fertility, and the lack of fertilizer during the plant production cycle is one factor responsible for low banana yield (Moreira, 1999). Local banana fertilizer recommendations do not specify the cultivar, so there is no recommendation for the different types or banana subgroups. According Teixeira, Nomura, Damatto Jr., and Fuzitani (2014), nitrogen fertilization recommendations range from 120 to 500 kg ha^{-1} in accordance with expected productivity, and potassium varies from 90 to 730 kg ha^{-1} in accordance with the potassium content in the soil.

Potassium (K) and nitrogen (N) are the most extracted macronutrients by banana plants, as they are directly related to plant growth, yield and banana fruit quality (Moreira, 1999). Potassium is the macronutrient extracted in greater amounts by banana plants (62% of the total macronutrient and 41% of the plant nutrients), which directly affects photosynthesis, the translocation of photosynthates, and the water balance in plants and fruits (Kumar & Kumar, 2008; Moreira, 1999). Nitrogen acts directly on vegetative growth, emissions and the development of shoot plants and increases the amount of dry matter in plants (Moreira, 1999).

This research aimed to identify the best fertilizer recommendation levels containing N and K and their effects on the phenological and productive characteristics of banana cultivars 'FHIA 17', 'Nanicão IAC 2001', and 'Grand Naine' grown in the soil and climatic conditions of the Ribeira Valley region, São Paulo State, Brazil.

Material and methods

This research was carried out at the experimental farm of Agência Paulista de Tecnologia dos Agronegócios (APTA) Regional Vale do Ribeira, in Pariquera-Açu County (Brazil), which has the following geographical coordinates: 24°36'31" South and 47°53'48" West, altitude of 25 m. The climate is tropical rainy with no dry season (Af), according to the Köppen classification. Data from a series of 10 years (2004-2014), recorded at APTA Regional Vale do Ribeira, shows that the average annual maximum and minimum temperatures were 26.8°C and 17.7°C, respectively, and that the average annual rainfall was 1,524.5 mm.

Soil samples were collected at the depth of 0-20 cm to determine the chemical properties of the samples and for lime and fertilizer recommendations for adequate banana production. The soil had the following characteristics: pH ($CaCl_2$) = 4.2; O.M. = 26.7 g dm^{-3}; P (resin) = 4.0 mg dm^{-3}; K = 0.6 mmol dm^{-3}; Ca = 12.7 mmol dm^{-3}; Mg = 4.7 mmol dm^{-3}; H + Al = 80.7 mmol dm^{-3}; SB = 18.0 mmol dm^{-3}; CTC = 98.7 mmol dm^{-3}; V = 19%; B = 0.24 mg dm^{-3}; Cu = 0.1 mg dm^{-3}; Fe = 111.7 mg dm^{-3}; Mn = 1.8 mg dm^{-3}; and Zn = 0.2 mg dm^{-3}. Dolomitic limestone was applied over the total area with the aim of elevating base saturation to 60% and increasing the magnesium content above 9.0 mmol dm^{-3}, as recommended by Teixeira et al. (2014). The phosphorus requirement was met with an application of 600 kg ha^{-1} of simple superphosphate to the total area.

'FHIA 17', 'IAC Nanicão 2001', and 'Grand Naine' banana production was evaluated for two production cycles (2013-2014). The 'Grand Naine' cultivar is considered the standard cultivar because it is widely used in commercial plantations in Brazil.

The standard fertilization with N and K (100%) was calculated in accordance to the soil chemical analysis in order to reach an expected yield of 40 to 50 ton ha^{-1} (as recommended by Teixeira et al., 2014). Fertilizer doses with N and K were as follows: NK0 = no fertilizer was applied; NK1 = an application of 50% of the standard recommendation of N and K (175 kg ha^{-1} of N and 285 kg K_2O ha^{-1} year^{-1}); NK2 = an application of 100% of the standard recommendation of N and K (350 kg ha^{-1} of N and 570 kg K_2O ha^{-1} year^{-1}); and NK3 = an application of 150% of standard recommendation of N and K (525 kg N ha^{-1} and 855 kg K_2O ha^{-1} year^{-1}). The N sources were urea and ammonium sulphate with interspersed applications, and the K source was potassium chloride. Fertilization in the 1st and 2nd production cycles were divided into five applications at the following percentages: 10, 25, 30, 20, and 15% of the total dose at each level of fertilization. The first fertilization application was applied 45 days after planting (in the 1st cycle) and 60 days after the harvest of the 1st bunch (in the 2nd cycle). The other fertilizer applications were applied in intervals of 60 days after the first application, preferably in the most favourable months for the development of banana (from August to May of each year).

Seedlings were produced by micropropagation, acclimatized in the nursery, and planted in the field when they had 5 to 6 leaves and were approximately 30 cm tall. Plant field spacing was determined based on cultivar average high: 2.0 m *x* 3.0 m (1,667 plants per hectare) for 'FHIA 17' (AAAA - subgroup Gros Michel) and 2.0 m *x* 2.5 m (2,000 plants per hectare) for 'Grand Naine' and 'Nanicão IAC 2001' (AAA - Cavendish). Crop management was based on Moreira (1999) recommendations.

Preventive fungicide applications were applied to the 'Grand Naine' cultivar because it is susceptible to black Sigatoka fungus. Fungicide applications were applied at intervals determined by plant monitoring, such as recommended by Fouré and Ganry (2008) and modified by Moraes, Modenese-Gorla, Fukuda, and Silva (2011). For the 'FHIA 17' and 'Nanicão IAC 2001' cultivars, fungicide applications were not performed because of their tolerance to black Sigatoka.

The plant development parameters evaluated were as follows: height of the pseudostem (measured from the ground level to the insertion of the last leaf), pseudostem diameter (measured at 30 cm above ground level), the number of active leaves at blooming and harvest time (only leaves that had more than half green area were counted), total cycle, and the daughter-plant height (measured at the mother-plant harvest time).

Bunches were harvested when the fruits of the 2nd hand had diameters between 34-36 mm, and the following was evaluated: the fresh mass of marketable bunches (kg); yield (estimated from the fresh mass of marketable bunches, planting density and the total cycle (ton ha^{-1} year^{-1})); the total number of fruits in a bunch; 2nd hand fruit fresh mass (g); and 2nd hand fruit length (measured on the convex face of the fruit (cm)).

The experimental design was a randomized block design, and treatments were organized in a 3 x 4 factorial scheme (cultivars x fertilization doses) with culture cycle as sub plots (2 production cycles). Individual treatments were replicated three times and contained four plants. The data were submitted to variance and F-test analyses using the statistical programme SISVAR (Ferreira, 2011). When the results were significant for the quantitative variable (fertilization), regression equations were adjusted, and when the results were significant only for the qualitative variables (cultivars and cycle), the averages were compared using Tukey's test at a 5% probability.

Results and discussion

Plant development

The leaf number at blooming, daughter-plant height, and the diameter and height of the pseudostem were influenced by cultivar and fertilizer dose. A triple interaction was not observed among cultivar, cycle and fertilizer dose.

The number of active leaves at blooming is an important parameter for banana bunch development because it reflects the potential yield, as these leaves are directly related to the plant photosynthetic rate (Soto-Ballestero, 2008) and promote starch accumulation in fruits. For the leaf number at blooming, it can be observed in Figure 1A that only the 'Grand Naine' cultivar regression equation was adjusted. Plant had more leaves with the application of 123% of the fertilizer recommendation. However, this fertilization level did not provide the highest fresh mass of marketable bunches and yield (Figure 2C and D). These results could be explained by the greater loss of leaves due to black Sigatoka fungus on plants fertilized with the highest levels. According to Moreira (1999), banana plants with excess N can show a high incidence of black Sigatoka. Nevertheless, this reduction in the number of leaves did not compromise yield. The leaf number in the other two cultivars did not present changes in response to fertilizer levels, and all of them had a sufficient number of leaves to reach suitable fruit development. According to Soto-Ballestero (2008),

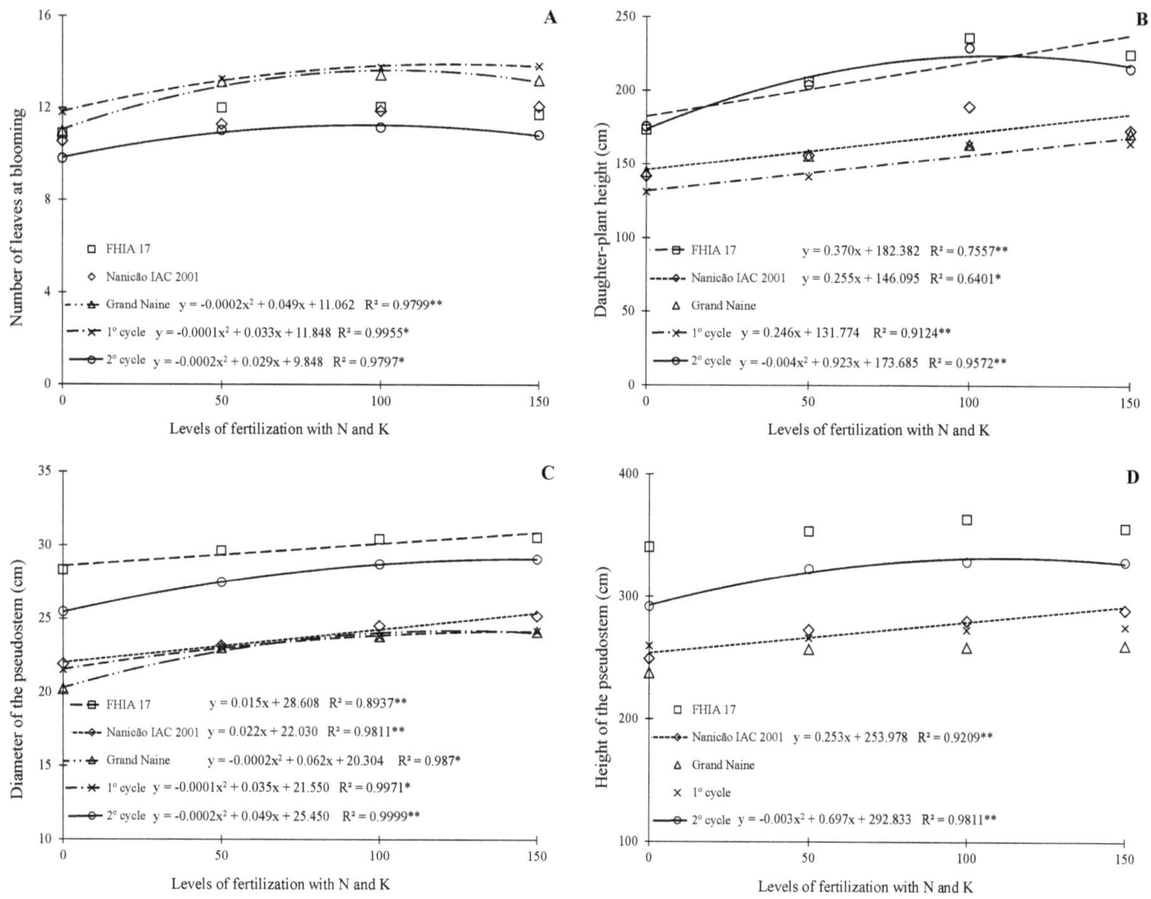

Figure 1. The development of banana cultivars 'Grand Naine', 'FHIA 17', and 'Nanicão IAC 2001' under different levels of fertilization with N and K (50%: 175 N + 285 K_2O; 100%: 350 N + 570 K_2O; and 150%: 525 N + 855 K_2O kg ha^{-1} $year^{-1}$) in two cycles. Pariquera-Açú, São Paulo State, Brazil, 2016. $^{**}p \leq 0.01$ and $^{*}p \leq 0.05$. R^2 = coefficient of determination.

banana plants of the Cavendish subgroup should have at least eight active leaves during bunch development. The maintenance of a larger number of leaves in the 'FHIA 17' and 'Nanicão IAC 2001' may be due to their resistance or tolerance to foliar diseases, especially black Sigatoka (Nomura et al., 2013b).

The active number of leaves at plant blooming presented an interaction between the production cycles and fertilizer levels in both production cycles, reaching a maximal response with the applications of 165% and 73% of the fertilizer recommendation in the 1st and 2nd cycles, respectively (Figure 1A).

The active leaf number at harvest time was different among cycles and among cultivars in the second cycle (Table 1). The cultivars maintained sufficient quantities of leaves in both cycles for suitable fruit. Teixeira, Ruggiero, and Natale (2001) recommended that the number of active leaves should be equal to the hands in the bunch. Rodrigues, Dias, and Pacheco (2009) recommended the maintenance of 10 to 12 leaves in the 'Prata-Anã' banana plant to achieve maximum yield and help maintain healthy banana plantations.

The 'Grand Naine' cultivar showed more leaves at harvest and were healthier due to the preventive applications of fungicides to control black Sigatoka. 'FHIA 17' and 'Nanicão IAC 2001', even with fewer leaves in the 2nd cycle (Table 1), had healthy and functional growth and yield without chemical disease control, proving their tolerance to black Sigatoka.

The fewer leaves at plant blooming until bunch harvest is mainly due to the translocation of assimilates to fruit filling, which becomes the preferred drain in the plant due to the natural senescence of leaves and because of the incidence *versus* tolerance to the black Sigatoka fungus (Silva, Pires, Pestana, Alves, & Silveira, 2006). Another factor that affects the durability of banana leaves after inflorescence emergence relates to the nutritional deficiency of N and K and is mainly due to the translocation of N and K from the older leaves to the fruits (Teixeira et al., 2001).

Table 1. The number of leaves at harvest, plant total cycle and fruit length of banana cultivars 'Grand Naine', 'FHIA 17' and 'Nanicão IAC 2001' under different levels of fertilization with N and K (50%: 175 N + 285 K_2O; 100%: 350 N + 570 K_2O; and 150%: 525 N + 855 K_2O kg ha^{-1} $year^{-1}$), Pariquera-Açú, São Paulo State, Brazil, 2016.

Number of leaves at harvest			
Cultivars	Cycles 1st	Cycles 2nd	Means
FHIA 17	8.2 A	4.9 b B	6.6 ab
Nanicão IAC 2001	7.9 A	3.7 c B	5.8 b
Grand Naine	8.0	7.1 a	7.6 a
Means	8.0 A	5.2 B	
V.C. 1 (%)	26.5		
V.C. 2 (%)	17.9		
Plant total cycle (days)			
Cultivars	Cycles 1st	Cycles 2nd	Means
FHIA 17	462.8 b	459.3 b	461.0 b
Nanicão IAC 2001	412.8 a B	359.2 a A	394.5 a
Grand Naine	406.7 a	382.3 a	386.0 a
Means	427.4 B	400.3 A	
V.C. 1 (%)	6.5		
V.C. 2 (%)	8.9		
Fruit length (cm)			
Cultivars	Cycles 1st	Cycles 2nd	Means
FHIA 17	18.7	23.4	21.0
Nanicão IAC 2001	23.3 A	18.7 B	21.0
Grand Naine	22.4	21.4	21.9
Means	23.0 A	19.6 B	
V.C. 1 (%)	13.6		
V.C. 2 (%)	13.0		

Means followed by different letters, lowercase among cultivars and uppercase between fertilization levels and production cycle, differ by Tukey's test ($p < 0.05$). V.C. = variation coefficient.

For the plant total cycle, a difference was observed between the qualitative variables (cultivars and cycles), as shown in Table 1. 'FHIA 17' had the longest total cycle among the evaluated cultivars, corroborating with Nomura et al. (2013a). The 'Grand Naine' and 'Nanicão IAC 2001' cultivars had shorter total cycles compared to 'FHIA 17'. In commercial banana production, reducing the period of fruit maintenance in the field is recommended to decrease pest and disease risk and to reduce pesticide use (Rodrigues, Souto, & Silva, 2006; Santos, Carneiro, Silveira Neto, Paniago Jr., & Peixoto, 2006).

The thinning of daughter-plants in banana production is aimed at continuing production over the years, and therefore, maintaining and developing daughter-plants in anticipation of the next production cycle ensures a rapid economic return on the invested capital. In evaluating daughter-plant height, it was observed that there was interaction between cultivars and fertilizer levels for this feature. 'FHIA 17' and 'Nanicão IAC 2001' reached maximum estimated values at the highest applied dose of N and K, regardless of the production cycle (Figure 1B). Despite that 'FHIA 17' displayed a larger daughter-plant at harvest, it did not promote the precocity in the 2nd cycle, which was possibly influenced by other factors such as genetics and climate.

There was interaction between cycles and fertilizer levels for banana daughter-plant height. In the 1st cycle, daughter-plants reached maximum height with the application of 150% of banana fertilizer recommendation, while in the 2nd cycle, daughter-plants reached the largest height with the application of 115% of the recommendation (Figure 1B). The low amount of fertilizers required in the 2nd cycle was because plants were well established in the field, the mother-plant translocated nutrients to the daughter-plant, and was due to the decomposition and nutrient release of the 1st cycle crop residues (Moreira & Fageria, 2009). Moreover, the successive applications of fertilizer increased soil K in the experimental field, as reported by Teixeira, Natale, Bettiol Neto, and Martins (2007), who found that successive applications of K in banana favoured the accumulation of this nutrient, most notably in the surface layer and in the areas closest to the plants.

The 'FHIA 17' and 'Nanicão IAC 2001' cultivars showed an increase in the pseudostem diameter due to the addition of N and K, obtaining the highest diameters with the application at the maximum dose (150%), whereas 'Grand Naine' reached an estimated maximum diameter by applying 155% of the recommendation (Figure 1C). The pseudostem is constituted by wrapped leaf sheaths, making it rigid, as the emergence of new leaves occurs (Soto-Ballestero, 2008). The proper nutrient supply furthers leaf development, and consequently, the pseudostem increases its height and diameter. Pseudostems stores large amounts of nutrients, especially N and K (67.9 and 233.1 kg ha^{-1}, respectively), which is used for fruit filling (Soto-Ballestero, 2008).

An interaction was observed between the production cycles and fertilizer levels for the pseudostem diameter. In the 1st cycle, the maximum value was estimated with an application of 175% of the fertilizer recommendation and in the 2nd cycle with the application of 123% of the recommendation (Figure 1C). The lower requirement of nutrients in the 2nd cycle is probably due to nutrient recycling from crop residue, especially that from the pseudostem, leaves and rhizomes of the banana plants. Hoffmann, Oliveira, Souza, Gheyi, and Souza Jr. (2010) found that the return of nutrients to the soil was between 497 to 305 kg ha^{-1} for K and 91 and 121 kg ha^{-1} for N for the 'Grand Naine' and 'Gros Michel' cultivars, respectively.

An interaction was found between cultivars and N and K fertilization levels in pseudostem height for 'Nanicão IAC 2001', reaching the minimum estimated height without fertilization (Figure 1D). For 'FHIA 17' and 'Grand Naine', plant height was similar among the N and K doses. This probably is due to previous nutrients stored in the soil, which was sufficient for proper plant development. There is also the possibility of nutrients being provided by crop residue mineralization, which mainly provides N and K.

Regardless of the cultivars, plant height in the 2nd cycle was simultaneously affected by fertilizer levels and by cycles. The highest pseudostem was obtained with the application of 116% of the fertilizer recommendation (Figure 1D). Farmers prefer shorter plants because there is no need to use anchors to support the plants and because it would be possible to increase planting density (Amorim, Santos-Serejo, Amorim, Ferreira, & Silva, 2013). According to the results obtained in this experiment, plants were smaller when they did not receive any fertilizer, but with this treatment, the production and yield were very low (Figure 2C and D). Despite the increase of pseudostem height by the fertilization levels, all cultivars were kept smaller than the band of 250 to 350 cm, which is preferable to the farmers (Santos et al., 2006). It was observed in the field that cv. 'FHIA 17' had a high number of fallen plants due to their susceptibility to banana weevil [*Cosmopolites sordidus* (Germ.)] and nematodes.

Banana plant height and pseudostem diameter are important parameters in crop management because they are related to plant vigour and they affect bunch harvesting and breaking/tipping adult plants, consequently causing fewer losses in yield (Farias, Donato, Pereira, & Silva, 2010). Generally, these overturned plants are caused by the direct action of strong winds and are related to large-sized plants with small pseudostem diameters (Teixeira et al., 2001). This is generally observed in cultivars of the Cavendish and Gros Michel subgroups (Moreira, 1999), which require anchors to support them to avoid yield and fruit quality loss.

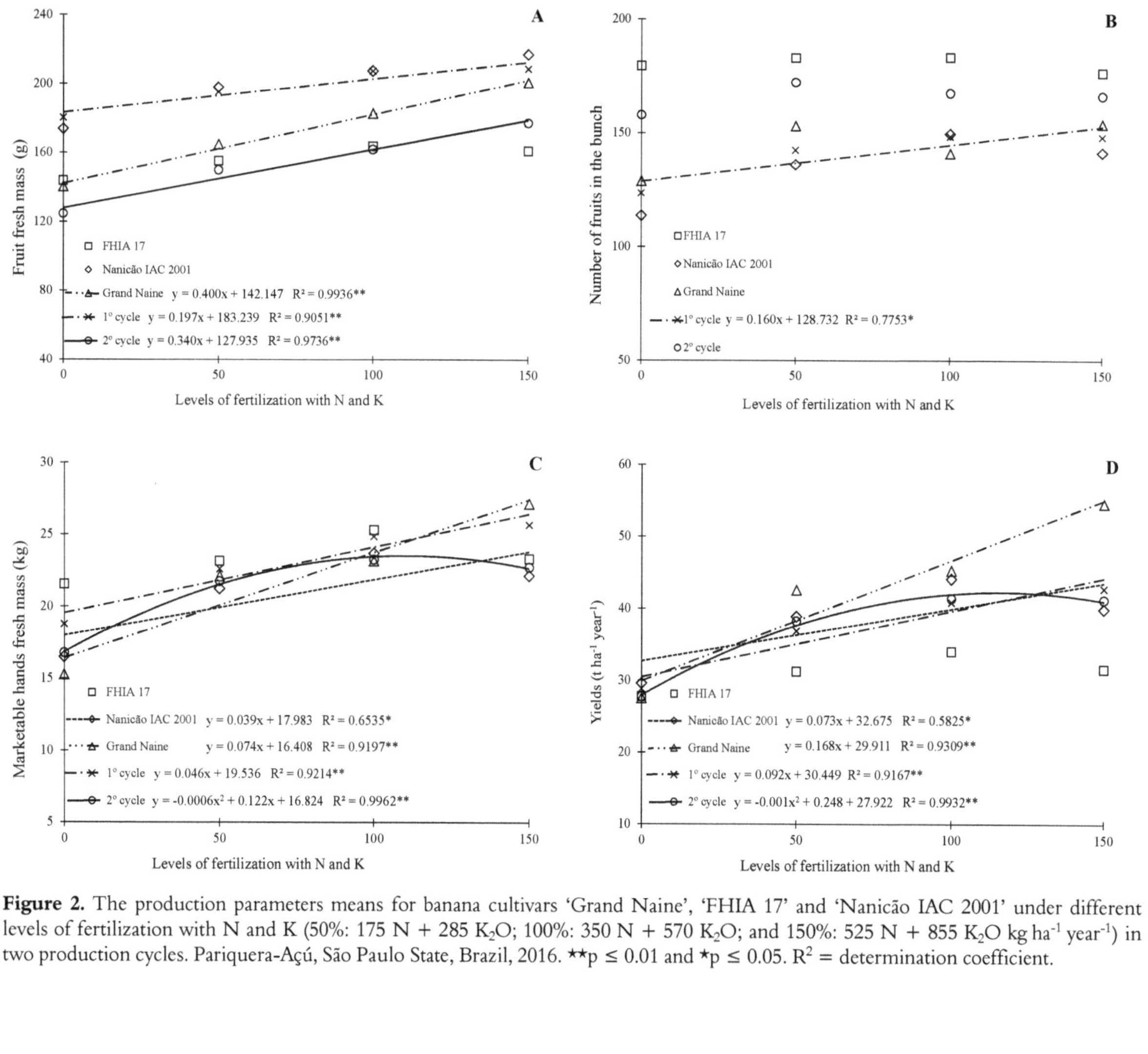

Figure 2. The production parameters means for banana cultivars 'Grand Naine', 'FHIA 17' and 'Nanicão IAC 2001' under different levels of fertilization with N and K (50%: 175 N + 285 K_2O; 100%: 350 N + 570 K_2O; and 150%: 525 N + 855 K_2O kg ha^{-1} $year^{-1}$) in two production cycles. Pariquera-Açú, São Paulo State, Brazil, 2016. **$p \leq 0.01$ and *$p \leq 0.05$. R^2 = determination coefficient.

Production

Fertilizer levels did not affect fruit length. Regardless of the production cycles, the 'FHIA 17', 'Nanicão IAC 2001' and 'Grand Naine' cultivars had fruit length averages, respectively, of 21.0 cm, 21.0 cm and 21.9 cm (Table 1). The fruit length of the Nanica type cultivars in this study were similar to the range (20.9 to 25.0 cm) found by Borges, Silva, Oliveira, and Roberto (2011), Donato et al. (2006), Mendonça et al. (2013), Njuguna et al. (2008), Nomura et al. (2013b), and Ramos, Leonel, Mischan, and Damatto Jr. (2009). The final size of banana fruit depends on several factors, such as temperature, the number of leaves and leaf area during bunch development, soil fertility, water availability, and maturity stage at harvest (Robinson & Gálan-Saúco, 2011). These authors found that banana shape and size are cultivar features, thus the reason why there was a low response to fertilization levels for this parameter in this study.

N and K levels increased fruit fresh mass in 'FHIA 17' and in both production cycles due to the significant interaction between cultivars and production cycles with fertilization levels. The highest fruit fresh mass was achieved at the highest level of fertilizer recommendation (150%) (Figure 2A). Fruit mass is an important trait for selecting cultivars and consumer preference, but fruit mass should not be considered an isolated trait because other components can reflect fruit quality, such as length and fruit diameter (Garruti et al., 2012).

The fruit number per bunch was not affected by cultivar and fertilizer levels (Figure 2B), but in the 2nd cycle, all cultivars had higher fruit numbers with an average of 181 fruits for 'FHIA 17', 144 fruits for 'Grand Naine', and 135 fruits for 'Nanicão IAC 2001'. The genomic group of each cultivar is a very important factor that controls production potential such as the number of fruit, hands in the bunch, and bunch mass. The cultivars that belong to the genomic group AAA (Cavendish and Gros Michel) and its hybrids ('FHIA 17') generally have a higher potential for fruit production, ranging from 10 to 30 fruits per hand and 5 to 16 hands per bunch, depending on the season and growing conditions (Robinson & Gálan-Saúco, 2011).

The total number of fruits in the 1st production cycle was affected by the fertilizer levels, where the estimated maximum response was achieved at the highest applied dose (150%) (Figure 1B). Regardless of fertilizer levels, there was a higher number of fruit in the 2nd cycle, with the average of 166 fruits per bunch. According to Robinson and Gálan-Saúco (2011) and Soto-Ballestero (2008), the number of hands and fruits in the bunch is mainly influenced by genetics, growing cycle, temperature, plant vigour, and by plant management at the floral differentiation stage. In this study, banana plants in the 2nd cycle had better nutritional conditions, as there was a connection between the rhizome of the mother-plant and daughter-plan, acting as an "umbilical cord", where sap and hormones could be exchanged (Moreira, 1999). It is recommended that the harvested plants remain in the field for at least one month to transfer nutrients from the mother-plant to the daughter-plant (Soto-Ballestero, 2008).

The increase of fertilizer for 'Nanicão IAC 2001' and 'Grand Naine' increased the marketable hands fresh mass, with the highest mass obtained by applying 150% of the fertilizer recommendation (Figure 2C). There was also a significant interaction between production cycles and fertilizer levels for marketable hands fresh mass, with maximum estimated values when 150% and 102% of the fertilizer recommendation was applied at the 1st and 2nd cycle, respectively (Figure 2C). The marketable hands mass for 'FHIA 17' and 'Nanicão IAC 2001' was reduced in the 2nd cycle due to a severe incidence of black Sigatoka in this cycle. The 'Grand Naine' did not had problems with the black Sigatoka fungus because it had been treated with preventive fungicides; therefore, 'Grand Naine' marketable hands produced higher fresh mass in the 2nd cycle. Furthermore, this cultivar in the 1st cycle had high productive potential, as also noted by Bolfarini, Javara, Leonel, and Leonel (2014), who observed bunches at 31.18 kg, grown in Botucatu (São Paulo State).

Regardless of production cycles, the fertilization levels resulted in increasing yields for 'Nanicão IAC 2001' and 'Grand Naine' (Figure 2D). The maximum estimated yields were achieved with the highest dose (150%), promoting an increase of 33.5 and 84.2% compared to the treatment without fertilization. These results corroborate the results of Soto-Ballestero (2008), who reported that Cavendish bananas are much more responsive to potassium fertilizer than the cultivars of the subgroup Gros Michel ('FHIA 17'). There were differences for yield among the cultivars, with banana cultivars 'Grand Naine' (42.5 ton ha^{-1} year^{-1}) and 'Nanicão IAC 2001' (38.1 ton ha^{-1} year^{-1}) having superior yield compared to 'FHIA 17' (31.1 ton ha^{-1} year^{-1}), regardless of the fertilizer levels and production cycles.

Despite that 'FHIA 17' presented marketable hands fresh mass very close to the other cultivars (Figure 2C), a low yield due to the lower planting density (1,667 plants ha^{-1}) and the highest total cycle (Table 1) was observed. Donato et al. (2006) found that cultivars from Cavendish and Gros Michel subgroups are potentially more productive than the Prata subgroup. These authors found a yield of 78.1 ton ha^{-1} year^{-1} for cv. 'Grand Naine' grown in Guanambi (BA), which is much higher than that found in this study. This increase is mainly because the region (Guanambi) has favourable climatic conditions for banana cultivation, the absence of the black Sigatoka fungus and appropriate water supplies by irrigation.

Variance analysis determined a significant effect in the interaction of cycles and fertilization levels on yield for both production cycles. In the 1st cycle, yield reached an estimated 37.3 ton ha^{-1} year^{-1} with the maximum applied dose (150%), while in the 2nd cycle, yield reached 37.1 ton ha^{-1} year^{-1} applying 124% of the fertilizer recommendation (Figure 2D). These values are above the international average yield (14.2 ton ha^{-1}) (Soto-Ballestero, 2008).

Conclusion

Given the experimental conditions of this study, the following was concluded: 'Nanicão IAC 2001' and 'Grand Naine' require an application of 150% of the fertilizer recommendation (525 kg N ha^{-1} year^{-1} and 855 kg K_2O ha^{-1} year^{-1}) to achieve maximum yield (43.6 and 55.1 ton ha^{-1} year^{-1}, respectively); the 'FHIA 17' banana showed a low response to nitrogen and potassium fertilization; and regardless of cultivar, to achieve maximum productivity (44.2 and 43.3 ton ha^{-1} year^{-1}, respectively), an application of 150 and 124% of the fertilizer recommendation in the 1st and 2nd production cycles is required, respectively.

Acknowledgements

We acknowledge the São Paulo Research Foundation (FAPESP), project 2012/50820-1, and Embrapa Cassava & Fruits (MP2 - MelhorMusa Project) for financial support and CAPES for a PhD scholarship to the first author.

References

Amorim, E. P., Santos-Serejo, J., Amorim, V. B. O., Ferreira, C., & Silva, S. (2013). Banana breeding at Embrapa cassava and fruits. *Acta Horticulturae, 986*(1), 171-176.

Bolfarini, A. C. B., Javara, F. S., Leonel, S., & Leonel, M. (2014). Crescimento, ciclo fenológico e produção de cinco cultivares de bananeira em condições subtropicais. *Revista Raízes e Amidos Tropicais, 10*(1), 74-89.

Borges, R. S., Silva, S. O., Oliveira, F. T., & Roberto, S. R. (2011). Avaliação de genótipos de bananeira no norte do estado do Paraná. *Revista Brasileira de Fruticultura, 33*(1), 291-296.

Castricini, A., Santos, L. O., Deliza, R., Coelho, E. F., & Rodrigues, M. G. V. (2015). Caracterização pós-colheita e sensorial de genótipos de bananeiras tipo Prata. *Revista Brasileira de Fruticultura, 37*(1), 27-37.

Donato, S. L. R., Silva, S. O., Lucca Filho, O. A., Lima, M. B., Domingues, H., & Alves, J. S. (2006). Comportamento de variedades e híbridos de bananeira (*Musa* spp.), em dois ciclos de produção no sudoeste da Bahia. *Revista Brasileira de Fruticultura, 28*(1), 139-144.

Farias, H. C., Donato, S. L. R., Pereira, M. C. T., & Silva, S. O. (2010). Avaliação fitotécnica de bananeiras tipo terra sob irrigação em condições semi-áridas. *Ciência e Agrotecnologia, 34*(4), 380-386.

Ferreira, D. F. (2011). SISVAR: A computer statistical analysis system. *Ciência e Agrotecnologia, 35*(6), 1039-1042.

Fouré, E., & Ganry, J. (2008). A biological forecasting system to control Black Leaf Streak. *Fruits, 63*(5), 311-317.

Garruti, D. S., Matias, M. L., Facundo, H. V. V., Silva, E. O., Costa, J. N., & Silva, M. A. A. P. (2012). Aceitação de cultivares de bananas resistentes à Sigatoka Negra junto ao consumidor da região Nordeste do Brasil. *Ciência Rural, 42*(5), 948-954.

Godoy, R. C. B., Waszczynskj, N., Santana, F. A., Silva, S. O., Oliveira, L. A., & Santos, G. G. (2016). Physico-chemical characterization of banana varieties resistant to black leaf streak disease for industrial purposes. *Ciência Rural, 46*(9), 1514-1520.

Hoffmann, R. B., Oliveira, F. H. T., Souza, A. P., Gheyi, H. R., & Souza Jr., T. F. (2010). Acúmulo de matéria seca e de macronutrientes em cultivares de bananeira irrigada. *Revista Brasileira de Fruticultura, 32*(1), 268-275.

Kumar, A. R., & Kumar, N. (2008). Potassium nutrition in banana. *The Asian Journal of Horticulture, 3*(2), 479-482.

Mendonça, K. H., Duarte, D. A., Costa, V. A. M., Matos, G. R., & Seleguini, A. (2013). Avaliação de genótipos de bananeira em Goiânia, estado de Goiás. *Revista Ciência Agronômica, 44*(3), 652-660.

Moraes, W. S., Modenese-Gorla da Silva, S. H., Fukuda, E., & Silva, C. M. (2011). Técnica de monitoramento da Sigatoka-negra na cultura da banana. *Pesquisa & Tecnologia, 8*(2), 9.

Moreira, R. S. (1999). *Banana: teoria e prática de cultivo* (2a ed., CD ROM). Campinas, SP: Fundação Cargill.

Moreira, A., & Fageria, N. K. (2009). Repartição e remobilização de nutrientes na bananeira. *Revista Brasileira de Fruticultura, 31*(2), 574-581.

Njuguna, J., Nguthi, F., Wepukhulu, S., Wambugu, F., Gitau, D., Karuoya, M., & Karamura, D. (2008). Introduction and evaluation of improved banana cultivars for agronomic and yield characteristics in Kenya. *African Crop Science Journal, 16*(1), 35-40.

Nomura, E. S., Damatto Jr., E. R., Fuzitani, E. J., Amorim, E. P., & Silva, S. O. (2013a). Avaliação agronômica de genótipos de bananeiras em condições subtropicais, Vale do Ribeira, São Paulo - Brasil. *Revista Brasileira de Fruticultura, 35*(1), 112-122.

Nomura, E. S., Moraes, W. S., Damatto JR., E. R., Fuzitani, E. J., Saes, L. A., Amorim, E. P., & Silva, S. O. (2013b). Evaluation of banana genotypes over two crop cycles under subtropical conditions in the Ribeira Valley, São Paulo, Brazil. *Acta Horticulturae, 986*(1), 61-70.

Oliveira, T. K., Lessa, L. S., Silva, S. O., & Oliveira, J. P. (2008). Características agronômicas de genótipos de bananeira em três ciclos de produção em Rio Branco, AC. *Pesquisa Agropecuária Brasileira, 43*(8), 1003-1010.

Ramos, D. P., Leonel, S., Mischan, M. M., & Damatto Jr., E. R. (2009). Avaliação de genótipos de bananeira em Botucatu-SP. *Revista Brasileira de Fruticultura, 31*(4), 1092-1101.

Reis, R. C., Viana, E. S., Jesus, J. L., Santos, T. S., & Oliveira, M. A. (2016). Physicochemical and sensorial quality of banana genotypes. *Pesquisa Agropecuária Tropical, 46*(1) 89-95.

Robinson, J. C., & Gálan-Saúco, V. (2011). *Bananas y plantains* (2a ed.). Madri, Spain: Mundi-Prensa.

Rodrigues, M. G. V., Dias, M. S. C., & Pacheco, D. D. (2009). Influência de diferentes níveis de desfolha na produção e qualidade dos frutos da bananeira 'Prata-Anã'. *Revista Brasileira de Fruticultura, 31*(3), 755-762.

Rodrigues, M. G. V., Souto, R. F., & Silva, S. de O. (2006). Avaliação de genótipos de bananeira sob irrigação. *Revista Brasileira de Fruticultura, 28*(3), 444-448.

Roque, R. L., Amorim, T. B., Ferreira, C. F., Ledo, C. A. S., & Amorim, E. P. (2014). Desempenho agronômico de genótipos de bananeira no recôncavo da Bahia. *Revista Brasileira de Fruticultura, 36*(3), 598-609.

Santos, S. C., Carneiro, L. C., Silveira Neto, A. M., Paniago Jr., E., & Peixoto, C. N. (2006). Caracterização morfológica e avaliação de cultivares de bananeira resistentes à Sigatoka-negra (*Mycosphaerella fijiensis* Morelet) no Sudoeste Goiano. *Revista Brasileira de Fruticultura, 28*(3), 449-453.

Silva, S., Pires, E. T., Pestana, R. K. N., Alves, J. S., & Silveira, D. C. (2006). Avaliação de clones de banana Cavendish. *Ciência e Agrotecnologia, 30*(5), 832-837.

Soto-Ballestero, M. (2008). *Bananos: técnicas de produción, poscosecha y comercialización* (3nd ed., CD-ROM). San José, CR: Litografía e Imprenta Lil.

Teixeira, L. A. J., Ruggiero, C., & Natale, W. (2001). Manutenção de folhas ativas em bananeira 'Nanicão' por meio do manejo das adubações nitrogenada e potássica e da irrigação. *Revista Brasileira de Fruticultura, 23*(3), 699-703.

Teixeira, L. A. J., Natale, W., Bettiol Neto, J. E., & Martins, A. L. M. (2007). Nitrogênio e potássio em bananeira via fertirrigação e adubação convencional-atributos químicos do solo. *Revista Brasileira de Fruticultura, 29*(1), 143-152.

Teixeira, L. A. J., Nomura, E. S., Damatto Jr., E. R., & Fuzitani, E. J. (2014). Banana. In A. T. E. Aguiar, C. Gonçalves, M. E. A. G. Paterniani, M. G. S. Tucci, & C. E. F. Castro (Eds.), *Instruções agrícolas para as principais culturas econômicas* (p. 46-51, Boletim Técnico, 200). Campinas, SP: Instituto Agronômico.

13

The influence of rootstock and time of harvest on the fruit quality during storage of in two grapefruit cultivars

Francisca Ligia de Castro Machado[1*], Janaína de Paula da Costa[2], Adunias dos Santos Teixeira[3] and José Maria Correia da Costa[4]

[1]*Instituto Federal de Educação, Ciência e Tecnologia, Campus Tianguá, Rod. CE-075, s/n, 62320-000, Aeroporto, Tianguá, Ceará, Brazil.* [2]*Departamento de Engenharia de Pesca, Universidade Federal do Ceará, Fortaleza, Ceará, Brazil.* [3]*Departamento de Engenharia Agrícola, Universidade Federal do Ceará, Fortaleza, Ceará, Brazil.* [4]*Departamento de Tecnologia de Alimentos, Universidade Federal do Ceará, Fortaleza, Ceará, Brazil. *Author for correspondence. E-mail: ligia.machado@ifce.edu.br*

ABSTRACT. The effects of the 'Cleopatra' mandarin and 'Swingle' citrumelo rootstocks and time of harvest on the quality of cold-stored (13°C and 90% R.H.) 'Ruby Red' and 'Star Ruby' grapefruit grown under the tropical conditions prevalent in Ceará state, Brazil, were evaluated. Fruit quality was assessed by mass loss and peel color, the percentage of juice, soluble solids, titratable acidity, pH, maturity index, and ascorbic acid content. The results suggested that 'Ruby Red' and 'Star Ruby' grapefruit presented similar amounts of soluble solids, pH, titratable acidity and ascorbic acid levels at harvest, and the trends over the course of storage followed similar patterns for both cultivars, regardless of rootstock. The influence of scion-rootstock combination on the studied variables was dependent on the time of harvest. The peel color of 'Ruby Red' grafted on 'Swingle' citrumelo is more vivid and redder when harvested in October than when harvested in August. 'Ruby Red' and 'Star Ruby' grapefruit harvested in October are sweeter, less acidic, juicier, and richer in ascorbic acid than those harvested in August. 'Cleopatra' mandarin rootstock might favor ascorbic acid levels at harvest, but the fruit may lose more weight during storage.

Keywords: *Citrus paradisi* Macf., postharvest, semi-arid conditions.

Introduction

Research has firmly established the association between fruit consumption and good health, such that purchasing fruit now involves considerations that go beyond visual quality. The regular consumption of citrus fruit has a beneficial effect on human health. The healthy properties of citrus are credited to its wide array of bioactive compounds, such as vitamin C and flavonoids (WU et al., 2007).

The bioactive compounds in citrus fruit vary according to the variety, environmental conditions, and stage of maturity (GIRENNAVAR et al., 2008). For example, vitamin C content, including ascorbic acid, is influenced by genotype differences, preharvest climate conditions, maturity, and postharvest handling (LEE; KADER, 2000). Among the many types of citrus, the grapefruit [*Citrus paradise* Mact. (Rutaceae)] has received much attention because of its well-known nutritional properties; for example, grapefruit juice has recently been found to protect against high blood pressure and elevated cholesterol (WU et al., 2007).

The stage of fruit maturity and time of harvest are important factors in citrus storability and directly affect fruit quality. Grapefruits, which do not continue to ripen after harvest, should be harvested ripe. Pailly et al. (2004) studied harvest time and storage conditions of 'Star Ruby' grapefruit for short distance summer consumption and concluded that the period during which fruit are picked may be a key factor because early or delayed harvest can lengthen or shorten storage life. Chen et al. (2010) studied chemical differences in grapefruit with respect to year, harvest time, and conventional versus organic farming and observed that 'Rio Red' grapefruit from early, mid, and late harvests had distinctive chemical compositions.

Another pre-harvest factor that affects fruit quality and postharvest behavior is rootstock. Machado et al. (2011) examined the effect of citrus rootstocks on fruit quality of grapefruit cv. 'Ruby Red' grafted on 'Swingle' citrumelo (*Citrus paradise* × *Poncirus trifoliata*) or 'Rangpur' lime (*Citrus limonia*) rootstock and observed that fruit quality at harvest and during storage was markedly influenced by rootstock. Cano and Bermejo (2011) observed that both citrus rootstock and scion cultivar influenced the bioactive constituents in the rinds of cultivars of 'Clementine' mandarin ('Fino', 'Loretina', and 'Marisol'), 'Satsuma' mandarin ('Owari'), 'Navel' orange ('Navelate' and 'Navelina') and common orange ('Valencia Late') grafted on different rootstocks ('Cleopatra' mandarin and 'Troyer' citrange).

The goal of this work was to study the influence of rootstock ('Swingle' citrumelo and 'Cleopatra' mandarin) and time of harvest (August and October) on the quality, at harvest and during storage, of 'Ruby Red' and 'Star Ruby' grapefruit grown under the tropical conditions prevalent in Ceará State, Brazil.

Material and methods

'Ruby Red' and 'Star Ruby' grapefruit scions grafted on 'Swingle' citrumelo or 'Cleopatra' mandarin rootstocks were harvested at two periods (August 15th 2010 and October 15th 2010) from plantings located at Limoeiro do Norte (5°13'S latitude and 37°54'W longitude), Ceará, Brazil. The climate of that region is classified as BS'w'h, according to the Koppen classification. The annual average temperature is 28.5°C, and the average precipitation is 772 mm year^{-1}, irregularly distributed during the months of January-June. The average relative humidity is 62%.

To minimize differences in chemical composition among cultivars and rootstocks, fruits were harvested from plants sharing the same environmental, cultural and soil conditions during growth. At each harvest date, two sets of healthy fruit at the same maturity stage (fruit exhibiting a yellow color on more than 2/3 of its surface) were taken at random from the field.

Fruit storage and analysis were conducted at Universidade Federal do Ceará, Fortaleza, Ceará State Brazil. Upon arrival to the laboratory, the fruits were sorted for shape, size, freedom from defects and sunburn. The fruits were then rinsed with water at room temperature, air-dried, and stored under refrigerated conditions (12±1°C, 90±2% R.H.). Assessment for non-destructive and destructive analyses was performed immediately prior to storage and every four days for twenty-four days of storage. The non-destructive measurements were cumulative mass loss and surface color. The destructive measurements included peel resistance to puncturing, thickness of the peel, pulp color, soluble solids, titratable acidity, pH, maturity index, ascorbic acid, and total soluble sugars.

Cumulative weight loss was determined by weighing each previously numbered fruit during storage and then calculating its mass loss relative to its mass at harvest. Peel color was determined using a Minolta Chroma Meter (Minolta Corporation Instrument Systems). Values were scored from the opposite sides of the fruit. The results were expressed as luminosity, chroma and hue angle. Peel resistance to puncturing was assessed with a force gauge model DD-200 fitted with a 5 mm cylindrical probe in the probe carrier. The grapefruits were placed upon a flat plate with the stem calyx axis parallel to the surface of the plate. Puncture resistance, expressed as N, was recorded as the minimum force required to puncture the fruit surface.

Juice was extracted from fruit halves using an automated squeezer with a rotating rod, filtered and homogenized before each analysis. Soluble solids (SS) were measured with a digital refractometer (0–45°Brix) (Palette 100, Atago, Co., Ltd.) and expressed as degree Brix. Acidity was determined by titration with 0.1 N NaOH with an endpoint of pH 8.2 and expressed as mg citric acid 100 g juice^{-1}; pH was measured with a pH meter directly in the juice. The maturity index was calculated by dividing SS by acidity. Ascorbic acid was measured following the methodology reported by Hernádez et al. (2006).

The experiment was designed as a randomized complete block with a split-plot arrangement. The classes in the ANOVA were scion-rootstock combination (main plot) and storage durations (sub-plot), with four replicates. Linear regressions were used to describe trends during storage ($R^2 \geq 70\%$) according to the procedure available in SAS. Mean separation was performed with Tukey test ($p \leq 0.05$). Comparisons between rootstocks at harvest were performed by the Student's t-test. Differences at $p \leq 0.05$ were considered significant.

Results and discussion

The weight of 'Ruby Red' fruit at harvest varied with the rootstock ($p \leq 0.05$), but this variation was dependent on the time of harvest; a significant difference was observed only for fruit harvested in August (Table 1). The weight of 'Star Ruby' grapefruit at harvest did not differ with rootstock at either harvest time ($p \leq 0.05$) (Table 2). In general, fruit harvested in October were heavier than those harvested in August.

Table 1. Differences in the quality parameters of freshly harvested 'Ruby Red' grapefruit grafted on 'Cleopatra' mandarin (CM) or 'Swingle' citrumelo (SC) rootstocks, harvested in August and October (Limoeiro do Norte, Ceará State, Brazil, 2010).

Quality parameters	Time of harvest					
	August			October		
	CM	SC	p-value	CM	SC	p-value
Weight	290.00	383.33	0.035**	438.67	448.0	0.875
Brightness	55.92	55.92	1.000	60.18	60.58	0.689
Chromaticity	26.40	26.50	1.000	29.71	35.08	0.001*
Hue angle	72.10	72.00	1.000	70.34	54.95	0.006*
Titratable acidity	1.36	1.65	0.052	1.50	0.93	0.177
Soluble solids	13.77	10.88	0.008*	11.50	13.63	0.109
SS:AT	10.08	6.58	0.001*	8.27	14.75	0.024**
pH	3.13	2.90	0.015**	3.10	3.51	0.027**
Ascorbic acid	51.74	54.65	0.194	61.19	66.48	0.642
Juice percentage	31.99	28.94	0.005*	32.52	38.74	0.097

* = Significant at 1% probability by Student´s Test; ** = Significant at 5% probability by Student´s Test.

Table 2. Differences in the quality parameters of freshly harvested 'Star Ruby' grapefruit grafted on 'Cleopatra' mandarin (CM) or 'Swingle' citrumelo (SC) rootstocks, harvested in August and October (Limoeiro do Norte, Ceará State, Brazil, 2010).

Quality parameters	Time of harvest					
	August			October		
	CM	SC	p-value	CM	SC	p-value
Weight	315.67	329.33	0.666	464.00	503.9	0.079
Brightness	54.63	53.68	0.507	58.84	6065	0.319
Chromaticity	27.37	29.52	0.132	30.29	29.83	0.540
Hue angle	70.90	60.62	0.276	65.89	67.03	0.665
Titratable acidity	1.69	1.23	0.004*	1.26	1.26	0.983
Soluble solids	11.37	9.67	0.073	12.47	12.13	0.581
SS:TA	6.76	7.54	0.295	9.99	8.82	0.132
pH	2.92	2.97	0.028**	3.27	3.16	0.227
Ascorbic acid	53.55	56.27	0.3019	67.11	58.28	0.084
Juice percentage	28.79	31.97	0.5841	32.94	31.67	0.7333

* = Significant at 1% probability by Student´s Test; ** = Significant at 5% probability by Student ´s Test.

During storage, cumulative weight loss increased with time, following a first-order polynomial equation for both cultivars harvested at either time (Figure 1).

For fruit harvested in August, rootstock played a role in cumulative weight loss. Grapefruits from both cultivars grafted on 'Cleopatra' mandarin lost a higher percentage of their weights during storage than did those grafted on 'Swingle' citrumelo rootstock ($p \leq 0.05$) (Figure 1). The cumulative weight loss of fruit harvested in October did not differ significantly ($p \leq 0.05$) between cultivars on both rootstocks, but it was affected over the course of storage. These results suggest that the effect of the scion-rootstock interaction on weight loss depends on the time of harvest.

In general, weight loss did not surpass 2.5% of the initial weight. This relatively low loss of weight may be due to the appropriate conditions during storage (12±1°C, 90±2% R.H.). The positive effects of low temperature and high relative humidity during storage on reducing weight loss in grapefruit have also been reported by Alferez et al. (2010).

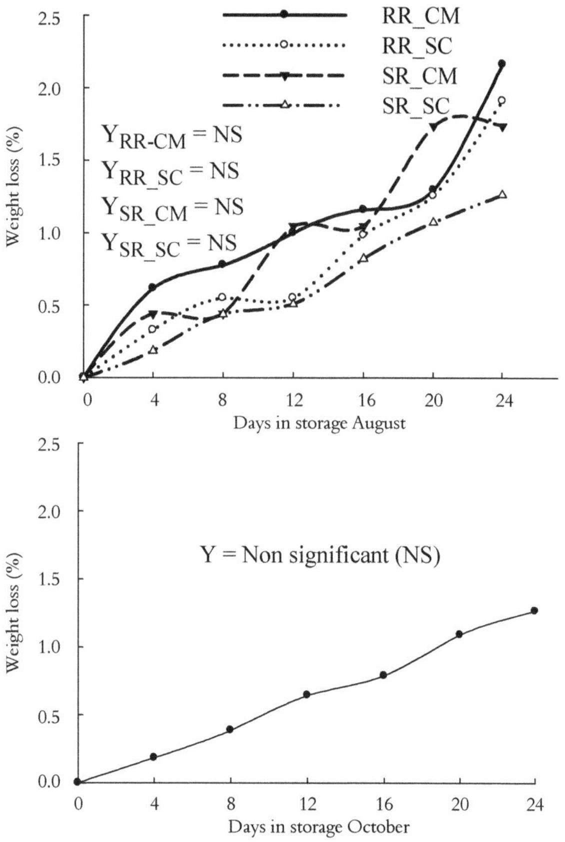

Figure 1. Weight loss (%) of 'Ruby Red' grapefruit grafted on 'Cleopatra' mandarin (RR-CM) and 'Swingle' citrumelo (RR-SC) rootstock and 'Star Ruby' grapefruit grafted on 'Cleopatra' mandarin (SR-CM) and 'Swingle' citrumelo (SR-SC) rootstock harvested in August and October and cold stored (Limoeiro do Norte, Ceará State, Brazil, 2010).

The influence of rootstock on the hue angle of the peel color of 'Ruby Red' fruit at harvest was dependent on the time of harvest, and a significant difference was observed only for 'Ruby Red' fruit harvested in October (Table 1). Fruit from 'Ruby Red' grafted on 'Cleopatra' mandarin had a yellow color, while fruit from 'Ruby Red' grafted on 'Swingle' citrumelo had a pink peel color (Figure 2). The peel color of 'Star Ruby' grapefruit at harvest did not differ with rootstock at either harvest date ($p \leq 0.05$) (Table 2).

During storage, the average hue angle values of fruit from both cultivars at both harvest times presented similar trends, with values decreasing linearly with time (Figure 2). The hue angle of the peel of fruit harvested in August did not differ significantly ($p \leq 0.05$) between cultivars on both rootstocks (Table 1 and Table 2) and decreased slightly with time. However, there were significant differences in hue angle for 'Ruby Red' fruit harvested in October (Table 1). Hue angle differences observed at harvest for fruit harvested in October were maintained during storage (Figure 2).

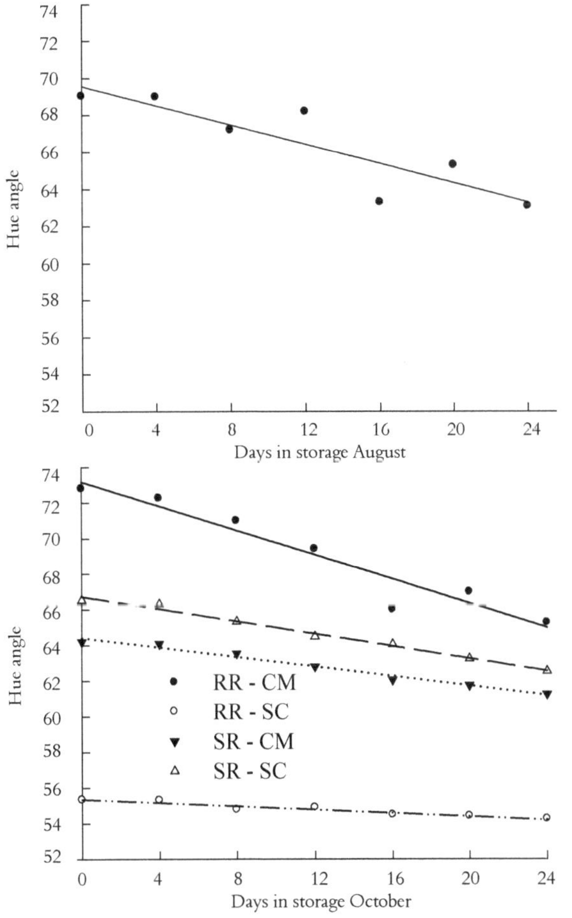

Figure 2. The hue angles of 'Ruby Red' grapefruit grafted on 'Cleopatra' mandarin (RR-CM) and 'Swingle' citrumelo (RR-SC) rootstock and 'Star Ruby' grapefruit grafted on 'Cleopatra' mandarin (SR-CM) and 'Swingle' citrumelo (SR-SC) harvested in August and October and cold stored (Limoeiro do Norte, Ceará State, Brazil, 2010).

Polynomial equations:

August: $Y = 69.55 - 0.26x$; $R^2 = 77.11\%$

October: $Y_{RR\text{-}CM} = 73.18 - 0.34x$; $R^2 = 92.61\%$; $Y_{RR\text{-}SC} = 55.35 - 0.47x$, $R^2 = 92.64\%$; $Y_{SR\text{-}CM} = 64.41 - 0.133x$; $R^2 = 97.91\%$; $Y_{SR\text{-}SC} = 66.71 - 0.17x$; $R^2 = 98.89\%$;

as observed for hue angle, the influence of the rootstock on the peel chromaticity of 'Ruby Red' fruit was dependent on time of harvest; a significant difference ($p \leq 0.05$) was observed only in fruit harvested in October (Table 1). The peel chromaticity of the 'Star Ruby' grapefruit at harvest did not differ between rootstocks at either harvest time ($p \leq 0.05$) (Table 2). During storage, the chromaticity in both cultivars at both harvest times exhibited similar trends, with values increasing linearly during storage (Figure 3). Peel chromaticity in August did not differ significantly ($p \leq 0.05$) between cultivars on both rootstocks, increasing with storage time. Fruit from both cultivars grafted on either rootstock and harvested in October showed significantly higher ($p \leq 0.05$) average chromaticity values at harvest, consequently exhibiting more vivid peel color (Figure 3).

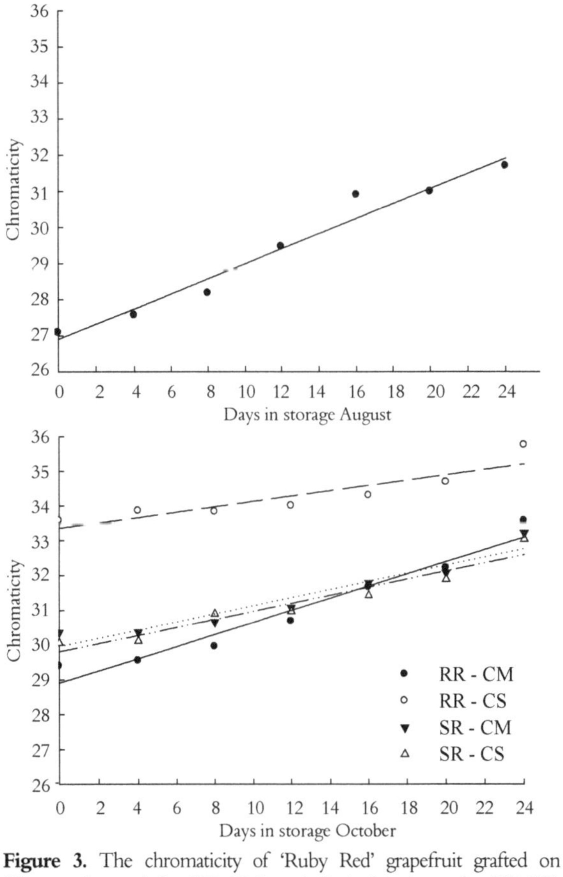

Figure 3. The chromaticity of 'Ruby Red' grapefruit grafted on 'Cleopatra' mandarin (RR-CM) and 'Swingle' citrumelo (RR-SC) rootstock and 'Star Ruby' grapefruit grafted on 'Cleopatra' mandarin (SR-CM) and 'Swingle' citrumelo (SR-SC) harvested in August and October and cold stored (Limoeiro do Norte, Ceará State, Brazil, 2010).

Polynomial equations:

August: $Y = 26.9 + 0.21x$; $R^2 = 96.51\%$

October: $Y_{RR-CM} = 28.9 + 0.17x$; $R^2 = 94.97\%$; $Y_{RR-SC} = 33.35 + 0.07x$, $R^2 = 81.25\%$; $Y_{SR-CM} = 29.95 + 0.11x$; $R^2 = 91.68\%$; $Y_{SR-SC} = 29.8 + 0.11x$; $R^2 = 92.68\%$;

peel brightness at harvest did not vary with rootstock or time of harvest for either cultivar (Tables 1 and 2). The pattern of change in peel brightness during storage (Figure 4) was similar to that observed for chromaticity (Figure 3) and the reverse of that of hue angle (Figure 2), indicating that the peel color became lighter as brightness values increased during storage for both harvest times.

Fruit harvested in August showed significantly lower ($p \leq 0.05$) average brightness values than those observed for fruit harvested in October (Figure 4). Thus, fruits harvested in August had a darker and grayer peel in comparison with those harvested in October. Considering that peel color is a major factor for consumer decisions, and assuming that a glowing, reddish color is more appealing to the human eye, 'Ruby Red' grafted on 'Swingle' citrumelo rootstock harvested in October stands out for its shining, vivid, pinkish colored fruit. These results corroborate our previous study (MACHADO et al., 2011), in which fruit from 'Ruby Red' or 'Star Ruby' grapefruit trees grafted on 'Swingle' citrumelo and harvested in October exhibited a bright, vivid, pinkish peel color.

Significant differences in peel color at harvest among the four scion-rootstock combinations observed in the October harvest may be the result of the interaction of the genotypes with climate factors during development and at time of harvest because all fruit shared the same environment, cultivation practices, and soil conditions. Additionally, the interactions between the genotypes and physiological factors, such as fruit load, may account for differences in peel color.

Rootstock and time of harvest affected the soluble solids content of 'Ruby Red' grapefruit, and Ruby Red' grafted on 'Cleopatra' mandarin harvested in August yielded the fruit with the highest soluble solids content (Table 1).

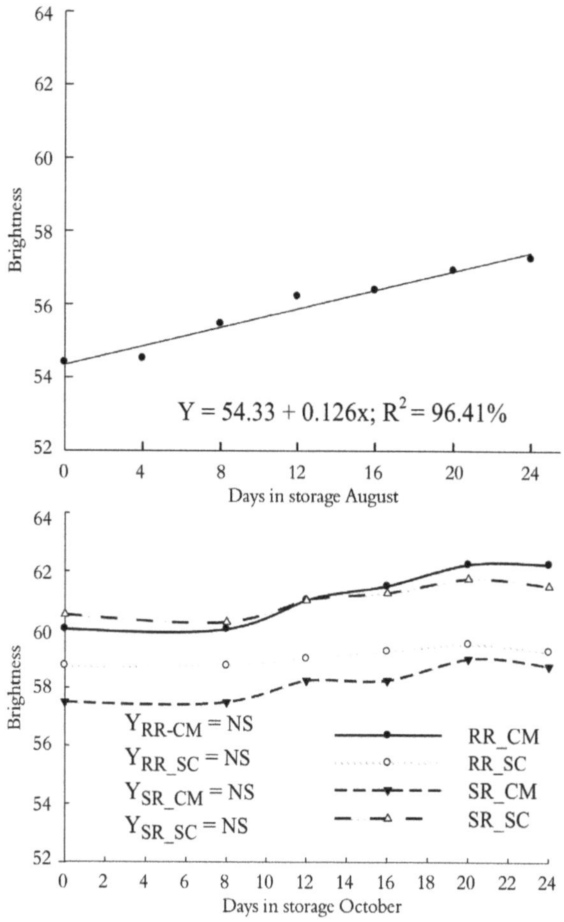

Figure 4. Brightness of 'Ruby Red' grapefruit grafted on 'Cleopatra' mandarin (RR-CM) and 'Swingle' citrumelo (RR-SC) rootstock and 'Star Ruby' grapefruit grafted on 'Cleopatra' mandarin (SR-CM) and 'Swingle' citrumelo (SR-SC) harvested in August and October and cold stored (Limoeiro do Norte, Ceará State, Brazil, 2010).

Both 'Ruby Red' and 'Star Ruby' grapefruit grafted on both 'Swingle' citrumelo and 'Cleopatra' mandarin showed similar trends in soluble solid content over the storage period. The rootstocks did not influence the soluble solids content, and no significant differences were observed along the storage period (Figure 5).

The rootstock affected the acidity content of 'Star Ruby' grapefruit harvested in August (Table 2). The most acidic fruits at harvest were produced by the 'Star Ruby' and 'Cleopatra' mandarin scion-rootstock combination, which exhibited 1.69 mg of citric acid 100 g-1 juice (Figure 6). Although the acidity of this scion-rootstock combination increased during storage, no significant difference was observed over the course of storage durations.

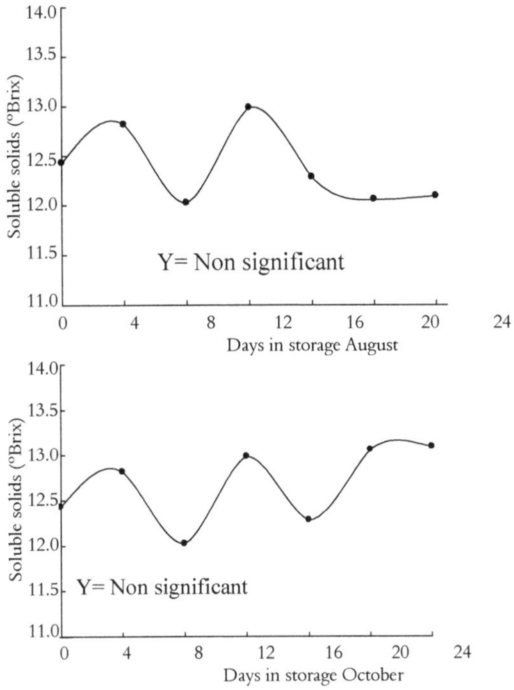

Figure 5. Soluble solids content (°Brix) in 'Ruby Red' grapefruit grafted on 'Cleopatra' mandarin (RR-CM) and 'Swingle' citrumelo (RR-SC) rootstock and 'Star Ruby' grapefruit grafted on 'Cleopatra' mandarin (SR-CM) and 'Swingle' citrumelo (SR-SC) harvested in August and October and cold stored. (Limoeiro do Norte, Ceará State, Brazil, 2010).

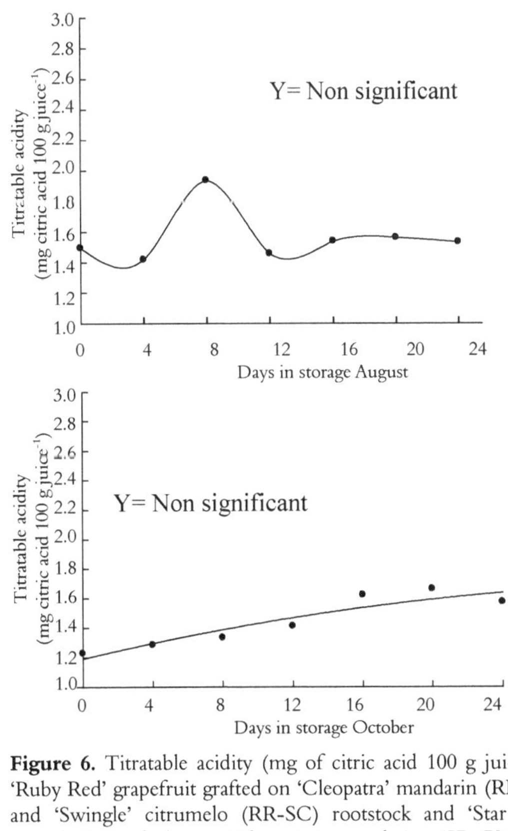

Figure 6. Titratable acidity (mg of citric acid 100 g juice^{-1}) of 'Ruby Red' grapefruit grafted on 'Cleopatra' mandarin (RR-CM) and 'Swingle' citrumelo (RR-SC) rootstock and 'Star Ruby' grapefruit grafted on 'Cleopatra' mandarin (SR-CM) and 'Swingle' citrumelo (SR-SC) harvested in August and October and cold stored. (Limoeiro do Norte, Ceará State, Brazil, 2010).

The fruit pH at harvest was affected by the scion-rootstock combination for both 'Ruby Red' and 'Star Ruby' cultivars (Tables 1 and 2). Fruit grown on 'Ruby Red' grapefruit grafted on 'Cleopatra' mandarin and harvested in August produced higher pH values at harvest and during storage than did fruit grown on 'Ruby Red' grapefruit grafted on 'Swingle' citrumelo. However, this pattern was not observed for fruit harvested in October (Figure 7).

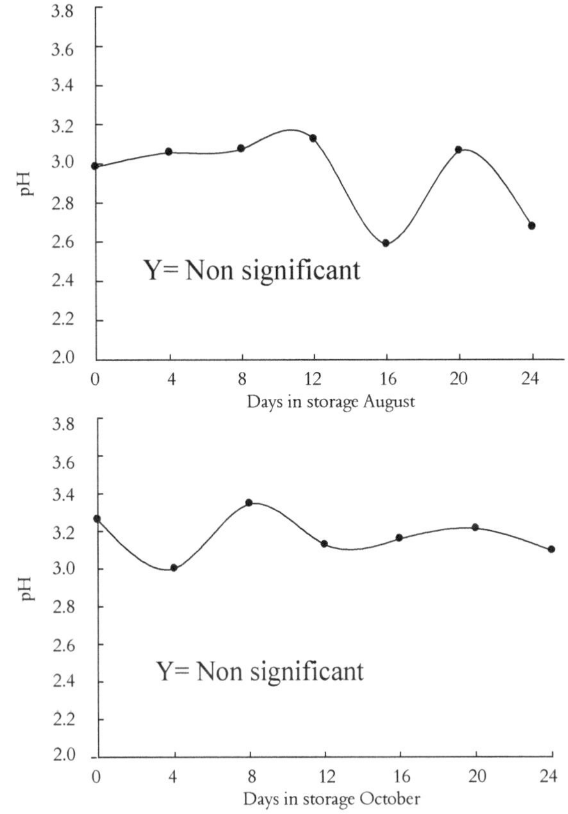

Figure 7. The pH of 'Ruby Red' grapefruit grafted on 'Cleopatra' mandarin (RR/CM) or 'Swingle' citrumelo (RR/SC) rootstock and 'Star Ruby' grapefruit grafted on 'Cleopatra' mandarin (SR/CM) and 'Swingle' citrumelo (SR/SC) harvested in August and October and cold stored. (Limoeiro do Norte, Ceará State, Brazil, 2010).

The maturity index of 'Ruby Red' fruit at harvest was affected by the rootstock and the time of harvest. The combination of 'Ruby Red' fruit and the 'Cleopatra' mandarin rootstock yielded a significantly higher ($p \leq 0.05$) soluble solids to acidity ratio in August. In October, however, the highest ratio was produced by the 'Ruby Red' and 'Swingle' citrumelo combination (Table 1). Indeed, October-harvested fruit exhibited higher acidity, while the soluble solids content was similar to that of August-harvested fruit. The maturity index of 'Star Ruby' grapefruit was not affected by rootstock or time of harvest (Table 2).

It is important to highlight that both cultivars had acceptable maturity indices at both harvests (Tables 1 and 2) and over the course of storage (Figure 8), meeting or surpassing existing international standards. In practical terms, both cultivars grafted on either rootstock produced excellent fruit, validated by the high maturity indices, which were similar to or higher than those reported for cultivars such as 'Rio Red,' 'Flame', and 'Ray Ruby' grown in Mexico (BECERRA-RODRÍGUEZ et al., 2008) and for 'Star Ruby' and 'Rio Red' grown in Pernambuco State, Brazil (LEDERMAN et al., 2005).

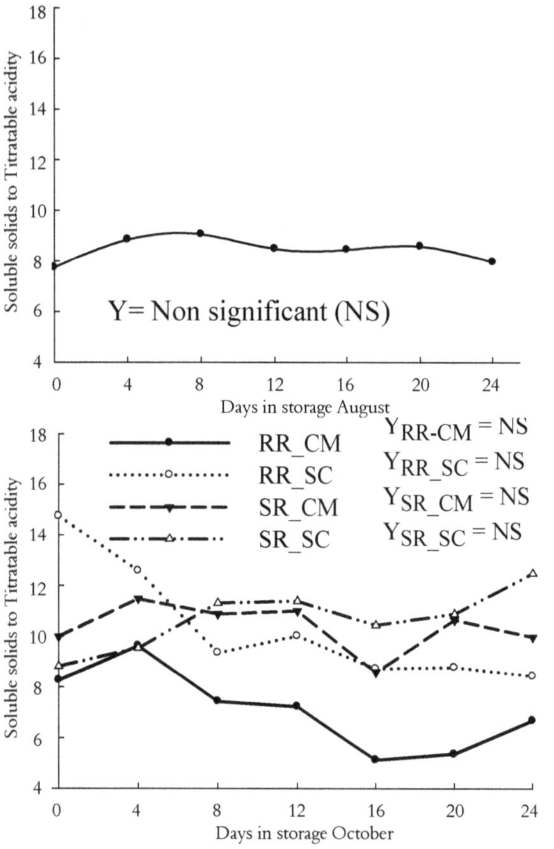

Figure 8. Maturity indices of 'Ruby Red' grapefruit grafted on 'Cleopatra' mandarin (RR-CM) and 'Swingle' citrumelo (RR-SC) rootstock and 'Star Ruby' grapefruit grafted on 'Cleopatra' mandarin (SR-CM) and 'Swingle' citrumelo (SR-SC) harvested in August and October and cold stored (Limoeiro do Norte, Ceará State, Brazil, 2010).

Although October-harvested fruit had higher ascorbic acid levels at harvest, no significant differences between rootstocks were observed for either cultivar at either time of harvest (Tables 1 and 2).

The influence of genotype on ascorbic acid content during storage was observed only in fruit harvested in October ($p \leq 0.05$). It is worth mentioning that fruit grown on 'Cleopatra' mandarin rootstock exhibited a greater increase in ascorbic acid during storage than that grown on Swingle citrumelo rootstock (Figure 9). Differences in ascorbic acid content have been shown to be the result of preharvest climatic conditions, genotype, and temperature.

Ascorbic acid content increased during storage for fruit harvested at either time. The average values at each time point during storage, as shown in Figure 9, surpassed the ascorbic acid values reported by Machado et al. (2011) for fruit harvested from 'Ruby Red' and 'Star Ruby' grafted on 'Swingle' citrumelo. That study also reported an increase in ascorbic acid during storage.

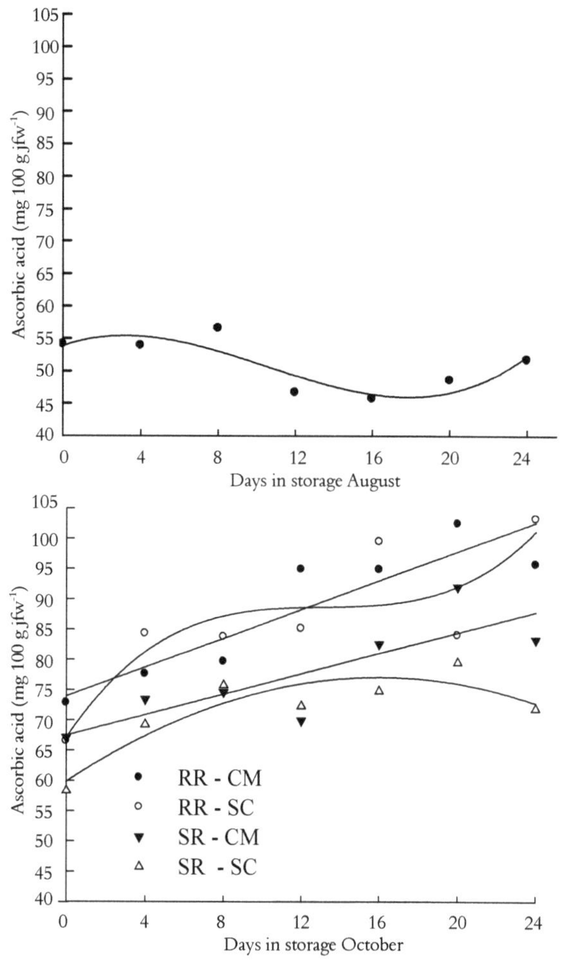

Figure 9. The ascorbic acid contents of 'Ruby Red' grapefruit grafted on 'Cleopatra' mandarin (RR-CM) and 'Swingle' citrumelo (RR-SC) rootstock and 'Star Ruby' grapefruit grafted on 'Cleopatra' mandarin (SR-CM) and 'Swingle' citrumelo (SR-SC) harvested in August and October and cold stored (Limoeiro do Norte, Ceará State, Brazil, 2010).

Polynomial equation:

August: $Y = 53.88 + 1.036x - 0.19x^2 + 0.006\,x^3$; $R^2 = 74.63\%$

October: $Y_{RR-CM} = 63.50 + 1.65x$; $R^2 = 95.69\%$; $Y_{RR-SC} = 67.15 + 4.82x - 0.36x^2 + 0.009\ x^3$, $R^2 = 75.71\%$; $Y_{SR-CM} = 65.33 + 1.16x$; $R^2 = 83.33\%$; $Y_{SR-SC} = 59.85 + 2.15x - 0.067x^2$; $R^2 = 82.89\%$;

'Ruby Red' grapefruit grafted on 'Cleopatra' mandarin rootstock produced significantly higher juice percentages in August (Table 1). Overall, the time of harvest influenced juice percentage, and fruits harvested in October were juicier than those harvested in August (Tables 1 and 2).

No decline in percent juice over the course of storage was observed from harvest to the end of storage for both harvests ($p \leq 0.05$). In general, the juice percentages found in this study were slightly lower than those reported for 'Ray Ruby' and 'March Red' grapefruits grown under dry tropical conditions in Colima, Mexico (BECERRA-RODRÍGUEZ et al., 2008).

Conclusion

October-harvested 'Ruby Red' grapefruit grafted on 'Swingle' citrumelo are redder with more vivid peel color.

The sweetest and least acid grapefruit is produced by the combination of 'Ruby Red' grapefruit grafted on 'Cleopatra' mandarin rootstock, harvested in August or 'Ruby Red' grapefruit grafted on 'Swingle' citrumelo rootstock, harvested in October.

October-harvested fruit had higher levels of ascorbic acid at harvest, and 'Cleopatra' mandarin rootstock produced the greatest increase in ascorbic acid over the course of storage.

October-harvested fruit are juicer than August-harvested fruit. Juice content does not decline during storage under the stated storage conditions.

Acknowledgements

The authors thank the Banco do Nordeste do Brasil (BNB) for financial support and the Fundação Cearense de Apoio a Pesquisa e Desenvolvimento (FUNCAP) for granting a scholarship to the first author. We also thank Anne Mary Gillen, PhD, for greatly contributing to the preparation of this manuscript and Professor Patrícia Verônica Pinheiro Sales for assistance with the statistical analyses.

References

ALFEREZ, F.; ALQUEZAR, B.; BURNS, J. K.; ZACARIAS, L. Variation in water, osmotic and turgor potential in peel of 'Marsh' grapefruit during development of postharvest peel pitting. **Postharvest Biology and Technology**, v. 56, n. 1, p. 44-49, 2010.

BECERRA-RODRÍGUEZ, S.; MEDINA-URRUTIA, V. M.; ROBLES-GONZÁLEZ, M. M.; WILLIAMS, T. Performance of various grapefruit (*Citrus paradisi* Macf.) and pummelo (*C. maxima* Merr.) cultivars under the dry tropic conditions of Mexico. **Euphytica**, v. 164, n. 1, p. 27-36, 2008.

CANO, A.; BERMEJO, A. Influence of rootstock and cultivar on bioactive compounds in citrus peels. **Journal of Science and Food Agriculture**, v. 91, n. 9, p. 1702-1711, 2011.

CHEN, P.; HARNLY, J. M.; LESTER, G. L. Flow Injection Mass Spectral Fingerprints Demonstrate Chemical Differences in Rio Red Grapefruit with Respect to Year, Harvest Time, and Conventional versus Organic Farming. **Journal of Agricultural and Food Chemistry**, v. 58, n. 8, p. 4545-4553, 2010.

GIRENNAVAR, B.; JAYAPRAKASHA, G. K.; JIFON, J. L.; PATIL, B. S. Variation of bioactive furocoumarins and Xavonoids in different varieties of grapefruits and pummelo. **European Food Research and Technology**, v. 226, n. 6, p. 1269-1275, 2008.

HERNÁDEZ, Y.; LOBO, M. G.; GONZÁLEZ, M. Determination of vitamin C in tropical fruits: A comparative evaluation of methods. **Food Chemistry**, v. 96, n. 4, p. 654-664, 2006.

LEDERMAN, I. E.; BEZERRA, J. E. F.; ALVES, M. A.; COELHO, Y. S.; SOBRINHO, A. P. C. Comportamento de seis variedades de pomelo (*Citrus paradisi Macfad.*) cultivadas sob irrigação na regiao semi-arida de Pernambuco. **Revista Brasileira de Fruticultura**, v. 27, n. 1, p. 245-247, 2005.

LEE, S. K.; KADER, A. A. Preharvest and postharvest factors influencing vitamin C content of horticultural crops. **Postharvest Biology and Technology**, v. 20, n. 3, p. 207-220, 2000.

MACHADO, F. L. C.; OLIVEIRA, V. S.; COSTA, J. M. C. Postharvest quality of grapefruit (*Citrus paradisi* Macf.) produced from initial plantings in Ceará state, Brazil. **Revista Ciência Agronômica**, v. 42, n. 4, p. 883-888, 2011.

PAILLY, O.; TISON, G.; AMOUROUX, A. Harvest time and storage conditions of 'Star Ruby' grapefruit (*Citrus paradisi* Macf.) for short distance summer consumption. **Postharvest Biology and Technology**, v. 34, n. 1, p. 65-73, 2004.

WU, T.; GUAN, Y.; YE, J. Determination of flavonoids and ascorbic acid in grapefruit peel and juice by capillary electrophoresis with electrochemical detection. **Food Chemistry**, v. 100, n. 4, p. 1573-1579, 2007.

14

Rhizogenic behavior of black pepper cultivars to indole-3-butyric acid

Welington Secundino[1], Rodrigo Sobreira Alexandre[2*], Edilson Romais Schmildt[2], Omar Schmildt[2], Gizele Cristina Magevski[2] and João Paulo Rodrigues Martins[3]

*[1]Instituto Capixaba de Pesquisa, Assistência Técnica e Extensão Rural, São Mateus, Espírito Santo, Brazil. [2]Centro Universitário Norte do Espírito Santo, Universidade Federal do Espírito Santo, Rod. BR-101 Norte, km 60, 29932-540, São Mateus, Espírito Santo, Brazil. [3]Departamento de Biologia, Universidade Federal de Lavras, Lavras, Minas Gerais, Brazil. *Author for correspondence. E-mail: rodrigosobreiraalexandre@gmail.com*

ABSTRACT. Little information is available regarding vegetative propagation of the species *Piper nigrum* L. to generate technical recommendations for the production of seedlings on a commercial scale. The purpose of this study was to investigate the rhizogenic behavior of cultivars of this species regarding indol-3-butyric acid (IBA). The experiment was performed at a vegetation house equipped with an intermittent nebulization irrigation system. The experimental site was located in the University Center of Northern Espírito Santo (CEUNES) of the Federal University of Espírito Santo (UFES), Brazil. The experimental design consisted of randomized complete blocks design arranged in a 3 x 5 factorial scheme: three cultivars (Bragantina, Iaçará and Guajarina) x five IBA concentrations (0; 1,500; 3,000; 4,500 and 6,000 mg kg^{-1}), with four repetitions of 16 cuttings each. Total immersion of the cuttings in IBA is recommended for the Iaçará and Guajarina cultivars, and immersion of only the basal region is recommended for cv. Bragantina. The recommended IBA concentration for these cultivars is 4,000 mg kg^{-1}.

Keywords: *Piper nigrum* L., genotypes, propagation, cutting.

Introduction

In Brazil, black pepper, in addition to being a valuable spice, has shown other applications in research as a larvicide to fight *Aedes aegypti* (SIMAS et al., 2007; GRZYBOWSKI et al., 2012) and for the bactericide properties of the essential oil of its seeds (TRAJANO et al., 2009).

Black pepper is very important both socially and economically for the northern region of the State of Espírito Santo, especially for its principal producing city, São Mateus. In this State, black pepper is grown on 2,725 ha, producing up to 6,371 ton, with an average yield of 2,723 kg ha^{-1}; this State is second in the national production ranking, surpassed only by the State of Pará, where black pepper production occupies an area of 17,605 ha, producing up to 38,472 ton, with an average yield of 2,185 kg ha^{-1} (IBGE, 2011). The position of the State of Espírito Santo relative the State of Pará, even though the planted area is significantly smaller, still indicates that we can improve in a few aspects, including the quality of seedlings. Improving the seedling quality starts by selecting the vegetal material and implies the maintenance of the mother plant characteristics, such as high productivity, vigor, absence of diseases, etc.

Black pepper is propagated through cuttings, which, unlike sexual reproduction, maintain an identical genetic constitution in the descendants of the mother plant, with essentially no signs of segregation and genetic combination (ARAÚJO;

BRUCKNER, 2008). Vegetative propagation is an important strategy to improve plants because it reduces the time spent in this process. Among the various methods of vegetative propagation, cutting is the most commonly used because it is an easy technique; thus, it is applied to multiply selected genotypes to maintain the characteristics of the matrix plant, reduce the juvenile period, and increase the plant's production (HARTMANN et al., 2011). Vegetative propagation process through cutting occurs via the appearance of adventitious roots, defined as post embryonic, which appear from the stem and leaves. Various exogenous and endogenous factors regulate the formation of adventitious roots, including auxins, Ca^{2+}, sugars, polyamines, ethylene, nitric oxide, hydrogen peroxide, carbon monoxide, cyclic guanosine monophosphate (cGMP), mitogen-activated protein kinase (MAPKs), peroxidase, phenols, etc. Such mediators seem to function in the signalization route during adventitious root formation (QADDOURY; AMSSA, 2003; LI et al., 2009; LI; XUE, 2010; LIMA et al., 2011a).

Auxins have aided in clonal propagation in various situations, such as declines in the formation of adventitious roots with increasing plant age (HUNT et al., 2011; SINGH et al., 2011), recalcitrant rooting genotypes (LEMAY et al., 2009) and low rooting during the cold seasons of the year (LEITE; MARTINS, 2007).

Only a few scientific papers can be found regarding the vegetative propagation of *P. nigrum* species (FRANCIS et al., 1993; THANUJA et al., 2002; THANKAMAN et al., 2008; SHARANGI; KUMAR, 2011). Although cutting is a routine technique among producers, no scientific research was found regarding the vegetative propagation of cvs. Bragantina, Iaçará and Guajarina.

Scientific information is scarce regarding this cloning technique and cultivars of *P. nigrum* in Brazil. Thus, the purpose of this research was to identify the ideal IBA concentration for rhizogenesis in cuttings of three cultivars that are economically significant in Northern Espírito Santo.

Material and methods

The work was conducted at a vegetation house equipped with an intermittent nebulization irrigation system at a relative humidity of 85 ± 5°C. The site was part of the Biologic and Agricultural Sciences Department of the Northern Espírito Santo University Center (CEUNES) of the Federal University of Espírito Santo (UFES), São Mateus, Espírito Santo State. The experiment was conducted from 9/29/2011 to 11/30/2011.

Three-year-old orthotropic herbal branches from adult plants in the field were used for the cuttings, which were performed using a pruning shear. Each removed branch was immersed in a styrofoam box containing distilled water to keep the tissue hydrated and prevent desiccation and thus avoid compromising the quality of the grafts in the rooting process. After collecting a sufficient number of branches, the branches were immediately transported to the vegetation house. Next, the branches were sanitized; in this process, cuttings with two knots and one leaf were individualized, and a simple bezel was made in the basin of the selected cuttings. After preparing a sufficient number of cuttings, the cuttings, with the exception of the inert talc sheets, were completely soaked with different concentrations of indole-3-butyric acid (IBA). For each treatment, the auxin was weighed on an analytical balance and immediately mixed with inert talc. After each stake was treated for three seconds with different concentrations of the auxin and inert talc formulation, the cuttings were planted in polyethylene trays containing vermiculita (Dimy®, Paraná State, Brazil) substrate [particle size distribution (%): > 4.75 mm (1.32%), 2.00 at 4.75 mm (17.14%), 1.00 at 2.00 mm (26.10%), 0.5 at 1.00 mm (22.19%), 0.25 at 0.5 mm (16.05%) and < 0.25 mm (17.21%)]. The substrate was irrigated before the cuttings were planted and was subsequently irrigated by the irrigation system. The cuttings were cultivated in a covered vegetation greenhouse for a period of 60 days.

The characteristics assessed in this experiment were survival (%); nodal, basal and total rooting (%); number of nodal, basal and total roots; length of nodal, basal and total roots (cm); volume of nodal, basal and total roots (cm^3); and dry matter mass from the nodal, basal and total roots (mg).

The design of the experiments used randomized blocks with a 3 x 5 factor system: three cultivars (Bragantina, Iaçará and Guajarina) x five IBA concentrations (0; 1,500; 3,000; 4,500 and 6,000 mg kg^{-1}), with 4 repetitions of sixteen cuttings each.

The averages of the variables were statistically analyzed using the F test, regression test and Tukey average test. The analyses were performed using the GENES (CRUZ, 2013) computer program.

Results and discussion

The Bragantina, Iaçará and Guajarina cultivars presented a high survival capacity of more than 99.3% with the use of an estimated IBA concentration of 3561 mg kg^{-1} (Figure 1). Sharangi et al. (2010) achieved a higher survival rate in *P. nigrum* for median cuttings with two knots treated with 100 mg L^{-1} of IBA. Working with the *P. amplum*, *P. arboreum* and *P.* sp., Magevski et al. (2011) also observed high rates of survival (98, 100 and 89%, respectively) regardless of the IBA concentration. Various factors are responsible for the maintenance of live cuttings during the adventitious rooting process, including genotype; high rooting percentage; the quality of the radicular system, including the number, volume and dry matter mass; and other factors, such as tissue consistency and reduced contamination of the vegetal material from the field, its proper disinfection or even the physical structure and the irrigation system of the greenhouse. According to Mateja et al. (2007), the physiologic stress of the vegetal material can be minimized by using an intermittent nebulization irrigation system to enable a higher survival rate, as observed in this paper.

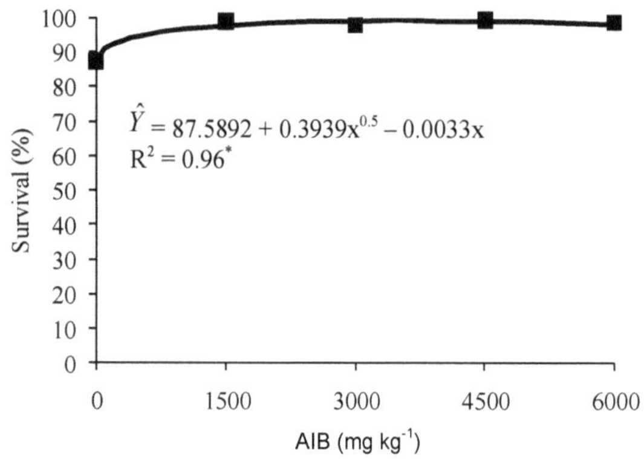

Figure 1. Survival (%) of *Piper nigrum* cvs. Bragantina, Iaçará and Guajarina cuttings treated with different concentrations of IBA. São Mateus, Espírito Santo State, 2012. *Significant at 5% probability by the F test.

Another factor that may be beneficial for the survival of *P. nigrum* cultivar cuttings is the presence of potential endodermal cells in the stem, as observed in other Piperaceous species, such as *Piper aduncum* (VIANNA; AKISUE, 1997), *Piper crassinervium* (ALBIERO et al., 2004), *Piper diospyrifolium*, *Peperomia dahlstedtii* (SOUZA et al., 2004) and *Peperomia parnassiflora* (OLIVEIRA et al., 2008). According to Mauseth (1988), these endodermic cells are important for controlling the water output cortex through foliar transpiration or water input provoked by the root pressure. This occurs mainly in the case of an abrupt loss of the leaves or during the preparation of the cuttings through the reduction of the foliar limbs to induce adventitious rooting, as observed in the present paper.

In *P. nigrum*, roots appear in the nodal area (leaf axil) (Figure 2Aa) and basin area slightly above the cut region (Figure 2Ab). In the Bragantina, Iaçará and Guajarina cultivars, the nodal regions of the cuttings present preformed roots, as was also observed by Souza et al. (2009) for the species *Piper amalago* and other species, such as *Citrus medica* and *Ribes* sp., which present latent preformed primitive roots that will form roots when placed in favorable conditions (FACHINELLO et al., 2005). The development of axillary gems is regulated by a positive signal produced by the roots, along with the inhibitive influences of the preexisting apex gems (apex dominance) (THOMAS; HAY, 2010). According to the authors, in the absence of nodal roots, the basal radicular system is unable to maintain the continuous growth of the sprout. According to Thomas and Hay (2008), stoloniferous plants present a physiologic link between the growth of nodal roots and the development of stem tissues.

Nodal rooting resulted in a crescent linear behavior for cv. Bragantina with maximum rooting (64.5%) achieved with 6,000 mg kg^{-1} of IBA (Figure 2B). This means that for this cultivar, higher IBA concentrations can increase the rooting in this region of the cutting. However, it would most likely be necessary to use even higher concentrations of the auxin to achieve maximum rooting, which may present an inconvenience. The Iaçará (98.7%) and Guajarina (91.8%) cultivars presented high percentages of nodal rooting that exhibited square root behavior and maxima at 3,351 and 3,866 mg kg^{-1} IBA, respectively (Figure 2B). Singh and Singh (1989) have also identified the node effect in *P. nigrum* cultivar Panniyur-1 cuttings, in which the presence of two nodes enabled maximum rooting. In other species, such as *Bambusa vulgaris,* nodal cuttings exhibited better rooting results when treated with 8,000 mg L^{-1} IBA (ISLAM et al., 2011). The presence of nodes in rose cuttings was also positive for adventitious rooting (PARK et al., 2010).

The basal rooting of the cuttings presented a square root behavior and maxima at 3,154; 3,166 and 3,351 mg kg^{-1}, with 88.5, 94.3 and 93.3% rooting for the Bragantina, Iaçará and Guajarina cultivars, respectively (Figure 2C). The most pronounced effect of IBA in the basal region of the Bragantina cultivar cuttings may have occurred due to the increased cellular plasticity in this location, possibly because of the increased biosynthesis of cellulase (RAVEN; CURTIS, 1975). This also means that if the IBA concentration is increased in the Bragantina cultivar to improve the effect in the nodal region, it can negatively interfere in the basal region, most likely due to phytotoxicity.

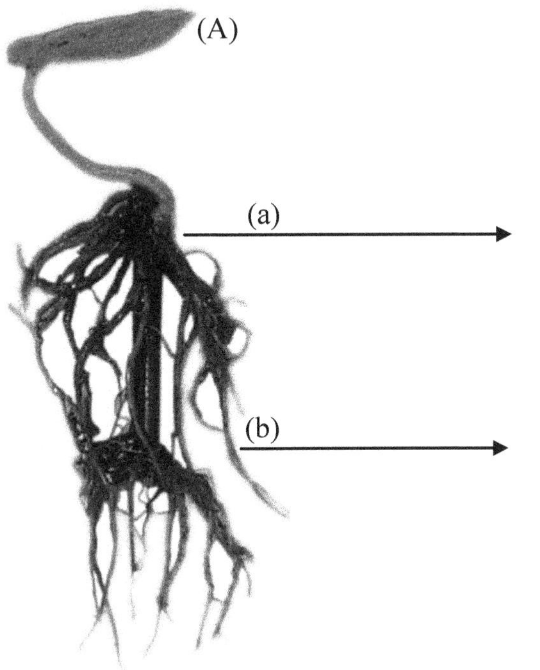

(B) Rooting nodal	R^2
$\hat{Y}_{Bragantina} = 10.7479 + 0.008963x$	0.73^{*}
$\hat{Y}_{Iaçará} = 39.9648 + 1.8609x^{0.5} - 0.0146x$	0.97^{*}
$\hat{Y}_{Guajarina} = 44.4511 + 1.7161x^{0.5} - 0.0138x$	0.97^{*}
(C) Rooting basal	R^2
$\hat{Y}_{Bragantina} = 10.5799 + 2.5613x^{0.5} - 0.0209x$	0.95^{*}
$\hat{Y}_{Iaçará} = 44.6172 + 1.6325x^{0.5} - 0.0133x$	0.95^{*}
$\hat{Y}_{Guajarina} = 2.3444 + 3.1052x^{0.5} - 0.0247x$	0.98^{**}
(D) Rooting total = (B) + (C)	R^2
$\hat{Y}_{Bragantina} = 13.1445 + 2.8224x^{0.5} - 0.0241x$	0.97^{*}
$\hat{Y}_{Iaçará} = 56.7599 + 1.4654x^{0.5} - 0.0126x$	0.96^{*}
$\hat{Y}_{Guajarina} = 42.6264 + 1.7527x^{0.5} - 0.0136x$	0.99^{**}

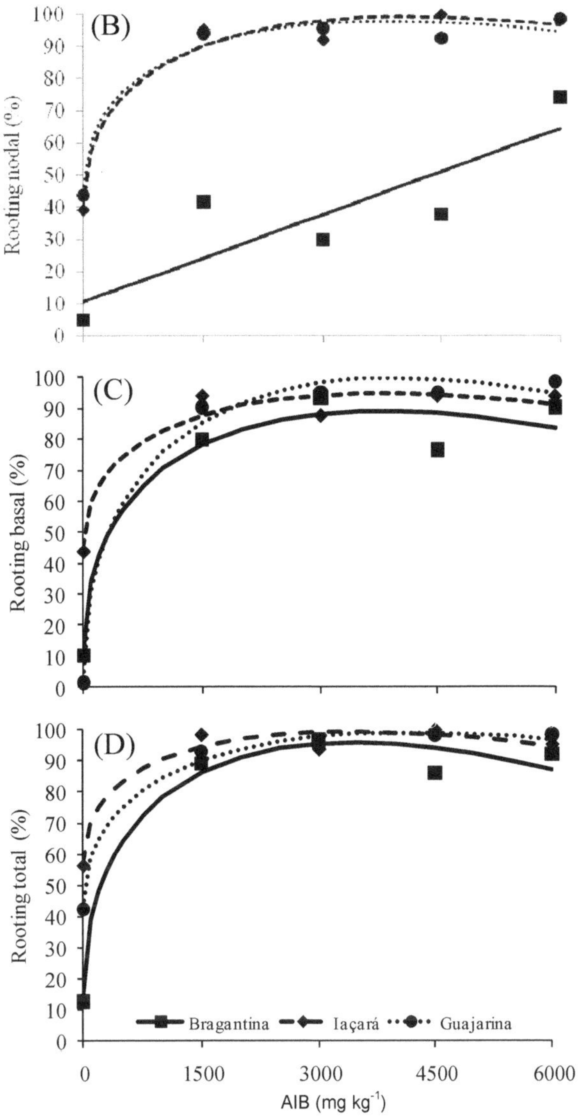

Figure 2. Rooted cutting of cv. Iaçará (A) with roots formed in the nodal (a) and basal (b) regions. Nodal (B), basal (C) and total (D=B+C) (%) rooting of cuttings of *Piper nigrum* cvs. Bragantina, Iaçará and Guajarina treated with different concentrations of IBA. São Mateus, Espírito Santo State, 2012. **,* Significant at 1 and 5% probability levels by the F test, respectively.

Total rooting in the absence of auxin was low for cvs. Bragantina (12%), Iaçará (56%) and Guajarina (42%) (Figure 2D and Table 1). Thanuja et al. (2002) also observed a low rooting value (25.75%) for the species *P. nigrum* cv. Panniyur-1 when growth regulators were not used. Cardoso and Santos (1981) reported that *P. nigrum* cuttings presented easy rooting but did not state whether growth regulators were needed.

In some Piperaceous plants, such as *Peperomia parnassifolia* species, vegetative propagation has been reported to occur naturally by stolon, defined as fresh plagiotropic stem aerial branches formed in the base of the plant. These stolons can be the origin of new individuals and may or may not detach from the mother plant (OLIVEIRA et al., 2008). In other species, vegetative propagation occurs through preexisting aerial adventitious roots in the nodal region of the vegetative branches, as observed by Souza et al. (2009) in the species *Piper amalago*. In such processes, the xylem and phloem, which are rich in amillaceous cellular content, are important for the emission and growth of the roots. The authors observed the same behavior for other Piperaceous plants, such as *Piper xylosteoides*, *Piper arboreum*, *Peperomia parnassiflora* and *Ottonia martiana*. A wide variety of amide grains have also been observed in the cells of the medullar parenchyma of *Piper crassenervium* (ALBIERO et al., 2004) and *Piper hispidum* (ALBIERO et al., 2006) stems.

It is observed that the use of IBA is important for the adventitious rooting of cuttings from all the studied cultivars of *P. nigrum*, with square root behavior and maxima at 3,428 (Bragantina), 3,381 (Iaçará) and 4,152 mg kg^{-1} IBA (Guajarina). The average rooting was 98.0% (Figure 2D). The smallest concentration of IBA employed (1,500 mg kg^{-1}) resulted in a high value of total rooting for Bragantina (89%), Iaçará (98.4%) and Guajarina (98.4%) cultivars (Table 1). There are relatively unsuccessful reports dated 50 years ago regarding the use of IBA-based products, such as Serradix® and Vigortone® (CARDOSO, 1961). Thanuja et al. (2002) achieved 60.58% rooting using 50 mg L^{-1} IBA. The IBA concentration depends on the species and also on the studied cultivar; for *Piper* sp., the adventitious rooting presented a linear behavior at the optimal concentration of 5,000 mg L^{-1} (87.5%) (SILVA et al., 2004). For *Piper mikanianum* var. *mikaniamum,* the ideal IBA concentration was 1,500 mg L^{-1} (65%) (PESCADOR et al., 2007), whereas for *P. nigrum* cv. Panniyur 1 (FRANCIS et al., 1993) and *P. longum* (SUNDHARAIYA et al., 2000), the ideal concentration was 1,000 mg L^{-1}, which resulted in a higher percentage of rooting and better quality of the radicular system.

In cocoa tree clones, different responses to IBA with regard to adventitious rooting have been reported (SANTOS JÚNIOR et al., 2008). Cuttings of *Arbutus unedo* and *Taxus baccata* treated with 8,000 and 10,000 mg L^{-1} (IBA, AIA and ANA) presented superior adventitious rooting in crescent order with IBA > AIA = ANA and IBA > AIA > ANA, respectively (METAXAS et al., 2004). *Pongamia pinnata* presented an efficiency of AIB > ANA > AIA (KESARI et al., 2009). The positive effect of IBA is most likely due to its higher chemical stability, and according to Fogaça and Fett-Neto (2005), this feature is related to rooting favoring, mainly in recalcitrant species. IBA was also superior to AIA and ANA in *Camellia sinensis* (ROUT, 2006), to ANA in *Tectona grandis* (HUSEN; PAL, 2007) and to AIA in *Cloezia buxifolia* and *Cloezia aquarum* (LEMAY et al., 2009). According to Husen and Pal (2007), IBA and ANA increased the total soluble sugar, amide, and protein levels and the peroxidase activity in the rooting zone. Sugar has also been observed to accumulate in the stem base of Ulmus vilosa (BHARDWAJ; MISHRA, 2005) and *Dianthus caryophyllus* (AGULLÓ-ANTÓN et al., 2011). According to Correa et al. (2011), the total levels of sugar and phenolic compounds and the peroxidase activity are important biochemical markers in the adventitious rooting process.

Table 1. Characteristics of the adventitious rooting of cuttings of *Piper nigrum* cvs. Bragantina, Iaçará and Guajarina treated with different concentrations of IBA. São Mateus, Espírito Santo State, 2012.

Caracteristics evaluated	Cultivars	AIB (mg kg^{-1})				
		0	1500	3000	4500	6000
Rootting nodal (%)	Bragantina	5.00 b	41.56 b	29.75 b	37.75 b	64.52 b
	Iaçará	39.25 a	95.25 a	92.00 a	100.00 a	98.50 a
	Guajarina	43.75 a	94.00 a	95.50 a	92.50 a	98.50 a
Rooting basal (%)	Bragantina	10.00 b	80.00 b	93.50 a	76.50 b	90.31 a
	Iaçará	43.75 a	93.75 ab	87.50 a	93.87 a	93.75 a
	Guajarina	1.50 b	98.50 a	98.50 a	100.00 a	100.00 a
Rooting total (%)	Bragantina	12.50 b	89.06 a	96.87 a	85.93 a	92.18 a
	Iaçará	56.25 a	98.43 a	93.75 a	100.00 a	95.31 a
	Guajarina	42.18 a	98.43 a	98.43 a	100.00 a	100.00 a
Number of roots nodal	Bragantina	0.05 b	2.07 b	0.70 c	1.03 c	1.01 c
	Iaçará	2.14 ab	7.27 a	13.20 a	16.12 a	14.69 a
	Guajarina	2.87 a	5.34 a	7.70 b	10.68 b	9.51 b
Number of roots basal	Bragantina	0.30 a	5.57 b	8.03 a	5.92 c	8.09 b
	Iaçará	1.31 a	10.30 a	10.95 a	11.23 b	10.57 ab
	Guajarina	0.01 a	8.93 a	9.30 a	15.23 a	12.52 a
Number of roots total	Bragantina	0.35 a	7.65 b	11.00 c	7.23 b	12.01 b
	Iaçará	3.45 a	17.57 a	24.15 a	27.35 a	25.26 a
	Guajarina	2.89 a	14.28 a	17.00 b	25.92 a	22.03 a
Length of the longest nodal root (cm)	Bragantina	0.12 b	3.21 b	8.85 b	3.01 b	5.93 c
	Iaçará	1.55 ab	7.33 a	9.04 a	9.47 a	7.86 b
	Guajarina	2.67 a	8.00 a	9.91 a	9.35 a	10.18 a
Total volume of roots (m^3)	Bragantina	0.0019 a	0.3384 b	0.5885 b	0.3865 b	0.9135 b
	Iaçará	0.0865 a	0.5577 b	0.7404 b	0.6539 b	0.5442 b
	Guajarina	0.0154 a	0.9327 a	1.2693 a	2.0096 a	1.5673 a
Dry mass of root nodal (mg)	Bragantina	0.0000 a	0.0150 b	0.0087 c	0.0177 c	0.0331 c
	Iaçará	0.0039 a	0.0566 a	0.1387 a	0.1435 a	0.1331 a
	Guajarina	0.0092 a	0.0419 ab	0.0665 ab	0.0735 b	0.0716 b
Dry mass of root basal (mg)	Bragantina	0.0021 a	0.0289 a	0.0614 a	0.0450 b	0.0594 a
	Iaçará	0.0039 a	0.0427 a	0.0519 a	0.0502 b	0.0704 a
	Guajarina	0.0000 a	0.0510 a	0.0510 a	0.0835 a	0.0612 a
Dry mass of root total (mg)	Bragantina	0.0021 a	0.0439 b	0.0700 c	0.0627 b	0.0925 b
	Iaçará	0.0077 a	0.0993 a	0.1906 a	0.1937 a	0.2035 a
	Guajarina	0.0092 a	0.0929 a	0.1175 b	0.1569 a	0.1327 b

*Averages followed by the same letter in the column between cultivars within each IBA concentration are not different by the Tukey Test at a 5% significance level.

However, species such as *Sapium glandulatum* exhibit difficulties in the adventitious rooting process even when large IBA concentrations (8,000 mg kg^{-1}) are used (FERREIRA et al., 2009).

The quality of the radicular system is very important to the effectiveness of the plantlet's growth; an indicator of the quality of the radicular system is the number of roots. It was, therefore, observed that very few roots formed in the nodal region of the cuttings of cv. Bragantina (1.82), with the maximum at 4,531 mg kg^{-1} IBA (Figure 3A), while only 8 basal roots were formed when the highest IBA concentration (6,000 mg kg^{-1}) was used (Table 1 and Figure 3B). In *Psidium guajava*, a larger number of roots were also produced when higher concentrations of IBA (8,000 mg L^{-1}) were used (ABDULLAH et al., 2006). The higher number of roots of cv. Bragantina formed in the basal region of the cuttings indicates less waste of the auxin, as the immersion of the cuttings in powder can be performed for this region only. The same cannot be said for cvs. Iaçará and Guajarina, for which full immersion of the cuttings in powder is needed because 10 to 12 roots are generally formed in the nodal region (Table 1 and Figure 3A).

The largest total numbers of roots were 12, 30 and 22 at estimated IBA concentrations of 3,937; 5,187 and 4,980 mg kg^{-1} for cvs. Bragantina, Iaçará and Guajarina, respectively (Figure 3C). It is observed that cv. Iaçará presented the largest number and the highest total dry matter mass of roots for all the studied IBA concentrations (Table 1), which indicated that this cultivar is very responsive with regard to the formation of roots; furthermore, this responsiveness can be influenced by various aspects, such as the presence of new leaves, degree of lignification of the cuttings, amide, hormonal balance, etc.

The nodal roots of cv. Bragantina grew at lower rates than those of the other studied cultivars, but the longest length (4 cm) was achieved at an IBA concentration of 6,000 mg kg^{-1}. IBA also influenced the growth of *Camellia sinensis* L. (LIMA et al., 2011b) roots. According to Corrêa et al. (2005), auxins regulate the rooting process and carbohydrates increase the root growth. The Iaçará (8.73 cm) and Guajarina (9.64 cm) cultivars presented a higher growth rate for their roots, with maxima at 3,639 and 5,374 mg kg^{-1} IBA, respectively (Figure 4A). The length of the roots in the basal region of the cuttings presented square root behavior in all cultivars, with a maximum at 3,117 mg kg^{-1} IBA (6.29 cm) (Figure 4B). The total length (average length in the nodal and basal regions) of the largest root was 10.1 cm when an estimated IBA concentration of 3,900 mg kg^{-1} was used (Figure 4C). A paper published by Ansari and Singh (2008) reported an interaction among half-siblings in the family of *Albizia procera* Benth with IBA, with a significant increase in the root length of four out of seven families. In *Dalbergia melanoxylon*, cuttings treated with IBA presented a higher rooting percentage and a superior number and length of roots compared to untreated cuttings.

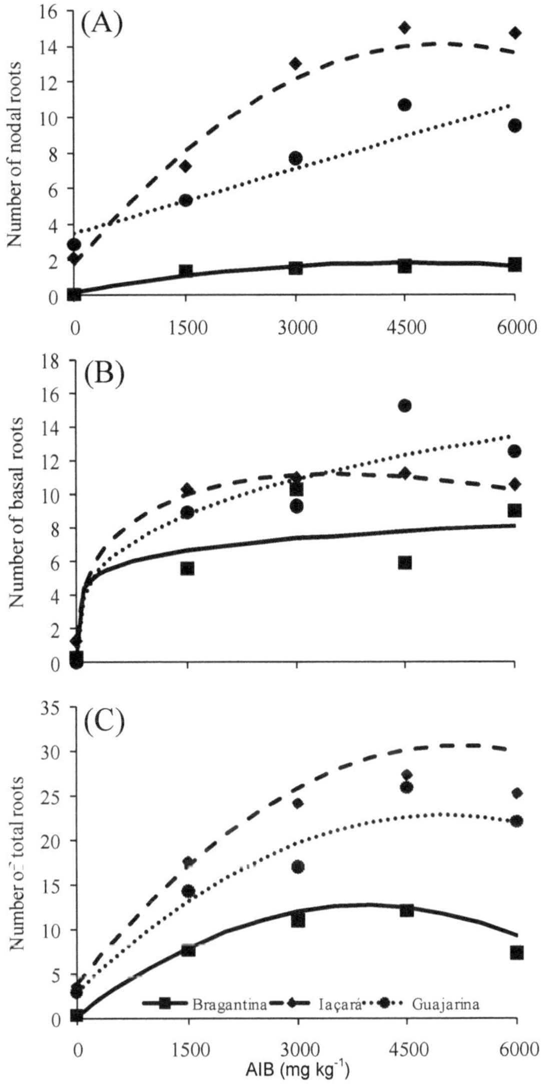

(A) Number of nodal roots	R^2
$\hat{Y}_{Bragantina} = 0.178286 + 0.000725x - 0.00000008x^2$	0.92^*
$\hat{Y}_{Iaçará} = 1.7668 + 0.004972x - 0.0000005x^2$	0.98^{**}
$\hat{Y}_{Guajarina} = 3.506 + 0.0012x$	0.86^*
(B) Number of basal roots	R^2
$Log\hat{Y}_{Bragantina} = 0.3555 + 0.1467logx$	0.97^{**}
$\hat{Y}_{Iaçará} = 1.3460 + 0.3320x^{0.5} - 0.0028x$	0.99^{**}
$Log\hat{Y}_{Guajarina} = -0.0130 + 0.3019logx$	0.99^{**}
(C) Number total roots = (A) + (B)	R^2
$\hat{Y}_{Bragantina} = 0.2406 + 0.0063x-0.0000008x^2$	0.99^*
$\hat{Y}_{Iaçará} = 3.772571 + 0.010375x - 0.000001x^2$	0.99^{**}
$\hat{Y}_{Guajarina} = 2.9600 + 0.007968x - 0.0000008x^2$	0.94^*

Figure 3. Nodal (B), basal (C) and total (C=A+B) root numbers of cuttings of *Piper nigrum* cvs. Bragantina, Iaçará and Guajarina treated with different concentrations of IBA. São Mateus, Espírito Santo, 2012. **,* Significant at 1 and 5% probability level by the F test, respectively.

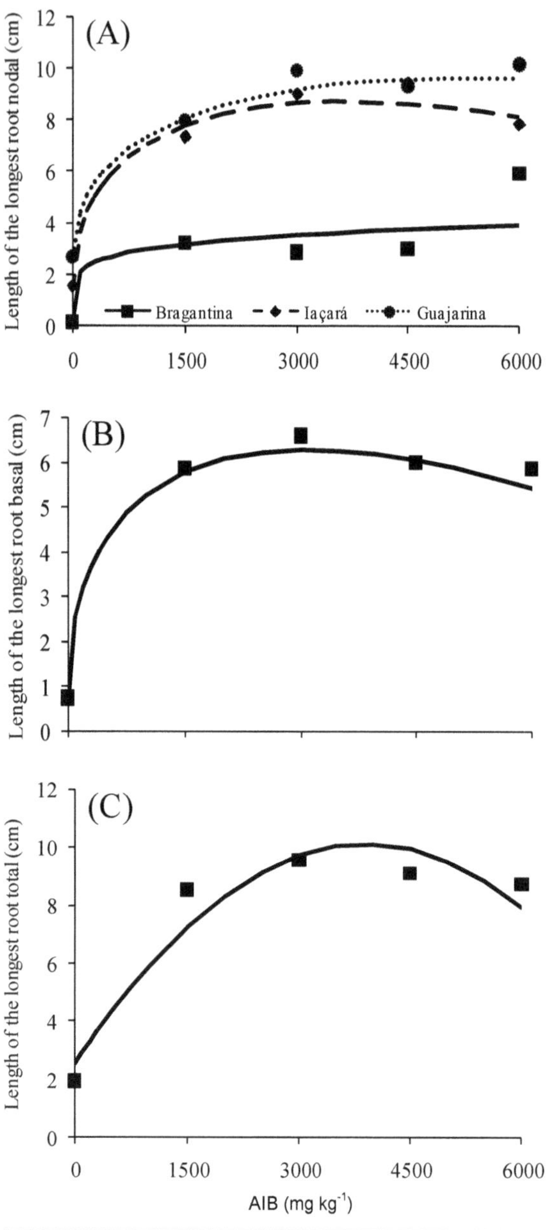

(A) Length of the longest root nodal	R^2
$\text{Log } \hat{Y}_{\text{Bragantina}} = 0.0142 + 0.1535\log x$	0.97^{**}
$\hat{Y}_{\text{Iaçará}} = 1.4609 + 0.2413x^{0.5} - 0.002x$	0.97^{*}
$\hat{Y}_{\text{Guajarina}} = 2.6631 + 0.1906x^{0.5} - 0.0013x$	0.98^{*}
(B) Length of the longest root basal	R^2
$\hat{Y} = 0.7343 + 0.2001x^{0.5} - 0.0018x$	0.99^{**}
(C) Length of the longest root total	R^2
$\hat{Y} = 2.532 + 0.0039x - 0.0000005x^2$	0.91^{**}

Figure 4. Length of the largest nodal (B), basal (C) and total (C=A+B) (cm) roots of cuttings of *Piper nigrum* cvs. Bragantina, Iaçará and Guajarina treated with different concentrations of IBA. São Mateus, Espírito Santo State, 2012. **,* Significant at 1 and 5% probability level by the F test, respectively.

The total volume of roots presented a quadratic behavior in all cultivars, with maxima at 4,000; 3,397 and 4,714 mg kg^{-1} IBA, achieving values of 0.828, 0.687 and 1.793 cm^3 for the Bragantina, Iaçará and Guajarina cultivars, respectively (Figure 5). The Guajarina cultivar presented a higher total volume of roots for all the IBA concentrations studied (Table 1), and it can be inferred that this cultivar can be used as rootstock, considering the higher exploitation of its radicular system and its consequently greater effectiveness of water absorption and use of mineral substrates; this was observed in a paper by Veloso et al. (2000), in which only the Guajarina cultivar responded to phosphorus application.

Root dry matter mass is associated with the rooting rate and the number of roots (HOLANDA et al., 2012). With regard to this characteristic, the Bragantina cultivar was 2.34 times superior in its basal region (67.38 mg) when 5,687 mg kg^{-1} IBA was used compared to the nodal region (28.78 g) when 6,000 mg kg^{-1} IBA was used (Figure 6A and B, respectively). For the other cultivars, the nodal dry matter mass was superior to the basal dry matter mass, and the Iaçará and Guajarina cultivars presented 164.3 and 69.61 mg of roots, respectively, with maxima at 6,000 and 4,500 mg kg^{-1} IBA (Figure 6A). The total dry matter masses of the roots were 75.71 mg for Bragantina, 195.5 mg for Iaçará and 153.82 mg for Guajarina, with maxima at 3,890; 4,650 and 4,316 mg kg^{-1} IBA, respectively (Figure 6C). Once again, it is observed that the rhizogenic capacity of the Bragantina cultivar is lower than those observed in the Iaçará and Guajarina cuttings.

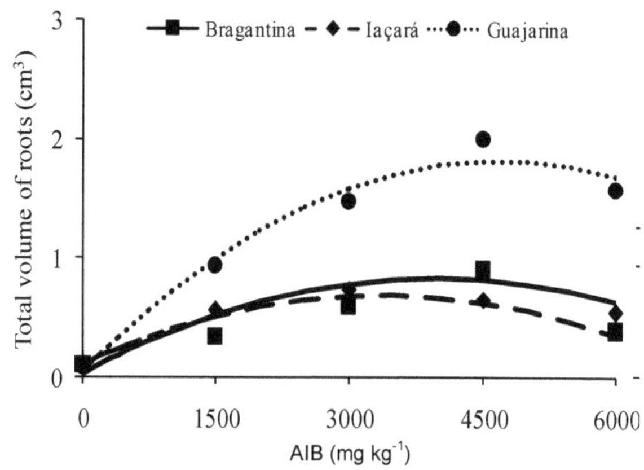

Total volume of roots	R^2
$\hat{Y}_{\text{Bragantina}} = 0.0289 + 0.0004x - 0.00000005x^2$	0.76^{*}
$\hat{Y}_{\text{Iaçará}} = 0.110314 + 0.0003397x - 0.00000005x^2$	0.96^{*}
$\hat{Y}_{\text{Guajarina}} = 0.041714 + 0.00075438x - 0.00000008x^2$	0.96^{*}

Figure 5. Total volume of roots (cm^3) of cuttings of *Piper nigrum* cvs. Bragantina, Iaçará and Guajarina treated with different concentrations of IBA. São Mateus, Espírito Santo, 2012. *Significant at 5% probability level by the F test.

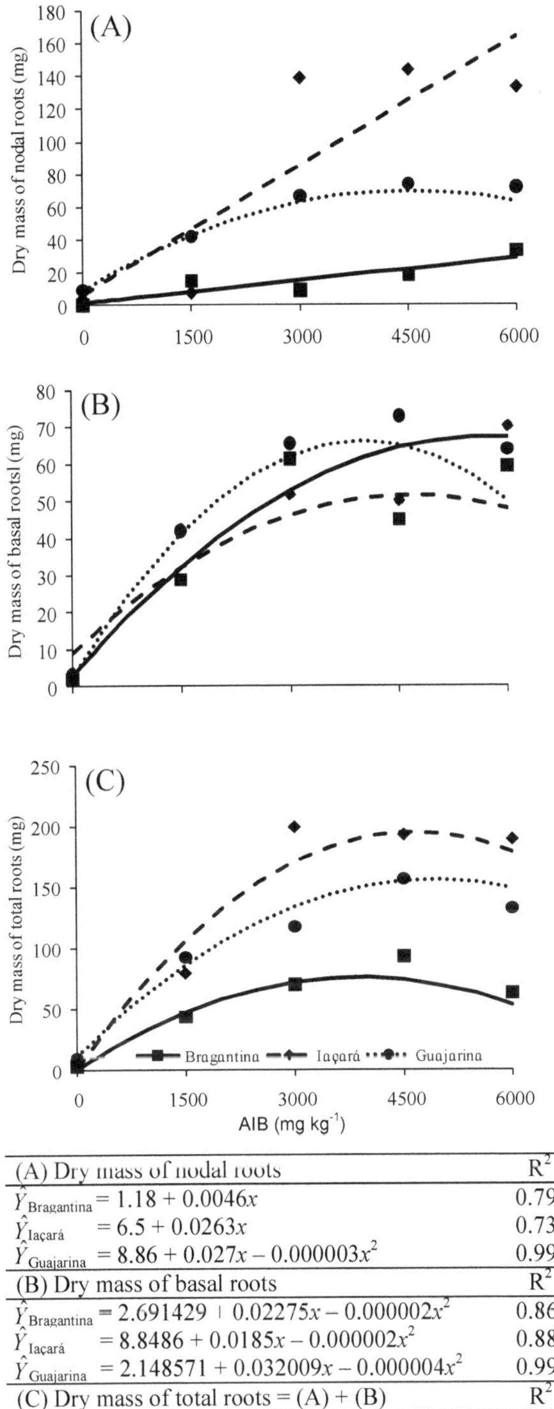

(A) Dry mass of nodal roots	R^2
$\hat{Y}_{Bragantina} = 1.18 + 0.0046x$	0.79^*
$\hat{Y}_{Iaçará} = 6.5 + 0.0263x$	0.73^*
$\hat{Y}_{Guajarina} = 8.86 + 0.027x - 0.000003x^2$	0.99^*
(B) Dry mass of basal roots	R^2
$\hat{Y}_{Bragantina} = 2.691429 + 0.02275x - 0.000002x^2$	0.86^*
$\hat{Y}_{Iaçará} = 8.8486 + 0.0185x - 0.000002x^2$	0.88^*
$\hat{Y}_{Guajarina} = 2.148571 + 0.032009x - 0.000004x^2$	0.99^*
(C) Dry mass of total roots = (A) + (B)	R^2
$\hat{Y}_{Bragantina} = 0.0549 + 0.0389x - 0.000005x^2$	0.95^*
$\hat{Y}_{Iaçará} = 0.9071 + 0.0837x - 0.000009x^2$	0.94^*
$\hat{Y}_{Guajarina} = 10.951 + 0.059x - 0.000006x^2$	0.97^*

Figure 6. Nodal (B), basal (C) and total (C=A+B) dry root mass (mg) of cuttings of *Piper nigrum* cvs. Bragantina, Iaçará and Guajarina treated with different concentrations of IBA. São Mateus, Espírito Santo State, 2012. *Significant at 5% probability level by the F test.

Conclusion

The total immersion of cuttings in IBA is recommended for the cvs. Iaçará and Guajarina, whereas immersion of only the basal region is recommended for cv. Bragantina because the nodal region produces virtually no roots. The recommended concentration for these cultivars is 4,000 mg kg^{-1}.

References

ABDULLAH, A. T. M.; HOSSAIN, M. A.; BHUIYAN, M. K. Clonal propagation of guava (*Psidium guajava* Linn.) by stem cutting from mature stockplants. **Journal of Forest Research**, v. 17, n. 4, p. 301-304, 2006.

AGULLÓ-ANTÓN, M. A.; SÁNCHEZ-BRAVO, J.; ACOSTA, M.; DRUEGE, U. Auxin of sugars: wat makes the difference in the adventitious rooting of stored carnation cuttings? **Journal of Plant Growth Regulation**, v. 30, n. 1, p. 100-113, 2011.

ALBIERO, A. L. M.; PAOLI, A. A. S.; SOUZA, L. A.; MOURÃO, K. S. M. Morfoanatomia dos órgãos vegetativos de *Piper crassinervium* H.B. & K. (Piperaceae). **Acta Botânica Brasílica**, v. 19, n. 2, p. 305-312, 2004.

ALBIERO, A. L. M.; PAOLI, A. A. S.; SOUZA, L. A.; MOURÃO, K. S. M. Morfoanatomia dos órgãos vegetativos de *Piper hispidum* Sw. (Piperaceae). **Revista Brasileira de Farmacognosia**, v. 16, n. 3, p. 379-391, 2006.

ANSARI, S. A.; SINGH, S. Genetic difference in adventitious rhizogenesis in *Albizia procera* Benth. with IBA treatment. **Journal of Forest Research**, v. 13, n. 2, p. 79-82, 2008.

ARAÚJO, R. C.; BRUCKNER, C. H. Biologia reprodutiva de fruteiras. In: BRUCKNER, C. H. (Ed.). **Fundamentos do melhoramento de fruteiras**. Viçosa: Editora UFV, 2008. p. 1-38.

BHARDWAJ, D. R.; MISHRA, V. K. Vegetative propagation of *Ulmus villosa*: effect of plant growth regulators, collection time, type of donor and position of shoot on adventitious root formation in stem cuttings. **New Forest**, v. 29, n. 2, p. 105-116, 2005.

CARDOSO, M. Sobre o enraizamento de estacas de pimenta-do-reino. **Bragantia**, v. 20, n. 16, p. 529-531, 1961.

CARDOSO, M.; SANTOS, R. R. Pimenteira-do-reino: produtividade segundo o tipo de muda. **Bragantia**, v. 40, n. 9, p. 221-224, 1981.

CORRÊA, L. A.; PAIM, D. C.; SCHWAMBACH, J.; FETT-NETO, A. G. Carbohydrates as regulatory factors on the rooting of *Eucalyptus saligna* Smith and *Eucalyptus globulus* Labill. **Plant Growth Regulation**, v. 45, n. 1, p. 63-73, 2005.

CORREA, L. R.; STEIN, R. J.; FETT-NETO, A. G. Adventitious rooting of detached *Arabidopsis thaliana* leaves. **Biologia Plantarum**, v. 56, n. 1, p. 25-30, 2011.

CRUZ, C. D. Genes: a software package for analysis in experimental statistics and quantitative genetics. **Acta Scientiarum. Agronomy**, v. 35, n. 3, p. 271-276, 2013.

FACHINELLO, J. C.; HOFFMANN, A.; NACHTIGAL, J. C.; KERSTEN, E. Propagação vegetativa por estaquia. In: FACHINELLO, J. C.; HOFFMANN, A.; NACHTIGAL, J. C. (Ed.). **Propagação de plantas frutíferas**. Brasília: Embrapa Informação Tecnológica, 2005.

FERREIRA, B. G. A.; ZUFFELLATO-RIBAS, K. C.; CARPANEZZI, A. A.; TAVARES, F. R.; KOEHLER, H. S. Metodologias de aplicação de AIB no enraizamento de estacas semilenhosas de *Sapium glandulatum* (Vell.) Pax. **Revista Brasileira de Plantas Medicinais**, v. 11, n. 2, p. 196-201, 2009.

FOGAÇA, C. M.; FETT-NETO, A. G. Role of auxin and its modulators in the adventitious rooting of *Eucalyptus* species differing in recalcitrant. **Plant Growth Regulation**, v. 45, n. 1, p. 1-10, 2005.

FRANCIS, G.; ARAVINDAKSHAN, M.; GOPIKUMAR, K.; VALSALAKUMARI, P. K. Effect of growth substances on rooting of planting materials in black pepper (*Piper nigrum* L.). **Journal of Tropical Agriculture**, v. 31, n. 1, p. 71-76, 1993.

GRZYBOWSKI, A.; TIBONI, M.; SILVA, M. A. N.; CHTOLINA, R. F.; PASSOS, M.; FONTANA, J. D. The combined action of phytolarvicides for the control of dengue fever vector, *Aedes aegypti*. **Revista Brasileira de Farmacognosia**, v. 1, n. 1, p. 15-23, 2012.

HARTMANN, H. T.; KESTER, D. E.; DAVIES JR., F. T.; GENEVE, R. L. **Plant propagation**: principles and practices. 8. ed. New Jersey: Prentice-Hall, 2011.

HOLANDA, F. S. R.; VIEIRA, T. R. S.; ARAÚJO FILHO, R. N.; SANTOS, T. O.; ANDRADE, K. V. S.; CONCEIÇÃO, F. G. Propagation through cutting technique of species occurring in the lower São Francisco in Sergipe Stage with different concentrations of indolbutiric acid. **Revista Árvore**, v. 36, n. 1, p. 75-82, 2012.

HUNT, M. A.; TRUEMAN, S. J.; RASMUSSEN, A. Indole-3-butyric acid accelerates adventitious root formation and impedes shoot growth of *Pinus elliottii* var. *elliottii* x *P. caribaea* var. *hondurensis* cuttings. **New Forests**, v. 41, n. 3, p. 349-360, 2011.

HUSEN, A.; PAL, M. Metabolic changes during adventitious root primordium development in *Tectona grandis* Linn. F. (teak) cuttings as affected by age of donor plants and auxin (IBA and NAA) treatment. **New Forest**, v. 33, n. 3, p. 309-323, 2007.

IBGE-Instituto Brasileiro de Geografia e Estatística. **Levantamento sistemático de produção agrícola**: pesquisa mensal de previsão e acompanhamento das safras agrícolas no ano civil. Rio de Janeiro: IBGE, 2011. v. 24, n. 11, p. 82.

ISLAM, M. S.; BHUIYAN, M. K.; HOSSAIN, M. M.; HOSSAIN, M. A. Clonal propagation of *Bambusa vulgaris* by leafy branch cuttings. **Journal of Forest Research**, v. 22, n. 3, p. 387-392, 2011.

KESARI, V.; KRISHNAMACHARI, A.; RANGAN, L. Effect of auxins on adventitious rooting from stem cuttings of candidate plus tree *Pongamia pinnata* (L.), a potential biodiesel plant. **Trees - Structural Function**, v. 23, n. 3, p. 597-604, 2009.

LEITE, J. B. V.; MARTINS, A. B. G. Efeito do ácido indolbutírico e época de coleta no enraizamento de estacas semi-lenhosas do cacaueiro. **Revista Brasileira de Fruticultura**, v. 29, n. 2, p. 204-208, 2007.

LEMAY, V.; GÂTEBLÉ, G.; McCOPY, S. Vegetative propagation of two endemic species of *Cloezia* Brongn. & Gris for conservation and mining revegetation activities in New Caledonia. **New Forest**, v. 37, n. 1, p. 1-8, 2009.

LI, S. W.; XUE, L. The interaction between H_2O_2 and NO, Ca^{2+}, cGMP, and MAPKs during adventitious rooting in mung bean seedlings. ***In Vitro* Cellular and Developmental Biology – Plant**, v. 46, n. 2, p. 142-148, 2010.

LI, S. W.; XUE, L.; XU, S.; FENG, H.; NA, L. Mediators, genes and signaling in adventitious rooting. **The Botanical Review**, v. 75, n. 2, p. 230-247, 2009.

LIMA, D. M.; BIASI, L. A.; ZANETTE, F.; ZUFFELLATO-RIBAS, K. C.; BONA, C.; MAYER, J. L. S. Capacidade de enraizamento de estacas de *Maytenus muelleri* Schwacke com a aplicação de ácido indol butírico relacionada aos aspectos anatômicos. **Revista Brasileira de Plantas Medicinais**, v. 13, n. 4, p. 422-438, 2011a.

LIMA, J. D.; LIMAS, A. P. S.; BOLFARINI, A. C. B.; SILVA, S. H. M. G. Enraizamento de estacas de *Camellia sinensis* L. em função da época de coleta de ramos, genótipos e ácido indolbutírico. **Ciencia Rural**, v. 41, n. 2, p. 230-235, 2011b.

MAGEVSKI, G. C.; CZEPAK, M. P.; SCHMILDT, E. R.; ALEXANDRE, R. S.; FERNANDES, A. A. Propagação vegetativa de espécies silvestres do gênero *Piper*, com potencial para uso como porta enxertos em pimenta-do-reino (*Piper nigrum*). **Revista Brasileira de Plantas Medicinais**, v. 1, n. especial, p. 559-563, 2011.

MATEJA, Š.; DOMINIK, V.; FRANCI, Š.; GREGOR, O. The effects of a fogging system on the physiological status and rooting capacity of leafy cuttings of woody species. **Trees - Structure and Function**, v. 21, n. 4, p. 491-496, 2007.

MAUSETH, J. D. **Plant anatomy**. Menlo Park: The Benjamin/Cummings Publishing Company, 1988.

METAXAS, D. J.; SYROS, T. D.; YUPSANIS, T.; ECONOMOU, A. S. Peroxidases during adventitious rooting in cuttings of *Arbutus unedo* and *Taxus baccata* as affected by plants genotype and growth regulator treatment. **Plant Growth Regulation**, v. 44, n. 3, p. 257-266, 2004.

OLIVEIRA, J. H. G.; SOUZA, L. A.; IWAZAKI, M. C. Estruturas de reprodução de *Peperomia parnassifolia* Miq. (Piperaceae). **Acta Scientiarum. Biological Sciences**, v. 30, n. 1, p. 1-7, 2008.

PARK, S. M.; WON, E. J.; PARK, Y. G.; JEONG, B. R. Effect of node position, number of leaflets left, and light intensity during cutting propagation on rooting and subsequent growth of domestic roses. **Horticulture, Environement, and Biotechnology**, v. 52, n. 4, p. 339-343, 2010.

PESCADOR, R.; VOLTONI, A. C.; GIRARDI, C. G.; ROSA, F. A. F. Estaquia de pariparoba-do-Rio Grande do Sul sob efeito do ácido indol-butírico em dois substratos. **Scientia Agraria**, v. 8, n. 4, p. 391-398, 2007.

QADDOURY, A.; AMSSA, M. Endogenous phenolic contents, peroxidase and polyphenoxidase activities in date palm (*Phoenix dactylifera* L.) off shoots related to rooting ability. **Acta Physiologiae Plantarum**, v. 25, n. 4, p. 417-421, 2003.

RAVEN, P. H.; CURTIS, H. **Integración del crescimento**: fitohormonas. Biología vegetal. Barcelona: Omega, 1975. p. 162–168.

ROUT, G. R. Effect of auxins on adventitious root development from single node cuttings of *Camellia sinensis* (L.) Kuntze and associated biochemical changes. **Plant Growth Regulation**, v. 48, n. 2, p. 111-117, 2006.

SANTOS JÚNIOR, A. J.; ALMEIDA, A. A. F.; SILVA, D. C.; FARIA, J. C.; MIELKE, M. S.; GOMES, F. P. Enraizamento de estacas, crescimento e respostas anatômicas de mudas clonais de cacaueiro ao ácido indol-3-butírico. **Revista Brasileira de Fruticultura**, v. 30, n. 4, p. 1071-1082, 2008.

SHARANGI, A. B.; KUMAR, R. Performace of rooted cuttings of black pepper (*Piper nigrum* L.) with organic substitution of nitrogen. **International Journal of Agricultural Research**, v. 6, n. 9, p. 673-681, 2011.

SHARANGI, A. B.; KUMAR, R.; SAHU, P. K. Survivability of black pepper (*Piper nigrum*) growing media. **Journal of Crop and Weed**, v. 6, n. 1, p. 52-54, 2010.

SILVA, J. M. M.; RAPOSO, A.; SOUZA, J. A.; MIRANDA, E. M. Indução de enraizamento em estacas de joão-brandinho (*Piper* sp.) com ácido indolbutírico. **Revista Ciência Agronômica**, v. 35, n. especial, p. 248-252, 2004.

SIMAS, N. K.; LIMA, E. C.; KUSTER, R. M.; LAGE, C. L. S.; OLIVEIRA FILHO, A. M. Potential use of *Piper nigrum* ethanol extract against pyrethroid-resistant *Aedes aegypti* larvae. **Revista da Sociedade Brasileira de Medicina Tropical**, v. 40, n. 4, p. 405-407, 2007.

SINGH, S. S.; SINGH, S. Effect of nodal cuttings and rooting media on the propagation of black pepper under South Andaman conditions. **Indian Cocoa Anecanut and Spices Journal**, v. 12, n. 4, p. 122-123, 1989.

SINGH, B.; YADAV, R.; BHATT, B. P. Effect of mother tree ages, different rooting mediums, light conditions and auxin treatment on rooting behaviour of *Dalbergia sissoo* branch cuttings. **Journal Forestry Research**, v. 22, n. 1, p. 53-57, 2011.

SOUZA, L. A.; MOSCHETA, I. S.; MOURÃO, K. S. M.; ALBIERO, A. L. M.; IWAZAKI, M. C.; OLIVEIRA, J. H. G.; ROSA, S. M. Vegetative propagation in Piperaceae species. **Brazilian Archieves of Biology Technology**, v. 52, n. 6, p. 1357-1361, 2009.

SOUZA, L. A.; MOSCHETA, I. S.; OLIVEIRA, J. H. G. Comparative morphology and anatomy of the leaf and stem of *Peperomia dahlstedtii*, *Ottonia martiana* and *Piper diospyrifolium* (Piperaceae). **Gayana Botánica**, v. 61, n. 1, p. 6-17, 2004.

SUNDHARAIYA, K.; PONNUSWAMI, V.; JASMINE, A. J. Effect of growth regulators in the propagation of sarkaraikolli (*Gymnema sylvestre*), medicinal coleus (*Coleus forskohlii*) and thippili (*Piper longum*). **Journal South Indian Horticulture**, v. 48, n. 1-6, p. 172-174, 2000.

THANKAMAN, C. K.; DINESH, R.; EAPEN, S. J.; KUMAR, A.; KANDIANNAN, K.; MATHEW, P. A. Effect of solarized potting mixture on growth of black pepper (*Piper nigrum* L.) rooted cuttings in the nursey. **Journal of Spices and Aromatic Crops**, v. 17, n. 2, p. 103-108, 2008.

THANUJA, T. V.; HEDGE, R. V.; SREENIVASA, M. N. Induction of rooting and root growth in black pepper cuttings (*Piper nigrum* L.) with the inoculation of arbuscular mycorrhizae. **Scientia Horticulturae**, v. 92, n. 3-4, p. 339-346, 2002.

THOMAS, R. G.; HAY, M. J. M. Adaptive variation in physiological traits underpinning stem elongation responses among nodally-rooting stoloniferous herbs. **Evolutionary Ecology**, v. 22, n. 3, p. 369-381, 2008.

THOMAS, R. G.; HAY, M. J. M. The role of nodal roots in prostate clonal herbs: 'phalanx' versus 'guerrilla'. **Evolutionary Ecology**, v. 24, n. 6, p. 1489-1504, 2010.

TRAJANO, V. N.; LIMA, E. O.; SOUZA, E. L.; TRAVASSOS, A. E. R. Propriedade antibacteriana de óleos essenciais de especiarias sobre bactérias contaminantes de alimentos. **Ciência e Tecnologia de Alimentos**, v. 29, n. 3, p. 542-545, 2009.

VELOSO, C. A. C.; CARVALHO, E. J. M.; MALAVOLTA, E.; MURAOK, T. Resposta de cultivares de pimenta-do-reino aos nutrientes NPK em um latossolo amarelo da Amazônia Oriental. **Scientia Agricola**, v. 57, n. 2, p. 343-347, 2000.

VIANNA, W. O.; AKISUE, G. Caracterização morfológica de *Piper aduncum* L. **Lecta**, v. 15, n. 1-2, p. 11-62, 1997.

15

Antioxidant activity and total phenol content of blackberries cultivated in a highland tropical climate

Mayara Neves Santos Guedes[1], Rafael Pio[2*], Luana Aparecida Castilho Maro[3], Fabíola Fonseca Lage[4], Celeste Maria Patto de Abreu[4] and Adelir Aparecida Saczk[4]

*[1]Departamento de Agricultura, Universidade Federal de São João del-Rei, Sete Lagoas, Minas Gerais, Brazil. [2]Departamento de Agricultura, Universidade Federal de Lavras, Cx. Postal 3037, 37200-000, Lavras, Minas Gerais,Brazil. [3]Empresa de Pesquisa Agropecuária e Extensão Rural de Santa Catarina, Itajaí, Santa Catarina, Brazil. [4]Departamento de Química, Universidade Federal de Lavras, Lavras, Minas Gerais, Brazil. *Author for correspondence. E-mail: rafaelpio@dag.ufla.br*

ABSTRACT. Blackberries are an important option for the diversification of fruit crops. However, there is currently no literature regarding plant cultivation in high-altitude tropical climates. Knowledge of the phenolic composition of blackberries is essential because variations in the levels of these components may exist between cultivars and may depend on environmental conditions. High performance liquid chromatography (HPLC) was used to evaluate the total phenol content of different blackberry cultivars (Arapaho, Brazos, Cainguangue, Cherokee, Choctaw, Comanche, Ébano, Guarani, Tupy and Xavante). Free radical scavenging activity in these cultivars was assayed using a DPPH test. The HPLC-UV chromatogram of blackberry fruit extracts at 280 nm revealed the presence of phenolic compounds. The results showed significant differences in the levels of phenolic compounds in the blackberry cultivars tested. Antioxidant activity was evaluated using the ABTS free radical and ranged from 2.7 ± 0.1 to 19 ± 2 µmole of Trolox equivalents per gram of sample (b.u.). These results are in good correlation with the phenolic contents of the blackberries tested. The Xavante blackberry cultivar had the highest levels of polyphenols that could be individually identified. Catechin polyphenols were found to be the main component in the blackberry varieties tested.

Keywords: total phenols, free radicals, HPLC, *Rubus* spp.

Introduction

Small red fruits orberries are characterized by the presence of high concentrations of a variety of bioactive compounds, including anthocyanins, phenolic compounds, organic acids, tannins and flavonoids (Szajdek & Borowska, 2008; Zozio, Dominique, & Dornier, 2011; Maro, Pio, Guedes,Abreu, & Curi, 2013).These compounds exhibit potent antioxidant activity and are widely necessary to prevent the deterioration of oxidizable products such as cosmetics and food products. In addition, these compounds are beneficial to human health (Curi, Pio, Moura, Lima, & Valle, 2014).

In this group of small fruits, blackberries (*Rubus*spp.) are considered excellent sources of phenolic compounds. However, there is a great diversity between blackberry cultivar results in fruits with different characteristics, both in flavor and staining; these differences are associated with the polyphenol content and profile of the fruit (Guedes et al., 2013; Curi et al., 2015).In addition, the chemical composition of the fruit is influenced by multiple genetic factors, which limits quality and consumer acceptance (Scalzo, Battino, Costantini, & Mezzetti, 2005).

Blackberries come from deciduous fruit-bearing trees of temperate climates. However, recently new orchards have been established in subtropical and tropical regions (Campagnolo & Pio, 2012; Curi et al., 2015; Caproni et al., 2016).The fruit chemical composition of a specific cultivar may be affected by several factors, and environmental conditions are among the most important factors that influence chemical composition (Maro et al., 2014).

The blackberry fruits grown in high-altitude tropical climates that are characterized by milder conditions require study to evaluate the chemical quality of the fruits. Therefore, the purpose of this study is to assess the quality and content of phenolic compounds in different cultivars of blackberries being grown in a high-altitude tropical climate.

Material and methods

Fruits from 10 blackberry cultivars were collected manually and randomly from different positions and orientations on thirty blackberry plants. The plants were at the commercial maturity stage. The orchard experiences a high-altitude tropical climate. The city is located at 21°14' S 45°00' W at an average altitude of 918 m in Brazil. According to Köeppen's classification, the climate is type Cwb, mild temperate (mesothermal). This climate is characterized by a dry winter and rainy summer (Dantas, Carvalho, & Ferreira, 2007).

For each cultivar, a total of 2,000 g of fruit was collected randomly from thirty plants. Of this amount, we selected 1,200 g. The fruits were transported to the biochemistry laboratory, where fruits were selected to form experimental units.Four replicates of 300 g were analyzed for each cultivar. Fruits were stored in a freezer at -18°C with 90-95% relative humidity until the time of analysis. The experimental design was completely randomized and contained four replications. The independent variable was the blackberry cultivar(Arapaho, Brazos, Cainguangue, Cherokee, Choctaw, Comanche, Ébano, Guarani, Tupy and Xavante), and the total phenol content and antioxidant of each cultivar was analyzed.

Reagents and standards

Sodium hypochlorite, ethanol, hydrochloric acid and acetone were purchased from Vetec Chemistry. Methanol was purchased from JT Baker Chemical Co. (Phillipsburg, NJ), and Trolox was purchased from Sigma - Aldrich (St. Louis, MO, USA). The mobile phase used for HPLC analysis consisted of ultrapure Milli-Q water (Millipore, Billerica, MA, USA), acetic acid and methanol (Merck, Darmstadt, Germany). The HPLC standards for gallic acid, *p*-coumaric acid, ferulic acid, ellagic acid, 3,4-dihydroxybenzoic acid, salicylic acid and syringic acid, monomers of condensed tannins gallocatechin, catechin and epigallocatechin gallate and resveratrol were acquired from Sigma-Aldrich (St. Louis, MO, USA). The standards for m- and o-cumaric acid and vanillic acid were obtained from Fluka (St. Louis, MO, USA). Stock standard solutions were prepared in dimethyl sulfoxide and/or methanol (Merck).

Extraction

Natural extractions of blackberry samples were performed as described by Arabbi, Genovese, and Lajolo (2004) with some modifications. Samples were extracted with a mixture of methanol/water/acetic acid (70:30:5) in a 1:15 (m:v) ratiowith ultrasonication for 10 minutes in an ice bath. The homogenized sample was filtered through paper filter (Whatman n° 1). The filtrate was then evaporated in a plate at 80°C until a volumeof 10 mL remained. This final volume was diluted with Milli-Q water to a volume of 15 mL. A 2 mL aliquot was filtered through a polyethylene filter with a 22μm pore size membrane (Millipore) and collected in a 5 ml vial. Extractions were performed in quadruplicate.

Separation, identification and quantification of phenolic compounds

Chromatography was performed using an Agilent Model 1100 HPLC.Compounds were best detected at a wavelength of 280 nm. Phenolic compound extracts and standards were injected into an Ascentis C18 column (25 cm x 4.6 mm x 5 μm) connected to a Supelguard Ascentis C18 pre-column (2 cm x 4.0 mm x 5 μm). The mobile phase consisted of 2% acetic acid (A) and methanol: water: acetic acid (70:28:2 v/v/v) (B). The flow rateused in all analyses was 100 mL min.$^{-1}$, and the injection volume was 20 μL. Eachrun was 65 minutes longwith a gradient as follows: 100% of solvent A for 5 minutes, 70% of solvent A for 20 minutes, 60% of solvent A for 18 minutes, 55% of solvent A for 7 minutes, and 0% of solvent A for 10 minutes. Solvent A (100%) was then added to equilibrate the column until the end of the run. The column was maintained at a temperature of 45°C.

Phenolic compounds present in blackberry fruit were identified by comparing the retention times of peaks in the extracts with those of standard compounds. Compound identities were confirmed co-chromatography of extracts and standards. To quantify phenolic compounds, an external standard was used. Standard curves of the compounds found in samples were obtained by diluting concentrated solutions to produce a concentration range of $5.0x10^{-7}$ to $1.5x10^{-3}$ mol L^{-1}. Calibration curves were constructed by plotting the average detector response (n = 3) for the peak area according to concentration.

The addition of standards toextracts was also used for identification. Quantification was performed using external standards with the following stock concentrations: galic acid (85.06 mg L^{-1}), gallocatechin (153.13 mg L^{-1}), 3,4 dihydroxybenzoic acid (77.06 mg L^{-1}), catechin (145.14 mg L^{-1}), chlorogenic acid (177.15 mg L^{-1}), epigallocatechin gallate (229.18 mg L^{-1}), vanillic acid (84.07 mg L^{-1}), epicatechin (145.13 mg L^{-1}), meta and ortho-coumaric (82.08 mg L^{-1}), ferulic acid (97.09 mg L^{-1}), resveratrol (114.12 mg L^{-1}), ellagic acid (151.09 mg L^{-1}) andsalicylic acid (69.06 mg L^{-1}). Each solution was injected three times into the HPLC system to obtain the average concentration and retention time of each compound.

Determination of antioxidant activity

Total antioxidant activity was measured according to the ABTS$^+$ [2,2azinobis-(3-ethylbenzthiazoline-6-sulfonic acid)] method established by Maro et al. (2013). Extracts were obtained according to the method described by Arabbi et al. (2004). The resultsare expressed as μmol Trolox equivalent per gram of sample.

Statistical Analysis

The data were subjected to an analysis of variance, and means were compared using the Scott-Knott test with a 5% probability of error. Data were analyzed to identify correlations between individual phenolic compounds and antioxidant activity.

Results and discussion

Figure 1 shows the UV chromatogram at 280 nm obtained from HPLC analysis of a 20 μL aliquot of a 5 x 10^{-5} mol L^{-1} standard solution of phenolic compounds.

Phenolic acids (syringic acid and salicylic acid), acids derived from the hydrolysis of hydrolysable tannins (gallic acid and ellagic acid) and condensed tannins monomers (catechin and gallocatechin), were identified and quantified in fruits from various blackberry cultivars using HPLC. These analyses showed high reproducibility (Figure 2).

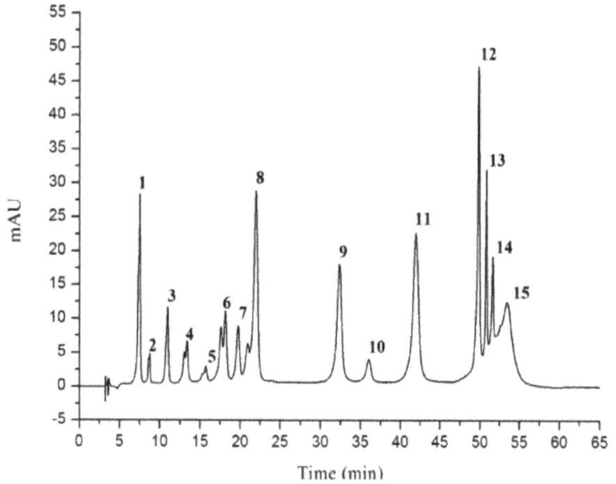

Figure 1. Chromatogram of the phenolic acid standard solution with spectrophotometric detection at 280 nm. Identified peaks: 1 = Gallic acid; 2 = gallocatechin; 3 = 3,4 dihydroxy phenolic 4 = catechin; 5 = chlorogenic acid; 6 = epigallocatechin; 7 = vanillic acid; 8= syringic acid; 9 = P-coumaric acid; 10 = ferulic acid; 11 = M-coumaric acid; 12 = O-coumaric acid; 13 = Resveratrol; 14 = ellagic acid and 15 = Salicylic acid.

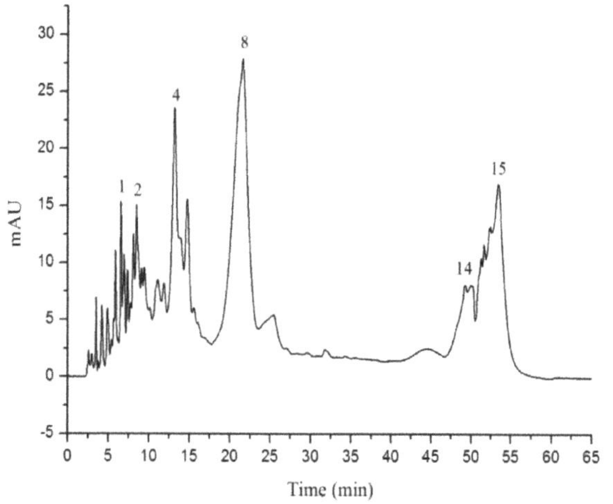

Figure 2. HPLC-DAD chromatograms of different cultivars of blackberry.Identified peaks: 1 = gallic acid; 2 = gallocatechin; 4 = catechin; 8 = syringic acid; 14 = ellagic acid and 15 = salicylic acid.

The most abundant phenolic compound was catechin, which was found in 92.38% of the blackberry fruits evaluated. A different result was obtained by Jacques, Pertuzatti, Barcia, Zambiazi, and Chim (2010), who reported that gallic acid was the most abundant phenolic compound. The content of catechin found in the Xavante cultivar was 258.58 mg per 100 g, which was higher than theother cultivars analyzed in this study (Table 1).

Table 1. Phenolic compound contents (mg per 100 gfresh weight) and antioxidant capacity (Trolox equivalent per g sample) of blackberries cultivars.

Blackberry cultivars	Phenolic compounds[1]						ABTS
	Ellagic acid	Gallic acid	Catechin	Gallocatechin	Syringic acid	Salicylic acid	
Arapaho	1.39c	0.58 g	156.95 c	1.06 c	2.01 d	17.14 d	51.35 a
Brazos	1.10 e	0.84 a	56.35 f	0.98 e	1.01 h	17.08 d	18.55 e
Caingangue	1.17 d	0.60 f	60.32 e	1.00 d	1.51 e	20.32 a	15.15 h
Cherokee	0.98 f	0.53 h	23.06 h	0.87 h	1.99 d	17.12 d	29.79 d
Choctaw	0.97 f	0.63 e	10.35 i	0.89 g	1.39 f	19.49 b	17.50 f
Comanche	1.16 d	0.53 h	38.28 g	0.92 f	2.80 b	16.46 d	33.05 c
Ébano	1.44 b	0.65 d	222.56 b	1.14 b	3.19 a	18.15 c	33.17 c
Guarani	0.91 g	0.70 c	23.77 h	0.86 i	1.23 g	19.20 b	17.10 f
Tupy	0.78 h	0.54 h	64.58 d	0.75 j	1.38 f	14.78 e	16.06 g
Xavante	3.08 a	0.77 b	258.58 a	1.20 a	2.59 c	16.73 d	50.19 b
Average	1.29	0.64	92.38	0.96	1.91	17.64	28.19
CV %	2.52	1.39	1.47	0.34	0.72	2.91	1.28
R^2	0.99	0.99	0.99	0.99	0.96	0.99	

[1] Means followed by the same small letter in the column do not differ from one another by the Scott-Knott test($p \leq 0.05$).

Catechin is a flavonoid compound. Flavonoids are present in fruits and vegetables and are considered to be therapeutic agents due to their beneficial effects on health. The protective properties of flavonoids against certain cancers, cardiovascular diseases and aging are all potentially beneficial for human health (Hidalgo, Sánchez-Moreno, & Pascual-Teresa, 2010; Carvalho, Cavaco, & Brodelius, 2011). The ellagic acid content of the10 black berry cultivars studied in this work ranged from 0.78 to 3.08 mg per 100 g fresh weight. The Xavante cultivar showed the highest levels of ellagic acid, followed by the Ebano cultivar. The lowest level of ellagic acid was observed in the Tupy cultivar (Table 1). Ellagic acid values in this study were lower than those obtained by Siriwoharn, Wrolstad, and Durst (2005) (50 mg per 100 g).

Recent interest in ellagic acid has been due to the compound's phytonutrient properties, which include disease prevention and antioxidant, antimutagenic, and anticarcinogenic effects (Maro et al., 2013; Souza et al., 2014).

Of the phenolic compounds identified and quantified, gallic acid was found at the lowest concentration, ranging from 0.53 to 0.84 mg per 100 g fresh weight. The highest amount of gallic acid was found in the Brazos cultivar, followed by the Shavante cultivar (Table 1). A higher gallic acid content (350.49 mg per 100 g fresh weight) was observed by Jacques et al. (2010) in Tupy black mulberry fruits cultivated in temperate climates. These discrepancies may be related to differences in species and cultivars as well as differences in extraction methodology.

Difference in gallocatechin contents were observed between cultivars, varying from 0.75 mg per 100 gin the Tupy cultivar to 1.20 mg per 100 g^{-1} in the Xavante cultivar. An average of 0.96 mg 100 g (Table 1) was observed. Previous studies have shown gallocatechinto aid in preventing hypoglycemia and in lowering cholesterol (Lee, Kim, Kim, Shin, & Baik, 2008).

The phenolic content of syringic and salicylic acids also varied among the blackberry cultivars studied. Syringic and salicylic acid contents averaged 1.91 and 17.64 mg per 100 g, respectively (Table 1). Phenol concentration sare influenced by factors such as the cultivar type, pre-harvest environmental and climatic conditions, plant diseases, harvesting procedure, point of maturity at harvest, geographic location, exposure to sunlight, post-harvest storage conditions, fruit processing, extraction method and quantification method (Cordenunsi, Nascimento, Genovese, & Lajolo, 2002; Vizzotto & Pereira, 2011; Souza et al., 2015).

The quantification data of individual phenolic compounds in domestic blackberry cultivars in highland tropical climate has not been reported previously; therefore, these results are important for the selection of varieties with superior properties for consumption and breeding of superior cultivars.

The antioxidant activity of blackberry fruits was determined using the ABTS method and reported in equivalents of Trolox, a water-soluble analog of vitamin E. Fruits from the Arapaho and Xavante cultivarsexhibited antioxidant activities of 51.35 and 50.19 μmol of Trolox equivalent per g of sample, respectively. These activities were 34-70% higher than the activities of the other cultivars.

These results corroborate the results obtained by Silva, Vendruscolo, and Toralhes (2011) for blackberry fruits cultivated in temperate conditions: the cultivar that showed the highest antioxidant activity (Xavante) also showed the highest total phenol content.

Correlations between antioxidant activity (ABTS) and individual polyphenol contents in blackberry fruits are shown in Table 2. Mean antioxidant activities correlate positively with the average valuesof most of individual polyphenols evaluated in this study. Correlations between ellagic acid, catechin, gallocatechin and syringic acid contents with antioxidant activity were observed (Table 2). On the other hand,there was no significant correlation (5% probability) between gallic acid content and antioxidant activity (Table 2). These results indicate that variations in phenolic compound contents can lead to different biological responses. Although the catechins represent potential constituents, other compounds may act synergistically, increasing the beneficial effects associated with the consumption of fruits and their blackberry derivatives.

Table 2. Pearson Correlation of ellagic acid, galic acid, catechin, gallocatechin, syringic acid, salicylic acid with antioxidant activity.

Variables	Correlation
Ellagic acid	0.71*
Galic acid	0.00ns
Catechin	0.73*
Gallocatechin	0.69*
Syringic acid	0.67*
Salicylic acid	0.32ns

* = F significant at 5%; ns = F not significant at 5%.

Conclusion

Fruits from blackberry cultivars grown under tropical high-altitude conditions exhibit varying levels of phytochemicals, which may contribute to their natural antioxidant properties.

The polyphenol catechin was the major component of the blackberry cultivars studied. In addition, the Xavante cultivar had higher individual levels of the polyphenols studied in this work.

References

Arabbi, P. R., Genovese, M. I., & Lajolo, F. M. (2004). Flavonoids in vegetable foods commonly consumed in Brasil and estimated ingestion by the Brazilian population. *Journal of Agricultural and Food Chemistry*, *52*(5), 1124-1131. doi: 10.1021/jf0499525

Campagnolo, M. A., & Pio, R. (2012). Phenological and yield performance of black and redberry cultivars in western Paraná State. *Acta Scientiarum. Agronomy*, *34*(4), 439-444. doi: 10.4025/actasciagron.v34i4.15528

Caproni, C. M., Curi, P. N., Moura, P. H. A., Pio, R., Gonçalves, E. D., & Pasqual, M. (2016). Blackberry and redberry production in crop and intercrop in Pouso Alegre, southern Minas Gerais, Brazil. *Ciência Rural, 46*(10), 1723-1728. doi: 10.1590/0103-8478cr 20150623

Carvalho, I. S., Cavaco, T., & Brodelius, M. (2011). Phenolic composition and antioxidant capacity of six artemisia species. *Industrial Crops and Products*, *33*(2), 382-388. doi: 10.1016/j.indcrop.2010.11.005

Cordenunsi, B. R., Nascimento, J. R. O., Genovese, M. I., & Lajolo, F. M. (2002). Influence of cultivar on quality parameters and chemical composition of strawberry fruits grown in Brazil. *Journal Agriculture of Food Chemistry*, *50*(9), 2581-2586. doi: 10.1021/jf011421i

Curi, P. N., Pio, R., Moura, P. H. A., Lima, L. C. O., & Valle, M. H. R. (2014). Qualidade de framboesas sem cobertura ou cobertas sobre o dossel e em diferentes espaçamentos. *Revista Brasileira de Fruticultura*, *36*(1), 199-205. doi: 10.1590/0100-2945-234/13

Curi, P. N., Pio, R., Moura, P. H. A., Tadeu, M. H., Nogueira, P. V., & Pasqual, M. (2015). Produção de amora-preta e amora-vermelha em Lavras-MG. *Ciência Rural*, *45*(8), 1368-1374. doi: 10.1590/0103-8478cr20131572

Dantas, A. A. A., Carvalho, L. G., & Ferreira, E. (2007). Classificação e tendências climáticas em Lavras, MG. *Ciência e Agrotecnologia*, *31*(6), 1862-1866. doi: 10.1590/S1413-70542007000600039

Guedes, M. N. S., Abreu, C. M. P., Maro, L. A. C., Pio, R., Abreu, J. R., & Oliveira, J. O. (2013). Chemical characterization and mineral levels in the fruits of blackberry cultivars grown in a tropical climate at an elevation. *Acta Scientiarum. Agronomy*, *35*(2), 191-196. doi: 10.4025/actasciagron.v35i2.16630

Hidalgo, M., Sánchez-Moreno, C., & Pascual-Teresa, S. (2010). Flavonoid-flavonoid interaction and its effect on their antioxidant activity. *Food Chemistry*, *121*(3), 691-696. doi: 10.1016/j.foodchem.2009.12.097

Jacques, A. C., Pertuzatti, P. B., Barcia, M. T., Zambiazi, R. C., & Chim, J. F. (2010). Estabilidade de compostos bioativos em polpa congelada de amora-preta (*Rubus fruticosus*) cv. Tupy. *Química Nova*, *33*(8), 1720-1725. doi: 10.1590/S0100-40422010000800019

Lee, S. M., Kim, C. W., Kim, J. K., Shin, H. J., & Baik, H. J. (2008). GCG-Rich tea catechins are effective in lowering cholesterol and triglyceride concentrations in hyperlipidemic rats. *Lipids*, *43*(5), 419-429. doi: 10.1007/s11745-008-3167-4

Maro, L. A. C., Pio, R., Guedes, M. N. S., Abreu, C. M. P., & Curi, P. N. (2013). Bioactive compounds, antioxidant activity and mineral composition of fruits of raspberry cultivars grown in subtropical areas in Brazil. *Fruits*, *68*(3), 209-217. doi: 10.1051/fruits/2013068

Maro, L. A. C., Pio, R., Guedes, M. N. S., Abreu, C. M. P., & Moura, P. H. A. (2014). Environmental and genetic variation in the post-harvest quality of raspberries in subtropical areas in Brazil. *Acta Scientiarum. Agronomy*, *36*(3), 323-328. doi: 10.4025/actasciagron.v36i3.18050

Scalzo, J., Battino, M., Costantini, E., & Mezzetti, B. Breeding and biotechnology for improving berry nutritional quality. (2005). *BioFactors*, *23*(4), 213-220.

Silva, R. S., Vendruscolo, J. L., & Toralhes, R. P. (2011). Avaliação da capacidade antioxidante em frutas produzidas na região Sul do RS. *Revista Brasileira de Agrociência*, *17*(3-4), 392-400.

Siriwoharn, T., Wrolstad, R. E., & Durst, R. W. (2005). Identification of ellagic acid in blackberry juice sediment. *Journal of Food Science*, *70*(3), 189-197. doi: 10.1111/j.1365-2621.2005.tb07124.x

Souza, V. R., Pereira, P. A. P., Silva, T. L. T., Lima, L. C. O., Pio, R., & Queiroz, F. (2014). Determination of the bioactive compounds, antioxidant activity and chemical composition of Brazilian blackberry, red raspberry, strawberry, blueberry and sweet cherry fruits. *Food Chemistry*, *156*(8), 362-368. doi: 10.1016/j.food chem.2014.01.125

Souza, V. R., Pereira, P. A. P., Teixeira, T. R., Silva, T. L. T., Pio, R., & Queiroz, F. (2015). Influence of processing on the antioxidant capacity and bioactive compounds in jellies from different blackberry cultivars. *International Journal of Food Science & Technology*, *50*(7), 1658-1665. doi: 10.1111/ijfs.12819

Szajdek, A., & Borowska, E. J. (2008). Bioactive compounds and health-promoting properties of berry fruits: Review. *Plant Foods for Human Nutrition*, *63*(4), 147-156.

Vizzotto, M., & Pereira, M. C. (2011). Amora-preta (*Rubus* sp.): otimização do processo de extração para determinação de compostos fenólicos antioxidantes. *Revista Brasileira de Fruticultura*, *33*(4), 1209-1214. doi: 10.1590/S0100-29452011000400020

Zozio, S., Dominique, P., & Dornier, M. (2011). Evaluation of anthocyanin stability during storage of a coloured drink made from extracts of the Andean blackberry (*Rubus glaucus* Benth.), açai (*Euterpe oleracea* Mart.) and black carrot (*Daucus carota* L.). *Fruits, 66*(3), 203-215. doi: 10.1051/fruits/2011030

16

Production and mass partitioning in Peruvian carrot plants grown under artificial shading period and intensity

Tatiana Pires Barrella[1], Mário Puiatti[1*], Ricardo Henrique Silva Santos[1] and Paulo Roberto Cecon[2]

[1]Departamento de Fitotecnia, Centro de Ciências Agrárias, Universidade Federal de Viçosa, Av. Peter Henry Rolfs, s/n, 36570-000, Viçosa, Minas Gerais, Brazil. [2]Departamento de Informática, Universidade Federal de Viçosa, Viçosa, Minas Gerais, Brazil.
**Author for correspondence: E-mail: mpuiatti@ufv.br*

ABSTRACT. Due to its long production cycle, the Peruvian carrot has been cultivated on marginal land, resulting in low yields. Intercropping is an option for planting in better areas, but there is a lack of research on crop shade tolerance. This study aimed to evaluate the effect of period and intensity of artificial shading on mass partitioning and root yield on Peruvian carrot 'Amarela de Carandaí'. The treatments were four shading intensities (18, 30 and 50% of shading with Sombrite® cover and full sun), imposed either from 30 to 120, or from 30 to 210, or from 120 to 210 days after planting (DAP), totaling 10 treatments at harvest (210 DAP), with four repetitions. Light restriction above 18% from 30–210 DAP and from 120–210 DAP reduced marketable root yield. The marketable root yield was less reduced when shading was imposed earlier (30–120 DAP) than later (120-210 DAP) on the growth cycle. Regarding light competition, Peruvian carrot should be intercropped with crops that promoted up to 50% of shading from 30-120 DAP, or up to 18% of shading from 120 to 210 or 30-120 DAP.

Keywords: *Arracacia xanthorrhiza*, arracacha, intercropping, light effect.

Introduction

Peruvian carrot (*Arracacia xanthorrhiza* Bancroft) is a tuberous vegetable of long cultivation cycle that occupies the land for a period of 10 to 12 months (CASALI, 1984; GRANATE et al., 2007; SANTOS, 1997; SANTOS; CARMO, 1998). In most cases, it is cultivated by producers with limited area, who are forced to implant the culture in marginal areas, usually soils with low fertility and high inclination, thus achieving low yields (CASALI, 1984; GIL LEBLANC et al., 2008). It is estimated that the area cultivated with Peruvian carrot in Brazil is 13,000 ha, with an average productivity of 9.2 t ha^{-1} (REGHIN et al., 2001). In spite of the high perishability of tuberous or storage roots (MENOLLI et al., 2008), the high prices achieved and the rusticity of the culture have encouraged their culture, mainly in the familiar agriculture system (GRANATE et al., 2007; HEREDIA ZÁRATE et al., 2007; SANTOS, 1997; SANTOS; CARMO, 1998; TOLENTINO JUNIOR et al., 2002; VIEIRA et al., 2003), due to the great involvement of manpower, especially at harvest (SANTOS; CARMO, 1998).

In situations of limited available areas, plant intercropping becomes an excellent alternative because it allows better soil exploration and an increase in production and income per unit of explored area and time unit (VANDERMEER, 1989). Due to the long cultivation cycle, the intercropping of Peruvian carrot with other plant species could be a solution for properties with area limitations, allowing its cultivation in better areas, an increment in culture yield and the diversification in production and income (HEREDIA ZÁRATE et al., 2007; TOLENTINO JÚNIOR et al., 2002; VIEIRA et al., 2003).

Some research works have been carried out on the intercropping of the Peruvian carrot and other vegetables of smaller size, such as lettuce (VIEIRA et al., 2003), lettuce and beet (TOLENTINO JÚNIOR et al., 2002) and carrot and coriander (HEREDIA ZÁRATE et al., 2007), with positive results in income for the intercropping of Peruvian carrot with lettuce (TOLENTINO JUNIOR et al., 2002; VIEIRA et al., 2003). However, although some producers cultivate Peruvian carrot in the row spacing of coffee plants (CASALI, 1984), its intercropping with coffee plants and other species of larger size lacks scientific information.

When establishing a plant intercropping system, it is very important to know the real light need or the tolerance cultures have to shading (VANDERMEER, 1989). Studies using some clones of Peruvian carrot demonstrated that the productivity of storage roots correlates directly and positively with the rate of CO_2 assimilation and the specific leaf area (JAIMEZ et al., 2008). In this context, studies on the use of artificial shading are desirable for a better evaluation of the effects of light by itself, since, without any other intercropped culture, other interferences such as competition for nutrients and water are not present.

Therefore, knowledge on the behavior of the Peruvian carrot culture as for light demand, as well as to the knowledge on the phenology of the cultures that will be associated with it, makes it easier to decide which cultures and periods would be more appropriate for intercropping implantation. Thus, the achievement of information that will help in the decisions related to the establishment of intercropping involving Peruvian carrots and other plant species has become a basic need.

The objective of the present work was to evaluate the effect of intensities and periods of artificial shading on the production and mass partitioning in the organs of the Peruvian carrot plant 'Amarela de Carandaí'.

Material and methods

The experiment was carried out in the Research Station of the Department of Plant Science, of the Universidade Federal de Viçosa, Viçosa, Minas Gerais State, from 5/15/2002 to 12/20/2002. The soil, Argissolo Vermelho Amarelo Câmbico (Ultisol), with mild topography, with approximately 1-2% of inclination, presented the following values in the chemical analysis (depth of 0-20 cm): pH in water (1:2.5) = 5.5; P = 89.3* and K = 162.0* mg dm^{-3}; Ca = 3.4**; Mg = 0.6**; Al = 0.0**; H + Al = 4.3***; SB = 4.2; CTC_t = 4.2 and CTC_T = 8.5 $cmol_c$ dm^{-3}; V = 49.0 and m = 0.0%. (*Mehlich 1 Extractor; ** KCl Extractor 1 mol L^{-1}; ***Calcium Acetate Extractor 0.5 mol L^{-1} pH 7.0). The soil was prepared with plowing and harrowing. Then, the planting furrows were made, with about 0.20 m of height, and spacing of 0.80 m.

Shoots from the clone 'Amarela de Carandaí' (BGH 5746) were used. They were achieved from the Vegetable Germplasm Bank of the Universidade Federal de Viçosa (BGH-UFV). The aerial and basal parts of the shoots were cut, maintaining 1.0 cm of the petiole, inclined cut at the base (simple bevel) and average length of 2.5 cm.

After the cuts were healed, the shoots were planted on the furrows, with a distance of 0.40 m between them, with enough depth to cover about 2.0 cm of the apex. No fertilization procedures were used because the fertility levels were fit for the culture. At 60 days after planting (DAP) one single cover nitrogen fertilization was performed, with ammonium sulfate, in the dose of 40 kg ha^{-1} of N, according to recommendations of Santos and Carmo (1998). Manual weeding and irrigation by infiltration were performed, when necessary. Pulverizations with pesticides were not necessary.

The treatments presented three shading intensities (18, 30 and 50%), implemented in three periods (30 to 120; 30 to 210 and 120 to 210 days after planting - DAP), and the control (control under full sun during all the cycle) (Table 1). The intensities of the shading were achieved with the use of Sombrite® covers, starting with 30 DAP when the plants were producing the second leaf, except for the control and shading of 120 to 210 DAP. Following this treatment, the covers were changed 120 DAP. Therefore, in the evaluation performed 120 DAP, there was a total of four treatments, while in the evaluation performed 210 DAP, there were 10 treatments (Table 1).

Table 1. Treatments of artificial shading intensities and periods used during the cultivation of the Peruvian carrot 'Amarela de Carandaí'. UFV, Viçosa, Minas Gerais State, 2002.

Treatment	Intensity	Period (Days)		
		0-30	30-120	120-210
1 (control)	SUN	SUN	SUN	SUN
2	18%	SUN	18%	18%
3	30%	SUN	30%	30%
4	50%	SUN	50%	50%
5	18%	SUN	18%	SUN
6	30%	SUN	30%	SUN
7	50%	SUN	50%	SUN
8	18%	SUN	SUN	18%
9	30%	SUN	SUN	30%
10	50%	SUN	SUN	50%

A randomized experimental design was used, with four replications. The experimental unit presented eight rows with 2.8 m of length, each one with seven plants, totaling 56 plants and an area of 17.92 m^2. The four central lines were considered useful, while the 0.80 m in the extremities were excluded. The Sombrite® covers were stretched about 0.50 m above the canopy of the plants, and were fixed by extensors in the lateral areas. During cultivation, the covers were lifted as plants grew.

Two harvests were carried out at 120 and 210 DAP; for each evaluation, six plants per plot were harvested, corresponding to two rows inside the useful area of each plot. The plants were separated into blade (limb), petiole, shoot, crown, absorbent roots and storage roots. All the roots with a diameter larger than 1.0 mm were considered storage roots; among them, those with a diameter larger or equal to 15.0 mm in the thicker portion were considered marketable. The dry matter (DM) of each organ was obtained after drying with forced air ventilation at 60 ± 5°C, until the constant mass was achieved. The DM ratio of the aerial part (blade + petiole + shoot + crown)/subterranean part (storage roots + absorbent roots) and the harvest index (HI = DM of marketable roots / plant total DM) were calculated.

The data achieved in the evaluations carried out on 120 and 210 DAP were separately submitted to the variance analysis. The averages of the treatments were compared to the average of the control treatment (full sun) using the Dunnett test at 5% probability. The descriptive analysis was carried out through the percentage proportion of DM in each plant organ in the treatments.

Results and discussion

In the evaluation carried out at 120 DAP, the accumulation of DM in the plant organs, except for the petiole and absorbent roots, was significantly lower in plants under shading than under full sun (Table 2). These results demonstrate that the shading in the initial period of cultivation (30 at 120 DAP) influenced the growth of the plants, since they accumulated less DM in their organs and in the whole plant; additionally, although non-significant in number, plants under higher intensities of shading presented, proportionally, a higher DM accumulation in the aerial part (DMAP/DMSP ratio), evidencing that there was a higher investment in the aerial structures, in search for light capture, a typical response of sun plants to light restriction (LARCHER, 2000). These results are in accordance with those found by Jaimez et al. (2008), who achieved a linear and positive relation between the specific leaf area and CO_2 assimilation rate and the production of storage roots, while working with five cultivars of Peruvian carrot in two locations with different altitudes in Venezuela.

Table 2. Dry matter (g $plant^{-1}$) of blade, petiole, shoot, crown, storage roots (RR), marketable (MR) and absorbent (AR), total of the plant, dry matter of aerial part / subterranean part ratio (AP/SP) and harvest index (HI = DM of marketable roots / plant total DM) observed at 120 and 210 days after planting, in Peruvian carrot plants cultivated under intensities (levels) and periods (days) of artificial shading. UFV, Viçosa, Minas Gerais State, 2002.

Artificial shading		Aerial part				Roots			Plant	Ratio AP/ SP	HI
Level (%)	Period (DAP)	Blade	Petiole	Shoot	Crown	RR	MR	AR	whole		
Evaluation 120 days after plantation											
Sun (Control)	-	162.4	66.7	38.2	17.0	24.3	-	3.4	312.0	10.3	-
18		64.9 *	33.5 ns	10.7 *	8.2 *	10.6 *	-	3.6 ns	131.5 *	8.3 ns	-
30	30-120	103.7 *	62.8 ns	13.3 *	10.9 *	10.1 *	-	3.3 ns	204.1*	14.2 ns	-
50		98.9 *	60.2 ns	15.0 *	10.4 *	7.9 *	-	3.3 ns	195.7 *	16.5 ns	-
Evaluation 210 days after plantation											
Sun (Control)	-	197.7	746.0	344.0	76.0	255.7	233.35	8.8	1628.2	5.7	0.14
18	30-210	256.2 ns	748.7 ns	588.0 ns	81.0 ns	255.0 ns	230.5 ns	5.4 ns	1934.3 ns	6.4 ns	0.12
30		263.2 ns	550.7 ns	532.0 ns	47.0 ns	170.6 ns	109.6 ns	11.4 ns	1574.9 ns	7.7 ns	0.07*
50		299.0 ns	533.9 ns	362.0 ns	47.0 ns	50.4 *	28.3 *	10.2 ns	1302.5 ns	20.5 *	0.02*
18		236.7 ns	543.7 ns	455.0 ns	79.0 ns	320.0 ns	308.3 ns	10.0 ns	1644.4 ns	4.0 ns	0.18
30	30-120	328.7 ns	618.5 ns	568.0 ns	76.0 ns	256.8 ns	205.3 ns	11.5 ns	1859.5 ns	5.9 ns	0.11
50		237.7 ns	626.5 ns	524.0 ns	92.0 ns	353.4 ns	272.3 ns	15.3 ns	1848.9 ns	4.2 ns	0.15
18		249.2 ns	545.2 ns	458.0 ns	80.0 ns	296.2 ns	265.0 ns	7.8 ns	1636.4 ns	4.4 ns	0.16
30	120-210	235.2 ns	631.0 ns	436.0 ns	54.0 ns	136.8 *	98.2 ns	9.4 ns	1502.4 ns	9.3 ns	0.07*
50		204.4 ns	561.0 ns	335.0 ns	45.0 ns	58.7 *	42.4 *	8.8 ns	1212.9 ns	17.0*	0.03*

*; ns: in each evaluation, significant and non-significant, respectively, at the level of 5%, in relation to the control (full sun), by the Dunnett test.

However, the results found in the evaluation at 210 DAP were different from those achieved in the evaluation at 120 DAP, for the same shading period (30-120), in which no difference for the dry matter of the organs was detected between the treatments of shading and the control (Table 2). It demonstrates that, although the young plants (120 DAP) present growth restriction for several organs due to shading, when exposed to full light (sun) during the three following months, these plants recovered that decrease, accumulating, at 210 DAP, dry matter in similar quantities to those plants that developed under full sun during the whole cycle. Such recovery is also evidenced by the total plant dry matter and the balance of the DMAP/DMSP ratio, as well as by the harvest index, which, inclusively, presented a numeric value higher than that of the control for the treatment of 18% of light restriction.

The harmful effect of the restriction of light on the growth of Peruvian carrot plants depends on the stage in which the plant is submitted to the stress and the level of restriction of light applied. So, levels of light restriction of up to 50% applied in the initial stage (30-120 DAP), and of up to 18% during the whole cycle (30-210 days) or in the final stage (120-210 days) were not harmful to the culture in terms of production of marketable root mass and harvest index. However, levels of light restriction above 18% during the whole cycle (30-210 days) or in the final stage (120-210 days) were harmful to the formation of marketable roots, causing a lower harvest index (Table 2).

Although the dry matter of the whole plant at 210 DAP for all the treatments of intensities and periods of shading was not different from the results for the control (Table 2), accumulation rate of the DM in its parts (partitioning) varied among the treatments (Figure 1). Plants submitted to the shading intensity of 30% and, mainly, of 50% during the whole cycle (30-210) or in the final period (120-210 DAP), proportionally allocated a higher percentage of mass for the aerial part (especially blade) than for the subterranean part, mainly the storage roots. Consequently, in these periods and intensities of shading, the DMAP/DMSP ratio was higher and the harvest index was lower (Table 2).

Therefore, plants under a higher restriction during the whole cycle (30-210 DAP) or in the final period (120-210 DAP) produced photo assimilates in amounts only enough for their maintenance and the investment in aerial parts, providing a restricted quantity of assimilates to be allocated for the storage roots. This harmful effect was proportional to the intensity of the restriction of light (Table 2, Figure 1). Thus, it could be used an intercropping with other species that provides for the Peruvian carrot culture shading intensities of up to 18% during the whole cycle (30-210) or in the final period (120-210), or up to 50% in the initial period (30-120 DAP), because it would less harmful to the production of marketable storage roots.

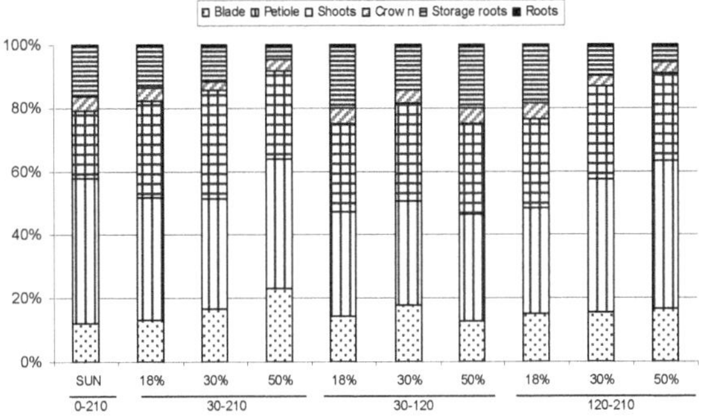

Figure 1. Partitioning of dry matter (in percentage) among the organs of the Peruvian carrot 'Amarela de Carandaí' plants cultivated under intensities and periods of artificial shading, 210 days after planting. UFV, Viçosa, Minas Gerais State, 2002.

Although LER (Land Equivalent Ratio) was higher than 1.0, a productivity decrease was observed for Peruvian carrot during the intercropping of Peruvian carrot with other vegetables of smaller size, such as lettuce (VIEIRA et al., 2003), lettuce and beet (TOLENTINO JÚNIOR et al., 2002) and carrot and coriander (HEREDIA ZÁRATE et al., 2007), but the results were positive for income for the intercropping of Peruvian carrot with lettuce (TOLENTINO JÚNIOR et al., 2002; VIEIRA et al., 2003). In cases like this, besides competition from other growth factors other than light (water, nutrients etc.) among plants of different cultures, there may have been a restriction of light for the Peruvian carrot plants in the initial stage of growth, since this species presents a slow initial growth when compared to other vegetables. Therefore, the harmful effect of the restriction of light, besides the level of restriction, depends on the stage in which the plant is submitted to the light restriction stress.

The plants in the treatments under restriction of light invested, proportionally, more mass in the aerial part (mainly blade and petiole), which suggests a morpho-adaptive response of the plant to the light environment. However, perhaps Peruvian carrot plants do not present more refined adaptive physiological mechanisms as to the restriction of available radiation, such as the diminishing of the respiratory standard and the light compensation point, usually observed in plants under shading, since in these treatments with higher restriction of light there was probably a restriction of assimilates for the storage roots. The low productivity of storage roots in some Peruvian carrot cultivars in Venezuela was also ascribed to the restriction of assimilates due to the rainy season, according to Jaimez et al. (2008).

The plants submitted to the highest level of light restriction during the whole cycle presented, proportionally, a great mass of aerial part (blade and petiole), but small production of storage roots (Figure 1 and Table 2). Therefore, a great leaf area is not enough, but it must be exposed to a level of radiation that allows to a higher liquid photosynthesis, according to observations by Jaimez et al. (2008) in Peruvian carrot cultivars in Venezuela, which demonstrates the significance of the leaf area and its efficiency in CO_2 assimilation.

The accumulation of dry matter in storage roots of the Peruvian carrot plant starts around the sixth month after planting (CÂMARA; CASALI, 1985); but, although already presenting marketable roots, they exhibit little mass until the seventh month (CÂMARA et al., 1985). In this work, although storage roots have already been observed at 120 DAP, their mass values were very low, below the commercial standard. Nevertheless, they presented a great increase of dry matter during the three following months, except for the treatments with restriction above 18% at 30-210 DAP and 120-210 DAP (Table 2), which proves that there had been restriction of assimilates for the storage roots in these treatments.

Although the culture was conducted until 210 DAP, it was evident that the negative effect of the light restriction on the production of storage roots is higher when implemented at 120 DAP, in comparison to the previous period (30 - 120 DAP). Somehow, these results are in accordance with Câmara and Casali (1985), who observed, in plants cultivated under full sun, that the restriction of the leaf area of these plants in the seventh month did not affect the following production of storage roots. However, since the plants had been cultivated under full sun, without light restriction, they could have stored reserves in the crown, as, in accordance with Gil Leblanc et al. (2008), the crown may play multiple roles, among which, it may work as a temporary drain, under certain circumstances.

Considering the population of 31,250 plants ha^{-1} and the accumulation of dry matter in the marketable roots of 0.233 kg $plant^{-1}$ of the plants under full sun, the estimated production of marketable roots, in the seventh month was 7.3 t ha^{-1} of dry matter; in the treatments that received the shading of 50% at 30-210 DAP and 120-210 DAP, these were only 0.88 and 1.31 t ha^{-1}, respectively. Câmara et al. (1985) obtained marketable root production of 4.44 t ha^{-1}, in the eighth month, for the cultivation under full sun. Therefore, the production under full sun at 210 DAP was very satisfactory.

The production of roots of Peruvian carrot cultivars in two locations in Venezuela was higher where the Photon Flux Density from 8:30 to 10:30 a.m. was about 30% lower (JAIMEZ et al., 2008). Therefore, although being a sun plant, Peruvian carrot seems to benefit from the partial reduction of incident radiation as for the production of tuberous roots, maybe because it is a C_3 species, which means that it is subject to photorespiration when it is under high irradiance.

Therefore, in accordance with the results achieved, the shading at the level of up to 50% could be used in the initial stage of cultivation, which corresponds to up to 120 DAP, without the decrease of the production of marketable roots; similarly, shading of up to 18% during the whole cycle (30-210 DAP) or from the beginning of the formation of storage roots could also be used. Nevertheless, during the whole cycle or starting at 120 DAP, shading above 18% would not be recommended because it causes the decrease of the production of marketable storage roots. So, during the implementation of the intercropping system of Peruvian carrot with other larger species, the associated culture should not promote light restriction to the Peruvian carrot culture above 18% during the whole cycle or after 120 DAP.

Conclusion

Intensities of shading above 18% during the whole cultivation cycle (30-210 days after plantation - DAP) or starting from the beginning of the formation of tuberous roots (after 120 DAP) were harmful to the production of marketable storage roots of Peruvian carrot.

The production of marketable storage roots was not affected when light restriction occurred in the initial stage (30-120 DAP), regardless of shading intensity.

Considering only the light effect, the intercropping of the Peruvian carrot must be used with plant species that promote shading of the culture without exceeding the maximum intensity of 50% in the initial stage of cultivation (30-120 DAP) or up to 18% during the whole cycle (30-210 DAP) or in the final period (after 120 DAP), without causing significant reductions in the production of the marketable storage roots.

Acknowledgements

To CNPq, for the Masters Scholarship granted to the first author.

References

CÂMARA, F. L. A.; CASALI, V. W. D. Época de plantio, ciclo e amassamento dos pecíolos da mandioquinha-salsa. **Horticultura Brasileira**, v. 3, n. 2, p. 25-28, 1985.

CÂMARA, F. L. A.; MAFFIA, L. M.; CASALI, V. W. D. Curva de crescimento e utilização da mandioquinha-salsa na alimentação. **Horticultura Brasileira**, v. 3, n. 2, p. 29-33, 1985.

CASALI, V. W. D. Escolha da área de plantio e preparo do solo para mandioquinha-salsa. **Informe Agropecuário**, v. 10, n. 120, p. 24-26, 1984.

GIL LEBLANC, R. E.; PUIATTI, M.; SEDIYAMA, M. A. N.; FINGER, L. F.; MIRANDA, G. V. Influência do pré-enraizamento e de tipos de mudas sobre a população, crescimento e produção da mandioquinha-salsa 'Roxa de Viçosa'. **Revista Ceres**, v. 55, n. 1, p. 74-82, 2008.

GRANATE, M. J.; SEDIYAMA, M. A. N.; PUIATTI, M. Batata-baroa ou mandioquinha-salsa (*Arracacia xanthorrhiza* Bancroft). In: PAULA JÚNIOR, T. J.; VENZON, M. (Ed.). **101 culturas**: manual de tecnologias agrícolas. Belo Horizonte: Epamig, 2007. p. 137-142.

HEREDIA ZÁRATE, N. A.; VIEIRA, M. C.; PONTIM, B. C. A.; FIGUEIREDO, P. G.; QUEVEDO, L. F.; ALMEIDA, S. O. Produção e renda bruta de mandioquinha-salsa, solteira e consorciada com cenoura e coentro. **Acta Scientiarum. Agronomy**, v. 29, n. 4, p. 549-553, 2007.

JAIMEZ, R. E.; SANTOS, N.; AÑEZ, B.; VÁSQUEZ, J.; ESPINOZA, W. Photosyntesis of field-grown Arracacha (*Arracacia xanthorrhiza* Bancroft) cultivars in relation to root-yield. **Scientia Horticulturae**, v. 118, n. 2 p. 100-105, 2008.

LARCHER, W. **Ecofisiologia vegetal**. São Carlos: RiMa, 2000.

MENOLLI, L. N.; FINGER, L. F.; PUIATTI, M.; BARBOSA, J. M.; BARROS, R. S. Atuação de enzimas oxidativas no escurecimento causado pela injúria por frio em raízes de batata-baroa. **Acta Scientiarum. Agronomy**, v. 30, n. 1, p. 57-63, 2008.

REGHIN, M. Y.; OTTO, R. F.; SILVA, J. B. C. "Stimulate Mo" e proteção com tecido 'não tecido' no pré-enraizamento de mudas de mandioquinha-salsa. **Horticultura Brasileira**, v. 18, n. 1, p. 49-52, 2001.

SANTOS, F. F. A cultura da mandioquinha-salsa no Brasil. **Informe Agropecuário**, v. 19, n. 190, p. 5-7, 1997.

SANTOS, F. F.; CARMO, C. A. S. **Mandioquinha-salsa**: manejo cultural. Brasília: CNPH, 1998.

TOLENTINO JÚNIOR, C. F.; HEREDIA ZÁRATE, N. A.; VIEIRA, M. C. Produção de mandioquinha-salsa consorciada com alface e beterraba. **Acta Scientiarum. Agronomy**, v. 24, n. 5, p. 1447-1454, 2002.

VANDERMEER, J. H. **The ecology of intercropping**. Cambridge: Cambridge University Press, 1989.

VIEIRA, M. C.; HEREDIA ZÁRATE, N. A.; GOMES, H. E. Produção e renda bruta de mandioquinha-salsa e alface, solteiras e consorciadas, com adubação nitrogenada e cama-de-frango em cobertura. **Acta Scientiarum. Agronomy**, v. 25, n. 1, p. 201-208, 2003.

17

Organic cultivation of onion under castor cake fertilization and irrigation depths

Gabriel Alves Botelho de Mello[1*], Daniel Fonseca de Carvalho[1], Leonardo Oliveira Medici[2], Aldir Carlos Silva[1], Daniela Pinto Gomes[1] and Marinaldo Ferreira Pinto[1]

[1]Departamento de Engenharia, Universidade Federal do Rio de Janeiro, BR-465, km 7, 23897-000, Seropédica, Rio de Janeiro, Brazil.
*[2]Departamento de Ciências Fisiológicas, Universidade Federal do Rio de Janeiro, Seropédica, Rio de Janeiro, Brazil. *Author for correspondence. Email: gabrirjrural@gmail.com*

ABSTRACT. Two experiments were carried out to evaluate the effects that different doses of castor bean (0, 200, and 300 g m^{-2}) and irrigation levels (204, 224, 278, and 321 mm in 2014, and 278, 302, 397, and 444 mm in 2015) have on both the productivity and quality of onion bulbs. In the experiments, the experimental design was randomized blocks (4 x 3 factorial scheme), with five replicates. Irrigation management was performed using a Simplified Irrigation Device (SID) in response to soil water tension in the treatment of highest irrigation depth. The following variables were evaluated: plant dry biomass (PDB), bulb dry biomass (BDB), total yield (TY), mean bulb fresh weight (MBFW), mean bulb dry weight (MBDW), mean bulb diameter (MBD) and water use efficiency (WUE). The highest irrigation depths positively influenced the mean production of onion bulbs, regardless of the applied dose of castor cake. The doses of castor cake positively influenced the production of onion bulbs when higher irrigation depths were applied.

Keywords: *Allium cepa* L., water deficit, organic fertilization, irrigation automation.

Introduction

Onion occupies a prominent position among bulbous vegetables and has the great economic importance in the world (El Balla, Hamid, & Abdelmageed, 2013). It is considered the third most important vegetable in economic value in Brazil (Cecílio Filho, May, Pôrto, & Barbosa, 2009). It is the main agricultural activity of more than 60 thousand families, and almost 100% of its production area is subjected to irrigation, except in the south region of the country (Ribeiro, Carvalho, Santos, & Guerra, 2016).

Because the cultivated area of onion is increasing worldwide, irrigation systems must be optimized, especially in arid and semi-arid regions (Zheng et al., 2013). In this context, studies to characterize the adequate water depth for the onion crop have been conducted, and their results indicate that higher yields occur when the soil is maintained with moisture close to field capacity (Channagoudra, Prabhudeva, & Kamble, 2009; Martin de Santa Olalla, Domínguez-Padilla, & López, 2004; Shock, Feibert, & Saunders, 2000).

Applying moderate water restriction to an onion crop aims to anticipate the beginning of bulb formation and the end of the crop cycle, which may be beneficial, depending on the purpose of the cultivation. However, the yields decrease when the water deficiency coincides with the beginning of bulb formation, which is considered the critical period for crop development (Zheng et al., 2013). The management of nutrients and irrigation directly influences the onion yield, the maturity stage of the plants and the harvest (Tekalign, Abdissa, & Pant, 2012; Leskovar, Yoo, & Pascual-Seva, 2012), and it interferes with the size, quality, sprouting percentage and total mass of the bulbs. However, few studies have analyzed the effects that water availability and application have on the performance of organic fertilizers. The soil moisture affects the mineralization process of these materials, resulting in a greater release of nutrients, especially nitrogen (N) (Lima et al., 2011), which increases the productivity of onion bulbs (Boyhan, Torrance, & Hill, 2007; Resende & Costa, 2008).

The organic production of onion is technically feasible when using organic materials, such as castor cake, that provide all of the nutrients required for its growth (Costa, Araujo, Santos, Resende, & Lima, 2008). Castor cake is a byproduct of castor oil extraction (Magriotis et al., 2014); it contains 4.15, 0.61, and 0.96% of N, P, and K, respectively, and has a C/N ratio of approximately 11.6 (Zapata, Vargas, Reyes, & Belmar, 2012). Thus, it is considered a high-quality fertilizer (Lima et al., 2011). This material has been widely used in the production of vegetables in organic systems (Santos, Espíndola, Guerra, Leal, & Ribeiro, 2012).

This study was undertaken to further evaluate the interactive effects that different irrigation levels and doses of castor bean cake have on the productivity and quality of onion bulbs in an organic production system.

Material and methods

Two experiments were carried out from May to October in both 2014 and 2015, at the experimental field of the "Fazendinha Agroecológica", km 47, in the municipality of Seropédica, Baixada Fluminense of Rio de Janeiro State, Brazil (22°45` S; 43°41`W; 33.0 m). The climate of the region is tropical and is classified as Aw according to Köppen's classification. During the experiments, the mean values of temperature and relative air humidity were 22.0°C and 72.9% in 2014 and 23.4°C and 77.2% in 2015. The cumulative rainfall was 128 mm in 2014 and 160 mm in 2015 (Figure 1).

The experiments were conducted in Red Yellow Argisol, with a sandy loam texture (Embrapa, 2006), which showed, in the 0 - 0.2 m layer, a soil density of 1.42 g cm^{-3} and contents of clay, sand and silt of 14, 70, and 13 g kg^{-1}, respectively. The soil chemical characteristics were the following: pH: 6.2; Al, 0.0 $mmol_c\ dm^{-3}$, Ca: 37 $mmol_c\ dm^{-3}$, Mg: 16 $mmol_c\ dm^{-3}$, K: 120 mg dm^{-3}, and P: 54 mg dm^{-3} (mean of 2014 and 2015). Minimum cultivation was adopted as the soil tillage.

The sowing of onion (cv. Alfa tropical) and its transplanting were performed in May and late June, respectively, in both 2014 and 2015. Basal fertilization was performed using aged bovine manure (1.0 kg m^{-2} of dry matter per plot), whose chemical characteristics were the following: N: 18.5 g kg^{-1}, P: 4.2 g kg^{-1}, K: 14.5 g kg^{-1}, Ca: 9.4 g kg^{-1}, and Mg: 7.7 g kg^{-1} (mean of 2014 and 2015).

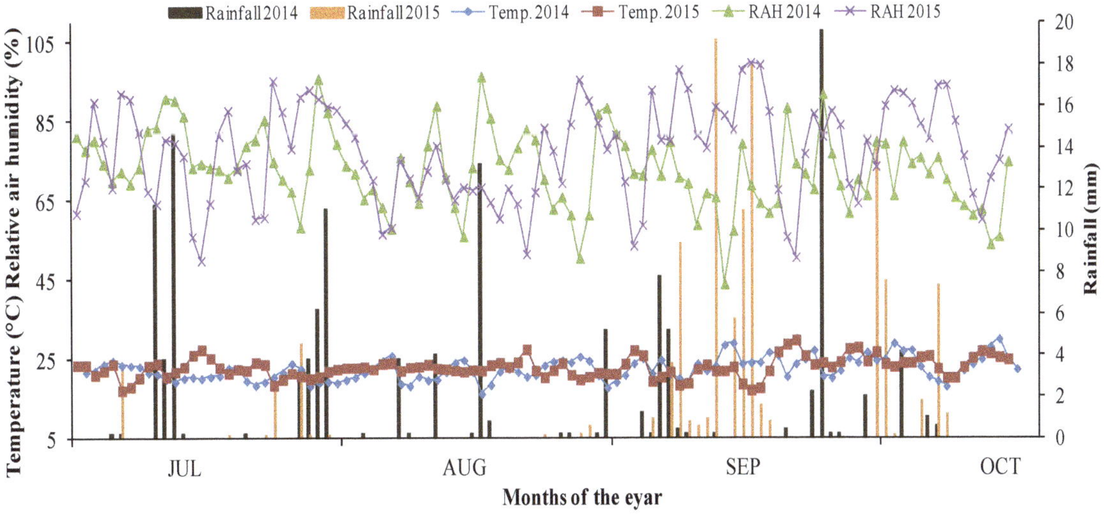

Figure 1. Mean temperature, relative air humidity and rainfall in the cultivation periods in 2014 and 2015.

The plots were arranged in 1-m^2 beds, spaced by 0.50 m of border and cultivated with 40 plants each, with a spacing of 0.10 m between plants and 0.25 m between rows. Cultivation practices consisted of manual weedings and phytosanitary control using Bordeaux mixture and roguing.

The different irrigation depths were applied through a drip system, using emitters ("iDrop normal"/Irritec) with nominal flow rates of 2.0 and 4.0 L h^{-1} that were adjusted at different spacings. Irrigation management was performed using the "Simplified Irrigation Device" (SID) (Medici, Rocha, Carvalho, Pimentel, & Azevedo, 2010), which operates in response to soil water tension and is regulated by the unevenness between a porous capsule and a pressure switch.

The adopted experimental design was in randomized blocks, in a 4 x 3 factorial scheme with five replicates. The treatments consisted of irrigation depths (L1- irrigation depth applied by the SID, L2 - 83% of L1, L3 - 50% of L1, and L4 - 42% of L1) and doses of castor cake as top-dressing (A - 0 g m^{-2}, B - 200 g m^{-2}, C - 300 g m^{-2}), equivalent to 0, 136.6, and 204.9 kg of N ha^{-1}. The chemical analysis of the castor cake showed the following nutrient contents: N: 68.3 g kg^{-1}, P: 9.4 g kg^{-1}, K: 14 g kg^{-1}, Ca: 5.4 g kg^{-1}, and Mg: 6.1 g kg^{-1} (mean of 2014 and 2015). Top-dressing fertilizations were performed at 30 and 55 days after transplanting (DAT) in 2014 but only at 30 DAT in 2015.

From transplanting to harvest, the SID was regulated to actuate the irrigation system when soil water tension was approximately 12 kPa in the L1 treatment. The soil water tension was monitored daily by using puncture tensiometers.

Irrigation actuation was simultaneously performed for each block, and the volume applied in the plots was daily monitored with hydrometers (Alpha mnf /FAE). The irrigation depths for each of the evaluated treatments were obtained based on the ratio between the water volume and either the wetted area (until 45 DAT) or the area shaded by the plant (from 46 DAT to harvest).

The accumulations of dry biomass in the plant (PDB) and in the bulb (BDB) were evaluated from 40 DAT onward, at 20-day intervals. Shoots, bulbs and roots were dried in a forced-air oven (70°C) until reaching a constant weight. To evaluate the total yield (TY), the mean bulb fresh weight (MBFW), mean bulb dry weight (MBDW) and mean bulb diameter (MBD) were measured transversely, and 16 bulbs per plot were collected and classified into four categories (> 60 mm, 60 to 41 mm, 40 to 30 mm and < 30 mm) (Kumar, Imtiyaz, Kumar, & Singh, 2007).

Water use efficiency (WUE) was obtained by calculating the ratio between TY and the water depth received by the crop (irrigation + rainfall), and the irrigation water use efficiency (IWUE) was obtained by calculating the ratio between TY and the irrigation depth (Lovelli, Perniola, Ferrara, & Tommaso, 2007).

The data were subjected to analysis of variance and regression; when significant, the fits were presented.

Results and discussion

In response to the development of culture and the soil and climatic conditions of the region, the irrigation system was activated by the SID 48 times during the onion crop cycle in 2014 and 53 times in 2015, applying irrigation levels plus rainfall of 204 (L4), 224 (L3), 278 (L2), and 321 mm (L1) in 2014, and 278 (L4), 302 (L3), 397 (L2), and 444 mm (L1) in 2015.

Evaluating the onion yield under the same edaphoclimatic conditions but with different soil covers and irrigation depths, as estimated through the soil water balance with a TDR, Ribeiro et al. (2016) obtained water depths of 372 and 351 mm (100% of crop evapotranspiration) in the absence and presence of soil cover, respectively. Marouelli, Abda, Madeira, Silva, and Oliveira (2010), in a study conducted in Brasília during the dry season of 2007, applied water depths of 525 and 625 mm in onion cultivation (cv. Baia Periforme) under no-tillage and conventional systems, respectively.

The growth dynamics of the crop in response to the different irrigation depths exhibited the same tendency in both cultivation years. The PDB was influenced from 100 DAT (Figure 2A), whereas the irrigation depths L1, L2, and L3 (9.0 g plant^{-1}) did not exhibit significant differences among themselves but did differ from L4 (6.3 g plant^{-1}). At 120 DAT, differences in PDB were observed among the highest (L1 and L2) and lowest (L3 and L4) irrigation depths (9.9 and 6.7 g plant^{-1}, respectively).

At the highest irrigation depths (L1 and L2), the accumulation of PDB increased during the cultivation period, but the same did not occur at the lowest irrigation depths (L4 and L3), which promoted a slight stagnation at 80 and 100 DAT, respectively, due to the reduction in the dry matter of leaves, bulbs and roots. These results demonstrate that the application of smaller irrigation depths activates the natural process of leaf senescence (Kumar et al., 2007, Zheng et al., 2013).

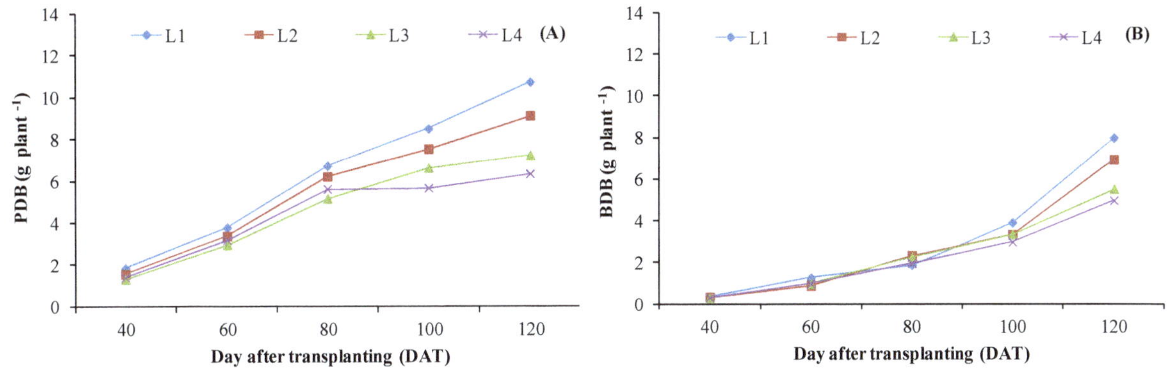

Figure 2. Plant dry biomass (PDB) (A) and bulb dry biomass (BDB) (B) in different development stages for irrigation depths at each dose of castor cake in both cultivation years (means of 2014 and 2015).

From the highest to the lowest irrigation depth, the PDB varied from 5.6 to 8.4 g plant^{-1} and from 6.3 to 10.7 g plant^{-1} at 100 and 120 DAT, respectively. In the cultivation of onion in Northern Minas Gerais, southeastern Brazil, Vidigal, Moreira, and Pereira (2010) obtained PDB values of 6.3 and 9.1 g plant^{-1}, respectively, at 101 and 122 DAT. Neto et al. (2014) obtained 14.5 g plant^{-1} at 90 DAT, for the cultivar IPA 11, when evaluating the growth and accumulation of macronutrients in the onion crop in Baraúna (Rio Grande do Norte State) and Petrolina (Pernambuco State). These variations in PDB are possibly attributed to the different cultivars and edaphoclimatic conditions of the cultivation regions.

BDB varied progressively during the cultivation period, regardless of the applied irrigation depth (Figure 2B). At 120 DAT, significant differences were observed, especially between L1 (higher BDB) and L3 and L4 (lower BDB). Bulb diameter is directly related to the amount of water applied (Kumar et al., 2007; Martin de Santa Olalla, Juan Valero, & Fabeiro Cortés, 1994). At 100 DAT, the BDB values obtained for different irrigation depths were similar to those found by Neto et al. (2014) (4 g) for the cultivar Texas Grano 502 and were lower than those of the cultivar IPA 11 (12.5 g).

In both cultivation years, regardless of the dose of castor cake applied as top-dressing, the irrigation depths significantly influenced the TY, MBFW, MBDW and MBD of the onion (Figure 3). Igbadun and Oiganji (2012), Ramalan, Nega, and Oyebode (2010) and Ribeiro et al. (2016) observed that irrigation depths had similar effects on the yield of onion bulbs.

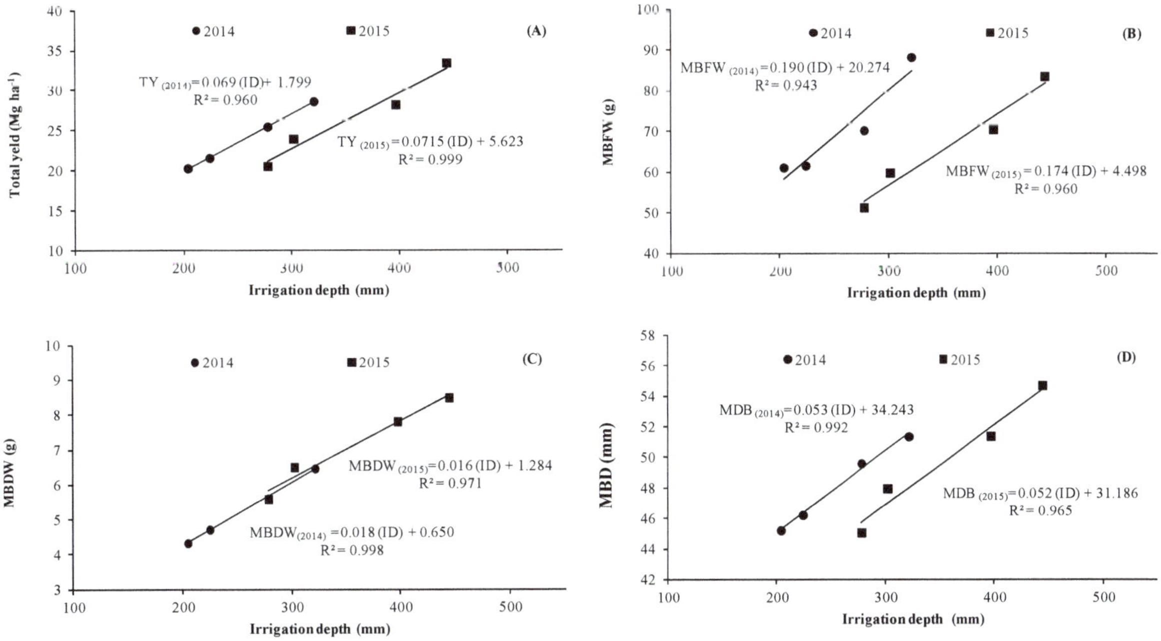

Figure 3. Total yield (TY) (A), mean bulb fresh weight (MBFW) (B), mean bulb dry weight (MBDW) (C) and mean bulb diameter (MBD) (D) for the irrigation depths at each dose of castor cake in both cultivation years (2014-2015). ID: Irrigation depth.

The reduction in irrigation depth of 36% (321 to 204 mm) in 2014 and of 37% (444 to 278 mm) in 2015 resulted in losses of 29, 30, 33, and 11% and 38, 38, 34, and 17% in TY, MBFW, MBDW and MBD, respectively, in each year. These results indicate that any degree of water stress can damage onion yield. A water deficit promotes numerous physiological alterations, such as a reduction in root mass, leaf area, number of leaves and, consequently, yield (Mao et al., 2003).

The castor cake significantly influenced TY (31.9 Mg ha^{-1}), MBFW (96.4 g), MBDW (7.4 g) and MBD (53.4 mm) in the cultivation of 2014, but only at the applied irrigation depth of 321 mm. A similar effect was observed in 2015 for TY (on average, 35.1 Mg ha^{-1}) and MBFW (on average, 87.9 g) at the highest irrigation depths (444 and 397 mm), for MBDW (9.5 g) at the irrigation depth of 444 mm and for MBD (52.2 mm), regardless of the applied irrigation depth (Figure 4). The significant effects of fertilization with castor cake as top-dressing only occurred at the highest irrigation depths because the moisture affects its mineralization, which results in the greater release of nutrients, particularly N (Lima et al., 2011). Boyhan et al. (2007) report that N fertilization has a strong impact on onion production and is necessary to obtain higher bulb yields.

The application of 200 g m^{-2} of castor cake did not promote the expected responses for TY, MBFW, MBDW or MBD (Figure 4), due to the high incidence of two leaf fungal diseases (Anthracnosis *Colletotrichum gloeosporioides f.sp. cepae* and Purple spot *Alternaria porri*) during this treatment, especially in 2014, when rainfalls occurred with higher frequency during the cultivation period (Figure 1). The statistical analysis showed low coefficients of determination (R^2) for the variable MBDW (Figure 4C), possibly because the source-sink relationship was damaged by the loss of photosynthetic apparatus caused by the diseases (Taiz & Zeiger, 2009). The degree of susceptibility of onion to diseases may be related to the mineral nutrition of the crop (Kurtz & Ernane, 2010).

In the present study, the highest TY values were 31.9 (2014) and 39.8 (2015) Mg ha^{-1} for irrigation depths of 321 (2014) and 444 (2015) mm, respectively, with the application of 300 g m^{-2} of castor cake. Costa et al. (2008) reported a yield of 44.2 Mg ha^{-1} in an organic production system, a value close to those found in the present study. Evaluating the yields of different onion cultivars, Ricci, Almeida, Guerra, Cocheto Junior, and Ribeiro (2014) obtained a value of 24.09 Mg ha^{-1} for the cultivar Alfa tropical in an organic production system under the same edaphoclimatic conditions used in this study.

The difference in TY obtained in both cultivation years can be explained, at least partially, by two reasons: a) in 2014, the rainfalls were better distributed (Figure 1), which promoted an environment that was more favorable to the occurrence of diseases; and b) there was variation in the installment of the top-dressing fertilization. In 2014, the application of castor cake was divided and resulted in non-mineralized residual material, thus not providing part of the nutrients to the crop. In 2015, the top-dressing fertilization was applied in a single dose (at 30 DAT), which favored higher mineralization.

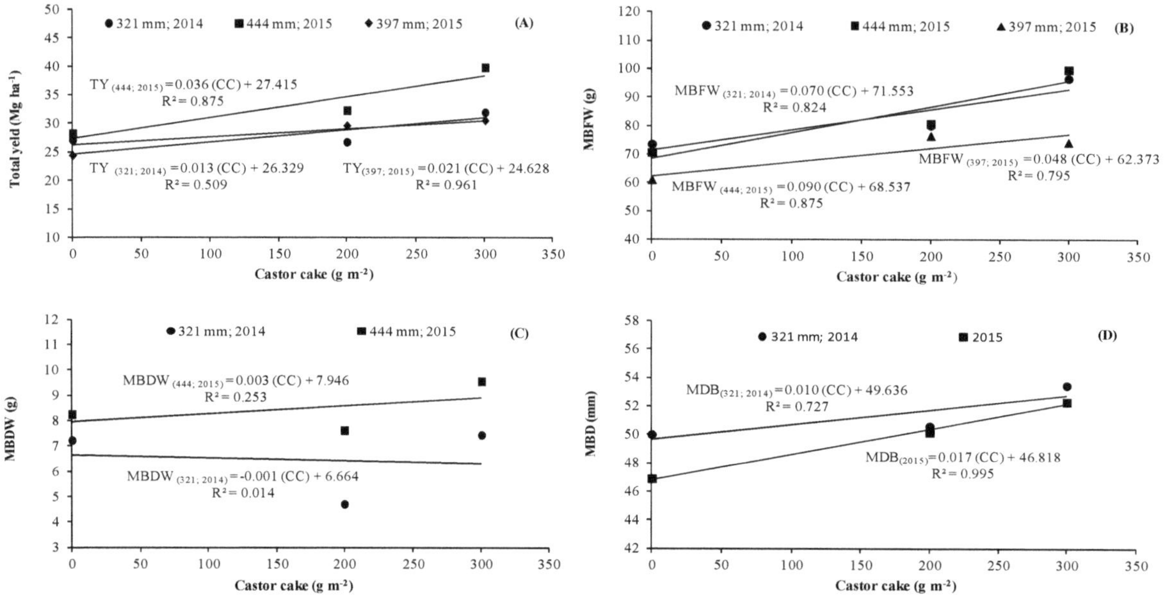

Figure 4. Total yield (TY) (A), mean bulb fresh weight (MBFW) (B), mean bulb dry weight (MBDW) (C) and mean bulb diameter (MBD) (D) for the doses of castor cake at each irrigation depth in both cultivation years (2014-2015). CC: Castor cake.

The percentage of bulbs with diameters greater than 60 mm was positively influenced by the increase in the irrigation levels in both the absence of fertilization and the addition of 300 g m^{-2} dose of castor bean cake in 2014, but it was independent of fertilization in 2015. The increase in irrigation levels favored the percentage of bulbs with a diameter between 41 and 60 mm when 200 g m^{-2} of castor bean was applied in 2014. In 2015, irrigation had an opposite effect, as a larger percentage of bulbs had diameters greater than 60 mm. The highest number of bulbs was included in this classification, as found by Ricci et al. (2014) and Ribeiro et al. (2016). The increase in water deficit favored an increase in the percentage of bulbs with diameters between 30 and 40 mm (smaller class) and with diameter smaller than 30 mm (without commercial value) (Figure 5).

Kumar et al. (2007) attributed this response pattern to the variation in soil moisture because higher water levels promote better water comfort to the plants. Shock, Feibert, and Saunders (1998), Martin de Santa Olalla et al. (2004) and Enciso, Weidenfeld, Jifon, and Nelson (2009) also observed this response pattern in their studies with different irrigation depths in onion cultivation.

Larger percentages of bulbs with diameters greater than 60 mm were found with the application of the 300 g m^{-2} dose of castor bean cake and at irrigation depths of 397 and 444 mm (larger depths) in 2015. A similar effect was observed with the application of this dose at depths of 397 and 278 mm for the percentage of bulbs with diameters between 41 and 60 mm. However, with the application at 444 mm, the percentage of bulbs with diameters between 41 and 60 mm decreased with increasing doses of castor bean cake. This is explained by the high percentage of fruits with diameters greater than 60 mm observed with the increase of castor bean dose, which reduced the percentage of bulbs with both diameters between 30 and 40 mm (smaller class) and diameters smaller than 30 mm (without commercial value) (Figure 6).

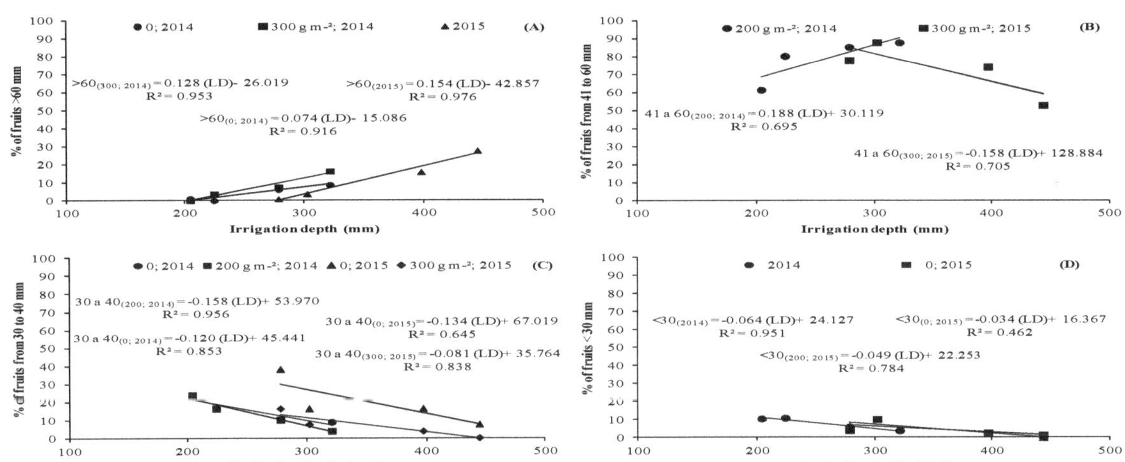

Figure 5. Percentage of transverse bulb diameter in the different bulb classes for irrigation depths in both cultivation years (2014-2015).

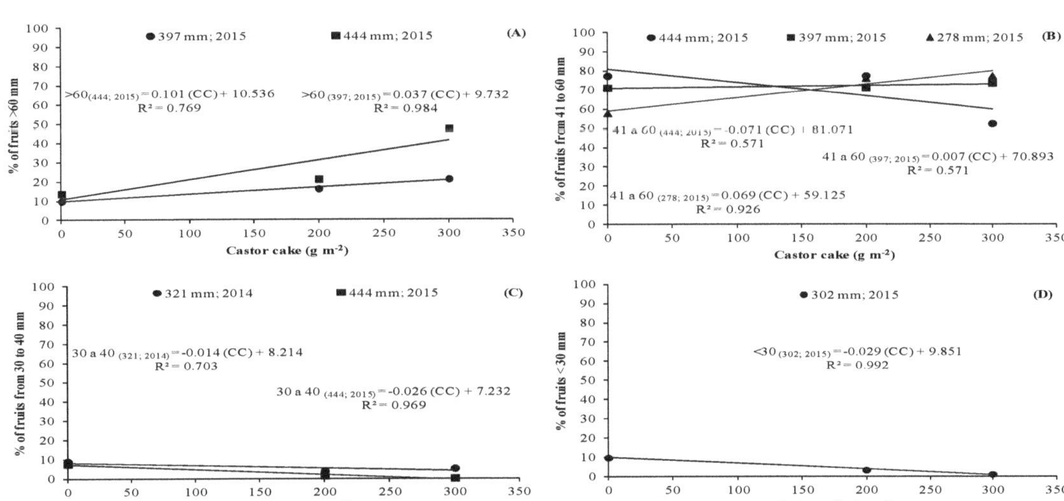

Figure 6. Percentage of transverse bulb diameter in the different bulb classes for the castor cake in both cultivation years (2014-2015).

Water use efficiency (WUE) varied from 99.30 (L4) to 88.86 (L1) kg ha^{-1} mm^{-1} in 2014 and from 100.60 (L4) to 103.85 (L1) kg ha^{-1} mm^{-1} in 2015. WUE was not influenced significantly by the irrigation depths, unlike the irrigation water use efficiency (IWUE), which showed a significant response, with the highest (266.2 kg ha^{-1} mm^{-1} in 2014, 269.7 kg ha^{-1} mm^{-1} in 2015) and lowest (147.5 kg ha^{-1} mm^{-1} in 2014, 172.4 kg ha^{-1} mm^{-1} in 2015) values observed, respectively, for L4 and L1 (Figure 7). Under water deficit conditions, the plant utilizes a survival strategy to develop its root system and create water reserves in the bulb (Zheng et al., 2013), which indicates that even with low water availability in the soil, onion plants can develop bulbs, although they may be of lower quality (Figure 5).

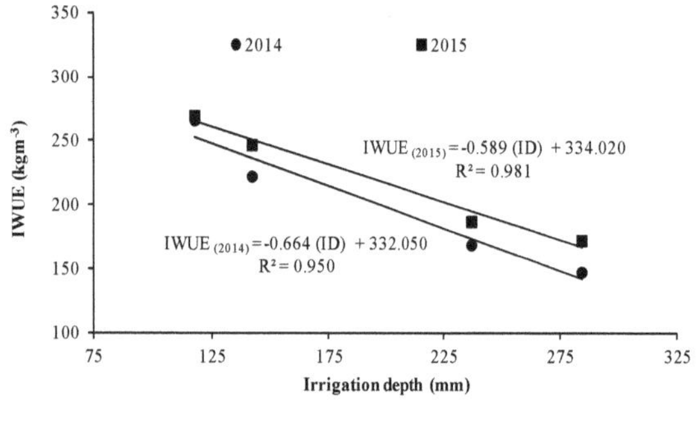

Figure 7. Irrigation water use efficiency (IWUE) for the irrigation depths at each dose of castor cake in both cultivation years (2014-2015).

Conclusion

The irrigation depths used with the automatic device positively and significantly influenced the production of onion bulbs, regardless of the applied dose of castor cake.

The doses of castor cake positively and significantly influenced the production of onion bulbs when higher irrigation depths were used.

References

Boyhan, G. E., Torrance, R. L., & Hill, C. R. (2007). Effects of nitrogen, phosphorus, and potassium rates and fertilizer sources on yield and leaf nutrient status of short-day onions. *HortScience*, *42*(3), 653-660.

Cecílio Filho, A. B., May, A., Pôrto, D. R. Q., & Barbosa, J. C. (2009). Accumulation of dry mass in onion depending on nitrogen, potassium and plant population in direct sowing. *Horticultura Brasileira, 27*(1), 49-54. doi:10.1590/S0102-05362009000100010

Channagoudra, R. F., Prabhudeva, A., & Kamble, A. S. (2009). Response of onion (*Allium cepa* L.) to different levels of irrigation and sulphur in alfisols of northern transitional tract of Karnatak. *The Asian Journal of Horticulture*, *4*(1), 152-155.

Costa, N. D., Araujo, J. F., Santos, C. A. F., Resende, G. M., & Lima, M. A. C. (2008). Evaluation of onion cultivars under organic cultivation in two soil types in the São Francisco Valley, Brazil. *Horticultura Brasileira, 26*(4), 476-480. doi:10.1590/S0102-05362008000400010

El Balla, M. M. A., Hamid, A. A., & Abdelmageed, A. H. A. (2013). Effects of time of water stress on flowering, seed yield and seed quality of common onion (*Allium cepa* L.) under the arid tropical conditions of Sudan. *Agricultural Water Management, 121*(C), 149-157. doi:10.1016/j.agwat.2013.02.002

Embrapa - Empresa Brasileira de Pesquisa Agropecuária. (2006). *Brazilian system of soil classification*. 2ªed. Brasília, DF: Embrapa Produção de informação; Rio de Janeiro, RJ: Embrapa Solos.

Enciso, J., Weidenfeld, B.; Jifon, J., & Nelson, S. (2009). Onion yield and quality response to two irrigation-scheduling strategies. *Scientia Horticulturae, 120*(3), 301-305. doi:10.1016/j.scienta.2008.11.004

Igbadun, H. E., & Oiganji, E. (2012). Crop coefficients and yield response factors for onion (*Allium cepa* L.) under deficit irrigation and mulch practices in Samaru, Nigeria. *African Journal of Agricultural Research*, 7(36), 5137-5152. doi:10.5897/AJAR12.689

Kumar, S., Imtiyaz, M., Kumar, A., & Singh, R. (2007). Response of onion (*Allium cepa* L.) to different levels of irrigation water. *Agricultural Water Management, 89*(1-2), 161-166. doi:10.1016/j.agwat.2007.01.003

Kurtz, C., & Ernani, P. R. (2010). Onion yield influenced by micronutrient application. *Revista Brasileira de Ciência do Solo, 34*(1), 133-142. doi:10.1590/S0100-06832010000100014

Leskovar, D. I., Agehara, S., Yoo, K., & Pascual-Seva, N. (2012). Crop coefficient-based deficit irrigation and planting density for onion: growth, yield, and bulb quality. *HortScience*, *47*(1), 31-37.

Lima, R. L. S., Severino, L. S., Sampaio, L. R., Sofiatti, V., Gomes, J. A., & Beltrão, N. E. M. (2011). Blends of castor meal and castor husks for optimized use as organic fertilizer. *Industrial Crops and Products*, *33*(2), 364-368. doi:10.1016/j.indcrop.2010.11.008

Lovelli, S., Perniola, M., Ferrara, A., & Tommaso, T. D. (2007). Yield response factor to water (ky) and water use efficiency of *Carthamus tinctorius* L. and *Solanum melongena* L. *Agricultural Water Management*, *92*(1-2), 73-80. doi:10.1016/j.agwat.2007.05.005

Magriotis, Z. M., Carvalho, M. Z., Sales, P. F., Alves, F. C., Resende, R. F., & Saczk, A. A. (2014). Castor bean (*Ricinus communis* L.) presscake from biodiesel production: An efficient low cost adsorbent for removal of textile dyes. *Journal of Environmental Chemical Engineering*, *2*(3), 1731-1740. doi:10.1016/j.jece.2014.07.005

Mao, X., Liu, M., Wang, X., Liu, C., Hou, Z., & Shi, J. (2003). Effects of deficit irrigation on yield and water use of greenhouse grown cucumber in the North

China Plain. *Agricultural Water Management*, *61*(3), 219-228. doi:10.1016/S0378-3774(03)00022-2

Marouelli, W. A., Abdalla, R. P., Madeira, N. R., Silva, H. R., & Oliveira, A. S. (2010). Water use and onion crop production in no-tillage and conventional cropping systems. *Horticultura Brasileira, 28*(1), 19-22. doi:10.1590/S0102-05362010000100004

Martin de Santa Olalla, F., Domínguez-Padilla, A., & López, R. (2004). Production and quality of the onion crop (*Allium cepa* L.) cultivated under controlled deficit irrigation conditions in a semi-arid climate. *Agricultural Water Management*, *68*(1), 77-89. doi:10.1016/j.agwat.2004.02.011

Martin de Santa Olalla, F., Juan Valero, J. A., & Fabeiro Cortés, C. (1994). Growth and production of onion crop (*Allium cepa* L.) under different irrigation schedulings. *European Journal of Agronomy*, *3*(1), 85-92. doi: 10.1016/S1161-0301(14)80113-5

Medici, L. O., Rocha, H. S., Carvalho, D. F., Pimentel, C., & Azevedo, R. A. (2010). Automatic controller to water plants. *Scientia Agricola*, *67*(6), 727-730. doi:10.1590/S0103-90162010000600016

Neto, P. A., Grangeiro, L. C., Mendes, A. M. S., Costa, N. D., Marrocos, S. T. P., & Sousa, V. F. L. (2014). Growth and accumulation of macronutrients in onion crop in Baraúna (RN) and Petrolina (PE). *Revista Brasileira de Engenharia Agrícola e Ambiental*, *18*(4), 370-380. doi:10.1590/S1415-43662014000400003

Ramalan, A. A., Nega, H., & Oyebode, M. A. (2010). Effect of deficit irrigation and mulch on water use and yield of drip irrigated onions. *WIT Transactions on Ecology and the Environment, 134*, 39-50. doi:10.2495/SI100041

Resende, G. M., & Costa, N. D. (2008). Nitrogen and potassium levels on the onion yield and storage with different planting times. *Pesquisa Agropecuária Brasileira*, *43*(4), 221-226. doi:10.1590/S0100-204X2008000200010

Ribeiro, E. C., Carvalho, D. F., Santos, L. A. F., & Guerra, J. G. M. (2016). Onion yield under agroecological farming system using distinct irrigation depths and soil covers. *Ciência Rural*, *46*(5), 783-789. doi:10.1590/0103-8478cr20150342

Ricci, M. S. F., Almeida, F. F. D., Guerra, J. G., Cocheto Junior, D. G., & Ribeiro, R. L. D. (2014). Organic cultivation of onion cultivars in the conditions of the Lowlands of Rio de Janeiro state, Brazil. *Horticultura Brasileira* *32*(1), 120-124. doi:10.1590/S0102-05362014000100021

Santos, S. S., Espíndola, J. A. A., Guerra, J. G. M., Leal, M. A. A., & Ribeiro, R. L. D. (2012). Production of organically grown onions depending on the use of mulch and castor bean cake. *Horticultura Brasileira*, *30*(3), 549-552. doi:10.1590/S0102-05362012000300032

Shock, C. C., Feibert, E. B. G., & Saunders, L. D. (2000). Irrigation criteria for drip-irrigated onions. *HortScience*, *35*(1), 63- 66.

Shock, C. C., Feibert, E. B. G., & Saunders, L. D. (1998). Onion yield and quality affected by soil water potential as irrigation threshold. *HortScience*, *33*(7), 1181-1191.

Taiz, L., & Zeiger, E. (2009). *Fisiologia vegetal.* Porto Alegre, RS: Artmed.

Tekalign, T., Abdissa, Y., & Pant, L. M. (2012). Growth, bulb yield and quality of onion (Allium cepa L.) as influenced by nitrogen and phosphorus fertilization on vertisol. II: Bulb quality and storability. *African Journal of Agricultural Research*, 7(45), 5980-5985. doi:10.5897/AJAR10.1025

Vidigal, S. M., Moreira, M. A., & Pereira, P. R. G. (2010). Growth and nutrients uptake by onion plants cultivated in the summer by direct sow and transplanting seedlings. *Bioscience Journal*, *26*(1), 59-70.

Zapata, N., Vargas, M., Reyes, J. F., & Belmar, G. (2012). Quality of biodiesel and press cake obtained from *Euphorbia lathyris*, *Brassica napus* and *Ricinus communis*. *Industrial Crops and Products*, *38*, 1-5. doi:10.1016/j.indcrop.2012.01.004

Zheng, J., Huang, G., Wang, J., Huang, Q., Pereira, L. S., Xu, X., & Liu, H. (2013). Effects of water deficits on growth, yield and water productivity of drip-irrigated onion (*Allium cepa* L.) in an arid region of Northwest China. *Irrigation Science*, *3*, 995-1008. doi:10.1007/s00271-012-0378-5

18

In vitro cultivation of *Anacardium othonianum* Rizz.: effects of salt concentration and culture medium volume

Kerlley Cristina de Assis[1], Flávia Dionísio Pereira[2], Juliana Silva Rodrigues Cabral[3], Fabiano Guimarães Silva[3*], José Waldemar Silva[4] and Silvia Correa dos Santos[5]

*[1]Empresa Brasileira de Pesquisa Agropecuária, Embrapa Clima Temperado, Pelotas, Rio Grande do Sul, Brazil. [2]Cooperativa Agroindustrial dos Produtores Rurais do Sudoeste Goiano, Rio Verde, Goiás, Brazil. [3]Instituto Federal de Educação, Ciência e Tecnologia Goiano, Rod. Sul Goiana, km 1, Rio Verde, Goiás, Brazil. [4]Universidade Federal de Uberlândia, Uberlândia, Minas Gerais, Brazil. [5]Universidade Federal de Goiás, Jataí, Goiás, Brazil. *Author for correspondence. E-mail: fabianocefetrv@yahoo.com.br*

ABSTRACT. *Anacardium othonianum* Rizz. is a medicinal plant species that is native to the Brazilian savannah. Adult plants are different from other genus members in this ecosystem due to their size, and efforts to locate the plants may lead to their extraction from the savannah and, frequently, plant death. Micropropagation has played a significant role in the propagation and preservation of specimens of several plant species; therefore, this study evaluated the effect of various salt concentrations and culture medium volumes on the *in vitro* cultivation of *A. othonianum* Rizz. Trial (I) evaluated two culture media (MS and WPM) and three salt concentrations (100, 50 and 25%) in a completely randomized design. Trial (II) evaluated two culture media and salt concentrations, MS (50%) and WPM (100%), and five medium volumes (10, 15, 20, 25 and 30 mL per test tube) as a 2 x 5 factorial in a completely randomized design. After 30 and 60 days of growth, the percentage of plantlet survival, average plantlet and leaf lengths and the average number of leaves and buds per explant were evaluated. The MS (50 and 25%) media and WPM (100 and 50%) media were the most effective for plantlet regeneration. The best responses were observed in 15- and 25-mL volumes of the MS (50%) medium. Therefore, the use of a 15-mL volume is suggested for greater medium economy.

Keywords: cashew, tissue culture, *in vitro* propagation.

Introduction

The native fruit species of Brazil are genetically diverse and are an important part of the agricultural sector of the country, playing important roles in the country's economic, social and agricultural systems (BASTOS, 2007).

Anacardium othonianum Rizz. is native to the Brazilian Savannah and is commonly known as the caju-de-árvore-do-cerrado. Adult plants differ from genus members in the ecosystem because of their size. Plant height and canopy diameter vary from three to four meters. The pseudo-fruit is pear-shaped and varies in color from yellow to red with a whitish-yellow interior. The species is regionally important, having medicinal properties and widespread acceptance as a food product. However, efforts to locate the species may lead to their extraction from the savannah and, frequently, plant death (SILVA et al., 2001).

Micropropagation has contributed significantly to the propagation and preservation of specimens of several plant species. Although this process involves a number of stages, the micropropagation protocol, once defined for a given species, can be optimized to obtain high-quality plantlets at a low cost of production. Micropropagation studies of little-known species must define the culture medium to be used because culture media, besides supplying essential elements for growth, also control the pattern of *in vitro* development (GRATTAPAGLIA; MACHADO, 1998).

A variety of culture media have been used for the regeneration of species from various genera. Some of these media were specifically developed to supply the particular demands of the species studied, such as the basic culture medium MS of Murashige and Skoog (1962), which was initially developed for *Nicotiana tabacum* pith tissue, and the Woody Plant Medium (WPM), which was made by Lloyd and McCown (1980) for the propagation of woody plants.

The salt and sugar concentrations within the media have a nutritive effect during *in vitro* growth, but also affect cell growth and morphogenesis, due to their osmotic properties (GEORGE, 1996).

Several formulations of basic media have been used for *in vitro* cultivation. There is no standard formula, but MS medium (MURASHIGE; SKOOG, 1962) has been successfully used, with various modifications and dilutions, for several species. However, for woody species, MS medium does not have the same effect, and compositions with more dilute macronutrients yield better growth and development (GRATTAPAGLIA; MACHADO, 1998). The growth medium WPM (LLOYD; McCOWN, 1980), for example, contains 25% of the nitrate and ammonia ions found in MS medium, but contains more potassium and sulfate ions, and has been widely and successfully used for the micropropagation of woody species (PASQUAL, 2001).

This work evaluated the effect of various salt concentrations and culture volumes on *in vitro* growth of *A. othonianum* Rizz. in preparation for basic micropropagation studies to preserve the species and clone its elite genotypes.

Material and methods

Vegetable Material

The vegetable material used for *in vitro* propagation was removed from *A. othonianum* Rizz. fruits that were collected in October 2008 at the Gameleira farm, which is located in Montes Claros county, GO, with geographical coordinates of 16° 06' 20" S, 51° 17' 11" W at 592 m above sea level. The plant exsiccate was deposited at the Herbário Jataiense of the Universidade Federal de Goiás, Campus Jataí, under the collection number 3793. The trials were performed in the Savannah Vegetable Tissue Culture Laboratory (Laboratório de Cultura de Tecidos Vegetais Cerrados) of the Instituto Federal Goiano, Campus Rio Verde, Goiás State.

The fruits were hand-pulped and were subsequently washed in tap water. Surface moisture was removed by blotting the seeds with paper towels at room temperature, and malformed seeds were culled. The seeds used were treated with the fungicide Vitavax-Thiram® [Active ingredient (carboxin + thiram): 200 + 200 g L^{-1}] at 300 mL fungicide per 100 kg of seeds and were dried in direct contact with silica gel in plastic trays (35 x 30 x 8 cm) until reaching a moisture level of 13%. Subsequently, seeds were packaged in plastic bags and stored at 18°C.

Stored seeds were germinated in groups of 100 in plastic trays (50 x 35 x 8 cm) containing washed sand as a substrate. The trays were maintained in a controlled environment with an average temperature of 25.6°C and a relative moisture of 58.2%. Phytosanitary control was maintained by spraying the systemic fungicide Derosal® at a 0.2% concentration 24 hours before collecting the explants. The seedlings were fertilized every two weeks with a nutrient solution prepared with salts derived from the medium MS (MURASHIGE; SKOOG, 1962).

Thirty days after sowing, when the seedlings were approximately 4 cm long, the nodal segments were removed and used as the explant source.

In vitro establishment

Nodal segments were removed from elite plants and immersed for 20 minutes in water containing three drops of neutral detergent. Subsequently, the nodal segments were immersed for 30 seconds in 70% alcohol (v v^{-1}) and for 15 minutes in 20% sodium hypochlorite. A triple rinse was performed with distilled and autoclaved water in a flow hood.

The medium used for establishment of explants was MS (50%), solidified with 4 g L^{-1} agar and modified with 3% sucrose, 2 g L^{-1} active charcoal and 30 μM BAP (6-benzylaminopurine). Preliminary studies (data not shown) indicated that this concentration of BAP resulted in better explant growth. The pH was adjusted to 5.7 ± 0.3 before autoclaving. Approximately 2 cm of each nodal segment were excised and inoculated in test tubes (25 x 150 mm) containing 10 mL culture medium. Inoculated tubes were kept in the dark in a growth room at 25 ± 3°C for 30 days. After the period in darkness, the explants were transferred to fresh culture medium and maintained under 16 hours of light per day for 30 days at 25 ± 3°C with photosynthetic active radiation of 45-55 μmol m^{-2} s^{-1}. The growth evaluations described in Trials I and II were made after 60 days of *in vitro* growth.

Trial (I) Evaluation of the effect of various growth-media salt concentrations on the regeneration of *A. othonianum* Rizz. nodal segments.

Nodal segments that were approximately 2 cm long and contained two buds were used as explant sources. These segments were grown in test tubes (25 x 150 mm) containing 20 mL of growth media. MS and WPM culture media (LLOYD; McCOWN, 1980) containing various salt concentrations (100, 50 or 25%) were used. The media were modified with 2 g L^{-1} active charcoal and 30 μM BAP; the pH was adjusted to 5.7 ± 0.3 before autoclaving, and the nodal segments were planted after cooling and solidification of the media. The cultures were maintained under 16 hours of light at 25 ± 3°C with photosynthetic active radiation of 45-55 μmol m^{-2} s^{-1}.

Thirty days later, the survival percentage, the average number of buds and the number and average lengths of leaves and plantlets were evaluated, and the plantlets were excised and inoculated in fresh media. The same variables were evaluated after another 30 days of growth. Therefore, evaluations were made at 30 and 60 days.

The experimental design was completely randomized, with six treatments and 25 repetitions of one test tube each, for a total of 150 experimental units at each growth time point (30 and 60 days). The data for the number of buds and leaves were submitted to analysis of variance using the software R (R DEVELOPMENT CORE TEAM, 2009), and the averages were compared by contrast using non-generalized linear models. The average lengths of plantlets and leaves and the survival percentages were evaluated by analysis of variance, and the averages were compared by the Scott Knott test at 5% probability with the software SISVAR (FERREIRA, 2003).

Trial (II) Evaluation of the effect of growth medium volume on regeneration of *A. othonianum* Rizz. nodal segments

The two best media, based on the results of Trial (I), were selected. The media pH was adjusted to 5.7 ± 0.3 before autoclaving. Five volumes (10, 15, 20, 25 or 30 mL) were evaluated for each culture medium.

Two-centimeter-long nodal segments containing two buds each were used as explants and were grown in test tubes (25 x 150 mm). The cultures were kept under 16 hours of light at 25 ± 3°C with photosynthetic active radiation of 45-55 μmol m^{-2} s^{-1}.

Thirty days later, the survival percentage, the average number of buds and the number and average lengths of leaves and plantlets were evaluated, and the plantlets were excised and inoculated in fresh media. The same variables were evaluated again after another 30 days of growth. Therefore, evaluations were made at 30 and 60 days.

The experimental design was completely randomized as a 2 x 5 factorial (culture media x media volume), and each treatment consisted of 25 repetitions of one test tube, for a total of 250 experimental units for each growth period. The data for the number of buds and leaves were submitted to analysis of variance using the software R (R DEVELOPMENT CORE TEAM, 2009), and the averages were compared by contrast using non-generalized linear models. The average lengths of plantlets and leaves and the survival percentages were evaluated by analysis of variance, and the averages were compared by the Scott Knott test at 5% probability with the software SISVAR (FERREIRA, 2003).

Results and discussion

Trial (I) Evaluation of the effect of various growth media salt concentrations on the regeneration of *A. othonianum* Rizz. nodal segments

After 30 days of growth, the nodal segments regenerated less vigorous plantlets (with a yellowish color) in the MS (100%) medium than did nodal segments regenerated in the other media, which presented well-formed, vigorous, dark-green plantlets. No oxidation, root or calli formation was observed. The MS (100%) medium was significantly different from all other conditions, as determined by the Scott Knott test, with 84% plantlet survival (Table 1).

Differences in plantlet lengths and leaves were observed at 30 days (Figure 1). The MS (50 and 25%) media and WPM (100 and 50%) yielded greater growth with lengths varying from 3.17 to 3.72 cm. The average leaf lengths were 0.97, 1.00 and 0.88 cm in the media MS (50%), MS (25%) and WPM (100%), respectively.

Table 1. Percentage of *A. othonianum* Rizz. plantlet survival after 30 and 60 days of *in vitro* growth under various concentrations of MS and WPM media.

Culture medium	Plantlet survival (%)	
	30 Days	60 Days
MS-100%	84 B*	12 B
MS-50%	100 A	96 A
MS-25%	100 A	100 A
WPM-100%	100 A	100 A
WPM-50%	100 A	100 A
WPM-25%	100 A	100 A

*Averages followed by the same letter in each column do not differ by the Scott Knott test at 5% probability.

After 60 days of growth, most plantlets grown in MS (100%) were dehydrated. The survival percentage was only 12%, differing from all other experimental conditions (Table 1).

Differences among the variables were observed after 60 days of growth (Figures 1 and 2). Media MS (50%), MS (25%) and WPM (50%) yielded the greatest average plantlet lengths (3.94, 4.36 and 4.34 cm, respectively) and leaf lengths (1.94, 2.13 and 2.36 cm, respectively).

A significant effect on the number of leaves was determined by contrast analysis at 30 days (Table 2). MS (50 and 25%) and WPM (100 and 50%) media had a greater effect than did other media concentrations. No bud formation was observed.

Significant differences were observed by contrast analysis in leaf number and buds after 60 days of growth (Table 2). The WPM medium had a greater effect than the MS medium for the three concentrations evaluated. The MS (50 and 25%) media had a greater effect than did MS (100%). In contrast, the WPM (100%) medium yielded a significant difference only in the number of leaves.

The MS (100%) medium was the least effective medium with respect to the characteristics evaluated in Trial (I). The MS medium has a higher salt concentration than the WPM culture medium, which appears to have a negative effect on the regeneration of *A. othonianum* Rizz. nodal segments. However, several authors have observed that the best regeneration and *in vitro* growth of *A. occidentale* L. occurs in the MS (100%) medium (ALIYU; AWOPETU, 2005; ANANTHAKRISHNAN et al., 2002; DAS et al., 1996).

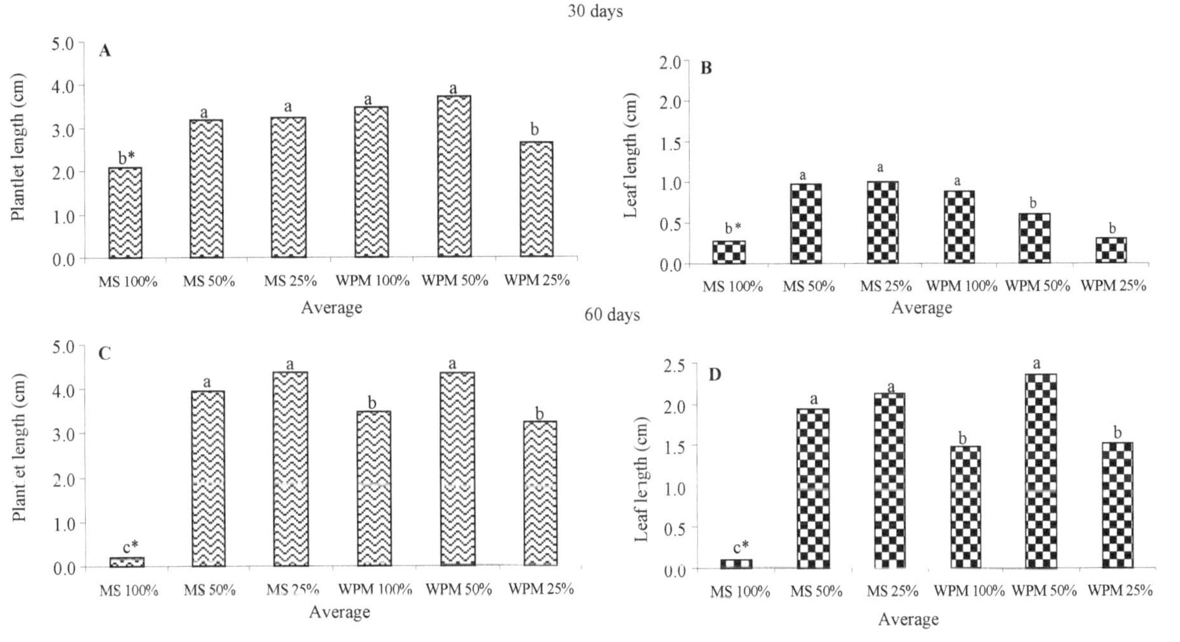

Figure 1. Average plantlet lengths (A and C) and average leaf lengths (B and D) of *A. othonianum* Rizz. after 30 and 60 days, respectively, in various concentrations of MS and WPM media.
*Averages followed by the same letter do not differ by the Scott Knott test at 5% probability.

Figure 2. Plantlets of *A. othonianum* Rizz. after 60 days of *in vitro* growth in the indicated concentrations of MS and WPM media. 2009. Bar = 10 mm.

Table 2. Contrast analyses for the average number of leaves and buds of *A. othonianum* Rizz. after 30 and 60 days of growth in various concentrations of MS and WPM media. 2009.

30 days of growth			
Contrast	Estimate	Standard Error	Pr (>|z|)
		Number of leaves	
MS-100+MS-50+MS-25 vs. WPM-100+WPM-50+WPM-25	0.16	0.18	0.37 n.s
MS-100 vs. MS-50+MS-25	-2.08	0.62	0.00 *
WPM-100 vs. WPM-50+WPM-25	1.26	0.52	0.01 *
MS-50 vs. MS-25	0.00	0.24	1.00 n.s
WPM-50 vs. WPM-25	1.20	0.38	0.00 *
60 days of growth			
Contrast	Estimate	Standard Error	Pr(>|z|)
		Number of leaves	
MS-100+MS-50+MS-25 vs. WPM-100+WPM-50+WPM-25	-0.82	0.26	0.00 *
MS-100 vs. MS-50+MS-25	-2.99	0.71	0.00 *
WPM-100 vs. WPM-50+WPM-25	0.41	0.20	0.04 *
MS-50 vs. MS-25	-0.20	0.22	0.37 n.s
WPM-50 vs. WPM-25	0.47	0.26	0.07 n.s
		Number of buds	
MS-100+MS-50+MS-25 vs. WPM-100+WPM-50+WPM-25	-1.07	0.39	0.00 *
MS-100 vs. MS-50+MS-25	-2.29	1.02	0.02 *
WPM-100 vs. WPM-50+WPM-25	-0.18	0.34	0.59 n.s
MS-50 vs. MS-25	-0.20	0.44	0.65 n.s
WPM-50 vs. WPM-25	-0.20	0.37	0.57 n.s

*significant at 5% probability; NS not significant.

Salt reduction is generally beneficial for the culture growth of woody species. This was confirmed in Trial (I) by reducing the salt concentration of the MS medium or by using the WPM medium; however, the WPM (25%) medium was not effective for supplying the plantlets' needs and promoting satisfactory growth. Other authors have also observed that reducing the salt concentrations of culture media favored *in vitro* growth of *A. occidentale* (LÉDO et al., 2007; RADMANN et al., 2009a and b; THIMMAPPAIAH et al., 2002).

Our experiments have shown that the MS (50 and 25%) media and the WPM (100 and 50%) media were the most effective in Trial (I) for the regeneration and growth of nodal segments of *A. othonianum* Rizz.

Trial (II) Evaluation of the effect of the growth-medium volume on the regeneration of *A. othonianum* Rizz. nodal segments

Consequently, Trial (II) evaluated the MS (50%) medium and the WPM (100%) medium, each supplemented with 2 g L^{-1} active charcoal and 30 μM BAP.

The type of media and media volume each had significant effects on the average leaf length, but these effects were only observed when the nodal segments were grown for 30 days (Table 3). MS medium volumes of 15, 20 and 25 mL promoted greater leaf growth (3.03, 3.29 and 3.17 cm, respectively) than did volumes of 10 and 30 mL (2.19 and 2.54 cm, respectively). The 10-mL volume promoted significant leaf growth in the WPM medium (3.46 cm), while in the MS medium, the average was 2.19 cm (Table 3).

The culture media alone had no significant effect on plantlet and leaf lengths after 30 days (Table 3). The average plantlet length was 4.70 cm for the MS medium and 4.50 cm for the WPM medium. The average leaf length was 2.85 cm for MS and 2.89 cm for WPM. Significant differences were found after 60 days for plantlet lengths only, with averages of 5.46 cm for medium MS and 4.69 cm for WPM. Volume alone did not have a significant effect on either plantlet or leaf lengths at 30 and 60 days (Table 3).

Contrast analysis of the varying conditions (Table 4) indicated a significant difference between the two media at each volume after 30 days of growth. The effect of the MS medium on the number of leaves was greatest at 30 mL. The MS medium also had a greater effect on the number of buds at volumes of 10, 15 and 25 mL. None of the other contrasts were significant at 30 days.

After 60 days, significant effects were observed when the culture media were contrasted at each volume. The MS medium mediated an increase in the number of leaves at volumes of 15, 25 and 30 mL. A greater number of buds were found in 10, 15 and 25 mL volumes of MS. None of the other contrasts were significant (Table 4).

Both culture media mediated the formation of vigorous plantlets at all volumes tested, with no calli and no root formation from the nodal segments (Figure 3).

The internal space available for explant growth has a significant effect on the *in vitro* development of explants. In addition to the medium volume, this space varies with container shape and volume. These variables are infrequently analyzed, although they affect the air volume within the container and the depth of the growth medium in it. These factors also affect the gas composition within the container and, consequently, culture growth and development (PEREIRA et al., 2006).

Table 3. Average plantlet and leaf lengths of *A. othonianum* Rizz. after 30 and 60 days of growth in various volumes of culture media. 2009.

30 days growth							
Variable (cm)	Media	Volume (mL)					
		10	15	20	25	30	Average
Plantlet length	MS	4.18*	5.22	4.98	5.02	4.12	4.70 a
	WPM	4.80	4.60	4.54	4.44	4.16	4.50 a
Average		4.49 A	4.91 A	4.76 A	4.73 A	4.14 A	
Leaf length	MS	2.19 b B	3.03 a A	3.29 a A	3.17 a A	2.54 a B	2.85 a
	WPM	3.46 a A	2.91 a A	2.67 a A	2.83 a A	2.57 a A	2.89 a
Average		2.83 A	2.97 A	2.98 A	3.00 A	2.55 A	
60 days growth							
Variable (cm)	Media	Volume (mL)					
		10	15	20	25	30	Average
Plantlet length	MS	4.90	6.16	5.68	5.64	4.94	5.46 a
	WPM	5.14	4.92	5.10	4.36	3.96	4.69 b
Average		5.02 A	5.54 A	5.39 A	5.00 A	4.45 A	
Leaf length	MS	2.72	3.48	3.26	3.08	2.70	3.05 a
	WPM	3.05	3.02	2.95	2.67	2.33	2.81 a
Average		2.89 A	3.25 A	3.10 A	2.87 A	2.52 A	

*Averages followed by the same uppercase letter in each row and the same lowercase letter in each column are not different by the Scott Knott test at 5% probability.

Table 4. Contrast analysis of the numbers of leaves and buds of *A. othonianum* Rizz. after 30 and 60 days of growth in various volumes of culture media. 2009.

30 days growth				
Fixed Contrast	Contrast	Estimate	Standard Error	Pr(>|z|)
			Number of leaves	
Volumes (mL)	MS-10 vs. WPM-10	0.22	0.24	0.34 ns
	MS-15 vs. WPM-15	0.36	0.23	0.13 ns
	MS-20 vs. WPM-20	0.08	0.24	0.71 ns
	MS-25 vs. WPM-25	0.33	0.26	0.19 ns
	MS-30 vs. WPM-30	0.55	0.25	0.02 *
			Number of buds	
Volumes (mL)	MS-10 vs. WPM-10	1.38	0.55	0.01 *
	MS-15 vs. WPM-15	1.16	0.51	0.02 *
	MS-20 vs. WPM-20	0.53	0.47	0.25 ns
	MS-25 vs. WPM-25	1.87	0.75	0.01 *
	MS-30 vs. WPM-30	0.69	0.46	0.13 ns
60 days growth				
Fixed Contrast	Contrast	Estimate	Standard Error	Pr(>|z|)
			Number of leaves	
Volumes (mL)	MS-10 vs. WPM-10	0.24	0.18	0.19 ns
	MS-15 vs. WPM-15	0.45	0.19	0.01*
	MS-20 vs. WPM-20	0.16	0.18	0.36 ns
	MS-25 vs. WPM-25	0.53	0.19	0.00*
	MS-30 vs. WPM-30	0.72	0.21	0.00 *
			Number of buds	
Volumes (mL)	MS-10 vs. WPM-10	1.84	0.62	0.00 *
	MS-15 vs. WPM-15	2.14	0.74	0.00 *
	MS-20 vs. WPM-20	0.40	0.52	0.44 ns
	MS-25 vs. WPM-25	2.19	1.05	0.03 *
	MS-30 vs. WPM-30	0.31	0.46	0.49 ns

*significant at 5% probability; NS not significant.

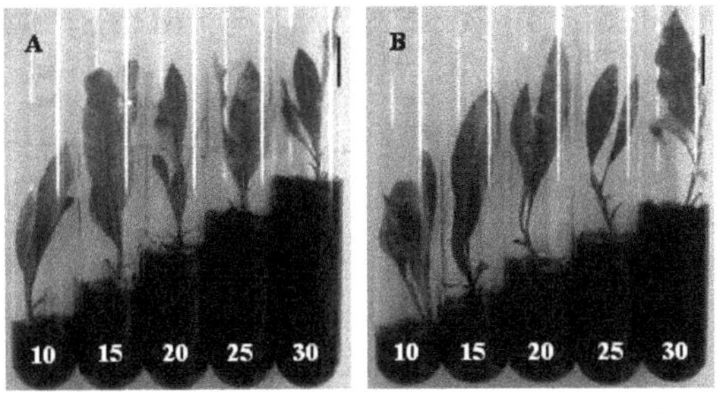

Figure 3. Plantlets of *A. othonianum* Rizz. after 60 days of *in vitro* growth. Numbers indicate media volume in mL. MS (50%) medium (A); WPM medium (B). 2009. Bar = 10 mm.

Previous studies have analyzed the ideal volume of culture medium for the growth of various species (CARVALHO et al., 1995; PEREIRA et al., 2006; REIS et al., 2004, 2007). Other studies have also evaluated various container volumes for their effect on plant growth (CAMOLESI et al., 2010; NICOLOSO; ERIG, 2002).

In summary, the MS (50%) medium showed greater potential than the WPM medium for supporting plantlet growth. Volumes of 15 and 25 mL yielded the best responses; the use of the 15 mL medium volume would, therefore, be the most economical.

Conclusion

The MS (50 and 25%) media and the WPM (100 and 50%) media were most effective for

Anacardium othonianum Rizz. plantlet regeneration. The best plantlet regeneration was observed in the MS (50%) medium in volumes of 15 and 25 mL; therefore, the use of the 15-mL volume is suggested for maximal economy.

References

ANANTHAKRISHNAN, G.; RAVIKUMAR, R.; GIRIJA, S.; GANAPATHI, A. *In vitro* adventitious shoot formation from cotyledon explants of cashew (*Anacardium occidentale* L.). **Scientia Horticulturae**, v. 93, n. 3-4, p. 343-355, 2002.

ALIYU, O. M.; AWOPETU, J. A. *In vitro* regeneration of hybrid plantlets of cashew (*Anacardium occidentale* L.) through embryo culture. **African Journal of Biotechnology**, v. 4, n. 6, p. 548-553, 2005.

BASTOS, L. P.; MOREIRA, M. J. S.; COSTA, M. A. P. C.; ROCHA, M. C.; HANSEN, D. S.; SILVA, S. A.; DANTAS, A. C. V. L.; SOUSA, C. S. Cultivo *in vitro* de Mangabeira (*Hancornia speciosa*). **Revista Brasileira de Biociências**, v. 5, supl. 2, p. 1122-1124, 2007.

CAMOLESI, M. R.; FARIA, R. T.; NEVES, C. S. V. J.; MARTINS, A. N. Volume do frasco e consistência do meio de cultura na multiplicação *in vitro* da bananeira 'Maçã'. **Ciência Rural**, v. 4, n. 2, p. 255-260, 2010.

CARVALHO, R.; FAVARETTO, N.; PINTO, J. E. B.; DESCHAMPS, C.; INNECCO, R. Influência de fatores físicos no desenvolvimento e crescimento *in vitro* de batata-doce [*Ipomoea batatas* (L.) Poir]. **Revista Ciência e Prática**, v. 19, n. 2, p. 158-164, 1995.

DAS, S.; JHA, T. B.; JHA, S. *In vitro* propagation of cashew nut. **Plant Cell Reports**, v. 15, n. 8, p. 615-619, 1996.

FERREIRA, D. F. **SISVAR**. Versão 4.3. Lavras: DEX/UFLA, 2003.

GRATTAPAGLIA, D.; MACHADO, M. A. Micropropagação. In: TORRES, A. C.; CALDAS, L. S.; BUSO, J. A. (Ed). **Cultura de tecidos e transformação genética de plantas**, v. 1. Brasília: Embrapa-SPI/Embrapa-CNPH, 1998. p. 183-260.

GEORGE, E. F. **Plant propagation by tissue culture**: the technology. Edington: Exegetics, 1996. (part 1).

LÉDO, A. S.; SECA, G. S. V.; BARBOZA, S. B. S. C.; SILVA JÚNIOR, J. F. Crescimento inicial de mangabeira (*Hancornia speciosa* Gomes) em diferentes meios de germinação *in vitro*. **Ciência e Agrotecnologia**, v. 31, n. 4, p. 989-993, 2007.

LLOYD, G.; McCOWN, B. Commercially feasible micropropagation of mountain laurel, *Kalmia latifolia*, by use of shoot tip culture. **Combined Proceedings International Plant Propagators Society**, v. 30, p. 421-327, 1980.

MURASHIGE, T.; SKOOG, F. A revised medium for rapid growth and bioassays with tobacco tissue culture. **Physiologia Plantarum**, v. 15, n. 3, p. 473-497, 1962.

NICOLOSO, F. T.; ERIG, A. C. Efeito do tipo de segmento nodal e tamanho do recipiente no crescimento de *Pfaffia glomerata in vitro*. **Ciência e Agrotecnologia**, Edição Especial, p. 1499-1506, 2002.

PASQUAL, M. **Textos acadêmicos**: Meios de Cultura. Lavras: FAEPE/UFLA, 2001.

PEREIRA, F. D.; PINTO, J. E. B. P.; RODRIGUES, H. C. A.; ROSADO, L. D. S.; BEIJO, L. A.; LAMEIRA, O. A. Proliferação *in vitro* de brotos de curauá utilizando diferentes volumes de meio. **Plant Cell Culture and Micropropagation**, v. 2, n. 2, p. 53-106, 2006.

RADMANN, E. B.; BIANCHI, V. J.; OLIVEIRA, R. P.; FACHINELLO, J. C. Multiplicação *in vitro* e alongamento das brotações micropropagadas do porta-enxerto 'Tsukuba 1' (*Prunus persica* L.). **Revista Brasileira de Fruticultura**, v. 31, n. 3, p. 656-663, 2009a.

RADMANN, E. B.; BIANCHI, V. J.; SOUZA, T. M.; FACHINELLO, J. C.; OLIVEIRA, R. P. Influência da composição do meio de cultivo e do tipo de explante na micropropagação do porta-enxerto de *Prunus* sp. 'GXN-9'. **Scientia Agraria**, v. 10, n. 2, p. 95-101, 2009b.

R DEVELOPMENT CORE TEAM. **R**: A language and environment for statistical computing. Vienna: R Foundation for Statistical Computing , 2009.

REIS, E. S.; PINTO, J. E. B. P.; ROSADO, L. D. S.; CORRÊA, R. M. Tipos de explantes e volumes de meio de cultura no cultivo *in vitro* de *Melissa officinalis* L. **Plant Cell Culture and Micropropagation**, v. 3, n. 2, p. 83-88, 2007.

REIS, E. S.; PINTO, J. E. B. P.; CORRÊA, R. M.; BERTOLUCCI, S. K. V.; LAMEIRA, O. A. Tamanhos e posições de explantes e volumes de meio de cultivo na multiplicação de ipeca (*Psychotria ipecacuanha* (Brot.) Stokes) *in vitro*. **Ciência e Agrotecnologia**, v. 28, n. 3, p. 703-709, 2004.

SILVA, D. B.; SILVA, J. A.; JUNQUEIRA, N. T. V.; ANDRADE, L. R. M. **Frutas do cerrado**. Brasília: Embrapa Informações Tecnológicas, 2001.

THIMMAPPAIAH; SHIRLY, R. A.; SADHANA, P. H. *In vitro* propagation of cashew from young trees. ***In Vitro* Cellular and Development Biology**, v. 38, n. 2, p. 152-156, 2002.

19

Evolution of nutrient availability in maturation phase of composting using proportions of different residues inoculated with *Beijerinckia indica*

Lusiene Barbosa de Sousa, Newton Pereira Stamford*, Wagner da Silva Oliveira, Emmanuella Vila Nova da Silva, Marllon dos Santos Martins and Carolina Etienne de Rosália e Silva Santos

*Departamento de Agronomia, Universidade Federal Rural de Pernambuco, Rua Dom Manuel de Medeiros, s/n, Dois Irmãos, 52171-900, Recife, Pernambuco, Brasil. *Author for correspondence. E-mail: newton.stamford@ufrpe.br*

ABSTRACT. Organic matter has low N content; however, organic matter may be enriched by inoculation with selected diazotrophic bacteria. The aim of the paper was to evaluate the effects of biofertilizer produced by mixing different types and proportions of organic matter inoculated with the diazotrophic bacteria *Beijerinckia indica*. The experiment consisted of plastic trays (6 L) containing 5 kg of three different types of organic matter (filter mud cake, earthworm compost, and crop residue) applied in the following proportions (v:v:v): 1-(5:0:0), 2-(4:1:0), 3-(4:0:1), 4-(3:2:0), 5-(3:1:1), 6-(3:0:2), 7-(2:3:0), 8-(2:2:1), 9-(2:1:2), 10-(2:0:3), 11-(1:4:0), 12-(1:3:1), 13-(1:2:2), 14-(1:1:3), 15-(1:0:4), 16-(0:5:0), 17-(0:4:1), 18-(0:1:4), 19-(0:3:2), 20-(0:2:3), and 21-(0:0:5). Samples were collected following inoculation with *B. indica* at 0, 10, 20, 30, and 40 days of the maturation phase. The chemical analyses were: pH (H_2O), total C and N, available P and K, and exchangeable Na^+, Ca^{+2}, Mg^{+2}. The treatments with the best results showed significant effects following the maturation phase. The exceptions were C and N, which presented only individual effects. In general, the organic matter proportions (5:0:0), (3:2:0), and (2:3:0) resulted in a significant increase in the availability of nutrients. Biofertilizers with optimal organic matter proportions may be used to produce organic substrates that are more effective and have the potential to be applied as alternatives to soluble NPK fertilizers.

Keywords: *Acidithiobacillus*; diazotrophic bacteria; earthworm compost; plant residue; sugarcane filter mud cake.

Introduction

Biofertilizer produced with organic matter (earthworm compost) has been shown to increase the availability of nutrients in various crops, especially phosphorus, potassium, calcium, magnesium, and sulfur (Lima, Stamford, Santos,

Lira Junior, & Dias, 2007; Stamford, Lima, Santos, & Dias, 2006; Stamford, Izquierdo, Fernández, & Moreno, 2008; Stamford et al., 2011). Nitrogen is also present in organic matter, although not in quantities that promote adequate nutrition and plant yield similar to an application of soluble fertilizer (Lima et al., 2010). However, organic matter inoculated with effective free-living diazotrophic bacteria can promote the production of N to acceptable levels (Lima et al., 2010).

The production of soluble NPK fertilizers first requires that they are chemically processed, resulting in high concentrations of readily available nutrients. Additionally, the production of soluble fertilizers often requires the addition of high temperatures or the use of strong acids (sulfuric acid, phosphoric acid or nitric acid), and these processes require high power consumption (Van Straaten, 2002).

Because of these requirements, organic matter seems to be a viable alternative for small, medium and large farms due to its low cost. Natural organic fertilizers can be produced without the use of corrosive products or high-tech equipment, using simple field techniques (Stamford et al., 2008).

The incorporation of organic matter generally improves the physical conditions of the soil and contributes to increased biological activity that provides nutrients to plants (Sousa et al., 2015). The primary materials used to produce organic biofertilizers include earthworm compost (Oliveira et al., 2015), crop compounds (Sousa et al., 2015), sugarcane filter mud cake (Oliveira et al., 2015), sugarcane bagasse, straw carbonized rice (Saideles, Winckler, Schirmer, & Sperandio, 2009), and others. Organic waste that has minimal environmental impact is preferred for biofertilizer production (Figueiredo & Tanamati, 2010). It is important to note that these materials must be evaluated to determine the appropriate proportion of each organic material for the production of biofertilizer to ensure nutrient availability and their usefulness in agriculture.

The aim of this paper was to evaluate the best proportions of different types of organic matter inoculated with effective free-living diazotrophic bacteria that would result in increased N and nutrient availability for the production of organic biofertilizer. The ability of the biofertilizers to be used as alternative fertilizer sources that could replace soluble fertilizers was also examined.

Material and methods

The experiment was performed at the Laboratory of Environmental Biotechnology of the Department of Agronomy, Federal Rural University of Pernambuco, Brazil. Using plastic trays (6 dm^3), 5 dm^3 of organic matter (sugarcane mud cake: earthworm compost: crop residue) was mixed in the following proportions (v:v:v): 1-(5:0:0), 2-(4:1:0), 3-(4:0:1), 4-(3:2:0), 5-(3:1:1), 6-(3:0:2), 7-(2:3:0), 8-(2:2:1), 9-(2:1:2), 10-(2:0:3), 11-(1:4:0), 12-(1:3:1), 13-(1:2:2), 14-(1:1:3), 15-(1:0:4), 16-(0:5:0), 17-(0:4:1), 18-(0:1:4), 19-(0:3:2), 20-(0:2:3), and 21-(0:0:5). The experiment was conducted as a complete randomized design using the proportions of the three types of organic matter as treatments and replicated four times.

The sugarcane mud cake was purchased from the "Petribu sugarcane industry", and the chemical analyzes showed: pH (H_2O) = 6.4, total C = 112 (g kg^{-1}), total N = 8.0 (g kg^{-1}), C:N = 14.0, available P = 150 (mg kg^{-1}), and available K = 119 (mg kg^{-1}); The earthworm compost was commercially available in the regional market (Febras Company): pH (H_2O)= 7.9, total C = 134 (g kg^{-1}), total N = 6.9 (g kg^{-1}), C:N = 19.4, available P = 120 (mg kg^{-1}), and available K = 425 (mg kg^{-1}); and the residue compost obtained at the District of Recife City, and the chemical analyzes showed: pH (H_2O) = 6.2, total C = 180 (g kg^{-1}), total N = 4.7 (g kg^{-1}), C:N = 38.2, available P = 80 (mg kg^{-1}), and available K = 159 (mg kg^{-1}).

After the respective treatments were mixed in the plastic trays, they were inoculated with the effective free-living bacteria *Beijerinckia indica* (strain NFB 10001), a strain selected from a previous study (Lima et al., 2010) and further characterized by the Macrogen Incorporation (Geumcheon-gu Seoul, Korea).

The pre-inoculum was initially prepared in a 250 mL Erlenmeyer flask, containing 100 mL of LG medium and incubated for two weeks under orbital agitation (500 gpm). After this period, the material was transferred to a 2,000 mL Erlenmeyer flask with 1,000 mL of LG medium to produce the liquid inoculum and incubated for 7 days with orbital agitation (500 gpm). Each tray received 200 mL of the liquid inoculum, which was homogenized with a sterile spatula. Treatments were incubated at room temperature (28 ± 2°C). The humidity control was set at maximum water retention capacity for each tray, and the daily addition of tap water passed through an activated carbon filter.

During the experiment, samples from each treatment were collected at various times during the maturation phase (0, 10, 20, 30, and 40 days) and chemical analyses of pH (H_2O), total C and N, available P and K, and exchangeable Ca, Mg and Na were all conducted following the Embrapa (2009) methodology.

The data for treatments deemed the most economically viable were statistically analyzed using SAS Institute (2011) software. Logarithmic and square root transformations of the data were not necessary to ensure the normality and homogeneity of the variance. For comparison between treatments, Tukey tests were carried ($p < 0.05$) and all statistical tests were evaluated at the 95% confidence level and values are given as mean ± standard error. Correlation analyses were also not carried out to examine the relationships.

Results and discussion

pH values

The pH values of the selected treatments with different proportions of the 3 types of organic matter (mud cake, earthworm compost and crop residue) at various incubation times are shown in Table 1. A low pH value at time zero (T_0) was observed in treatment 7 (1:1:3), and the other treatments with different proportions were not significantly different when compared at the beginning of the experiment (T_0). The low variation may be due to the amount of sugarcane mud cake and earthworm compost because these types of organic matter have higher pH values (6.7 and 7.9, respectively).

Table 1. The pH values of selected treatments with different proportions of the 3 types of organic matter (mud cake: earthworm compost: crop residue), inoculated with the diazotrophic bacteria *Beijerinckia indica* (strain NFB 10001), after incubation at various times (0, 10, 20, 30, and 40 days) at room temperature (28 ± 2°C).

Treatments	Incubation time (days)				
Proportions	0	10	20	30	40
			pH (H_2O)		
1 (5-0-0)	6.97a±0.12	7.06b± 0.09	7.06b± 0.01	7.16a± 0.06	7.15a± 0.00
2 (2-3-0)	6.61a± 0.01	7.25a± 0.12	7.29a± 0.06	7.31a± 0.09	7.25a± 0.12
3 (3-2-0)	6.90a± 0.31	7.28a± 0.11	7.29a± 0.09	7.31a± 0.02	7.31a± 0.02
4 (3-1-1)	6.81a± 0.02	7.06a± 0.15	7.19a± 0.17	7.25a± 0.07	7.17a± 0.05
5 (2-0-3)	6.68a± 0.15	7.38a± 0.06	7.06b± 0.20	7.14a± 0.12	7.30a± 0.17
6 (2-2-1)	6.73a± 0.04	6.98b± 0.40	7.19a± 0.01	7.27a± 0.09	7.21a± 0.07
7 (1-1-3)	6.41b± 0.24	6.82b± 0.06	6.82b± 0.01	6.76b± 0.02	6.81b± 0.01
8 (1-2-2)	6.63a± 0.00	6.87b± 0.04	7.00b± 0.06	7.02b± 0.09	7.08a± 0.4
9 (0-4-1)	6.75a± 0.02	6.92b± 0.01	6.98b± 0.03	6.80b± 0.22	6.83b± 0.02

Means with the same letter in the same column are not different according to Tukey's test ($p < 0.05$).

A slight increase in pH was observed with increased incubation time, especially from time zero to 30 days of incubation. In general, similar results were observed for the incubation times from 10 to 40 days, with an increase in pH when incubated for 30 days. Higher pH values were observed with treatments (2:3:0) and (3:2:0) at 30 days of incubation; these treatments had a high amount of earthworm compost (pH 7.9).

However, despite the different proportions of the 3 types of organic matter used, these pH values do not pose a problem for the production of organic biofertilizers because the pH values were between 6 and 7. Plant yield and development were not affected when phosphate and potassic rock biofertilizers that have low pH values (pH 3.0-3.5) were mixed with earthworm compost (pH 7.9). Santana et al. (2014) also evaluated the effects of rock biofertilizers mixed with earthworm compost on green pepper yield (*Capsicum annuum*) and observed increases in soil pH and soil nutrient availability after plant harvest.

The effect of mixing organic matter with a high pH, such as earthworm compost (pH 7.9), with rock biofertilizers, which results in a higher soil pH, is very important. Inoculation of rock biofertilizers with the oxidative bacteria *Acidithiobacillus* reduced the pH (3.0-3.5), and the addition of organic matter with a high pH neutralized the acidity, resulting in satisfactory plant growth and yield (Stamford, Stamford, Stamford, Barros Neto, & Takaki, 2007; 2009; Santana et al., 2014; Oliveira et al., 2015).

Exchangeable calcium

The analysis of exchangeable calcium in the treatments with different proportions of the 3 types of organic matter showed significant differences ($p < 0.05$) among the treatments (Table 2).

Table 2. Soluble calcium in the biofertilizers composed of different proportions of the 3 types of organic matter (sugarcane mud cake: earthworm compost: crop residue), inoculated with diazotrophic bacteria *Beijerinckia indica* (strain NFB 10001), after incubation for various maturation phases (0, 10, 20, 30, and 40 days) at room temperature (28 ± 2°C).

Treatments	Incubation time (days)				
(Proportion)	0	10	20	30	40
		Exchangeable calcium (mg dm^{-3})			
1 (5-0-0)	686a± 38	576a± 14	636a± 32	880a±39	570a± 11
2 (2-3-0)	310d±33	457a± 84	387c± 61	452c± 22	398c± 42
3 (3-2-0)	523b±66	378b± 54	423c± 24	480c± 20	448b± 42
4 (3-1-1)	501b±80	534a± 20	579a± 46	715b± 80	544a± 50
5 (2-0-3)	382c±19	333b± 3	486b± 35	369d± 55	503a± 24
6 (2-2-1)	416c±84	493a± 31	498b± 6	820a± 5	457b± 86
7 (1-1-3)	407c±57	456a± 71	542b± 15	475c± 9	424b± 53
8 (1-2-2)	397c±74	409b± 23	336c± 42	489c± 26	276d± 28
9 (0-4-1)	223d±38	324b± 27	315c± 13	371d± 49	304d± 5

Means with the same letter in the same column are not different according to Tukey's test ($p < 0.05$).

The addition of organic matter to the different treatments resulted in increased calcium in those biofertilizers with a high proportion of sugarcane mud cake (5:0:0), probably due to the calcium concentration in this natural substrate. The best results were observed in treatments (5:0:0) and (2:2:1) followed by treatment (3:1:1) when the material was in maturated phase for 30 days.

The results reported by Santana et al. (2014) showed that applying higher rates of mixed biofertilizer (phosphate and potassic rocks mixed with earthworm compost) on green pepper (*Capsicum annuum*) increased exchangeable calcium in the soil compared with a soluble fertilizer application. These results differ from those reported by Stamford et al. (2011), who showed that different fertilization treatments did not differ from one another (mixed biofertilizer, soluble fertilizer, and rock biofertilizer) and who found exchangeable calcium in the soil after two subsequent grapevine harvests.

Rossetto, Dias, and Vitti (2008), in describing the effects of sugarcane mud cake on crop yield, agreed that sugarcane mud cake increases crop yield with the addition of organic matter, phosphorus and calcium. Stamford et al. (2006) reported a significant increase in exchangeable calcium in the soil when applications of phosphate and potassic rock biofertilizer were utilized in sugarcane grown in a tableland Ultisol of northeastern Brazil. Similar results were reported by Stamford et al. (2011) when applying mixed biofertilizer and soluble fertilizer in grape (*Vitis vinifera*) grown in the semiarid region of the San Francisco Valley, Brazil.

Exchangeable magnesium

The treatments with different proportions of the 3 types of organic matter showed significant differences in exchangeable Mg^{+2} in the substrates (Table 3).

Table 3. Soluble magnesium in the biofertilizers with different proportions of the 3 types of organic matter (sugarcane mud cake: earthworm compost: crop residue), inoculated with diazotrophic bacteria *Beijerinckia indica* (strain NFB 10001), after incubation for the maturation phases with various times (0, 10, 20, 30, and 40 days) at room temperature (28 ± 2°C).

Treatments	Incubation time (days)				
(Proportions)	0	10	20	30	40
	Exchangeable magnesium (mg dm^{-3})				
1 (5-0-0)	93a±4	112a ± 9	130a ± 2	187a ± 4	132a ± 1
2 (2-3-0)	66b±9	66c ± 6	59c ± 2	84d ± 7	60d ± 3
3 (3-2-0)	67b±9	57d ± 7	110b ± 5	185a ± 9	115b ± 2
4 (3-1-1)	78b±5	109a ± 2	136a ± 6	167b ± 2	89c ± 9
5 (2-0-3)	56c±4	66c ± 6	60c ± 5	101b ± 9	112b± 5
6 (2-2-1)	62b±7	106a ± 9	107b ± 5	107b ± 7	116b± 9
7 (1-1-3)	47c±8	47d ± 3	98b ± 2	77d ± 1	100c ± 4
8 (1-2-2)	62b±7	57d ± 1	99b ± 9	62e ± 3	39e ± 3
9 (0-4-1)	68b±9	63c ± 1	101b ± 5	106b ± 5	108b ± 7

Means with the same letter in the same columns are not different according to Tukey's test ($p < 0.05$).

The (5:0:0) treatment had higher levels of exchangeable magnesium in the substrates incubated from time zero to 30 days in maturation phase and showed values that were very different compared with the other mixed proportions. The effects of the different proportions on exchangeable magnesium during 10 to 30 days of maturation phase were evident in the proportions (5:0:0), (3:2:0), and (3:1:1). It is important to note that these treatments used sugarcane mud cake in higher proportions. A reduction in exchangeable magnesium was observed during the 40-day of the incubation phase. Santana et al. (2014) did not report significant differences in exchangeable magnesium in the soil when different fertilization treatments were applied.

Other studies have reported an increase in exchangeable magnesium in soil when organic fertilizers are applied (Moura, Stamford, Duenhas, Santos, & Nunes, 2007; Stamford et al., 2006; Stamford et al., 2009; Stamford et al., 2014). The available magnesium may be released from minerals as biotite, which is used in the production of mixed fertilizers with PK rocks and organic matter, especially when biotite and sulfur are inoculated with the oxidative bacteria *Acidithiobacillus*, which produces sulfuric acid and releases magnesium due to acidity.

Available phosphorus

The available phosphorus in the different proportions of the 3 types of organic matter was significantly affected by the various incubation periods of the maturation phase (Table 4).

Table 4. Available phosphorus in the biofertilizers with different proportions of the 3 types of organic matter (sugarcane mud cake: earthworm compost: crop residue), inoculated with the diazotrophic bacteria *Beijerinckia indica* (strain NFB 10001), after incubation for various times in the maturation phase (0, 10, 20, 30, and 40 days) at room temperature (28 ± 2°C).

Treatments	Incubation time (days)				
(Proportions)	0	10	20	30	40
	Available phosphorus (mg dm^{-3})				
1 (5-0-0)	1553d±18	1578c ±20	2264b±90	2867a±38	2337a±33
2 (2-3-0)	1521d±10	1246c±16	1564d±80	2839a±95	1987b±93
3 (3-2-0)	2232a±35	2371a±90	2749a±50	3035a±40	2457a±93
4 (3-1-1)	2253a±46	2300a±31	1907c±40	1976b±63	1946b±66
5 (2-0-3)	1942b±85	1539c±16	2308b±90	1840b±53	1657c±75
6 (2-2-1)	1269d±58	1067c±95	2740a±96	2296b±56	2041a±63
7 (1-1-3)	698e±83	1689b±27	1575d±64	1723c±28	1901b±57
8 (1-2-2)	1795c±31	1214c±72	1214e±40	2278b±23	1308d±43
9 (0-4-1)	1143d±90	1457c±93	2274b±78	1996b±47	1951b±48

Means with the same letter in in the same column are not different according to Tukey's test ($p < 0.05$).

The available P significantly differed in the (3:2:0) and (3:1:1) treatments at time zero (T_0). After 10 and 20 days of incubation in the maturation phase, treatment effects were observed for proportions (3:2:0) and (3:1:1). The highest available P was obtained in the maturation phase correspondent to 30 days in proportions (3:2:0) and (2:3:0). At 40 days of incubation, similar to the tendency observed with the other nutrients, we observed a decrease in the P content.

The availability of phosphorus also depends on the ability of diazotrophic bacteria to fix N, as well as the activity of microorganisms that degrade organic materials, which contributes to an increase in the availability of phosphorus and other nutrients. Furthermore, the addition of organic material to the substrate increases microbial activity by promoting the mineralization of the material, thus allowing an increase in the availability of nutrients in these systems, soil, or organic substrates.

Stamford et al. (2014) evaluated mixed biofertilizers that were produced with PK rocks and organic matter with high N content due to inoculation with diazotrophic bacteria and found an increase in available P in the soil. Similar results were reported by Silva, Stamford, Amorim, Almeida Junior, and Silva (2011), who evaluated melon growth and phosphorus availability in the soil by applying phosphate and potassium biofertilizers after three subsequent cycles.

Available potassium

Available potassium was significantly affected by the different proportions of the 3 types of organic matter inoculated with diazotrophic bacteria during various incubation times of the maturation phase (Table 5).

Table 5. Available potassium in the biofertilizers with different proportions of the 3 types of organic matter (sugarcane mud cake: earthworm compost: crop residue), inoculated with the diazotrophic bacteria *Beijerinckia indica* (strain NFB 10001), after incubation for various times of the maturation phase (0, 10, 20, 30, and 40 days) at room temperature (28 ± 2°C).

Treatment	Incubation time (days)				
(Proportions)	0	10	20	30	40
	Available phosphorus (mg dm^{-3})				
1 (5-0-0)	109e±15	232c±62	201d±44	215d±38	130d±19
2 (2-3-0)	710a±12	639a±63	602b±114	404b±19	380c±84
3 (3-2-0)	519b±53	489b±31	497b±46	303c±73	339c±114
4 (3-1-1)	320d±16	368c±76	350c±12	351c±25	312c±54
5 (2-0-3)	433c±31	521b±32	530b±75	390b±82	339c±15
6 (2-2-1)	445c±43	512b±25	376c±62	394b±88	357c±22
7 (1-1-3)	445c±43	426b±51	450c±50	597a±207	511a±7
8 (1-2-2)	463c±24	485b±50	431c±25	411b±65	323c±198
9 (0-4-1)	701a±31	624a±6	623a±6	410b±38	449b±12

Means with the same letter in the same column are not different according to Tukey's test ($p < 0.05$).

Available potassium was affected by the different proportions of the 3 types of organic matter at the beginning of the experiment (time zero), specifically with proportions (2:3:0) and (0:4:1) likely due to the addition of earthworm compost. After 10 days of incubation, available potassium increased in several treatments with similar results. In general, the available potassium increased during the incubation time; however, the available potassium decreased after 40 days of the maturation phase.

Increases in available potassium in soil and in organic substrates may occur in mixed biofertilizers produced with phosphate, potassic rocks and organic matter (earthworm compost) and may have residual effects as reported by Stamford et al. (2014). The effects of the mixed biofertilizer may be caused by sulfuric acid release by the sulfur oxidative bacteria *Acidithiobacillus*, which increases the availability of phosphorus and potassium contained in the natural rocks used to produce the mixed biofertilizers.

Santos, Tiritan, Foloni, and Fabris (2010) reported that sugarcane mud cake is an organic compound that is also rich in potassium but that varies in its composition according to the variety of sugarcane and plant maturation.

Soluble sodium

Significant differences in soluble sodium were observed in the treatments with different proportions of the 3 types of organic matter (Table 6).

Table 6. Soluble sodium in the biofertilizers with different proportions of organic matter (sugarcane mud cake: earthworm compost: crop residue), inoculated with the diazotrophic bacteria *Beijerinckia indica* (strain NFB 10001), after incubation in the maturation phase with various times (0. 10, 20, 30, and 40 days) at room temperature (28 ± 2°C).

Treatment	Incubation time (days)				
(Proportions)	0	10	20	30	40
	Exchangeable sodium (mg dm^{-3})				
1 (5-0-0)	84c ±10	196b±32	102b±9	104d±19	93d±4
2 (2-3-0)	513b±167	366a±12	398a±58	224c±6	178c±58
3 (3-2-0)	494b±133	265a±77	338a±91	485a±259	379a±32
4 (3-1-1)	187c±71	219a±25	292a±25	160d±19	272b±101
5 (2-0-3)	792a±45	347a±2	309a±132	471a±175	255b±19
6 (2-2-1)	260c±84	334a±32	275a±102	228c±64	250b±25
7 (1-1-3)	753a±130	256a±25	283a±25	352b±175	139d±48
8 (1-2-2)	233c±38	288a±45	352a±65	201c±51	294b±64
9 (0-4-1)	270c±16	316a±6	339a±103	361b±218	234b±38

Means with the same letter in the same column are not different according to Tukey's test ($p < 0.05$).

At the initial time (time zero), higher levels of soluble sodium were observed in treatments (2:0:3) and (1:1:3); those treatments contained greater amounts of the organic matter crop culture. The other treatments contributed less soluble sodium to the substrate. The increase in soluble sodium during the period from 10 to 20 days was similar and did not significantly differ. This finding is not of great significance because the increase in soluble sodium in soil and in organic substrates is problematic; excess $Na+$ can increase the osmotic potential of the soil and reduce the availability of water, which reduces water infiltration into the soil and can have negative consequences on seed germination and plant growth.

Total N and C

The total N and C did not significantly differ ($p > 0.05$) between the incubation periods in the maturation phase, and the different treatments only had individual effects of the organic matter proportions on these nutrients (Figures 1 and 2).

The total N increased an average of 8.58 to 11.61 g kg^{-1} in the treatments and proportions 2:0:3, 1:2:2, and 0:4:1 were significantly different (Figure 1) and superior to the other treatments. These results differed from those of Stamford et al. (2014), who observed significant differences, in total N in soil when various fertilizer treatments (rock biofertilizers, mixed organic biofertilizers, and soluble fertilizer) were applied compared to earthworm compost (20 t ha^{-1}).

All applied treatments increased total C, showing a maximum increase in total C of 80 to 112 g kg^{-1} (Figure 2). The use of different proportions of the 3 types of organic matter to produce the biofertilizers may have promoted an increase in nutrients, particularly N and C because these substrates normally have a higher quantity of these elements that are released by the decomposition and mineralization of organic matter.

Increased total C in soil was reported by Stamford et al. (2011) using different fertilizer treatments (mixed biofertilizer, rock biofertilizers and soluble fertilizer) in grapevine (*Vitis vinifera*) grown in the San Francisco Valley, Brazil (Brazilian semiarid region).

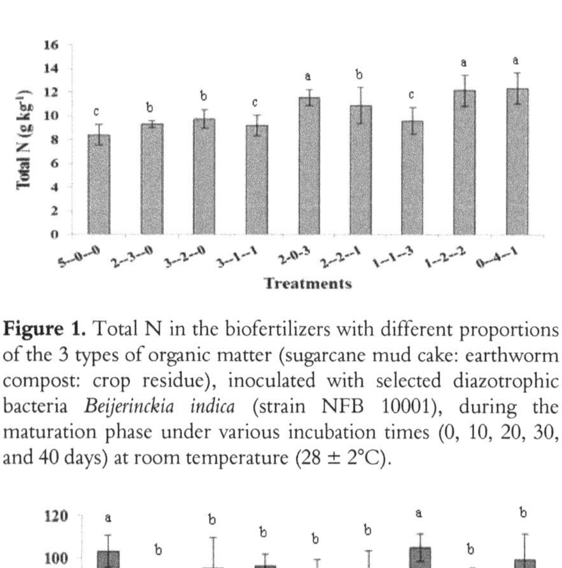

Figure 1. Total N in the biofertilizers with different proportions of the 3 types of organic matter (sugarcane mud cake: earthworm compost: crop residue), inoculated with selected diazotrophic bacteria *Beijerinckia indica* (strain NFB 10001), during the maturation phase under various incubation times (0, 10, 20, 30, and 40 days) at room temperature (28 ± 2°C).

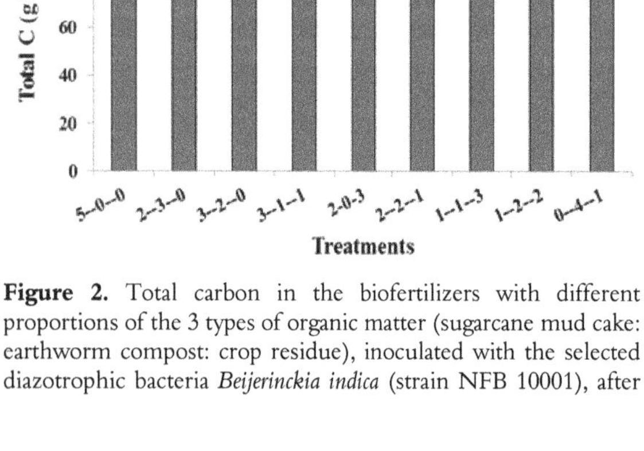

Figure 2. Total carbon in the biofertilizers with different proportions of the 3 types of organic matter (sugarcane mud cake: earthworm compost: crop residue), inoculated with the selected diazotrophic bacteria *Beijerinckia indica* (strain NFB 10001), after incubation for various times (0, 10, 20, 30, and 40 days) at room temperature (28 ± 2°C).

Conclusion

The different proportions of the organic matter used to produce organic biofertilizer contributed to increase nutrients availability in the substrates (biofertilizers), especially with proportions (5:0:0) and (3:2:0) for Ca, Mg and P; proportions (2:3:0) and (2:0:3) for K and N, respectively. The study confirmed the estimative of nutrient availability after residue addition which increased with the incubation period of 30 days. Biofertilizers with optimal organic matter proportions may be used as alternatives to soluble NPK fertilizers.

Acknowledgements

The authors are grateful to CNPq "Brazilian Scientific and Technological Development Council), FACEPE "Foundation for the Support of Science and Technology of the State of Pernambuco), and CAPES "Coordination to the Improvement of Education Personnel", for the financial support and scholarships.

References

Empresa Brasileira de Pesquisa Agropecuária [Embrapa] (2009). *Manual de análises químicas de solos, plantas e fertilizantes* (2a ed.). Brasília, DF: Embrapa.

Figueiredo, P. G., & Tanamati, F. Y. (2010). Adubação orgânica e contaminação ambiental. *Revista Verde, 5*(3), 1-4.

Lima, F. S., Stamford, N. P., Sousa, C.S., Lira Junior, M. A., Malheiros, S. M. M., & Van Straaten, P. (2010). Earthworm compound and rock biofertilizer enriched in nitrogen by inoculation with free living diazotrophic bacteria. *World Journal of Microbiology and Biotechnology, 26*(10), 1769-1777. doi: 10.1007/s11274-010-0357-z

Lima, R. C. M., Stamford, N. P., Santos, C. E. R. S., Lira Junior, M. A., & Dias, S. H. (2007). Effectiveness and residual effect of PK rock biofertilizers with sulfur and *Acidithiobacillus* in lettuce. *Brazilian Journal of Horticulture, 25*(3), 402-407. doi: 10.1590/S0102-05362007000300016

Moura, P., Stamford, N. P., Duenhas, L. H., Santos, C. E. R. S., & Nunes, G. H. A. S. (2007). Effectiviness of rock biofertilizers with *Acidithiobacillus* on melon grown in the San Francisco Valley. *Brazilian Journal of Soil Science, 2*(1), 1-7.

Oliveira, F. L. N., Stamford, N. P., Neto, D. S., Oliveira, E. C. A., Oliveira, W. S., & Santos C. E. R. S. (2015). Effects of biofertilizers produced from rocks and organic matter, enriched by diazotrophic bacteria inoculation on growth and yield of sugarcane. *Australian Journal of Crop Science, 9*(6), 504-508.

Rossetto, R., Dias, F. L. F., & Vitti, A. C. (2008). Problemas nutricionais dos solos nas novas fronteiras canavieiras. *Idea News*, *8*(94), 78-90.

Saideles, F. L. F., Winckler, M. V. C., Schirmer, W. N., & Sperandio, H. V. (2009). Carbonized rice hull as substratum to produce tamboril da mata and garapeira seedlings. *Semina: Ciências Agrárias*, *30*(4), 1173-1186. doi: 10.5433/1679-0359.2009v30n4Sup1p1173

SAS Institute (2011). The SAS 10.2 *software, Statistical Analysis System for Windows*. Cary, NC: Procedure guide for personal computer.

Santana, R. S., Stamford, N. P., Silva Junior, S., Santos, C. E. R. S., Freitas, A. D. S., & Arnaud, T. M. S. (2014). Influence of Bioprotector with Microbial Inoculation on green pepper yield and improvement on soil nutrients availability. *International Journal of Agricultural Innovations Research*, *2*(6), 2319-2473.

Santos, D. H., Tiritan, C. S., Foloni, J. S. S., & Fabris, L. B. (2010). Produtividade de cana-de-açúcar sob adubação com torta de filtro enriquecida com fosfato solúvel. *Pesquisa Agropecuária Tropical*, *40*(4), 454-461.

Silva, M. O., Stamford, N. P., Amorim, L. B., Almeida Junior, A. B., & Silva, M. O. (2011). Diferentes fontes de P no desenvolvimento do meloeiro e disponibilidade de fósforo no solo. *Revista Ciência Agronômica*, *42*(2), 268-277. doi: 10.1590/S1806-66902011000200003

Sousa, L. B., Nóbrega, R. S. A., Lustosa, J. F., Amorim, S. P. N., Ferreira, L. V. M., & Nóbrega, J. C. A. (2015). Cultivo de *Sesbania virgata* (Cav. Pers) em diferentes substratos. *Revista Ciências Agrárias*, *58*(3), 240-247. doi: 10.4322/rca.1942

Stamford, N. P., Izquierdo, C. G., Fernández, M. T. H., & Moreno, M. C. M. (2008). Biofertilizante de rochas fosfatadas e potássicas com enxofre e *Acidithiobacillus*. In M. V. Figueiredo, H. A. Burity, N. P. Stamford, & C. E. R. S. Santos (Eds.), *Microrganismos e Agrobiodiversidade*: o novo desafio para a agricultura (p. 401-421). Guaíba, RS: Ed. Agrolivros.

Stamford, N. P., Lima, R. A., Santos, C. E. R. S., & Dias, S. H. L. (2006). Rock biofertilizers with *Acidithiobacillus* on sugarcane yield and nutrient uptake in a Brazilian soil. *Geomicrobiology Journal*, *23*(5), 261-265. doi: 10.1080/0149 0450600760658

Stamford, N. P., Moura, P. M., Lira Junior, M. A., Santos, C. E. R. S., Duenhas, L. H., & Gava, C. A. T. (2009). Chemical attributes of an Argisol of the Vale do São Francisco after melon growth with phosphate and potash rocks biofertilizers. *Brazilian Journal of Horticulture*, *27*(4), 447-452. doi: 10.1590/S0102-05362009000400008

Stamford, T. C. M., Stamford, T. L. M., Stamford, N. P., Barros Neto, B., & Takaki, G. M. C. (2007). Growth of *Cunninghamella elegans* UCP 542 and production of chitin and chitosan using yam bean medium. *Electronic Journal of Biotechnology*, *10*(1), 61-68. doi: 10.2225/vol10-issue1-fulltext-1

Stamford, N. P., Andrade, I. P., Silva Junior, S., Lira Junior, M. A., Santos, C. S., Freitas, A. D. S., & Van Straaten, P. (2011). Soil properties and grape yield affected by rock biofertilizers with earthworm compound. *Journal of Soil Science and Plant Nutrition*, *11*(4), 15-25. doi: 10.4067/S0718-95162011000400006

Stamford, N. P., Silva Junior, S., Santos, C. E. R. S., Freitas, A. D. S., Santos, C. M. A., Arnaud, T. M. S., & Soares, H. R. (2014). Yield of grape (*Vitis labrusca* cv. Isabel) and soil nutrients availability affected by biofertilizer with diazotrophic bacteria and fungi chitosan. *Australian Journal of Crop Science*, *8*(2), 301-306.

Van Straaten, P. (2002). *Rocks for crops: Agrominerals of Sub-Saharan Africa*. Nairobi, KE: ICRAF.

20

Organic cherry tomato yield and quality as affect by intercropping green manure

Edmilson José Ambrosano[1], Gabriela Cristina Salgado[2*], Ivani Pozar Otsuk[3], Patrícia Patri[1], Celina Maria Henrique[1] and Paulo César Tavares de Melo[2]

[1]*Agência Paulista de Tecnologia dos Agronegócios, Pólo Regional Centro Sul, Rodovia SP-127, km 30, 13400-970, Piracicaba, São Paulo, Brazil.*
[2]*Departamento Produção Vegetal, Escola Superior de Agricultura "Luiz de Queiroz", Universidade de São Paulo, Piracicaba, São Paulo, Brazil.*
[3]*Agência Paulista de Tecnologia dos Agronegócios, Instituto de Zootecnia, Nova Odessa, São Paulo, Brazil. *Author for correspondence. E-mail: salgado.gc@gmail.com*

ABSTRACT. The aim of the study was to determine the yield and quality parameters of organic cherry tomatoes cultivated by intercropping with green manure in two successive years. The experimental design was a randomized block with eight treatments and five replicates as follows: two controls with single cherry tomato crop, one with straw addition and the other without corn straw as a cover crop; cherry tomato intercropped with jack bean (*Canavalia ensiformis* DC); tomato intercropped with white lupine (*Lupinus albus* L.); tomato intercropped with sun hemp (*Crotalaria juncea* L.); tomato intercropped with velvet bean-dwarf [*Mucuna deeringiana* (Bort) Merrill]; tomato intercropped with mung bean [*Vigna radiata* (L.) Wilczek]; and tomato intercropped with cowpea [*Vigna unguiculata* (L.) Walp.]. All the treatments with green manure received corn straw as mulch. The treatment group with cowpea had a lower number and weight of commercial fruits compared to the velvet bean-dwarf and jack bean treatments. The number and weight of the commercial fruits were lower in the second year. The fruits in the first year exhibited a higher average weight, a lower nutrient content and a lower soluble solids total TSS/AT ratio than the second year fruits. Green manures did not negatively affect the quality or nutrient content of the tomato fruit.

Keywords: organic agriculture; *Solanum lycopersicum*; production; intercropping.

Introduction

Concern about food insecurity and the environmental impact of the production of foods has led consumers to seek food grown via alternative systems over those grown conventionally (Rembiałkowska, 2007; Hidalgo-Baz, Martos-Partal, & González-Benito, 2017). Organic agriculture has grown by 75% in the last 15 years, reaching a planted area of 43.7 million hectares in 142 countries in 2014; of this total area, 0.5% is intended for the production of vegetables (legumes, hardwoods and fruit) (Lernoud & Willer, 2016).

Tomato is one of the most produced vegetables in Brazil, with the 2015 harvest reaching a production of

4.1 million tons (IBGE, 2016). For organic producers, the commercialization of tomatoes has a better economic return due to demand and the higher price compared to other vegetables, thus reducing the external input cost (Jouzi et al., 2017). The demand for healthy, nutritious products and the market for gourmet products have increased the value of organic foods (Willer & Lernoud, 2016). One study has shown that organic fruits have a higher content of vitamins, phenolics and antioxidants than conventionally produced fruits (Mditshwa, Magwaza, Tesfay, & Mbili, 2017).

However, organic producers face challenges due to the scarcity of inputs such as fertilization. According to Seufert, Ramankutty, and Foley (2012), when organic agriculture received higher amount of N than conventional agriculture, its performance improved, showing that organic agriculture is limited by the N input. The sources of nitrogen used in organic agriculture are essentially composting, biofertilizer and green manure. The practice of green manuring has been shown to be important for different crops, whether they use intercropping or crop rotation systems. This is due to the innumerable benefits provided by this practice, such as the improvements in the physical, chemical and biological aspects of soil, increased productivity, and weed control (Ambrosano et al., 2011; Dantas et al., 2015; Tao et al., 2017). However, few studies have shown the influence of green manures on the yield and quality of organic foods such as tomatoes.

In this context, the aim of this study was to determine the yield and quality parameters of organic cherry tomatoes grown in concert with leguminous cover crops in two successive years.

Material and method

The experiment was conducted in an agroecological area under protected cultivation (greenhouse) for two successive years (2012/2013 and 2013/2014) in the municipality of Piracicaba, São Paulo State, Brazil (latitude 22°43'S and longitude 47°38'W) with an altitude of 540 m, an average monthly temperature of over 22°C and a temperature of the coldest month of 16.9°C. The average temperature and mean relative humidity in the period corresponding to the cherry tomato crop are presented in Figure 1.

The soil in this area corresponds to kandiudox with the following physical and chemical characteristics: clay: 36%; silt: 23%; fine sand: 28%; coarse sand: 13%; pH: 5,8; OM: 20 g dm^{-3}, 29 mg dm^{-3}; K: 8.4 mmol dm^{-3}; Ca: 35 mmol dm^{-3}; Mg: 11 mmol dm^{-3}; H + Al: 20 mmol dm^{-3}; Al: 0 $mmol_c$ dm^{-3}; S: 5 mg dm^{-3}; B: 0.42 mg dm^{-3}; Cu: 4.5 mg dm^{-3}; Fe: 51 mg dm^{-3}; Mn: 45 mg dm^{-3}; Zn: 3,1 mg dm^{-3}; SB: 54.4 mmol dm^{-3}; CTC: 74.4 mmol dm^{-3}; V: 69%; and m: 0%.

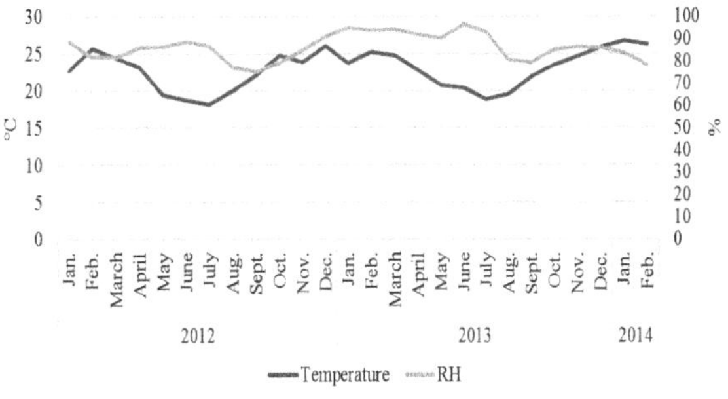

Figure 1. Average temperature (°C) and RH - Relative Humidity (%) for the years 2012/2013/2014.

The greenhouse used was a metal structure of the arch type with a 75 µm-thick transparent low-density polyethylene (LDPE) plastic film and its open sides covered by only sombrite. The dimensions of the structure were 6.0 x 50.0 m, and ceiling height was 2.5 m, giving a total area of 300 m^2.

The cherry tomato seeds (*Solanum lycopersicum* L.), Access 21 of the Instituto Agronômico (IAC) (Filho & Melo, 2001), were seeded in polystyrene trays with 128 cells and transplanted into pits (10^{-4} m^3) at the density of 3.3 m^2 in single lines (0.9 x 0.6 m). The green manure was seeded in the experimental area between the tomato lines. Tomato transplanting and sowing of the green manure occurred simultaneously in the month of August in the two years. The tomatoes were fertilized in the pit with 25 g of thermophosphate, equivalent to 75 kg ha^{-1} of P_2O_5, and 2.7 g of potassium sulphate, equivalent to 25 kg ha^{-1} of K_2O. After transplanting, the soil was covered with corn straw in all plots, except for the control without corn straw.

The cherry tomatoes and green manure plants were irrigated according to their water requirement using drip irrigation. The tomatoes were tutored, where the apical pruning was performed 120 days after transplanting (DAT), which corresponds to the eighth racemic.

Tomatoes harvested in the mature stage were separated into commercial fruits and damaged fruits and were later counted and weighed to determine the number of commercial and damaged fruits per plant (NCF pl^{-1}, NDF pl^{-1}), commercial and damaged fruit weight per plant (CFW pl^{-1} and DFW pl^{-1}) and average commercial and damaged fruit weight (ACFW, ADFW). A sample of the fruits was taken to determine the chemical parameters of the cherry tomato.

In the chemical analysis of cherry tomato quality, the following parameters were determined: pH using a potentiometer according to method No. 981.12 – AOAC (1997); soluble solids content (°Brix) using a refractometer according to method No. 932.12 – AOAC (1997); titratable total acidity (% citric acid) using potentiometric titration according to the second method No. 942.15- AOAC (1997); total sugars and reducing sugars (%) using the method of Eynon and Lane described by Carvalho, Mantovani, Carvalho, and Moraes (1990); and the ratio obtained by dividing the soluble solids content (°Brix) by the value of the total titratable acidity (%).

To analyse the macro- and micronutrients in the cherry tomato fruit, they were sampled, taken to the laboratory, dried at 65°C in a drying oven with forced air circulation for 72 hours, and passed through a Wiley type mill, and then their contents of N, P, K, Ca, Mg, Cu, Fe, Mn, and Zn were determined according to the methodology described in Malavolta, Vitti, and Oliveira (1997).

The experimental design was a randomized complete block design with eight treatments and five replicates. The treatments were as follows: two controls with a single cherry tomato crop, one with added straw and the other without added straw in addition as a cover crop; cherry tomato intercropped with jack bean (*Canavalia ensiformis* DC); cherry tomato intercropped with lupine (*Lupinus albus* L.); cherry tomato intercropped with sun hemp (*Crotalaria juncea* L.); cherry tomato intercropped with velvet bean-dwarf [*Mucuna deeringiana* (Bort) Merrill]; cherry tomato intercropped with mung bean [*Vigna radiata* (L.) Wilczek]; and cherry tomato intercropped with cowpea beans [*Vigna unguiculata* (L.) Walp.]. All the treatments with green manure intercropped with cherry tomato received maize straw as mulch. Green fertilizers were sown at different densities according to recommendations (Table 1).

Table 1. Planting density of the green manures used in the experiment.

Treatments	Density of sowing ----- Seeds m^{-1}----
1- Check with corn straw (CCS1)	--
2- Check without corn straw (CCS2)	--
3- Intercropping cherry tomato with jack beans	5
4- Intercropping cherry tomato with sun hemp	11
5- Intercropping cherry tomato with velvet bean-dwarf	5
6- Intercropping cherry tomato with mung bean	20
7- Intercropping cherry tomato with white lupine	7
8- Intercropping cherry tomato with cowpea beans	20

Statistical analysis was performed using the concept of time-repeated measures, and data were analysed using the MIXED procedure of SAS software (Statistical Analysis System, 9.3). The level of significance adopted for analysis of variance was $p < 0.1$.

Result and discussion

There were differences ($p < 0.1$) in the number of commercial fruits (NCF) between the velvet bean-dwarf and bean cowpea treatments and in commercial fruit weight (CFW) between the jack bean and cowpea bean treatments; these differences were independent of the year (Table 2). The treatment of cherry tomatoes intercropped with velvet bean-dwarf (129.0 NFC plant^{-1}) produced 27% more NCF than the treatments with cowpea (94.4 NCF plant^{-1}). The CFW of treatments with jack bean (1,410.8 g plant^{-1}) was 30% higher than that of the treatments with bean cowpea (985.3 g plant^{-1}).

Table 2. Number and weight of commercial fruits per cherry tomato plant.

Treatments	Number of Commercial Fruits (NCF)			Weight of Commercial Fruits (WCF)		
	Year 1	Year 2	Average	Year 1	Year 2	Average
	-- NCF plant^{-1}--			---------- g plant^{-1}---------		
CCS1	116.3	122.5	119.4 AB	1317.0	1313.1	1315.0 AB
CCS2	139.3	100.9	120.1 AB	1601.3	994.5	1297.9 AB
Jack beans	139.1	111.5	125.3 AB	1667.4	1154.1	1410.8 A
Sun hemp	128.2	94.0	111.1 AB	1440.3	958.8	1199.5 AB
Velvet bean-dwarf	128.4	129.6	129.0 A	1465.6	1199.9	1332.8 AB
Mung beans	127.0	100.3	113.7 AB	1465.5	1014.2	1239.9 AB
White lupine	124.3	113.5	118.9 AB	1383.5	1083.8	1233.7 AB
Cowpea beans	104.7	84.0	94.4 B	1150.0	820.6	985.3 B
Average	125.9 a	107.0 b		1436.33a	1067.4 b	
*CV (%)	29.93			31.47		

Averages followed by different lowercase letters in the row differ ($p < 0.1$) from each other based on the F test, and means followed by different capital letters in the columns differ ($p < 0.1$) from each other based on the Tukey-Kramer test. *coefficient of variation; *CCS1 - check with corn straw; *CCS2 - check without corn straw.

According to Ambrosano, Rossi, Dias, Tavares, and Ambrosano (2014), cowpea bean intercropped with cherry tomato in the field produced 1.6 Mg ha^{-1} dry mass, which was 63% higher than the dry mass produced when velvet bean-dwarf was used (0.6 Mg ha^{-1}). In this way, it seems that there was an inter specific competition between the cherry tomato and the cowpea bean for resources such as space and light, which affected the yield of the cherry tomato. Liu, Cheng, Meng, Ahmad, and Zhao (2014) observed that tomato-garlic intercropping had a negative impact on tomato productivity because of the competition between plants for water, space and nutrition. According to Sakai et al. (2011) lettuce intercropped with green manure had lower yields compared to single lettuce cultivation. According to the same author, competition can be established for different factors (water, light, and nutrients) that occur simultaneously, in which one competition factor can lead to another.

On average, the cherry tomato plants in the first year had 15 and 26% more NCF (125.9 NCF plant^{-1}) and WCF (1,436.3 g plant^{-1}), respectively, compared to those in the second year (107.0 NCF plant^{-1}; 1,067.37 g plant^{-1}) (Table 2). The number of damaged fruits (NDF) and weight of damaged fruit (WDF) were also higher in the first year (46.7 NDF plant^{-1}, 430.6 g plant^{-1}) than in the second year (23.1 NDF plant^{-1}; 179.8 g plant^{-1}) (Table 3). These productivity values were similar to those found by Sari et al. (2017) for a cherry tomato hybrid.

There was no difference ($p > 0.1$) in the NDF and WDF between the treatments, indicating that the consortium with green manures did not interfere negatively in the cherry tomato productivity compared to single tomato cultivation (Table 3).

Table 3. Number and weight of damaged fruits per cherry tomato plant.

Treatments	Number of Damaged Fruits (NDF)			Weight of Damaged Fruits (WDF)		
	Year 1	Year 2	Average	Year 1	Year 2	Average
	NDF plant^{-1}			g plant^{-1}		
CCS1	43.43	23.07	34.64 A	401.87	213.26	307.57 A
CCS2	48.75	20.73	34.74 A	442.12	155.20	298.66 A
Jack beans	51.65	26.42	39.04 A	499.15	210.20	354.68 A
Sun hemp	51.25	22.05	36.65 A	473.62	162.16	317.89 A
Velvet bean-dwarf	45.52	20.25	32.88 A	408.53	155.29	281.91 A
Mung beans	40.50	20.85	30.67 A	376.05	161.05	268.55 A
White lupine	42.27	21.20	31.73 A	380.43	160.24	270.33 A
Cowpea beans	50.27	27.24	38.76 A	463.20	220.73	341.96 A
Average	46.70 a	23.07 b		430.62 a	179.77 b	
*CV (%)	23.13			24.58		

Averages followed by different lowercase letters in the rows differ ($p < 0.1$) from each other based on the F test, and means followed by different capital letters in the columns differ ($p < 0.1$) from each other based on the Tukey-Kramer test. *coefficient of variation; *CCS1 - check with corn straw; *CCS2 - check without corn straw.

The intercropping treatments with jack beans had an average weight of commercial fruits (WACF) that was on average 8% higher than with that of the velvet-bean dwarf, white lupine and cowpea treatments (Table 4). For the same variable, there was also a difference ($p < 0.1$) between the years (Table 4). The WACF found in year 1 (11.38 g fruit^{-1}) was 14% higher than that found in year 2 (9.77 g fruit^{-1}). Proportionally, the WADF was also higher in year 1 (9.25 g fruit^{-1}) than in year 2 one (7.56 g fruit^{-1}) (Table 4).

Table 4. Average weight of commercial fruits (WACF) and average weight of damaged fruits (WADF) of cherry tomato.

Treatments	WACF			WADF		
	Year 1	Year 2	Average	Year 1	Year 2	Average
	g fruit^{-1}					
CCS1	11.21	10.48	10.84 AB	9.21	7.82	8.52 A
CCS2	11.51	9.62	10.56 AB	9.11	7.23	8.17 A
Jack beans	11.96	10.31	11.14 A	9.70	7.83	8.77 A
Sun hemp	11.21	9.86	10.53 AB	9.32	7.24	8.28 A
Velvet bean-dwarf	11.43	9.13	10.28 B	9.09	7.46	8.27 A
Mung beans	11.60	9.85	10.73 AB	9.34	7.49	8.41 A
White lupine	11.12	9.33	10.23 B	8.99	7.51	8.25 A
Cowpea beans	10.98	9.57	10.27 B	9.20	7.88	8.54 A
Average	11.38 a	9.77 b		9.25 a	7.56 b	
*CV (%)	7.59			10.45		

Averages followed by different lowercase letters in the rows differ ($p < 0.1$) from each other based on the F test, and means followed by different capital letters in the columns differ ($p < 0.1$) from each other based on the Tukey-Kramer test. *coefficient of variation; *CCS1 - check with corn straw; *CCS2 - check without corn straw.

The successive cultivation of tomato in the same area may have favoured, possibly, the propagation of the late blight (*Phytophthora infestans*), which, due to the favourable conditions (lower temperatures and high relative humidity), contributed to a greater rate of infection of this disease in the leaves of the cherry tomato plants. This may be the cause of the lower productivity of the cherry tomato in the second year. According to Lima, Maffia, Barreto, and Mizubuti (2009), successive tomato cultivation threatens new crops because of the presence in that area of sporangia produced in crop residues and the presence of host plants in the region. The late blight is considered one of the major diseases affecting organic tomato agriculture. According to Bruggen and Finckh (2016), multiple cycle leaf diseases such as *P. infestans* can be a serious problem for organic farmers because they lack effective control measures. The late blight occurs mainly in environments with high relative humidity and low temperatures. However, there is a study that indicates that tomato plants can also be infected by this disease at higher temperatures (27°C) (Danies et al., 2013; Maziero, Maffia, & Mizubuti, 2009).

Regarding the macro- and micronutrients in the fruit, there was no difference ($p > 0.1$) in the phosphorus (P), potassium (K), magnesium (Mg), calcium (Ca), copper (Cu), iron (Fe), manganese (Mn) and zinc (Zn) content between treatments, independent of the year (Figure 2). The results showed that the access cherry tomato fruits exhibited 22.15 g kg^{-1} of nitrogen (N), 3.39 g kg^{-1} of P, 33.28 g kg^{-1} of K, 1.95 g kg^{-1} of Mg, 1.38 g kg^{-1} of Ca, 7.0 mg kg^{-1} of Cu, 54.36 mg kg^{-1} of Fe, 8.90 mg kg^{-1}of Mn, and 27.71 mg kg^{-1} of Zn. These values are close to the levels of N, P, K, Ca, and Mg found by Kumar, Rouphael, Cardarelli, and Colla (2015). By contrast, the levels of Cu, Fe, and Zn found in this study are larger than those found by Kumar et al. (2015) and Schmautz et al. (2016).

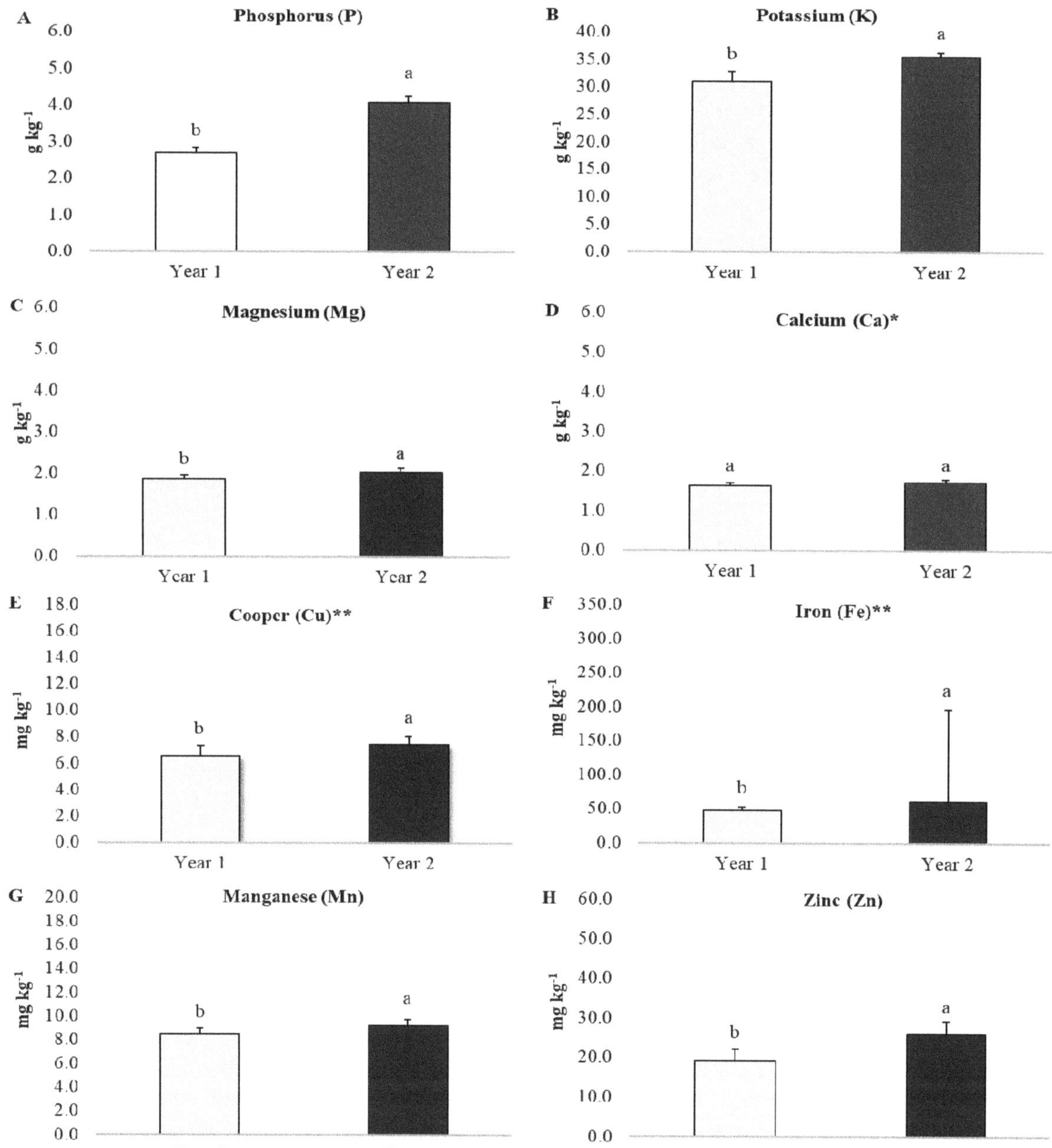

Figure 2. Nutrient content of the cherry tomato fruit: (A) phosphorus, (B) potassium, (C) magnesium, (D) calcium, (E) cooper, (F) iron, (G) manganese, and (H) zinc; Averages followed by different lowercase letters on the bars differ ($p < 0.1$) from each other based on the F test; *Statistic referring to the square root transform of the data; **Statistic referring to the log transform of the data.

However, for all the nutrient contents presented in this study, except for Ca, there were differences in the macro- and micronutrient contents in the fruit between the cultivated years ($p < 0.1$) (Figure 2). The content of P, K, Mg, Cu, Fe, Mn, and Zn in the fruit from the second year was 34, 13, 8, 13, 20, 8, and 26% higher, respectively, than that in the fruit from the first one. In the second year, the productivity and WACF were lower; however, the macro- and micronutrient contents of the fruit were higher. Most likely, the lower WACF in the second year led to a higher concentration of nutrients in the fruit.

Among the quality parameters, the cherry tomato did not exhibit a difference ($p > 0.1$) between the treatments with green manure. However, there was a difference ($p < 0.1$) between the years for all variables, except for the soluble solids content (Figure 3). The average values of the chemical characteristics of this study on cherry tomato SST 6.5 Brix, pH 4.35, were similar to the values found by Jorge, Nascimento, Junior, Silva, and Barbosa (2017) (Figure 3).

The SSC/TTA ratio found in the cherry tomatoes in year 2 (19.5) is above the range reported in the work of (Rossi et al., 2013) (Figure 3C). In year 2, the increase in the SSC/TTA ratio was due to the 17% decrease in TTA; thus, the cherry tomato was less acidic and consequently sweeter in the second year (Figure 3D). Despite the decrease in TTA in year 2, the pH was only 1% higher in year 1 compared to year 2 on average (Figure 3A). In year 2, the total sugar was 24% lower than that in year 1, and the reducing sugar was 25% higher than that in year 1 (Figure 3E and F). The present variety of cherry tomato had a lower SSC/TTA than cultivars of commercial cherry tomato; therefore, this variety of cherry tomatoes is less sweet than the commercial cultivars (Bhandari, Chae, & Lee, 2016).

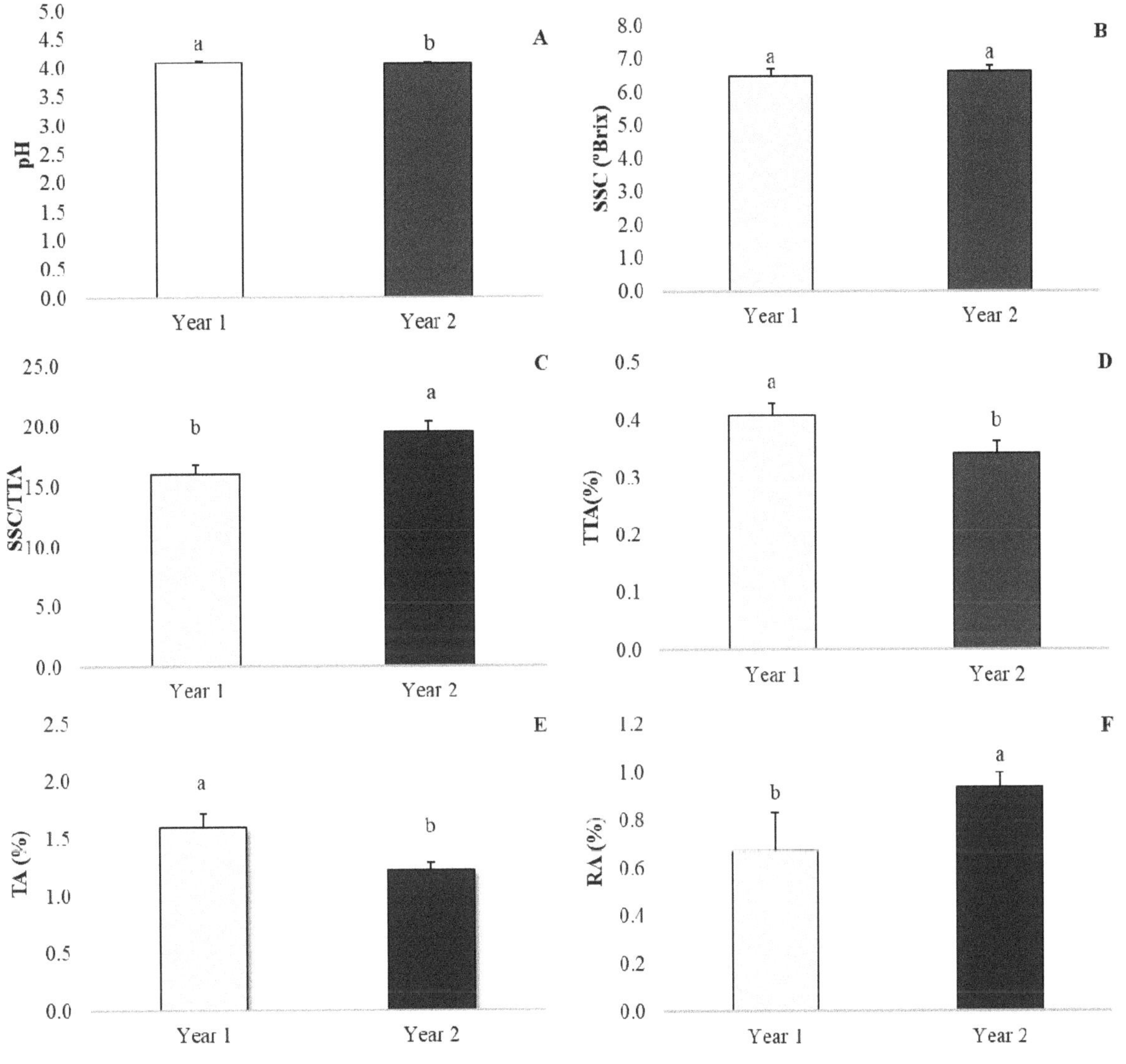

Figure 3. Chemical parameters of the cherry tomato fruit: pH (A), SSC - soluble solids content (B), SSC/TTA - ratio of soluble solids content to titratable acidity (C), TTA - titratable acidity (D), TA - total sugars (E), RA - reducing sugars (D); Means followed by the same lowercase letters do not differ from each other (Test F, $p > 0.1$).

Pre-harvest factors such as soil properties, nutrition, temperature, irrigation, genetics and fruit maturity may influence fruit quality (Dorais, Ehret, & Papadopoulos, 2008; Luna, Selma, Tudela, & Gil, 2013). These factors were similar in the two years of the cherry tomato cultivation experiment, except for the incidence of late blight in the leaves of the cherry tomato plants in the second year.

Biotic or abiotic stress can cause changes in the sugar composition in the phloem sap (Gil, Ben-ari, Turgeon, & Wolf, 2012). In addition, some pathogens can alter the physiology of the host plant and acquire sugar from it for their nutrition (Chen et al., 2010). Thus, the reduction in total sugar in the second year may be related to the defence system of the tomato plant.

Conclusion

The cherry tomatoes intercropped with green manures did not have reduced NCF, WCF, or WACF compared to the controls. In addition, green manure did not negatively interfere with the chemical parameters of the quality of the cherry tomatoes.

Acknowledgements

We would like to acknowledge the technical support of the research of APTA, the CNPq productivity scholarship to the first author, the Professor Glaucia Maria Bovi Ambrosano in support of statistical analysis, FAPESP and Piraí Seeds.

References

Ambrosano, E. J., Cantarella, H., Ambrosano, G. M. B., Schammas, E. A., Dias, F. L. F., Rossi, F., … Azcón, R. (2011). Produtividade da cana-de-açúcar após o cultivo de leguminosas. *Bragantia*, *70*(4), 810–818. doi: 10.1590/S0006-87052011000400012

Ambrosano, E. J., Rossi, F., Dias, F. L. F., Tavares, S., & Ambrosano, G. M. B. (2014). Desempenho do Tomate-Cereja e Milho-Verde Após Cultivo de Fabaceas e Seu Efeito no Solo. *Cadernos de Agroecologia*, *9*(4), 1–12.

Association of Official Analytical Chemists [AOAC]. (1997). *Official methods of analysis* (16th ed.). Washington D.C.: AOAC.

Bhandari, S. R., Chae, Y., & Lee, J. G. (2016). Assessment of Phytochemicals, Quality Attributes,and Antioxidant Activities in Commercial Tomato Cultivars. *Korean Journal of Horticultural Science & Technology*, *34*(5), 677–691. doi: 10.12972/kjhst.20160071

Bruggen, A. H. C. Van, & Finckh, M. R. (2016). Plant Diseases and Management Approaches in Organic Farming Systems. *Annual Review of Phytopathology*, *54*, 25–54. doi: 10.1146/annurev-phyto-080615-100123

Carvalho, C. R. L., Mantovani, D. M. B., Carvalho, P. R. N., & Moraes, R. M. (1990). *Análises químicas dos alimentos*. Campinas, SP: ITAL.

Chen, L., Hou, B., Lalonde, S., Takanaga, H., Hartung, M. L., Qu, X., … Frommer, W. B. (2010). Sugar transporters for intercellular exchange and nutrition of pathogens. *Nature*, *468*(7323), 527–532. doi: 10.1038/nature09606

Danies, G., Small, I. M., Myers, K., Childers, R., & Fry, W. E. (2013). Phenotypic characterization of recent clonal lineages of *Phytophthora infestans* in the United States. *Plant Disease*, *97*(7), 873–881. doi: 10.1094/PDIS-07-12-0682-RE

Dantas, R. A., Carmona, R., Carvalho, A. M., Rein, T. A., Malaquias, J. V., Santos Júnior, J. D. G., … Santos Júnior, J. D. G. (2015). Produção de matéria seca e controle de plantas daninhas por leguminosas consorciadas com cana-de-açúcar em cultivo orgânico. *Pesquisa Agropecuária Brasileira*, *50*(8), 681–689. doi: 10.1590/S0100-204X2015000800006

Dorais, M., Ehret, D. L., & Papadopoulos, A. P. (2008). Tomato (*Solanum lycopersicum*) health components: from the seed to the consumer. *Phytochemistry Reviews*, 7(2), 231–250. doi: 10.1007/s11101-007-9085-x

Filho, J. A. A., & Melo, A. M. T. (2001). Avaliação de tomate silvestre do tipo cereja. *Horticultura Brasileira*, *19*, 1–5.

Gil, L., Ben-ari, J., Turgeon, R., & Wolf, S. (2012). Effect of CMV infection and high temperatures on the enzymes involved in raffinose family oligosaccharide biosynthesis in melon plants. *Journal of Plant Physiology*, *169*(10), 965–970. doi: 10.1016/j.jplph.2012.02.019

Hidalgo-Baz, M., Martos-Partal, M., & González-Benito, Ó. (2017). Attitudes vs. purchase behaviors as experienced dissonance: The roles of knowledge and consumer orientations in organic market. *Frontiers in Psychology*, *24*(8), 248. doi: 10.3389/fpsyg.2017.00248

Instituto Brasileiro de Geografia e Estatística. [IBGE] (2016). *Levantamento sistemático da produção agrícola* (v. 29). Rio de Janeiro, RJ: IBGE.

Jorge, M. F., Nascimento, K. O., Junior, J. L. B., Silva, L. D. B., & Barbosa, M. I. M. J. (2017). Physicochemical Characteristics, Antioxidant Capacity and Phenolic Compounds of Tomatoes Fertigated With Different Nitrogen Rates *Revista Caatinga*, *30*(1), 237–243. doi: 10.1590/1983-21252017v30n126rc

Jouzi, Z., Azadi, H., Taheri, F., Zarafshani, K., Gebrehiwot, K., Van Passel, S., & Lebailly, P. (2017). Organic farming and small-scale farmers: Main opportunities and challenges. *Ecological Economics*, *132*(C), 144–154. doi: 10.1016/j.ecolecon.2016.10.016

Kumar, P., Rouphael, Y., Cardarelli, M., & Colla, G. (2015). Effect of nickel and grafting combination on yield, fruit quality, antioxidative enzyme activities, lipid peroxidation, and mineral composition of tomato.*Journal of Plant Nutrition and Soil Science*, *178*(6) 848–860. doi: 10.1002/jpln.201400651

Lenoud, J., & Willer, H. (2016). Statistics on selected crop: vegetables. In H. Willer, & J. Lenoud (Ed.), *The world of organic agriculture. Statistics and Emerging Trends* (p. 114-116). Switzerland; Germany: Research Institute of Organic Agriculture (FiBL) and IFOAM- Organics International.

Lima, M. A., Maffia, L. A., Barreto, R. W., & Mizubuti, E. S. G. (2009). *Phytophthora infestans* in a subtropical region : survival on tomato debris, temporal dynamics of airborne sporangia and alternative hosts. *Plant Pathology*, *58*(1), 87–99. doi: 10.1111/j.1365-3059.2008.01951.x

Liu, T., Cheng, Z., Meng, H., Ahmad, I., & Zhao, H. (2014). Growth, yield and quality of spring tomato and physicochemical properties of medium in a tomato/garlic intercropping system under plastic tunnel organic medium cultivation. *Scientia Horticulturae*, *170*, 159–168. doi: 10.1016/j.scienta.2014.02.039

Luna, C., Selma, V., Tudela, J. A., & Gil, I. (2013). Influence of nutrient solutions in an open-field soilless system on the quality characteristics and shelf life of fresh-cut red and green lettuces (*Lactuca sativa* L.) in different seasons. *Journal of the Science of Food and Agriculture*, *93*(2), 415–421. doi: 10.1002/jsfa.5777

Malavolta, E., Vitti, G. C., & Oliveira, S. A. (1997). *Evaluation of the nutritional status of plants:* principles and applications. Piracicaba, SP: POTAFOS.

Maziero, J. M. N., Maffia, L. A., & Mizubuti, E. S. G. (2009). Effects of temperature on events in the infection cycle of two clonal lineages of *Phytophthora infestans* causing late blight on tomato and potato in Brazil. *Plant Disease*, *93*(5), 459–466. doi: 10.1094/PDIS-93-5-0459

Mditshwa, A., Magwaza, L. S., Tesfay, S. Z., & Mbili, N. (2017). Postharvest quality and composition of organically and conventionally produced fruits: A review. *Scientia Horticulturae*, *216*, 148–159. doi: 10.1016/j.scienta.2016.12.033

Rembiałkowska, E. (2007). Quality of plant products from organic. *Journal of the Science of Food and Agriculture*, *87*(15), 2757–2762. doi: 10.1002/jsfa3000

Rossi, F., Ambrosano, E. J., Schammass, E. A., María, G., & Ivani, E. (2013). Intercropping influence on the quality of fruit. *Espamciencia*, *4*(2), 77–82.

Sakai, R. H., Ambrosano, E. J., Clarissa, A., Negrini, A., Cesar, P., Trivelin, O., … César, P. (2011). N transfer from green manures to lettuce in an intercropping cultivation system. *Acta Scientiarum. Agronomy*, *33*(4), 679–686. doi: 10.4025/actasciagron.v33i4.6766

Sari, B. G., Lúcio, A. D., Santana, C. S., Lopes, S. J., Sari, B. G., Lúcio, A. D., … Lopes, S. J. (2017). Linear relationships between cherry tomato traits. *Ciência Rural*, *47*(3) e20160666. doi: 10.1590/0103-8478cr20160666

Schmautz, Z., Loeu, F., Liebisch, F., Graber, A., Mathis, A., Griessler Bulc, T., & Junge, R. (2016). Tomato productivity and quality in aquaponics: Comparison of three hydroponic methods. *Water*, *8*(11), 1-21. doi: 10.3390/w8110533

Seufert, V., Ramankutty, N., & Foley, J. A. (2012). Comparing the yields of organic and conventional agriculture. *Nature*, *485*(7397), 229 232. doi: 10.1038/nature11069

Tao, J., Liu, X., Liang, Y., Niu, J., Xiao, Y., Gu, Y., … Yin, H. (2017). Maize growth responses to soil microbes and soil properties after fertilization with different green manures. *Applied Microbiology and Biotechnology*, *101*(3), 1289–1299. doi: 10.1007/s00253-016-7938-1

Willer, H., & Lenoud, J. (2016). The world of organic agriculture: summary. In H. Willer, & J. Lenoud, (Ed.), *The world of organic agriculture. Statistics and Emerging Trends* (p. 114-116). Switzerland; Germany: Research Institute of Organic Agriculture (FiBL); IFOAM - Organics International.

21

'Navelina' oranges submitted to pre-harvest resistance inducers

Marines Batalha Moreno Kirinus[1*], Caroline Farias Barreto[1], Pricila Santos da Silva[2], Roberto Pedroso de Oliveira[3], Marcelo Barbosa Malgarim[1] and José Carlos Fachinello[1†]

[1]*Departamento de Fitotecnia, Faculdade de Agronomia Eliseu Maciel, Universidade Federal de Pelotas, Centro Agropecuário, Avenida Eliseu Maciel, s/n, Cx. Postal 354, 96050-500, Capão do Leão, Rio Grande do Sul, Brazil.* [2]*Departamento de Fitotecnia, Faculdade de Agronomia, Universidade do Estado de Santa Catarina, Centro Agroveterinário, Lages, Santa Catarina, Brazil.* [3]*Empresa Brasileira de Pesquisa Agropecuária, Embrapa Clima Temperado, Pelotas, Rio Grande do Sul, Brazil.* †*In memoriam. *Author for correspondence. E-mail: marinesfaem@gmail.com*

ABSTRACT. The objective of this study was to evaluate the physical-chemical characteristics, rot index and bioactive compounds of 'Navelina' oranges under postharvest refrigerated storage conditions after pre-harvest resistance induction in crops from 2015 and 2016. The field experimental design was completely randomized blocks. The treatment factors were composed of the following resistance inducers: noresistance inducer (control), selenium (Se), silicon (Si), acibenzolar-s-methyl (ASM), methyl jasmonate (MeJa), thiamethoxam (TMT) and imidacloprid (IMI). In the laboratory, the experimental design was the same as that in the field, but it used a two-factor scheme instead of a unifactorial scheme. In the two-factor scheme, factor A was composed of the abovementioned resistance inducers, and factor B was composed of the refrigerated storage periods (zero and 30 days). The analyses investigated the coloration, fresh mass loss, rot index, soluble solids, pH, titratable acidity, SS/TA ratio, ascorbic acid, total phenolic compounds and antioxidant capacity of the oranges. The application of pre-harvest resistance inducers was efficient in maintaining the physical-chemical characteristics of the 'Navelina' oranges in postharvest, increasing their bioactive compounds in comparison to the control. The resistance inducers Se, Si, MeJa, and IMI reduced rot rates, while ASM and MeJa prevented fresh fruit mass loss.

Keywords: *Citrus sinensis*; navel orange; elicitors; refrigerated storage; rotting.

Introduction

The orange production area of Brazil is the largest in the world; however, compared to other countries, it has the 10th highest productivity. The orange sector is highly organized and competitive, accounting for 30% of the global production and comprising one of the largest centers of orange juice production; in addition, over 19 million tons of oranges are harvested per productive cycle (FAO, 2016). Orange trees are susceptible to various diseases, primarily citrus canker. These diseases cause economic damage to production, increasing costs. The most frequent procedure for disease control consists of the use of agrochemicals and resistant cultivars and the encouragement of positive cultural practices and crop management.

The induction of systemic acquired resistance (SAR) is a promising alternative for disease control, as it exploits a natural defense mechanism of plants. After application, SAR can confer long-term protection against a broad spectrum of microorganisms (David, Yinong, Cassiana, & Monica, 2010). Sensitive plants can acquire a greater ability to defend against pathogen attacks from primary infection. This process involves a series of biochemical and physiological reactions that trigger the production of several secondary metabolites (Hall, Kim, & DeLuca, 2011).

The substances that promote SAR induction that are most commonly reported in the literatureare: acibenzolar-s-methyl, methyl jasmonate, selenium, silicon, and neonicotinoids. Acibenzolar-s-methyl (ASM), a functional analog of salicylic acid that can activate plant defenses, such as pathogenesis-related proteins, is largely used in apples (Quaglia, Ederli, Pasqualini, & Zazzerini, 2011) and citrus (Graham & Myers, 2011; Neto, Maraschin, & DiPiero, 2015). Methyl jasmonate (MeJa) is the methyl ester of the phytohormone jasmonic acidand shows promising results in SAR induction, interfering in the physiological and/or biochemical processes, a sign of the endogenous molecules of loquats (Cai, Cao, Yang, & Zheng, 2011; Cao, Cai, Yang, Joyce, & Zheng, 2014), pomegranates(Sayyari et al., 2011)and bananas(Zhao et al., 2012). Silicon plays multiple roles in cell growth and development, combining physical and chemical barriers such as cell wall lignification and the induction of defense proteins against diseases (French-Monar, Rodrigues, Kordorfer, & Datnoff, 2010), as observed in avocados (Tesfay, Bertling, & Bower, 2011), cotton (Oliveira et al., 2012) and tomatoes (Andrade et al., 2013). Selenium is absorbed and transported by plants in the form of selenite, presenting a high antioxidant capacity and the induction of plant defense systems (Hasanuzzaman, Nahar, & Fujita, 2014). Recently, substances such as imidacloprid neonicotinoids (IMI) and thiamethoxam (TMT) have been used with success in inducing SAR in pomegranates (Graham & Myers, 2011; Bagio, Canteri, Barreto, & Júnior-Leite, 2016).

Therefore, the occurrence of diseases is one of the main factors of losses in orange production in all regions of Brazil, particularly in the southern region, where the amount and frequency of rainfall are high. Studies are necessary to investigate the application of resistance inducers during the pre-harvest period to promote fruit conservation, reduce pesticide applications and increase the levels of beneficial bioactive compounds in fruit to humans. This study aimed to evaluate the physical-chemical characteristics, rot indexes and bioactive compounds of 'Navelina' oranges under postharvest refrigerated storage after resistance induction in the pre-harvest 2015 and 2016 crops.

Material and method

Resistance inducers were applied in the 2015 and 2016 crops of a commercial orchard of 'Navelina' oranges (*Citrus sinensis* (L.) Osbeck) in Santa Silvana, the 6th district of the municipality of Pelotas, Rio Grande do Sul State, Brazil (31°25'58"S and 52°16'58" and 193 meters). The soil in the region, which is classified as red-yellow argisoil, is moderately deep, with medium texture in the A horizon and a clayey texture in the B horizon (Santos et al., 2006). The climate features a Cfa classification (Köppen & Geiger, 1928), i.e., a temperate or humid subtropical climate with hot summers and an average annual rainfall of 1,582 mm, average annual temperature of 17.7°C and average annual relative humidity of 78.8% (INMET, 2016).

The plants (4 years old) were installed under trifoliate rootstock (*Poncirus trifoliate* (L.) Raf.) with 6 m spacing between rows and 4 m between plants. The experimental field was handled in accordance with the requirements of integrated production for citrus (Marodin & Schafer, 2009). To the orchard, the fungicide Nativo® (trifloxystrobin and tebuconazole) was applied three times at an interval of 30 days, with the first application in the phenological stage of the newly formed fruit lets. In addition, the Bordeaux mixture (copper sulfate and lime) was used with six applications spaced 45 days, beginning during the flowering and finishing 60 days before harvest.

For the application of resistance inducers, the experimental design of the field used completely randomized blocks, with five replicates three plants per plot, and the evaluation of the central plant with a unifactorial scheme. The treatment was composed of resistance inducers [no resistance inducer (control, water), selenium (Se, 10 mg L^{-1}), silicon (Si, 400 mg L^{-1}), acibenzolar-s-methyl (ASM, 100 mg L^{-1}), methyl jasmonate (MeJa, 10 mg L^{-1}), thiamethoxam (TMT, 2000 mg L^{-1}) and imidacloprid (IMI, 714 mg L^{-1})].

The resistance inducers were applied during three applications in the orchard ata 15-day interval, with applications occurring 45, 30, and 15 days before harvest using the total dosage. The Si, ASM and MeJa products were applied by spraying with Coastal Spray (Guarani®) with a flat fan nozzle and fine droplet size (101-200 μ) in the entire plant canopy, avoiding run-off. A total of 0.1% non-ionic adhesive spreader Silwet L-77® was added. For the Se, TMT and IMI resistance inducers, syrups were prepared with water for each product and applied to the soil around the plant canopy.

When they reached commercial maturation, the oranges were collected randomly in four quadrants of the tree canopy, placed in plastic boxes, cleaned and sanitized, and transported to the Agronomy Laboratory, Department of Plant Science at the Universidade Federal de Pelotas (UFPel), where they underwent a standardized pre-screening by removing the damaged fruit. The fruit were then submitted to pre-cooling (15 ± 2°C) for 24 hours.

In the lab, the design used was the same established in the field but in a two-factorial scheme, where factor A was composed of the same resistance inducers described previously and factor B was composed by the storage periods(zero and 30 days). Time zero corresponded to the fruit that were not subjected to storage, and the 30-days to rage corresponded to refrigerator storage at 5 ± 1°C, under 85-95% relative humidity. After removal from the chamber, the fruit were submitted to a simulation of commercialization time, 7 days at 20±1°C. For each treatment, three replicates were used with 20 oranges each, and the same number of repetitions was used in the refrigerated storage, totaling 840 fruits.

The coloration was measured with a Minolta colorimeter CR-300, with the reading system CIE L★ a★ b★ and the chromatic tone represented by the hue angle (*h*°) through the arctangent formula b★/a★. The result of this equation, expressed in radians, was then converted to degrees (Minolta, 1994). The fresh fruit loss was obtained by the difference between the initial and final mass of fruit in the cold storage period, and the values were expressed in percentages (%). The rot index was established by the percentage of fruit attacked by pathogens through the visual inspection of fruit, where fruit with lesions greater than or equal to 5 mm were considered to have rot. Both evaluations were conducted after 30 days of refrigerated storage. Soluble solids (SS) were quantified with a digital Refractometer (Atago®) model PAL-1, and the results were expressed in °Brix. The hydrogen potential (pH) was measured with a digital pH meter (Digimed®). For titra table acidity (TA), 10 mL of orange juice was added to 90 mL of distilled water. The sample titration was made with the aid of a digital burette (Brand®) containing a sodium hydroxide solution (0.1 N) up to pH 8.1. The titra table acidity was expressed as the percentage of citric acid. The SS/TA ratio of oranges was expressed by the relationship between the soluble solids and titra table acidity (SS/TA) (Zenebon, Pascuet, & Tiglea, 2008). The ascorbic acid content was quantified through the official AOAC (1997) method by oxidative titration with 2.6-dichlorophenol indophenol (DCFI), in which the titration point is detected by the appearance of pink coloration, and the result is expressed in mg ascorbic acid per 100 g of the sample (Jacobs, 1958; Leme & Malavolta, 1950).

To analyze the phenolic compounds and antioxidant capacity of the fruit, the exocarp or epicarp (peel) was separated from the endocarp (pulp) and evaluated separately to monitor translocation in the fruit (Chitarra & Chitarra, 2005). Total phenolic compounds were quantified using the Folin-Ciocalteau reagent, as described by Swain and Hillis (1959), and expressed in mg of chlorogenic acid equivalent (CAE) per 100 g^{-1}. The antioxidant capacity was determined by spectrophotometry, according to a method adapted from Brand-Williams, Cuvelier, and Berset (1995), and the results were expressed as µg of Trolox equivalent antioxidant capacity (TEAC) g^{-1}.

The 2015 and 2016 crops were used as replicates. The data were analyzed for normality by the Shapiro-Wilk test and homoscedasticity by the Hartley test. Subsequently, the data were submitted to an analysis of variance ($p \leq 0.05$). To determine significance, the effects of the resistance inducers were analyzed by the Tukey test ($p \leq 0.05$), and the effects of the storage period were analyzed by the t test ($p \leq 0.05$). To compare the control with the resistance inducers, the Dunnett test ($p \leq 0.05$) was carried out. The presence of correlations between the dependent variables of this study was analyzed with the Pearson correlation coefficient (r) ($p < 0.0001$).

Result and discussion

For color variables (L★ and b★), soluble solids (SS) and ascorbic acid, there were interactions with the treatment factors tested (Tables 1 and 2), while the color expressed by a★ and the hue angle, the pH, the titratable acidity (TA) and the ratio of SS/TA only had significant interactions with the main effects of the storage period (Table 3). The applications of resistance inducers did not change the luminosity coloration of 'Navelina' oranges, as expressed in L★ coordinates, in either assessment period; however, L★ values decreased over the storage period for degradative processes in all but the TMT resistance inducer treatments. Compared to the control, all resistance inducers maintained the fruit luminosity levels (L★), except for the MeJa treatment at day zero, which had a higher level (Table 1). As observed in this study, the storage effect also reduced the luminosity parameters of

'Valencia Delta' fruit submitted to resistance inducers (Pereira, Machado, & Costa, 2014). An investigation on the effectiveness of the MeJa resistance inducer applied during pre-harvest in mangoes (*Mangifera indica* L.) showed uniform development of the red color in the peel after harvesting, with an increase in the L★ and a★ values (Muengkaew, Chaiprasart, & Warrington, 2016). This increase is possibly due to MeJa resistance inducer performance in accumulating certain proteins related to pathogenesis, thus promoting metabolic changes that keep color strength in oranges (Brinceño et al., 2012).

Regarding the coloration values of the b★ coordinate, the highest values determining the intensity of yellow-orange in oranges were produced by the resistance inducers Se, MeJa and IMI at day zero, with no significant resistance inducer effects observed at 30 days of refrigerated storage (Table 1). For storage purposes, ASM and TMT resistance inducers increased the b★ intensity in the fruit. When compared to the control, differences were found for the ASM, TMT and IMI treatments only at the end of the storage period.

Table 1. Coloration of 'Navelina' orange fruit with different resistance inducers applied during pre-harvest. Coloration is expressed by luminosity level (L★) and intensity of yellow-orange (b★). Storage period data represents refrigerated storage with a subsequent simulation of commercialization time (7 days at 20 ± 1°C) in the 2015 and 2016 crops. Ufpel, Pelotas, Rio Grande do Sul State, Brazil.

Resistance Inducers	L★				b★			
	Storage period							
	0		30		0		30	
Control	69.32		64.51		69.32		64.51	
Selenium	71.42	aA[1]ns	66.48	aBns	71.42	aA[1]ns	66.48	aBns
Silicon	71.19	aAns	66.88	aBns	71.19	aAns	66.88	aBns
Acibenzolar-s-methyl	70.74	aAns	67.17	aBns	70.74	aAns	67.17	aBns
Methyl Jasmonate	72.67	aA★	67.40	aBns	72.67	aA★	67.40	aBns
Thiamethoxam	70.30	aAns	68.41	aAns	70.30	aAns	68.41	aAns
Imidacloprid	71.39	aAns	67.62	aBns	71.39	aAns	67.62	aBns
C.V. (%)	3.0				3.0			

[1]Means followed by the same lowercase letter in the column do not differ by the Tukey test ($p \leq 0.05$) that compared the effects of the resistance inducers in each storage period. Means followed by the same uppercase letter in the row do not differ by the t test ($p \leq 0.05$) comparing the storage of each resistance inducer. ★ and ns mean significant and not significant, respectively, in relation to the control (no resistance inducer) by the Dunnett test ($p \leq 0.05$). C.V: coefficient of variation.

Regarding the soluble solids (SS) of the 'Navelina' orange fruit, there were no differences among the resistance inducers in either assessment period (Table 2). However, the oranges treated by the ASM resistance inducer showed an increase in sugar contents during the storage period. In addition, there were no differences between the treatments and control in either assessment period (zero and 30 days) and no effects on sugar metabolism throughout the storage period.

The applied resistance inducers did not affect the ascorbic acid levels in each period. However, for the ASM treatment, the reduction of these levels during refrigerated storage caused degradation during fruit ripening. Fruit treated with the resistance inducers Se and IMI showed higher levels of ascorbic acid compared to the control in both assessment times. In comparison, the ascorbic acid levels of the MeJa treated fruit were higher only at 30+7 days (Table 2). Other studies have shown that storage of 'Pera Bianchi' oranges was linked with increases in the fruit ascorbic acid levels from 48.89 mg 100 mL^{-1} at 15 days to 56.76 mg 100 mL^{-1} at 45 days of storage (Rosa, Clemente, Oliveira, Todisco, & Costa, 2016).

Table 2. Soluble solids (°Brix) and ascorbic acid (mg 100 g^{-1}) of 'Navelina' orange fruit treated by resistance inducers during pre-harvest. Data are shown for periods of refrigerated storage with a subsequent simulation of commercialization time (7 days at 20 ± 1°C) in the 2015 and 2016 crops. Ufpel, Pelotas, Rio Grande do Sul State, Brazil.

Resistance Inducers	Soluble Solids (°Brix)				Ascorbic Acid(mg 100 g^{-1})			
	Storage period							
	0		30		0		30	
Control	10.15		11.23		44.69		43.49	
Selenium	10.91	aAns	11.58	aAns	51.78	aA★	49.44	aA★
Silicon	10.75	aAns	11.06	aAns	48.21	aAns	46.72	aAns
Acibenzolar-s-methyl	10.16	aAns	10.48	aAns	48.36	aAns	45.72	aBns
Methyl Jasmonate	10.41	aAns	10.46	aAns	49.10	aAns	47.69	aA★
Thiamethoxam	10.40	aAns	11.33	aAns	47.91	aAns	46.48	aAns
Imidacloprid	10.73	aAns	11.18	aAns	52.23	aA★	48.86	aA★
C.V. (%)	5.5				6.0			

[1]Means followed by the same lowercase letter in the column do not differ by the Tukey test ($p \leq 0.05$) that compared the effects of the resistance inducers in each storage period. Means followed by the same uppercase letter in the row do not differ by the t test ($p \leq 0.05$) comparing the storage of each resistance inducer. ★ and ns mean significant and not significant, respectively, in relation to the control (noresistance inducer) by the Dunnett test ($p \leq 0.05$). C.V: coefficient of variation.

The a★ coordinate for orange fruit coloration intensified throughout the storage period, with the orange color becoming more reddish. Based on the hue angle, the fruit lost its typical yellowish coloring. Similarly, the pH of oranges increased throughout the storage period (Table 3), atypical feature of the cultivar studied (Koller, 2013). With ripening, oranges lost acidity, as shown by a rapid increase of the pH, the inverse of the hydrogen ion concentration used in respiration and ripening. There was a reduction in the levels of citric acid and the ratio of SS/TA with storage time (Table 3), which consequently reduced the fruit flavor. In studies conducted on 'Valencia Delta' oranges during storage at room temperature, the application of postharvest coating was associated with increased acidity (citric acid) and the SS/TA ratio in oranges, while the coloration tone (hue angle) decreased over the storage period (Pereira et al., 2014).

Table 3. Coloration(a* and hue angle), pH, titra table acidity (% citric acid) and SS/TA ratio in 'Navelina' oranges over a refrigerated storage period with the subsequent simulation of commercialization time (7 days at 20 ± 1°C) in the 2015 and 2016 crops. Ufpel, Pelotas, Rio Grande do Sul State, Brazil.

Variables	Storage period		C.V.
	0	30	(%)
a*	13.85 b[1]	21.12 a	29.29
Hue angle	78.35 a	73.31 b	5.18
pH	3.46 b	3.60 a	3.90
Titratable acidity (% citric acid)	1.02 a	0.93 b	12.48
SS/TA ratio	11.40 a	10.86 b	10.43

[1]Means followed by the same lowercase letter in the row do not differ by the t test ($p \leq 0.05$) comparing the storage periods. C.V: coefficient of variation

The ASM and MeJa resistance inducer treatments differed from the control after 30 days of refrigerated storage with subsequent simulation of commercialization time (7 days at 20 ± 1°C) (Table 4). In another study, the application of salicylic acid activated the synthesis of secondary metabolites, promoters of systemic resistance; however, it did not affect the biomass loss of fresh fruit (Borsatti, Mazaro, Danner, Nava, & Dalacosta, 2015), which corroborates the results of this work.

Regarding rot rate after 30 days of refrigerated storage, the ASM and TMT resistance inducer treatments did not differ from the control (Table 4). However, the treatments with other resistance inducers were efficient in rot control in the studied period, signaling defense responses and inducing biosynthesis substances generating physical and chemical barriers. In another study investigating 'Satsumas' tangerines stored at 14 ± 2°C, the application of resistance inducers in the postharvest period reduced rot significantly during the first six days of storage (Zhu et al., 2015).

The total phenolic compounds and antioxidant capacity of both peel and pulps howed interactions with the treatment factors tested (Tables 5 and 6). At day zero, the Si and ASM resistance inducer treatments showed higher levels of total content of phenolic compounds in the pulp than the others (Table 5). However, at 30 days, there was no significant difference between the resistance inducer treatments. When comparing the resistance inducers with control, only the IMI treatment did not differ in either assessment period evaluated. Previous studies have shown that resistance inducers increase the demand of enzymes for the biosynthesis of phenolic compounds needed to fight pathogens (Oliveira, Varanda, & Félix, 2016).

Table 4. Fresh mass loss (%) and rot index (%) of 'Navelina' oranges treated with resistance inducers in the pre-harvestperiod of the 2015 and 2016 crops. Ufpel, Pelotas, Rio Grande do Sul State, Brazil.

Resistance inducers	Fresh mass loss (%)	Rot index (%)
Control	8.43 ab[1]	6.66a
Selenium	9.75 A	1.66b
Silicon	8.45 Ab	0.83b
Acibenzolar-s-methyl	6.71 B	5.03ab
Methyl Jasmonate	6.46 B	0.83b
Thiamethoxam	7.91 ab	5.03ab
Imidacloprid	7.01 ab	0.83b
C.V. (%)	30.2	124.7

[1]Means followed by the same lowercase letter in the column do not differ by the Tukey test ($p \leq 0.05$). C.V: coefficient of variation.

In the case of the phenolic compounds in the pulp, there was no difference between the resistance inducer treatments at both zero and 30 days after cold storage (Table 5). Similarly, storage period had no effect on the phenolic compounds in the pulp. However, when compared with control, fruit from the Se and Si resistance inducer treatments showed higher values in the two assessment periods. The application of these resistance inducers in postharvest raises levels of phenolic compounds in plant tissues, which usually have antioxidant properties that are highly beneficial to humans (Romanazzi et al., 2016). The Se and Si resistance inducers confer tolerance to oxidative stress by strengthening the defense system in plants through increased antioxidant capacity (Hasanuzzaman, Nahar, & Fujita, 2014).

Table 5. Total phenolics (mg CAE 100 g^{-1}) in the pulp and peel of 'Navelina' oranges treated by resistance inducers applied in pre-harvest. The storage period represents a period of refrigerated storage with the subsequent simulation of commercialization time (7 days at 20 ± 1°C) in the 2015 and 2016 crops. Ufpel, Pelotas, Rio Grande do Sul State, Brazil.

Resistance inducers	Total phenolics in pulp		Total phenolics in peel	
	(mg CAE 100 g^{-1})			
	Storage period			
	0	30	0	30
Control	108.94	85.82	367.74	350.58
Selenium	129.02 dA[1] *	106.26 aB *	425.44 aA *	409.61 aA *
Silicon	161.38 aA *	103.95 aB *	424.32 aA *	411.98 aA *
Acibenzolar-s-methyl	147.22 abA *	104.37 aB *	403.28 aA ns	386.44 aA ns
Methyl Jasmonate	134.18 bcA *	102.57 aB *	395.01 aA ns	380.34 aA ns
Thiamethoxam	135.59 bcA *	113.02 aB *	389.13 aA ns	371.29 aA ns
Imidacloprid	117.89 dA ns	98.03 aB ns	375.56 aA ns	359.56 aA ns
C.V. (%)	7.2		7.4	

[1]Means followed by the same lowercase letter in the column do not differ by the Tukey test ($p \leq 0.05$) that compared the effects of the resistance inducers in each storage period. Means followed by the same uppercase letter in the row do not differ by the t test ($p \leq 0.05$) comparing the storage of each resistance inducer. * and ns mean significant and not significant, respectively, in relation to the control (noresistance inducer) by the Dunnett test ($p \leq 0.05$). C.V: coefficient of variation.

Table 6. Antioxidant capacity (DPPH, µg TEAC g^{-1}) in the pulp and peel of 'Navelina' oranges treated by resistance inducers applied in pre-harvest. The storage period represents a period of refrigerated storage with the subsequent simulation of commercialization time (7 days at 20 ± 1°C) in the 2015 and 2016 crops. Ufpel, Pelotas, Rio Grande do Sul State, Brazil.

Resistance inducers	DPPH in pulp (µg TEAC g^{-1}) Storage period 0	DPPH in pulp 30	DPPH in peel 0	DPPH in peel 30
Control	241.64	127.31	351.08	232.41
Selenium	389.23abA ★	153.00aB ns	455.54aA ★	281.54abB ★
Silicon	453.18aA ★	150.08aB ns	472.54aA ★	303.19aB ★
Acibenzolar-s-methyl	363.39bA ★	138.14aB ns	421.29aA ns	247.29bB ns
Methyl Jasmonate	397.58abA★	159.06aB ns	440.82aA★	268.49abB ns
Thiamethoxam	327.29bA ★	173.64aB ★	438.20aA ★	272.69abB ns
Imidacloprid	410.65abA ★	173.69aB ★	441.85aA ★	274.20abB ★
C.V. (%)	13.8		11.1	

[1]Means followed by the same lowercase letter in the column do not differ by the Tukey test ($p \leq 0.05$) that compared the effects of the resistance inducers in each storage period. Means followed by the same uppercase letter in the row do not differ by the t test ($p \leq 0.05$) comparing the storage of each resistance inducer. ★ and ns mean significant and not significant, respectively, in relation to the control (noresistance inducer) by the Dunnett test ($p \leq 0.05$). C.V: coefficient of variation.

The antioxidant capacity in the pulp was higher for the treatments with the resistance inducers Se, Si, MeJa, and IMI, differing from the others at day zero. At 30 days, there were no differences between the resistance inducer treatments (Table 6). A reduction in the antioxidant capacity of the oranges was observed during storage for all resistance inducers. However, the antioxidant capacity of the oranges treated with TMT and IMI differed from the control at day zero and at 30 days. Neonicotinoids induce a defense with increased bioactive compounds through the increased biosynthesis of enzymes primarily in young citrus plants, which keep this induction for a long period, (Graham & Myres, 2013).

The antioxidant capacity in orange peels, at day zero, showed no difference s among the resistance inducers; however, at 30 days, there was a reduction in the capacity with the application of the ASM resistance inducer (Table 6). Similar to the pulp, the antioxidant capacity in the peel decreased with storage time for all resistance inducers. Higher levels in the resistance inducer treatments than in the control were observed, mainly for the resistance inducers Se, Si and IMI at day zero. Induced resistance raises the synthesis of phenolic compounds in plant tissues through the stress caused by the resistance inducers that lead to changes in phenolic metabolism, as these compounds exhibit antioxidant properties (Wu et al., 2014; Orabi, Dawood, & Salman, 2015).

Regarding correlations between the treatments and measured variables, significant results were found for the total phenolics and antioxidant capacity variables, which showed the highest positive correlation coefficients (Table 7) for all resistance inducers used. These compounds confer an increase in receptors in the cell membrane, thus mimicking the inevitable phenomenon of electron leakage of chloroplasts, mitochondria and plasma membrane (Bhattacharjee, 2012; Sharma, Jha, Dubey, & Pessarakli, 2012). When there was an increase in the levels of total phenolics, there was also an increase in the antioxidant capacity of oranges. In this context, regarding the association between phenolic compounds and antioxidant capacity of orange pulp, the Si and ASM resistance inducers showed correlation coefficients that were higher than those in the control. A previous study showed that the application of resistance inducers to 'Fortune' tangerines in pre-harvest provided an increase in the gene expression and synthesis of phenolic compounds (Llorens, Scalschi, Fernández-Crespo, Lapeña, & García-Agustín, 2015).

The Se and Si resistance inducer treatments obtained correlation coefficients between the antioxidant capacities in the peel and pulp that were higher than those in the control, demonstrating that Se and Si promoted an increased antioxidant capacity that was transported from the peel to the pulp of the oranges (Table 7). This behavior is due to the powerful antioxidant capacity of phenolic compounds. In 'Valencia' and 'Lane Late' oranges, the application of resistance inducers in postharvest as a curative activity showed the positive effect of increasing bioactive compounds in citrus plants (Moscoso-Ramírez & Palou, 2013).

Table 7. Pearson correlation coefficients (r, $p < 0.0001$) among the total phenolic compounds (phenols) and antioxidant capacity (DPPH)in 'Navelina' oranges treated with the resistance inducers selenium (Se), silicon(Si), acibenzolar-s-methyl (ASM), methyl jasmonate (MeJa), thiamethoxam (TMT), and imidacloprido (IMI) that were applied in the pre-harvest period. Oranges were submitted to refrigerated storage with a subsequent simulation of commercialization time (7 days at 20 ± 1°C) in the 2015 and 2016 crops.Ufpel, Pelotas, Rio Grande do Sul State, Brazil.

	Control	Selenium	Silicon	ASM	MeJa	TMT	IMI
			DPPH in pulp				
Phenols inpulp	0.91844	0.75758	0.98504	0.95590	0.85551	0.78681	0.81085
			Phenols in pulp				
DPPH in peel	0.91922	0.69533	0.92491	0.90886	0.72638	0.70688	0.86271
			DPPH in pulp				
	0.96059	0.97714	0.96091	0.90057	0.91894	0.89481	0.95486

Conclusion

The application of resistance inducers in the pre-harvest period is an efficient method to maintain the physical-chemical properties of 'Navelina' oranges during postharvest, providing increased bioactive compounds in both the peel and pulp when compared to the control. The resistance inducers Se, Si, MeJa, and IMI reduce the rot index, while ASM and MeJa prevent the loss of fruit fresh mass.

References

Andrade, C. C. L., Resende, R. S., Rodrigues, F. A., Ferraz, H. G. M., Moreira, W. R., Oliveira, J. R., & Mariano, R. L. R. (2013). Silicon reduces bacterial speck developmento on tomato leaves. *Tropical Plant Pathology*, *38*(5), 436-442. doi: 10.1590/S1982-56762013005000021

Association of Official Analysis Chemists [AOAC]. (1997). *Official methods of analysis*. 18nd ed. Gaitherburg, MD: AOAC International.

Bagio, T. Z., Canteri, M. G., Barreto, T. P., & Júnior Leite, R. P. (2016). Activation of systemic acquired resistance in citrus to control huanglongbing disease. *Semina: Ciências Agrárias*, *37*(4), 1757-1766. doi: 10.5433/1679-0359.2016v37n4p1757

Bhattacharjee, S. (2012). The language of reactive oxygen species signaling in plants. *Journal of Botany*, *2012*, ID985298, 1-22. doi: 10.1155/2012/985298

Borsatti, F. C., Mazaro, S. M., Danner, M. A., Nava, G., & Dalacosta, N. L. (2015). Induction of resistance and postharvest quality of blackberry treated with salicylic acid. *Revista Brasileira de Fruticultura*, *37*(2), 318-326. doi: 10.1590/0100-2945-087/14

Brand-Willians, W., Cuvelier, M. E., & Berset, C. (1995). Use of a free radical method to evaluate antioxidant activity. *Food Science and Technology*, *28*(1), 25-30. doi: 10.1016/S0023-6438(95)80008-5

Brinceño, Z., Almagro, L., Sabater-Jara, A. B., Calderón, A. A., Pedreño, M. A., & Ferrer, M. A. (2012). Enhancement of phytosterols, taraxasterol and induction of extracellualar pathogenesis-related proteins in cell cultures of *Solanum lycopersicum* cv. Micro-Tom elecited with cyclodextrins an methyl jasmonate. *Journal of Plant Physiology*, *169*(11), 1050-1058. doi: 10.1016/j.jplph.2012.03.008

Cai, Y., Cao, S., Yang, Z., & Zheng, Y. (2011). MeJA regulates enzymes involved in ascorbic acid and glutathione metabolism and improves chilling tolerance in loquat fruit. *Postharvest Biology Technology*, *59*(3), 324-326. doi: 10.1016/j.postharvbio.2010.08.020

Cao, S., Cai, Z., Yang, Z., Joyce, D. C., & Zheng, Y. (2014). Effect of MeJa treatment on polyamine, energy status and anthracnose rot of loquat fruit. *Food Chemistry*, *145*(2014), 86-89. doi: 10.1016/j.foodchem.2013.08.019

Chitarra, M. I. F., & Chitarra, A. B. (2005). *Pós-colheita de frutos e hortaliças*: fisiologia e manuseio. 2. ed. Lavras, MG: UFLA.

David, V., Yinong, Y., Cassiana, V. C., & Monica, H. O. (2010). Abscisic Acid-Induced Resistance against the Brown Spot Pathogen Cochliobolusmiyabeanus in Rice Involves MAP Kinase-Mediated Repression of Ethylene Signaling. *Plant Physiology*, *152*(4), 2036-2052. doi: 10.1104/pp.109.152702

Food and Agriculture Organization of the United Nations [FAO]. (2016). *Faostat*: Production crops. Retrieved on Nov. 10, 2016, from http://faostat.fao.org/site/567/DesktopDefault.aspx?PageID=567#ancor

French-Monar, R. D., Rodrigues, F. A., Kordorfer, G. H., & Datnoff, L. E. (2010). Silicon suppresses Phytophthora blight development on bell pepper. *Journal of Phytopathology*, *158*(7-8), 554-560. doi: 10.1111/j.1439-0434.2009.01665.x

Graham, J. H., & Myers, M. E. (2011). Soil Application of SAR Inducers Imidacloprid, Thiamethoxam, and Acibenzolar-S-Methyl for Citrus Canker Control in Young Grapefruit Trees. *Plant Disease*, *95*(6), 720-729. doi: 10.1094/PDIS-09-10-0653

Graham, J. H., & Myers, M. E. (2013). Integration of soil applied neocicotinoid insecticides and acibenzolar-S-methyl for systemic acquired resistance (SAR) control of citrus canker on young citrus trees. *Crop Protection*, *54*(2013), 239-243. doi: 10.1016/j.cropro.2013.09.002

Hall, D., Kim, K. H., & DeLuca, V. (2011). Molecular cloning and biochemical characterization of three Concord grape (*Vitis labrusca*) flavonol 7-O-glucosyltransferases. *Planta*, *234*(1), 1201-1214. doi: 10.1007/s00425-011-1474-0

Hasanuzzaman, M., Nahar, K., & Fujita, M. (2014). Silicon and Selenium: Two vital trace elements that confer abiotic stress tolerance to plants. In P. Ahmad, & S. Rasool (Ed.), *Emerging techonologies and management of crop stress tolerance* (p. 377-422). Tokyo, JN: Biological Techniques.

Instituto Nacional de Meteorologia [INMET]. (2016). Retrieved on Dec. 12, 2016, from http://www.inmet.gov.br/portal/index.php?r=bdmep/bdmep

Jacobs, M. B. (1958). *The chemical analysis of foods and food products*. New York, US: Van Nostrand.

Koller, O. L. (2013). *Citricultura catarinense*. Florianópolis, SC: Epagri.

Köppen, W., & Geiger, R. (1928). *Klimate der Erde*. Munique, GE: Gotha, Verlag Justus Perthes.

Leme, J. J., & Malavolta, E. (1950). Determinação fotométrica de ácido ascórbico. *Anais da Escola Superior de Agricultura Luiz de Queiroz*, 7(1), 115-129. doi: 10.1590/S0071-12761950000100016

Llorens, E., Scalschi, L., Fernández-Crespo, E., Lapeña, L., & García-Agustín, P. (2015). Hexanoic acid provides long-lasting protection in 'Fortune' mandarin against *Alternaria alternata*. *Physiological and Molecular Plant Pathology*, *91*(2015), 38-45. doi: 10.1016/j.pmpp.2015.05.005

Marodin, G. A. B., & Schafer, G. (2009). Produção integrada de citros. In O. C. Koller (Ed.), *Citricultura* p. (269-316). Porto Alegre, RS: Rígel.

Minolta, K. (1994). *Precise color communication: color control from feeling to instrumentation*. Tokyo, JN: Minolta Co. Ltda.

Moscoso-Ramírez, P. A., & Palou, L. (2013). Evaluation of postharvest treatments with chemical resistance inducers to control green and blue molds on orange fruit. *Postharvest Biology and Technology*, *85*, 132-135. doi: 10.1016/j.postharvbio.2013.05.013

Muengkaew, R., Chaiprasart, P., & Warrington, I. (2016). Changing of physiochemical properties and color development of mango fruit sprayed methyl jasmonate. *Scientia Horticulturae*, *198*, 70-77. doi: 10.1016/j.scienta.2015.11.033

Neto, A. C. R., Maraschin, M., & DiPiero, R. M. (2015). Antifungal activity of salicylic acid against Penicilliumexpansum and its possible mechanisms of action. *International Journal of Food Microbiology*, *215*(215), 64-70. doi: 10.1016/j.ijfoodmicro.2015.08.018

Oliveira, J. C., Alburquerque, G. M. R., Mariano, R. L. R., Gondim, D. M. F., Oliveira, J. T. A., & Souza, E. B. (2012). Reduction of the severity of angular leaf spot of cotton mediated by silicon. *Journal Plant Pathology*, *94*(2), 297-304. doi: 10.4454/JPP.FA.2012.044

Oliveira, M. D. M., Varanda, C. M. R., & Félix, M. R. F. (2016). Induced resistance during the interaction pathogen x plant and the use of resistance inducers. *Phytochemistry Letters*, *15*, 152-158. doi: 10.1016/j.phytol.2015.12.011

Orabi, S. A., Dawood, M. G., & Salman, S. R. (2015). Comparative study between the physiological role of hydrogen peroxide and salicylic acid in alleviating the harmful effect of low temperature on tomato plants grown under sand-ponic culture. *Scientia Agriculturae*, *9*(1), 49-59. doi: 10.15192/PSCP.SA.2015.1.9.4959

Pereira, G. S., Machado, F. L. C., & Costa, J. M. C. (2014). Application of coating extends postharvest quality in the 'Valencia Delta' orange during ambient storage. *Revista Ciência Agronômica*, *45*(3), 520-527. doi: 10.1590/S1806-6902014000300012

Quaglia, M., Ederli, L., Pasqualini, S., & Zazzerini, A. (2011). Biological control agentes and chemicas induceres of resistance for postharvest control of *Penicillium expansum* on apple fruit. *Postharvest Biology and Technology*, *59*(3), 307-315. doi: 10.1016/j.postharvbio.2010.09.007

Romanazzi, G., Sanzani, S. M., Bi, Y., Tian, S., Martínez, P. G., & Alkan, N. (2016). Induced resistance to control postharvest decay of fruit and vegetables. *Postharvest Biology and Technology*, *122*(2016), 82-94. doi: 10.1016/j.postharvbio.2016.08.003

Rosa, C. I. L. F., Clemente, E., Oliveira, D. M., Todisco, K. M., & Costa, J. M. C. (2016). Effects of 1-MCP on the postharvest quality of the orange cv. Pera stored under refrigeration. *Revista Ciência Agronômica*, *47*(4), 624-632. doi: 10.5935/1806-6690.20160075

Santos, J. H., Jacomine, P. K.T., Anjos, L. H. C., Oliveira, V. A., Oliveira, J. B., Coelho, M. R., ... Cunha, T. J. F. (2006). *Sistema Brasileiro de Classificação de Solos* (2a ed.). Rio de Janeiro, RJ: Embrapa Solos.

Sayyari, M., Babalar, M., Kalantari, S., Martínes-Romero, D., Guillén, F., Serrano, M., & Valero, D. (2011). Vapour treatments with methyl salicylate or methyl jasmonate alleviated chilling injury and enhanced antioxidant potential during postharvest storage of pomegranates. *Food Chemistry*, *124*(3), 964-970. doi: 10.1016/j.foodchem.2010.07.036

Sharma, P., Jha, A. B., Dubey, R. S., & Pessarakli, M. (2012). Reactive oxygen species, oxidative damage, and antioxidative defense mechanism in plants under stressful conditions. *Journal of Botany*, *2012*, ID217037, 1-26. doi: 10.1155/2012/217037

Swain, T., & Hills, W. E. (1959) The phenolic constituents of *Punnus domestica*. The quantitative analysis of phenolic constituents. *Journal of the Science of Food and Agriculture*, *10*(1), 63-68. doi: 10.1002/jsfa.2740100110

Tesfay, S. Z., Bertling, I., & Bower, J. P. (2011). Effects of postharvest potassium silicate application on phenolics and other anti-oxidant systems aligned to avocado fruit quality. *Postharvest Biology and Technology*, *60*(2), 92-99. doi: 10.1016/j.postharvbio.2010.12.011

Wu, Z. L., Yin, X. B., Lin, Z. Q., Bañauelos, G. S., Yuan, L. X., Liu, Y., & Li, M. (2014). Inhibitory effect of selenium against *Penicillium expansum* and its possible mechanisms of action. *Current Microbiology*, *69*(2), 192-201. doi: 10.1007/s00284-014-0573-0

Zenebon, O., Pascuet, N. S., & Tiglea, P. (2008). *Métodos físico-químicos para análise de alimentos* (4a ed.). São Paulo, SP: Instituto Adolfo Lutz.

Zhao, M. L., Wang, J. N., Shan, W., Fan, J. G., Kuang, J. F., Wu, K. Q., ..., Lu, W. J. (2012). Induction of jasmonate signaling regulators MaMyC2s and their physical interactions with MaICE1 in methyl jasmonate-induced chilling tolerance in banana fruit. *Plant Cell Environment*, *36*(1), 1365-3040. doi: 10.1111/j.1365-3040.2012.02551.x

Zhu, F., Yun, Z., Ma, Q., Gong, Q., Zeng, Y., Xu, J., ..., Deng, X. (2015). Effects of exogenous 24-epibrassinolide treatment on postharvest quality and resistance of Satsuma mandarin (*Citrus unshiu*). *Postharvest Biology and Technology*, *100*, 8-15. doi: 10.1016/j.postharvbio.2014.09.014

22

Estimating soybean yields with artificial neural networks

Guiliano Rangel Alves[1*], Itamar Rosa Teixeira[1], Francisco Ramos Melo[1], Raniele Tadeu Guimarães Souza[1] and Alessandro Guerra Silva[2]

*[1]Departamento de Engenharia Agrícola, Universidade Estadual de Goiás, Campus Henrique Santillo, BR-153, Fazenda Barreiro do Meio, 75132-400, Anápolis, Goiás, Brazil. [2]Departamento de Agronomia, Universidade de Rio Verde, Rio Verde, Goiás, Brazil. *Author for correspondence. E-mail: guiliano@gmail.com*

ABSTRACT. The complexity of the statistical models used to estimate the productivity of many crops, including soybeans, restricts the use of this practice, but an alternative is the use of artificial neural networks (ANNs). This study aimed to estimate soybean productivity based on growth habit, sowing density and agronomic characteristics using an ANN multilayer perceptron (MLP). Agronomic data from experiments conducted during the 2013/2014 soybean harvest in Anápolis, Goiás State, B razil, were used to conduct this study after being normalized to an ANN-compatible range. Then, several ANNs were trained to choose the best-performing one. After training the network, a performance analysis was conducted to select the ANN with a performance most appropriate for the problem, and the selected network had a 98% success rate with training data and a 72% data validation accuracy. The application of the MLP to the data used in the experiment shows that it is possible to estimate soybean productivity based on agronomic characteristics, growth habit and population density through AI.

Keywords: *Glycine max* (L.) Merrill; agronomic characteristics; modeling; MLP; perceptron.

Introduction

Soybeans are currently considered the main agricultural commodity of Brazil, which is the second largest producer of this oilseed. During the 2015/2016 harvest, approximately 95.4 million tons of this grain were produced (Conab, 2017), but there are productivity gaps among regions that are due to several factors during the development of the cultivar in the field. The productivity potential of soybean is genetically determined (Homrich, Wiebke-Strohm, Weber, & Bodanese-Zanettini, 2012), but how this potential can be achieved depends on the effect of limiting factors that act at different stages during the production cycle (Kron, Souza, & Ribeiro, 2008).

The factors that influence soybean productivity include row spacing and plant density (Akond et al., 2013; Souza, Teixeira, Reis, & Silva, 2016). Soybean productivity was successfully estimated in studies by Monteiro and Sentelhas (2014) using an agrometeorological model. The main agronomic characteristics that are influenced by the different behaviors of each cultivar include the number of branches produced per plant, number of pods and seeds per plant, number of internodes, insertion of the first pod, stem diameter, plant height and, obviously, grain production (Liu et al., 2010; Passos et al., 2011).

The use of agronomic traits in soybeans at the R6 stage or later makes it possible to estimate productivity without data from previous stages (Lee & Herbek, 2005). At these later stages, the grains have completely filled the cavity near the pod and are similar to the pods collected at harvest (Oliveira, Silva, Mielezrski, Lima, & Edvan, 2016).

Artificial neural networks (ANNs) are widely applied in research due to their ability to model highly nonlinear systems in which the relationships between variables are unknown or very complex (Russell & Norvig, 2009; Goyal, 2013), and this capability has led to the use of ANNs in different fields of science. In the agricultural sciences, several studies have used ANNs (Zhang, Bai, & Liu, 2007; Alvarez, 2009; Gago, Martínez-Núñez, Landín, & Gallego, 2010; Erzin, Rao, Patel, Gumaste, & Singh, 2010; Soares, Pasqual, Lacerda, Silva, & Donato, 2013; Safa, Samarasinghe, & Nejat, 2015; Alves et al., 2017).

This work aimed to evaluate the possibility of using an artificial neural network as a tool to evaluate the main agronomic traits of soybean cultivars with different growth habits and subjected to different sowing densities to obtain estimates of productivity.

Material and methods

The development of a multilayer perceptron (MLP) artificial neural network (ANN) requires supervised training to adjust the weights of the synapses. Thus, the MLP used in this study to estimate soybean productivity was created using data from an experiment conducted during the 2013/2014 harvest at an experimental site belonging to Emater/Goiás, Anápolis, Goiás State, Brazil (48°18'23" W, 16°19'44" S). According to the Köppen classification, the climate is AW humid tropical, that is, characterized by a dry winter and a rainy summer. The soil of the area is classified as a Rhodic Hapludox (dystrophic Red Latosol).

The experiment employed a completely randomized, 3 x 3 factorial design with eight replications. The treatments consisted of three soybean cultivars with different growth habits and types (BRS Valiosa RR, BMX Potencia RR and NA 7337RR) and three plant densities (D1: 245,000 plants ha^{-1}, D2: 350,000 plants ha^{-1} and D3: 455,000 plants ha^{-1}).

At harvest, the following agronomic characteristics (variables) were evaluated in ten plants from each plot: plant height, number of branches per plant, number of pods per plant, number of grains per pod, weight of 1,000 seeds (WTS) and grain yield (productivity).

To conduct ANN training, independent variables (cultivars with different growth habits and population density) and dependent variables such as the agronomic characteristics of plant height (PH), number of branches per plant (B), number of pods per plant (P), number of seeds per pod (S), weight of 1,000 seeds (WTS) and grain yield (Prod. kg ha^{-1}) were selected. These variables were normalized to equalize the ANN input data (Leal, Miguel, Baio, Neves, & Leal, 2015) so that the initial weights of the variables were assumed to be equivalent at the beginning of the training, thus avoiding the difficulties posed by variables with different weights that can prevent the ANN from converging.

In addition to having different magnitudes, variables that are not numerical, such as growth habit variables and population density, should also be considered as categories. It is recommended that the same dummy treatment applied to the variables in multiple regression analysis be followed for these types of variables (Sharma, Sharma, & Kasana, 2007; Bohl, Diesteldorf, Salm, & Wilfling, 2016).

For the other input variables whose values are real numbers, a linear transformation was used. The maximum and minimum values of the variables used in this transformation are shown in Table 1.

Table 1. Values used for X_{min} and X_{max} by variable.

Variable	Unit	X_{min}	X_{max}
Plant height (PH)	cm	20	200
Number of branches per plant (B)	un	1	15
Number of pods per plant (P)	un	10	150
Number of seeds per pod (S)	un	1	4
Weight of 1,000 seeds (WTS)	kg	5	40
Productivity (Prod)	kg ha^{-1}	1000	6000

The values used as inputs and the expected results were normalized to values between minus one and one, but after the networks were trained and validated, the resulting value of the network was transformed back to its original quantity. To perform this transformation, Equation 1 was used, considering the minimum and maximum values for each variable (Table 1).

$$X = \frac{(scaledX - d2)(X_{max} - X_{min})}{d2 - d1} + X_{min} \qquad (1)$$

where: X = the value of the original quantity, scaledX = the transformed value, X_{max} = the maximum possible value for the variable, X_{min} = the minimum possible value for the variable, d1 = the value of the lower limit of the converted value (-1 in this study), and d2 = the value of the upper limit of the converted value (1 in this study).

The input layer was set to begin the development of the multilayer perceptron (MLP). One neuron was used for each of the input variables (Table 2), and the output layer contained one neuron representing productivity. To conduct the training and the validation processes, a program was developed using the Levenberg-Marquardt training algorithm (Schiavo, Prinari, Gronski, & Serio, 2015) and the mean squared error (MSE) performance function, Equation 2, to enable various ANN architecture settings to be explored.

$$MSE = \frac{1}{N}\sum_{i=1}^{N}(e_i)^2 = \frac{1}{N}\sum_{i=1}^{N}(t_i - \alpha_i)^2 \quad (2)$$

where: N = the number of data presented for the training, e = the difference between the expected value and the estimated value of the network, t = the value estimated by the network, and α = the expected value.

Table 2. Variables presented to the input neuron layer.

Variable	Transformation	Description
C2 C3	Category/type	Growth habit
D2 D3	Category/type	Density
PH	Linear transformation	Plant height
B	Linear transformation	Number of branches per plant
P	Linear transformation	Number of pods per plant
S	Linear transformation	Number of seeds per pod
WTS	Linear transformation	Weight of 1,000 seeds

C2, C3: Growth habit identification. D2, D3: Density identification.

After designing the program, a training was conducted with 20,000 MLP networks. One thousand networks were trained in each architecture by varying the number of neurons in the hidden layer between one and twenty, following the assumption that $2i+1$ neurons in the hidden layer are needed to map any continuous function with i entries.

After completing the neural network training, a file was generated for each training that contained the training data (the parameters used in the training, the content of the training data, and the test, validation and performance sets). Another file was generated containing the consolidated data for all the trainings.

The resulting file contained one line for each trained network (20,000). The columns one to sixty-five were the values estimated by the network, and some columns were added (added columns filled by the formula) (Table 3).

To determine the network with the best performance, some networks were selected using the following criteria: general performance of the experiment, performance of the training set, performance of the validation set, performance of the test set, training R^2, validation R^2, test R^2, and general R^2. Pearson's correlations (training, validation and tests) followed a decreasing order because the higher the value of the Pearson's correlation, the closer the estimated value to the observed value.

Table 3. Description of the columns in the file containing the information network training data.

Column index	Column name	Formula	Information
66	BN	-	Overall network performance
67	BO	-	Performance in joint training
68	BP	-	Performance in the validation set
69	BQ	-	Performance in the test set
70	BR	-	R linear training set
71	BS	$=(BRnn)^2$	R^2 training set
72	BT	-	R linear validation set
73	BU	$=(BTnn)^2$	R^2 validation set
74	BV	-	R linear training set
75	BW	$=(BVnn)^2$	R^2 training set
76	BX	-	R linear general set
77	BY	$=(BXnn)^2$	R^2 from general set
78	BZ	-	Pearson correlation of the training set
79	CA	-	Pearson correlation of the validation set
80	CB	-	Pearson correlation of the test set
81	CC	-	Epoch that was the training stop (convergence)
82	CD	-	Number of neurons in the hidden-layer training
83	CE	-	Training number held in the architecture

nn: the line number.

Results and discussion

The descriptive statistics of the variables are presented for the test, training and validation of the MLP network (Table 4), and the low number of samples (65 samples) that were used as inputs for the training, validation and testing can be seen. Considering that MLPs learn from examples, this hindered the training and validation of the network.

The coefficient of variation for all the characteristics was equal to or greater than 10 %, which contributed to the training since the variation in the values allows for a better adjustment of the synapse connection weights, but this high coefficient of variation may also represent outliers in the data. However, the MLP networks coped with these values using cross-validation while avoiding the influence of the adjusted synapse weights, ensuring that the network did not model the noise in the samples.

Table 4. Descriptive statistics of the variables used to train the ANN.

Characteristic	N	Average	Min	Max	Standard deviation	C.V.
Plant height	65	92.37361	78.80	112.18	8.87538	10 %
Branches per plant	65	4.81704	2.00	11.30	2.09585	44 %
Grains mass per plant	65	18.18143	10.00	30.57	5.32825	29 %
Nº pods per plant	65	59.63328	27.60	119.80	19.28325	32 %
Seeds per pod	65	2.18204	1.82	2.67	0.27201	12 %
Productivity (kg ha^{-1})	65	4347.56436	2958.72	5248.17	477.41305	11 %

N: number of samples; Min: minimum value; Max: maximum value.

The 65 obtained samples were randomly divided into three subsets: training (42 samples, 65%), validation (16 samples, 25%) and test (7 samples, 10%). For each new training, a new draw was performed, and this division strategy made the training difficult because by performing the random division without considering the treatments used to obtain the samples, the sets were formed with poor sample representation. This resulted in multiple networks with high training performance but a low validation network, with 19 neurons in training 806, line 3, and an observed network performance training with an R^2 of 0.999 and a validation R^2 of 0.145 (Table 5).

The development of the program was essential for determining the architecture that could perform adequately because the scaling problem involves the adjusting the complexity of the neural model based on the complexity of the problem. It was possible to vary the complexity of the architecture using the program, thereby enabling performance evaluation in different architectures and configurations.

Another device was used to repeat the training a thousand times in every architecture. This approach, by which each new training performed a new draw of the training, validation and testing sets, as well as the initialization of the synapse weightings allowed the common problem of tending to get stuck in minimal places when training MLP networks to be overcome using a backpropagation algorithm (Zweiri, Seneviratne, & Althoefer, 2005). The importance of the repetitions was observed in the network with the best performance (R^2 = 0.987 in the training and 0.727 in the validation), which was determined after 963 repetitions in the architecture training with nine neurons in line two of the hidden layers (Table 5).

The strategy of drawing new training, validation and test sets during training led to problems regarding the convergence and generalization of the networks because the treatments were not considered. This resulted in the grouping of samples with little representation during the training, test and validation sets and proved that the network began to specialize in the training set. This is illustrated by the high R^2 of 0.999 obtained during the network training with 19 neurons after 806 repetitions (Figure 1), at which the training set line near time 48 fell dramatically, indicating that the network had memorized the training set. At the same time, the lines for the validation and test sets did not follow this patterns, showing low R^2 values (0.145 and 0.045, respectively).

Table 5. ANN training data.

Training			Performance				R^2				Pearson's correlation			
NN	NT	C	Gen	Tra	Val	Test	Gen	Tra	Val	Test	Gen	Tra	Val	Test
14	124	18	0.009	0.001	0.03	0.03	0.743	0.980	0.240	0.203	0.862	0.990	0.490	0.451
9	963	19	0.03	0.001	0.033	0.155	0.611	0.987	0.727	0.033	0.782	0.993	0.852	0.181
19	806	48	0.044	0.000	0.138	0.092	0.357	0.999	0.145	0.045	0.598	0.999	0.381	0.212
2	212	926	0.028	0.022	0.046	0.021	0.239	0.423	0.000	0.991	0.489	0.650	0.013	0.996
6	336	42	0.029	0.034	0.009	0.048	0.195	0.283	0.264	0.151	0.441	0.532	0.513	0.389
1	340	119	0.033	0.038	0.032	0.003	0.107	0.244	0.041	0.090	0.328	0.494	0.203	0.300

C: cycles or epoch at which the architecture training was finalized; NN: -number of neurons in the hidden layer; NT: training number held in the NN architecture; Tra: training; Val: validation; Performance: mean squared error (MSE) = 7.

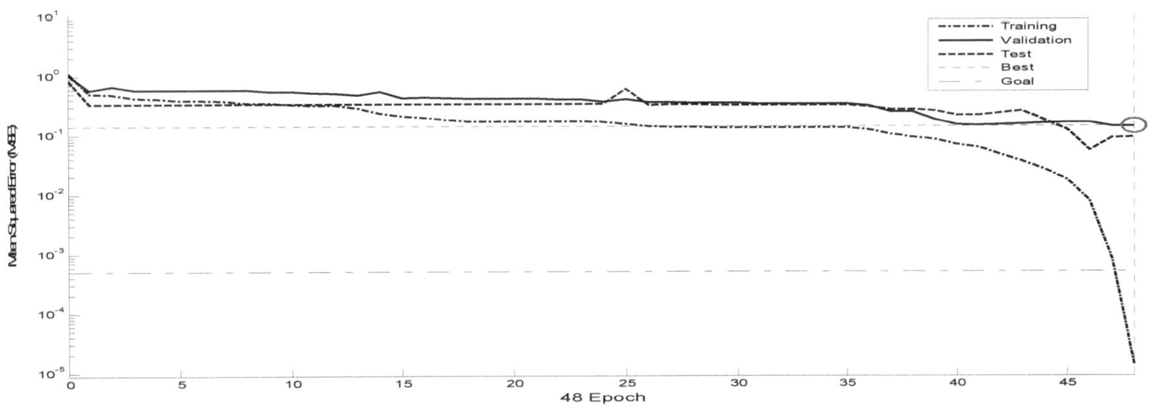

Figure 1. Graphical representation of network training with 19 neurons and 806 repetitions.

The network with two neurons and 212 repetitions (Figure 2) had the same problem with drawing the sets. The only difference was that the set with the best Pearson's correlation had a value of 99.55 %, but the validation set showed a correlation of only 13.45 %.

Various criteria were used to select the network that best managed to generalize the problem. Networks with fewer neurons in the hidden layer are more generalizable, but a network with few neurons in the hidden layer may not be able to solve a problem with a high degree of complexity and may result in under fitting (Patterson, 1996). This was observed for the estimated and observed values for the network with a neuron with 340 repetitions (Figure 3) that had an MSE of 0.0315, which was better than the network with nine neurons and 963 repetitions (network chosen as the most appropriate) that presented an MSE of 0.0334. Despite the low MSE, the R^2 and the Pearson's correlation values were not satisfactory (0.041 and -0.203, respectively) (Table 5), and the network failed to model the relationship between the input variables and the expected result.

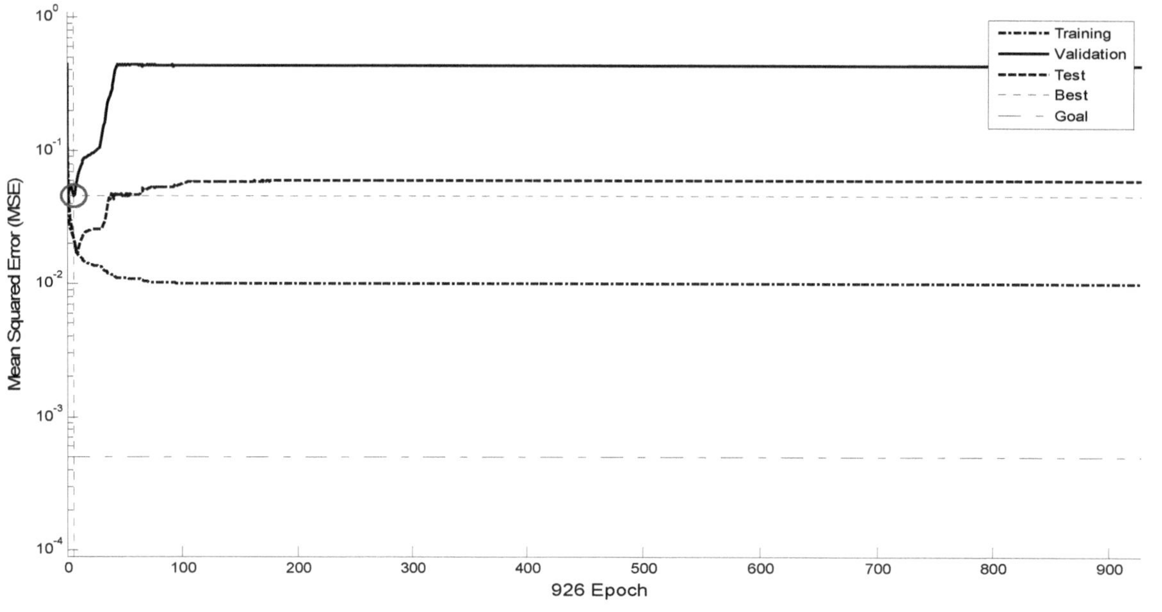

Figure 2. Graphical representation of network training with 2 neurons and 212 repetitions.

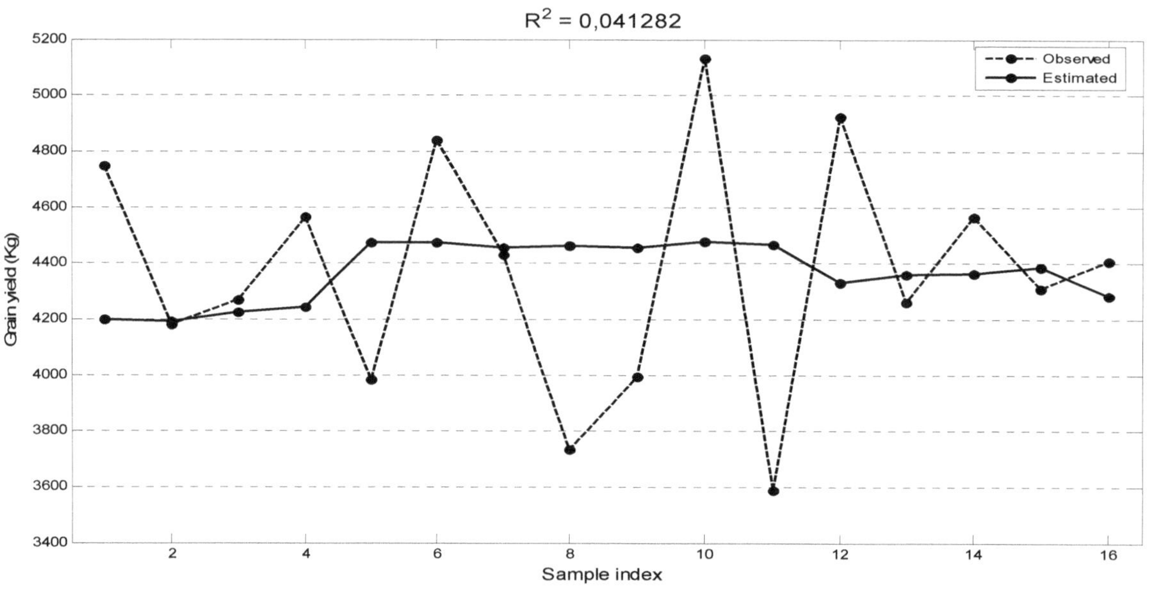

Figure 3. Comparative graphical representation of the estimated and observed values of the validation set for the network with one neuron and 340 repetitions.

With the validation of the network with nine neurons and 963 repetitions, it was possible to verify the tendency of the observed values to follow the estimated values (Figure 4), which was confirmed by the R^2 of 0.726 and the Pearson's correlation of 85 %. If samples 7 and 11 and indexes 3 and 4 (Figure 4) were removed from this validation, the R^2 would rise to 0.811, and the Pearson's correlation would rise to 90 %. This simulation was performed upon observing that samples 11 and 2 have agronomic characteristics with very similar values but very different productivities (a difference of nearly 800 kg), which raises the possibility that factors that were not recorded in the experiment that created the data influenced the productivity of the samples.

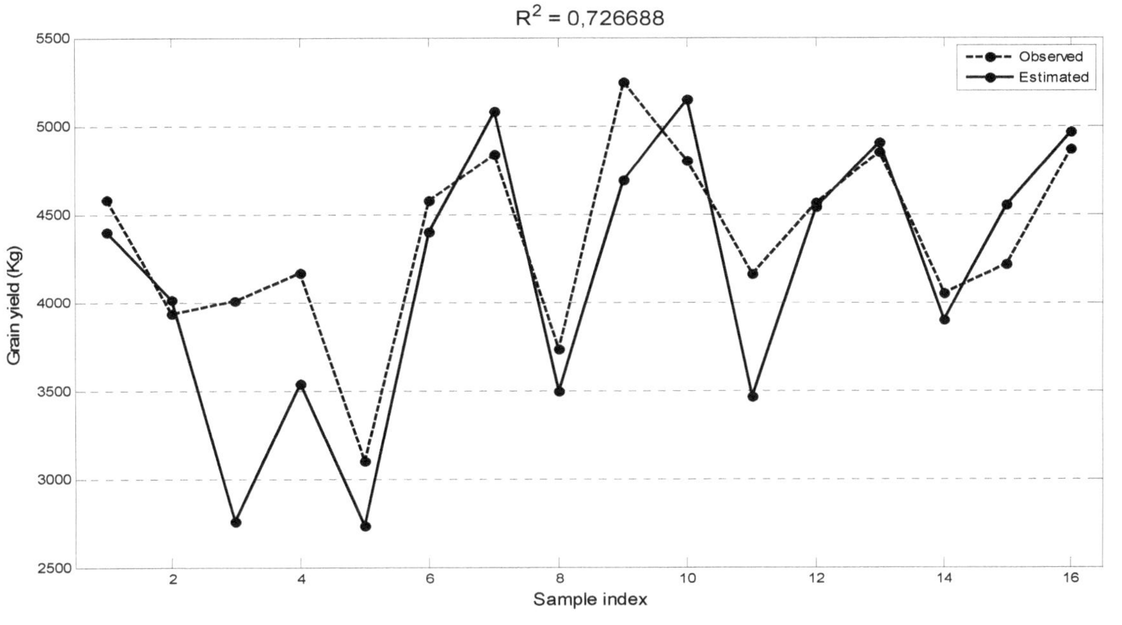

Figure 4. Comparative graphical representation of the estimated and observed values of the validation set for the network with 9 neurons and 963 repetitions.

Among all the trained networks, the one with nine neurons and 963 repetitions, which was selected as the best solution to the problem, did not present the best R^2 during training. Rather, it was selected because it presented the best validation, with an R^2 of 0.727 and a Pearson's correlation of 85 %. The estimated and observed values were observed to be very close during the training, which is evidence of the learning capacity of the MLP (Figures 5 and 6).

Despite achieving considerable performance during validation (R^2 = 0.727 and Pearson's correlation = 85%), the selected network had some values that were distant from the line. This can be attributed to the influence of the productivity of the selected samples on the validation set, which was not considered during training. Furthermore, the performance of the test set was very low (R^2 = 0.033 and Pearson's correlation = -18%), which reinforces the point that the selection of the sets was entirely random and clustered samples with low representation. The influence of non-registered factors should also be considered.

The difference between the observed and the estimated grain productivity values (error) of the network with the training set was 30.42 kg (Table 6), confirming the high Pearson's correlation of 99%. The absolute average error was evaluated instead of the average error (17.1 kg) to avoid masking the distance between the observed and estimated values because estimated values may be negative, indicating that the value estimated by the network is greater than the observed, which decreases the average error.

The performance of the MLP was considered good relative to other studies of soybean productivity estimation as represented by Fontana, Berlato, Lauschner, and Mello (2001). This author obtained a correlation of 0.85 between observed and estimated values using the Jensen model modified to estimate the soybean crop yield in the State of Rio Grande do Sul. Using an agrometeorological model, Monteiro and Sentelhas (2014) obtained an R^2 of 0.64 when estimating soybean productivity in different regions.

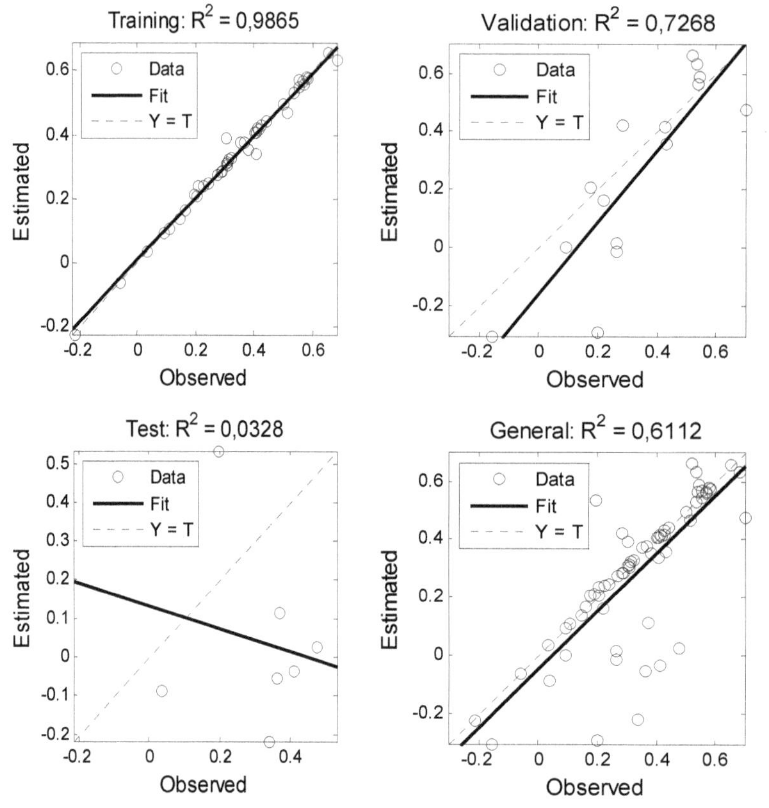

Figure 5. Graphical representation of linear R training, validation, testing and the overall network with 9 neurons and 963 trainings.

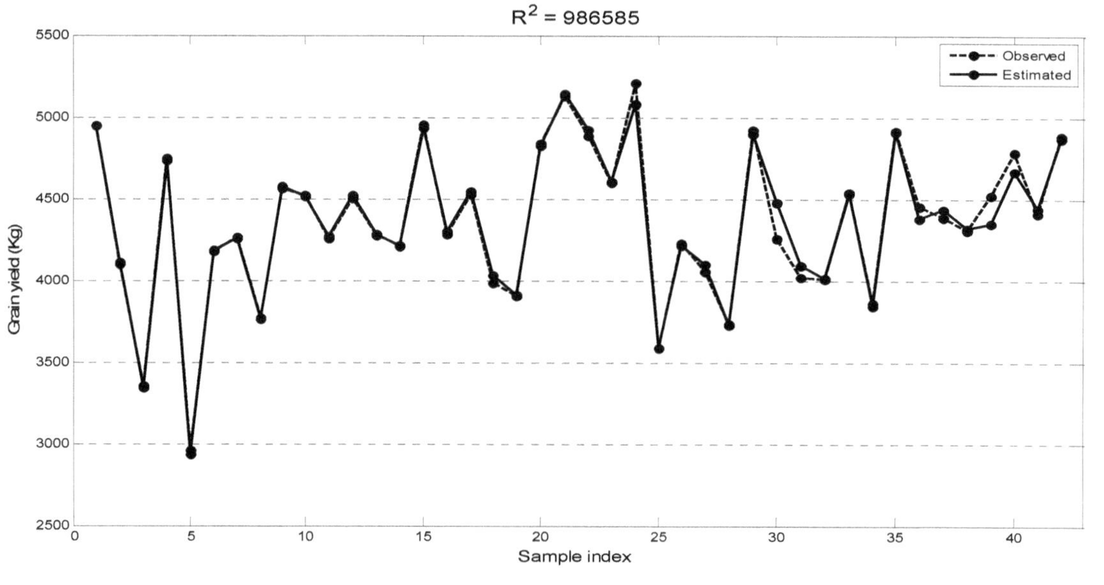

Figure 6. Comparative graphical representation of the observed and estimated training values for the network with 9 neurons and 963 repetitions.

Table 6. Analysis of network errors with nine neurons and 963 repetitions.

	Samples	Min	Max	Average	Mean Absolute Error	Deviation
General	65	-841.23	1398.14	122.28	203.78	380.75
Training	42	-216.42	167.81	1.17	30.42	55.85
Validation	16	-351.78	1246.33	193.32	340.56	427.76
Test	7	-841.23	1398.14	686.57	926.92	761.56

Min: minimum; Max: maximum; Mean Absolute Error: average of the absolute error values. Minimum values below zero indicate that the network was estimated above that value. All values are in kg.

Conclusion

The multilayer perceptron (MLP) ANN with supervised training could estimate (validation R^2 of 0.72 and training R^2 of 0.98) a yield with considerable assertiveness using information regarding the agronomic characteristics, growth habit and population density of the soybean crop.

The use of artificial neural networks to estimate soybean yield is viable, since the back propagation training technique allowed the relationship between the independent variables (soybean agronomic characteristics, growth habit and population density) and soybean yield to be identified with high precision (72%).

References

Akond, A. G. M., Bobby, R., Bazzelle, R., Clark, W., Kantartzi, S. K., Meksem, K., & Kassem, M. A. (2013). Effect to two row spaces on several agronomic traits in soybean [*Glycine max* (L.) Merr.]. *Atlas Journal of Plant Biology, 1*(2), 18-23. doi: 10.5147/ ajpb.2013.0073

Alvarez, R. (2009). Predicting average regional yield and production of wheat in the argentine pampas by an artificial neural network approach. *European Journal of Agronomy, 30*(1), 70-77. doi: 10.1016/ j.eja.2008.07.005

Alves, D. P., Tomaz, R. S., Laurindo, B. S., Laurindo R. C. F., Fonseca e Silva, F., Cruz, C. D.,... Silva, D. J. H. (2017). Artificial neural network for prediction of the area under the disease progress curve of tomato late blight. *Scientia Agricola*, *74*(1), 51-59. doi: 10.1590/1678-992x-2015-0309

Bohl, M. T., Diesteldorf, J., Salm, C. A., & Wilfling B. (2016). Spot market volatility and futures trading the pitfalls of using a dummy variable approach. *The Journal of Futures Markets*, *36*(1), 30-45. doi: 10.1002/fut.21723

[Conab] (2017). Companhia Nacional de Abastecimento. *Grãos*: 4° levantamento da safra 2016/17, janeiro/2017. Brasília, DF: Conab, 2017. Retrieved on Nov. 15, 2017 from: http://www.conab.gov.br/OlalaCMS/uploads/arquivos/17_01_11_11_30_39_boletim_graos_janeiro_2017.pdf

Erzin, Y., Rao, B. H., Patel, A., Gumaste, S. D., & Singh, D. N. (2010). Artificial neural network models for predicting electrical resistivity of soils from their thermal resistivity. *International Journal of Therm Sciences, 49*(1), 118-130. doi: 10.1016/ j.ijthermalsci.2009.06.008

Fontana, D. C., Berlato, M. A., Lauschner, M. H., & Mello, R. W. (2001). Modelo de estimativa de rendimento de soja no Estado do Rio Grande do Sul. *Pesquisa Agropecuária Brasileira, 36*(3), 399-403.

Gago, J., Martínez-Núñez, L., Landín, M., & Gallego, P. P. (2010). Artificial neural networks as an alternative to the traditional statistical methodology in plant research. *Journal of Plant Physiology*, *167*(1), 23-27. doi: 10.1016/j.jplph.2009.07.007

Goyal, S. (2013). Artificial neural networks in vegetables: a comprehensive review. *Scientific Journal of Crop Science*, *2*(7), 75-94. doi: 10.14196/sjcs.v2i7.928

Homrich, M. S., Wiebke-Strohm, B., Weber, R. L. M., & Bodanese-Zanettini, M. H. (2012). Soybean genetic transformation: a valuable tool for the functional study of genes and the production of agronomically improved plants. *Genetics and Molecular Biology*, *35*(4), 998-1010.

Kron, A. P., Souza, G. M., & Ribeiro, R. V. (2008). Water deficiency at different developmental stages of *Glycine max* can improve drought tolerance. *Bragantia*, *67*(1), 43-49. doi: 10.1590/S0006-87052008000100005

Leal, A. J. F., Miguel, E. P., Baio, F. H. B., Neves, D. C., & Leal, U. A. S. (2015). Redes neurais artificiais na predição da produtividade de milho e definição de sítios de manejo diferenciado por meio de atributos do solo. *Bragantia*, *74*(4), 436-444. doi: 10.1590/1678-4499.0140

Lee, C., & Herbek, J. (2005). Estimating soybean yield. *Cooperative Extension Service*, *188*(1), 7-8.

Liu, B, Liu, X. B., Wang, C., Li, Y. S., Jin, J., & Herbert, S. J. (2010). Soybean yield and yield component distribution across the main axis in response to light enrichment and shading under different densities. *Plant, Soil and Environment*, *56*(8), 384–392

Monteiro, L. A., & Sentelhas, P. C. (2014). Calibration and testing of an agrometeorological model for the estimation of soybean yields in different Brazilian regions. *Acta Scientiarum. Agronomy*, *36*(3), 265-272. doi: 10.4025/actasciagron.v36i3.17485

Oliveira, R. D., Silva, C. M., Mielezrski, F., Lima, J. S. B., & Edvan, R. L. (2016). Harvest growth stages in soybean cultivars intended for silage. *Acta Scientiarum. Animal Sciences*, *38*(4), 383-387. doi: 10.4025/actascianimsci.v38i4.31837

Passos, A. M. A., Rezende, P. M., Alvarenga, A. A., Baliza, D. P., Carvalho, E. R., & Alcântara, H. P. (2011). Yield per plant and other characteristics of soybean plants treated with kinetin and potassium nitrate. *Ciência e Agrotecnologia*, *35*(5), 965-972. doi: 10.1590/S1413-70542011000500014

Patterson D. W. (1996). *Artificial Neural Networks: Theory and Applications, Simon and Schuster (Asia) Pte. Ltd.* Singapore: Prentice Hall

Russell, S., & Norvig, P. (2009). *Artificial intelligence*: a modern approach. 3. ed. London, UK: Prentice Hall.

Safa, M., Samarasinghe, S., & Nejat, M. (2015). Prediction of wheat production using artificial neural networks and investigating indirect factors affecting it: case study in Canterbury province, New Zealand. *Journal of Agricultural Science and Technology*, *17*(4), 791–803.

Schiavo, M. L., Prinari, B., Gronski, J. A., & Serio, A. V. (2015). An artificial neural network approach for modeling the ward atmosphere in a medical unit. *Mathematics and Computers in Simulation*, *116*(1), 44-58. doi: 10.1016/j.matcom.2015.04.006

Sharma, A. K., Sharma, R. K., & Kasana, H. S. (2007).

Prediction of first lactation 305-day milk yield in karan fries dairy cattle using ann modeling. *Applied Soft Computing*, 7(3), 1112-1120. doi: 10.1016/j.asoc.2006.07.002

Soares, J. D. R., Pasqual, M., Lacerda, W. S., Silva, S. O., & Donato, S. L. R. (2013). Utilization of artificial neural networks in the prediction of the bunches' weight in banana plants. *Scientia Horticulturae, 155*(1), 24-29. doi: 10.1016/j.scienta.2013.01. 026

Souza, R. T. G.; Teixeira, I. R., Reis, E. F., & Silva, A. G. (2016). Soybean morphophysiology and yield response to seeding systems and plant populations. *Chilean Journal of Agricultural Research*, *76*(1), 3-8. doi:http://dx.doi:10.4067/S0718-58392016000100001

Zhang, W., Bai, C., & Liu, G. (2007). Neural network modeling of ecosystems: A case study on cabbage growth system. *Ecological Modelling*, *201*(3-4), 317-325. doi: 10.1016/j.ecolmodel.2006.09.022

Zweiri, Y. H., Seneviratne, L. D., & Althoefer, K. (2005). Stability analysis of a three-term backpropagation algorithm. *Neural Networks*, *18*(10), 1341-1347.

23

Phenological and yield performance of black and redberry cultivars in western Paraná State

Marcelo Angelo Campagnolo[1] and Rafael Pio[2*]

*[1]Universidade Estadual do Oeste do Paraná, Marechal Cândido Rondon, Paraná, Brazil. [2]Departamento de Agricultura, Universidade Federal de Lavras, Cx. Postal 3037, 37200-000, Lavras, Minas Gerais, Brazil. *Author for correspondence. E-mail: rafaelpio@dag.ufla.br*

ABSTRACT. Because grain cultivation in Paraná is becoming unsustainable on small farms, growing fruit should be explored on these production units. Thus, the objective of the present work was to quantify the phenology and production of cultivars of black and redberry plants in the western region of Paraná in the town of Marechal Cândido Rondon, Paraná State, Brazil. The experimental design was in randomized blocks, with four blocks and eleven treatments (blackberry cultivars Arapaho, Xavante, Ébano, Comanche, Caingangue, Choctaw, Tupy, Guarani, Brazos and Cherokee and the redberry plants). In each plot, composed of five plants, the phenological and production data for the 2009/10 and 2010/11 production cycles were collected, and the physicochemical quantification of the fruits in the last production cycle was performed. Most of the cultivars presented production cycles longer than 90 days, with the harvest beginning in late October and extending through late January. 'Brazos' presented the highest estimated yield. The Brazos, Guarani and Tupy cultivars produced fruits of higher masses. The redberry presented a high fruit yield and proved an excellent option for cultivation in this region.

Keywords: *Rubus* spp., *Rubus rosifolius*, phenology, fruit quality.

Introduction

The western region of the state of Paraná is known for the intensive farming of grains, such as soybean, corn and wheat, occupying the largest portion of cultivated land and, therefore, constituting the basis of the economy of this region. However, phytosanitary problems associated with monocultures, increased production costs and reduced profitability have made the cultivation of these crops unviable on small farms (FURLANETTO et al., 2010). An option for these farmers would be to grow fruit as an alternative to grains, which could lead to improved profitability in western Paraná. Some studies have already demonstrated the potential exploitation of the fig tree (CAMPAGNOLO et al., 2010; DALASTRA et al., 2009) and grape vine (SILVA et al., 2010; WERLE et al., 2008).

The expansion of fruit growing in this region would also include the cultivation of the blackberry tree (*Rubus* spp.), which has the advantage that the establishment and maintenance cost of the orchards is relatively low when compared with other perennial fruit-bearing trees already under cultivation, offering income that is not based on household farming in western Paraná (ATTILIO et al., 2009). Indeed, the cultivation of Rubus species in Brazil is promising (ANTUNES et al., 2010).

The interest in blackberries has increased in recent years due, in part, to the high concentrations of phenolic and carotenoid compounds, which can help to protect against degenerative diseases (FERREIRA et al., 2010; JACQUES et al., 2010). In addition, several other natural pigments, mainly anthocyanin, are used in the making of dairy products, jams and preserves (ANTUNES, 2002).

Being a deciduous fruit-bearing tree of temperate climates, the main blackberry producing regions lie in Rio Grande do Sul State, Brazil. Nevertheless, new orchards have been established in subtropical regions, such as Caldas, Minas Gerais State, Brazil, and around the higher elevations of Minas Gerais and São Paulo Mantiqueira.

Antunes et al. (2006) conducted a trial of blackberry cultivation in Caldas, Minas Gerais State, and found that some cultivars presented an excellent yield performance, with fruit production higher than 'Tupy', Brazil's and Paraná State main cultivar Campagnolo and Pio (2012b). As each cultivar is unique with regard to the temperature requirements, differences in the performances of cultivars in different growing regions can occur. Therefore, the evaluation of other potentially higher-yielding cultivars is a highly important factor in the expansion of berry crops in the western region of western region of Paraná state.

The objective of the present work was to quantify the phenology and yield of black and redberry cultivars in the western region of Paraná State.

Material and methods

The experiment was conducted in the town of Marechal Cândido Rondon, Paraná State, Brazil, situated at 472 m of elevation, 24°35'42" latitude South and 53°59'54" longitude West. According to the climate classification, the region possesses a Cfa climate – a humid subtropical zone, with rainforest and annual maximum and minimum average temperatures of 28.5 and 16.6°C, respectively. The soil in the experimental area is Eutroferric Red Latosol.

The seedlings of blackberry cultivars, thornless 'Arapaho', 'Xavante' and 'Ébano' and thorned 'Comanche', 'Caingangue', 'Choctaw', 'Tupy', 'Guarani', 'Brazos', and 'Cherokee', and redberry (*Rubus rosifolius*), species native to the Mantiqueira range, were grown according to the methodology by Villa et al. (2003) and Campagnolo and Pio (2012a). Seedlings were planted in the field in November of 2008 at a spacing of 3.0 x 0.5 m (a stand of 6,667 plants per hectare), and the plants were trained under single trellises of one thread in a "T" (parallel double threads), spaced 60 cm apart and 80 cm from the soil.

The experiment was conducted in randomized blocks containing 11 treatments (cultivars), with four blocks and five plants per experimental unit. The trees were cultivated according to the recommendations of Campagnolo and Pio (2012b). At each pruning, five liters of farmyard manure were distributed per linear meter in the direction of the wire threads. Four weedings of spontaneous plants were performed in the experimental area in the months of June, September and November of 2009 and 2010. In addition to the applications of farmyard manure (three liters per linear meter), single applications were performed in October and in December for each evaluation cycle.

In the production cycle of 2009/10 and 2010/11, the phenological stages (early budbreak, and duration of blooming and harvest) were evaluated from after the pruning, on April the first for the redberry plants and July the fifth for the blackberry cultivars, until the end of the harvest. The yield variables, number of fruits per plant, fresh mass of fruits (g), yield (g per plant) and estimated yield (kg ha^{-1}), in the two production cycles were evaluated from April to January. The fruits collected at each harvest, every three days, were counted and weighted with the aid of a digital semi-analytical scale. At the end of the production cycle, all of the values recorded for the determination of the yield per plant were summed, and the estimated yield was calculated by multiplying the yield by the population density (6,667 plants per hectare). The maximum and minimum average temperatures and accumulated rainfall for the months of April of 2009 to January of 2011 (Figure 1) were recorded.

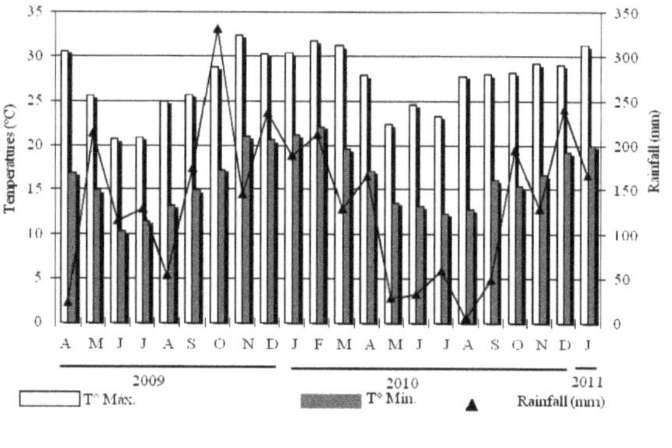

Figure 1. Maximum and minimum temperatures and accumulated rainfall (mm) for the months of April of 2009 to January of 2011 in the town of Marechal Cândido Rondon, Paraná State, Brazil. Marechal Cândido Rondon, Paraná State, Unioeste, 2011.

During the second harvest for each cultivar, in the month of December of 2010, 20 fruits per block were collected for the determination of the physical variables (length and average diameter of the fruits, with the aid of a digital pachymeter) and chemical variables, including the total titratable acidity (obtained through the titration of the samples with 0.1 Mol L^{-1} solutions of NaOH and expressed as a percentage of citric acid), total soluble solids (with the aid of a portable refractometer [ATAGO, model PAL^{-1}], at a temperature of 20°C, with the reading expressed in °Brix and the ratio of soluble solids and acidity.

The data were submitted to an analysis of variance, and the means were compared using the Scott-Knott test at a 5% of probability of error. The SISVAR (System for Variance Analysis- SISVAR) program was utilized for the statistical analysis.

Results and discussion

With regard to the early budbreak of the blackberry plants in the first production cycle (2009/10), the Comanche cultivar was the first to enter the budbreak stage. 'Cherokee', 'Brazos', 'Guarani' and 'Choctaw' initiated the budbreak stage in the second fortnight of August, and the others began in the first fortnight of September (Table 1). However, during the second production cycle, all cultivars initiated budbreak during the second fortnight of August, with the exception of 'Brazos' for which the budbreaking began nine days before the other cultivars. It is believed that, in the first year when the plants were younger, motive that caused the difference in the beginning of the budbreaks of the stems relative to the second production year. These results are important from the standpoint of the distribution of the blackberry plants in locations with climatic conditions similar to western Paraná State (Cfa – humid subtropical zone), but without the incidence of frost from the second fortnight of August and thereafter, which could cause decreased yield due to the injury caused to the new leaves.

It was found that blooming of the blackberry plants began in September in both of the production cycles, with the exception of the Brazos and Arapaho cultivars (Table 1). The duration of blooming in the second production cycle was longer, as the plants were more mature, an observation in agreement with the results of Antunes et al. (2000), who also found that the duration of blooming in young plants is shorter. The duration of blooming in the Ébano cultivar was shorter in relation to the others during both of the production cycles. The same trend occurred with the harvest duration, which was only 38 days during the second production cycle and 52 days during the second cycle. In addition, that cultivar was the latest, with harvests concentrated between December and January, in accordance with the results of Antunes et al. (2000) in Caldas, Minas Gerais State, Brazil (Cwa climate – humid subtropical zone). However, the harvest of the fruits of the other cultivars began during the second fortnight of October and lasted until the second fortnight of January, with the Brazos cultivar presenting the longest harvest duration in both of the production cycles.

Table 1. Phenological description of the production cycles 2009/10 and 2010/11 – early budbreak (EB), start, finish and duration of the blooming (BS, BF and DB) and harvest (SH, FH and DH) of plants of berries cultivars in Western Paraná, Brazil. Marechal Cândido Rondon, Paraná State, Unioeste, 2011.

Cultivar	EB	BS	BF	DB (days)	SH	FH	DH (days)
				Production cycle 2009/10			
Amora-vermelha	15/04	08/05	28/10	173	23/06	01/12	161
Xavante	15/09	22/09	10/12	79	18/11	07/01	50
Tupy	12/09	29/09	02/01	95	26/11	26/01	61
Caingangue	01/09	20/09	02/01	104	26/11	29/01	64
Comanche	03/08	06/09	21/12	106	30/10	18/01	80
Cherokee	29/08	07/09	18/12	102	24/10	18/01	86
Brazos	18/08	07/09	20/11	74	15/10	24/12	101
Arapaho	04/09	11/09	10/12	90	06/11	09/01	64
Guarani	16/08	18/09	05/12	78	06/11	02/01	57
Ébano	14/09	03/11	03/01	61	19/12	26/01	38
Choctaw	25/08	02/10	06/12	65	29/11	04/01	42
				Production cycle 2010/11			
Amora-vermelha	10/03	25/03	28/11	248	10/05	10/01	245
Xavante	19/08	10/09	09/01	122	01/11	10/01	70
Tupy	21/08	15/09	08/01	116	01/11	24/01	84
Caingangue	20/08	01/09	08/01	130	18/10	24/01	98
Comanche	19/08	01/09	08/01	130	23/10	19/01	88
Cherokee	22/08	01/09	07/01	135	26/10	18/01	84
Brazos	10/08	18/08	08/01	143	16/10	24/01	100
Arapaho	19/08	24/08	07/01	138	26/10	12/01	90
Guarani	19/08	01/09	06/01	128	28/10	08/01	84
Ébano	19/08	06/10	12/01	98	03/12	24/01	52
Choctaw	20/08	04/09	09/01	128	25/10	08/01	87

Regarding the performance of the yield of the blackberry cultivars, the greatest fruit yield in the first production cycle was recorded for 'Brazos' (194 fruits). In spite of the Tupy and Guarani cultivars having produced fruits of larger thickness, equal to 'Brazos', the number of fruits were not sufficient to amount to the yield of 7,523.8 kg ha^{-1} recorded for 'Brazos' (Table 2). However, for the second production cycle, the Brazos and Guarani cultivars bore increased number of fruits (493 and 476, respectively), and these two cultivars, along with 'Tupy', bore fruits with the largest thicknesses (larger than five grams). Consequently, the highest yield per plant occurred for the Guarani and Brazos cultivars, but, due to the number of fruits and slightly superior thickness, the greatest yield was found for 'Brazos' (18.602.5 kg ha^{-1}), followed by 'Guarani' (15.129.8 kg ha^{-1}) and then 'Tupy', 'Comanche' and 'Arapaho', but with yields of approximately 50% in relation to that obtained for 'Brazos'.

The results obtained in the present work corroborate the findings of Antunes et al. (2000) who also recorded increased yield with the Brazos cultivar, followed by Guarani, in Caldas, Minas Gerais State, Brazil. Differences were recorded with regard to the yield of the blackberry cultivars lacking thorns. In the present work, yields of 926.3 kg ha^{-1} in the first cycle and 1,427.2 kg ha^{-1} in the second cycle were recorded for the Ébano cultivar. In contrast, in a study conducted in Jundiaí, São Paulo State , Brazil (Cwa climate, mesothermal with dry winter, commonly named altitude tropical, according to Köppen classification), Martins and Pedro Junior (1999) recorded yields of 1,720 kg ha^{-1} in the first cycle and 2,893 kg ha^{-1} in the second cycle for the Ébano cultivar. Broetto et al. (2009) recorded a yield of 3,026 kg ha^{-1} in the first cycle using the Xavante cultivar at Guarapuava Paraná State, Brazil (Cfb climate, mesothermal with rainy winter); in the present study, we only recorded 62.2 kg ha^{-1} in the first cycle and 1,274.7 kg ha^{-1} in the second cycle. These differences may be related to the thermal requirements of each cultivar.

The greatest sizes (length and diameter) of fruits were reported for 'Brazos', followed by 'Guarani' and 'Tupy' (Table 3). These results were expected, as the fruits of those cultivars exhibited the greater masses (Table 2). The fruits of the Tupy and Caingangue cultivars presented lower acidity, whereas the fruits of the other cultivars proved to be more acidic, with the exception of the fruits of the Arapaho cultivar, which were of an intermediary classification (Table 3). As to the soluble solid contents, the fruits of the Tupy, Caingangue, Cherokee, Arapaho, Guarani and Choctaw cultivars recorded higher rates; however, due to the acidity of the fruits, the highest ratio between the concentration of total soluble solids and acidity were obtained for the Tupy and Caingangue cultivars (Table 3), suggesting palatability for fresh consumption.

Table 2. Average number of fruits, average fresh mass, production and estimated yield of the production cycles 2009/10 and 2010/11 of plants of berries cultivars in Western Paraná, Brazil. Marechal Cândido Rondon, Paraná State, Unioeste, 2011.

Cultivar	Number of fruits	Fresh mass (g)	Production (g)	Estimated yield (kg ha^{-1}) [(2)]
	Production cycle 2009/10 [(1)]			
Amora-vermelha	366.9 a	1.6 f	582.9 b	3,887.9 b
Xavante	4.7 f	2.0 e	9.4 f	62.2 f
Tupy	24.0 e	5.3 a	136.7 d	911.6 d
Caingangue	42.0 d	3.4 c	144.9 d	966.3 d
Comanche	77.6 c	3.5 c	277.6 c	1,851.7 c
Cherokee	23.3 e	2.5 d	60.6 e	404.3 e
Brazos	194.2 b	5.8 a	1,128.1 a	7,523.8 a
Arapaho	52.5 d	3.5 c	183.8 d	1,225.7 d
Guarani	76.6 c	5.3 a	412.4 b	2,750.2 c
Ébano	46.4 d	2.9 d	138.9 d	926.3 d
Choctaw	35.4 e	4.0 b	150.8 d	1,005.4 d
C.V. (%)	7.1	5.6	9.4	9.4
	Production cycle 2010/11 [(1)]			
Amora-vermelha	1,136.5 a	1.4 g	1,709.3 b	11,395.9 c
Xavante	76.6 e	3.5 d	222.7 e	1,274.7 f
Tupy	307.9 c	5.5 a	1,596.9 b	10,647.0 c
Caingangue	164.2 d	3.4 e	516.0 d	3,440.3 e
Comanche	386.2 c	3.9 c	1,411.2 b	9,412.5 c
Cherokee	159.0 d	3.1 e	451.5 d	3,012.3 e
Brazos	492.8 b	5.7 a	2,780.2 a	18,602.5 a
Arapaho	246.0 d	3.6 d	817.6 c	5,451.3 d
Guarani	475.8 b	5.4 a	2,269.3 a	15,129.8 b
Ébano	89.8 e	2.8 f	214.0 e	1,427.2 f
Choctaw	396.0 c	4.3 b	1,580.4 b	11,395.9 c
C.V. (%)	6.3	5.4	7.9	7.9

[(1)]Means followed by the same small letter in the column do not differ from one another by the Scott-Knott test ($p \leq 0.05$). [(2)]Calculation considering the spacing 3.0 x 0.5 m with the density of 6,667 plants ha^{-1}.

Table 3. Length, diameter, total titrable acidity, total soluble solids (STT, expressed in °Brix) and ratio SST/acidity in fruits from the production cycle 2009/10 of plants of berries cultivars in Western Paraná, Brazil. Marechal Cândido Rondon, Paraná State, Unioeste, 2011.

Cultivars	Length (mm)[1]	Diameter (mm)	Total titrable acidity	Total soluble solids (SST, °Brix)	Ratio (SST/acidity)
Amora-vermelha	18.0 d	18.9 d	1.7 c	9.8 a	5.8 a
Xavante	16.6 e	17.3 e	2.3 a	8.6 b	3.7 c
Tupy	24.5 b	23.6 b	1.8 c	10.0 a	5.6 a
Caingangue	18.5 d	17.5 e	1.8 c	10.1 a	5.6 a
Comanche	21.5 c	20.9 c	2.4 a	8.6 b	3.6 c
Cherokee	17.5 d	18.5 d	2.2 a	9.2 a	4.2 b
Brazos	27.3 a	25.1 a	2.4 a	7.5 c	3.1 c
Arapaho	21.0 c	20.3 c	2.0 b	9.8 a	4.9 b
Guarani	25.2 b	24.2 b	2.3 a	10.4 a	4.5 b
Ébano	17.8 d	18.7 d	2.5 a	7.6 c	3.0 c
Choctaw	21.5 c	20.6 c	2.6 a	9.5 a	3.7 c
C.V. (%)	2.8	2.5	13.8	7.0	8.9

[1]Means followed by the same small letter in the column do not differ from one another by the Scott-Knott test ($p \leq 0.05$).

It should be emphasized that variations can occur in the chemical composition due to the differences in the intensity of the solar radiation rather than to differences in the site where the plants are grown and to temperature variation, which affect the organolepetic characteristics of blackberry fruits (ALI et al., 2011; SIRIWOHARN et al., 2004).

As temperatures during the crop cycle are very elevated in that region (Figure 1) and blackberry fruits are very delicate and perishable (ANTUNES et al., 2003), the 'Brazos' and 'Guarani' blackberry cultivars can be cultivated in the western region of the Paraná state due to the good yields obtained.

Another option aiming at the production of fruits in western Paraná, in addition to other regions, would be the cultivation of redberry fruits. That species (Rubus rosifolius), native to the Mantiqueira range (LORENZI et al., 2006), has not yet been studied in detail, but the results obtained in the present study are quite promising. The budbreak of the plants began in April during the first production cycle and during March in the second cycle, with a blooming duration of 173 days in the first production cycle and 248 days in the second cycle. In the second cycle, the recorded blooming period ranged between late March and late November; such that the fruit production ranged from May to January in the second cycle (a 245-day duration) (Table 1). The plants bore an average of 367 fruits in the first cycle and 1,136.5 fruits in the second cycle, but, due to the small size of the fruit (average mass of 1.5 g and sizes smaller than 18 mm), the yield per plant per cycle was 582.9 g and 1,709.3 g, respectively, presenting a yield of 11,395.9 kg ha^{-1} for the second production cycle (Table 2). Furthermore, the fruits possess low acidity and a high content of soluble solids (Table 3).

Conclusion

Most of the blackberry cultivars examined presented a production cycle longer than 90 days, with a harvest beginning in late October and extending through late January. The Brazos, Guarani and Tupy cultivars produced fruits of the greatest mass, but the greatest estimated yield was recorded for 'Brazos'. Both 'Tupy' and 'Cainguangue' presented a good balance between the total soluble solids and acidity. The redberry tree presented a high fruit yield and was found to be an excellent option for processing.

References

ALI, L.; SVENSSON, B.; ALSANIUS, B. W.; OLSSON, M. E. Late season harvest and storage of *Rubus* berries-major antioxidant and sugar levels. **Scientia Horticulturae**, v. 129, n. 3, p. 376-381, 2011.

ANTUNES, L. E. C. Amora-preta: nova opção de cultivo no Brasil. **Ciência Rural**, v. 32, n. 1, p. 151-158, 2002.

ANTUNES, L. E. C.; CHALFUN, N. N. J.; REGINA, M. A.; DUARTE FILHO, J. Fenologia e produção de variedades de amora preta nas condições do planalto de Poços de Caldas-MG. **Revista Brasileira de Fruticultura**, v. 22, n. 1, p. 89-95, 2000.

ANTUNES, L. E. C.; DUARTE FILHO, J. D.; SOUZA, C. M. Conservação pós-colheita de frutos da amoreira-preta. **Pesquisa Agropecuária Brasileira**, v. 38, n. 3, p. 413-419, 2003.

ANTUNES, L. E. C.; GONÇALVES, E. D.; TREVISAN, R. Fenologia e produção de cultivares de amoreira-preta em sistema agroecológico. **Ciência Rural**, v. 40, n. 9, p. 1929-1933, 2010.

ANTUNES, L. E. C.; TREVISAN, R. GONÇALVES, E. D.; FRANZON, R. C. Produção extemporânea de amora-preta. **Revista Brasileira de Fruticultura**, v. 28, n. 3, p. 430-434, 2006.

ATTILIO, L. B.; BOLIANI, A. C.; TARSITANO, M. A. A. Custo de produção de amora-preta em região tropical. **Revista Brasileira de Fruticultura**, v. 31, n. 4, p. 1042-1047, 2009.

BROETTO, D.; BOTELHO, R. V.; PAVANELLO, A. P.; SANTOS, R. P. Cultivo orgânico de amora-preta cv. Xavante em Guarapuava-PR. **Revista Brasileira de Agroecologia**, v. 4, n. 2, p. 2208-2212, 2009.

CAMPAGNOLO, M. A.; PIO, R. Enraizamento de estacas caulinares e radiculares de amoreira-preta coletadas em diferentes épocas, armazenadas a frio e tratadas com AIB. **Ciência Rural**, v. 42, n. 2, p. 232-237, 2012a.

CAMPAGNOLO, M. A.; PIO, R. Produção da amoreira-preta 'Tupy' sob diferentes épocas de poda. **Ciência Rural**, v. 42, n. 2, p. 225-231, 2012b.

CAMPAGNOLO, M. A.; PIO, R.; DALASTRA, I. M.; CHAGAS, E. A.; GUIMARÃES. V. F.; DALASTRA, G. M. Sistema desponte na produção de figos verdes 'Roxo de Valinhos'. **Ciência Rural**, v. 40, n. 1, p. 25-29, 2010.

DALASTRA, I. M.; PIO, R.; CAMPAGNOLO, M. A.; DALASTRA, G. M.; CHAGAS, E. A.; GUIMARÃES, V. F. Épocas de poda na produção de figos verdes 'Roxo de Valinhos' em sistema orgânico na região oeste do Paraná. **Revista Brasileira de Fruticultura**, v. 31, n. 2, p. 447-453, 2009.

FERREIRA, D. S.; ROSSO, V. V.; MERCADANTE, A. Z. Compostos bioativos presentes em amora-preta (*Rubus* spp.). **Revista Brasileira de Fruticultura**, v. 32, n. 3, p. 664-674, 2010.

FURLANETTO, C.; SEIFERT, K. E.; FENSTERSEIFER, C. E.; PAGE, E. C.; DAVI, J. J. S.; GRABOWSKI, M. M. S. Desenvolvimento das culturas de soja, milho e trigo cultivadas em áreas infestadas com o nematoide Tubixaba tuxaua no Oeste do Paraná. **Tropical Plant Pathology**, v. 35, n. 5, p. 295-302, 2010.

JACQUES, A. C.; PERTUZATTI, P. B.; BARCIA, M. T.; ZAMBIAZI, R. C.; CHIM, J. F. Estabilidade de compostos bioativos em polpa congelada de amora-preta (*Rubus fruticosus*) cv. Tupy. **Química Nova**, v. 33, n. 8, p. 1720-1725, 2010.

LORENZI, H.; BACHER, L.; LACERDA, M.; SARTORI, S. **Frutas brasileiras e exóticas cultivadas (de consumo in natura)**. São Paulo: Instituto Plantarum de Estudos da Flora, 2006.

MARTINS, F. P.; PEDRO JUNIOR, M. J. Influência da produtividade da amora-preta, cv. Ébano, em Jundiaí. **Bragantia**, v. 58, n. 2, p. 317-321, 1999.

SILVA, T. P.; PIO, R.; SALIBE, A. B.; DALASTRA, I. M.; STANGARLIN, J. R.; KUHN, O. J. Avaliação de porta-enxertos de videira em condições subtropicais. **Bragantia**, v. 69, n. 1, p. 93-97, 2010.

SIRIWOHARN, T.; WROLSTAD, R. E.; FINN, C. E.; PEREIRA, C. B. Influence of cultivar, maturity, and sampling on blackberry (*Rubus L. Hybrids*) anthocyanins, polyphenolics and antioxidant properties. **Journal of Agricultural and Food Chemistry**, v. 52, n. 26, p. 8021-8030, 2004.

VILLA, F.; PIO, R.; CHALFUN, N. N. J.; GONTIJO, T. C. A.; COELHO, J. H. C.; DUTRA, L. F. Propagação de amoreira-preta utilizando estacas lenhosas. **Ciência e Agrotecnologia**, v. 27, n. 4, p. 829-834, 2003.

WERLE, T.; GUIMARÃES, V. F.; DALASTRA, I. M.; ECHER, M. M.; PIO, R. Influência da cianamida hidrogenada na brotação e produção da videira 'Niagara Rosada' na região oeste do Paraná. **Revista Brasileira de Fruticultura**, v. 30, n. 1, p. 20-24, 2008.

24

Influence of pollination on canola seed production in the Cerrado of Uberlândia, Minas Gerais State, Brazil

Leandro Fuzaro[1*], Nayara Luchini Xavier[1], Fábio Janoni Carvalho[2], Flávia Andrea Nery Silva[1], Stephan Malfitano Carvalho[3] and Vanessa Andaló[2]

*[1]Instituto de Ciências Agrárias, Universidade Federal de Uberlândia, Avenida Amazonas, s/nº, Bloco 2E, Sala 2E 152, Campus Umuarama, Uberlândia, Minas Gerais, Brazil. [2]Instituto de Ciências Agrárias, Universidade Federal de Uberlândia, Campus Monte Carmelo, Monte Carmelo, Minas Gerais, Brazil. [3]Departamento de Entomologia, Universidade Federal de Lavras, Campus Universitário, Lavras, Minas Gerais, Brazil. *Author for correspondence. E-mail: fuzaro.leandro@gmail.com*

ABSTRACT. The objective of this study was to evaluate the influence of pollinating insects on canola productivity for different commercial hybrids, Hyola 61 and Hyola 433, in Minas Gerais State, Brazil. The studies were conducted at the Fazenda Experimental Água Limpa of the Federal University of Uberlândia. To evaluate the influence of pollination by bees on the production of canola seeds, an experiment was carried out to exclude floral visitors and compare canola seed production following four treatments: (a) autogamy in Hyola 61, (b) autogamy in Hyola 433, (c) open pollination in Hyola 61, with insects having free access, and (d) open pollination in Hyola 433, with insects having free access. Flowers that were available for pollination had 25% and 18.8% heavier pods and 22.5% and 20.6% more seeds per pod in Hyola 61 and Hyola 433, respectively, than those bagged without pollination. The results show that the cross-pollination process, fostered by pollinator visits, positively influenced some parameters related to production, presenting an advantage in relation to canola autogamy in Cerrado Mineiro.

Keywords: *Apis mellifera*; *Brassica napus*; tropicalization; productivity.

Introduction

Canola (*Brassica napus* L.) is an oleaginous plant of the Brassicaceae family that was developed by conventionally breeding rapeseed cultivars; rapeseed is a grain with high levels of erucic acid and glucosinolates (antinutritional substances). From its seeds, excellent quality oil for human consumption and for bioenergy is produced, and it is considered the third most produced oilseed in the world, surpassed only by the production of palm and soybean (De Mori, Tomm, & Ferreira, 2014).

Worldwide, the cultivation of canola covered an area of approximately 34 million hectares in 2014. The world's largest producers and consumers of canola grains are in the European Union, with production of 21.1 million tons for the 2013/2014 harvest and a forecast of 24 million tons for the 2014/2015 harvest. Consumption of canola grains was 24.5 million tons in 2013/2014, with grain imports of 3.4 million tons. The second largest producer and consumer is China, with production of 14.5 million tons for the 2013/2014 harvest and grain imports of approximately 5 million tons to supply the domestic market (USDA, 2016).

In Brazil, canola cultivation started in the 1980s and currently involves only spring canola of the species *B. napus* var. *oleifera*, used in crop rotation systems mainly in the south of the country (Luz et al., 2012). The state of Rio Grande do Sul has been considered an outstanding state in terms of the production and planted area of canola, with values of 85.2% and 86.7%, respectively. The state of Paraná contributes 14.8% of the national production and 13.3% of the planted area of canola in Brazil (CONAB, 2016). Canola cultivation fits well with the crop rotation system adopted in Brazil, and it is rotated with soybean, both of which are sown before corn (Tomm et al., 2009a).

Although canola is a plant adapted to cold regions and cultivated mainly in temperate regions (latitudes between 35 and 55 degrees) (Tomm, Wiethölter, Dalmago, & Santos, 2009b), recent efforts and research have been carried out to introduce and adapt the crop to warmer climates at lower latitudes (tropicalization). Experiments and commercial crops in Goiás and Minas Gerais States have demonstrated that canola is a crop with a great potential to contribute to the expansion of Brazilian agribusiness, as it is perfectly suited to the crop production systems of the Central-West region of the country (Tomm, 2005). Moreover, this plant adapts to regions considered too adverse for the development of other cultivars (Ávila, Braccini, Scapim, Fagliari, & Santos, 2007).

Canola's flowers are hermaphrodites; they present both female (gyneum) and male (androecium) parts. They have a pistil with three parts: the ovary, which houses the eggs and which has 28 eggs per flower on average; the stylet, which is the medial portion of the pistil; and the stigma, which is the surface that receives the pollen grains. The androecium is formed by stamens. The canola flowers have six stamens, four long and two short. Each stamen contains a fillet that supports an anther. The anthers contain pollen grains, which are accessible after the opening of the anther and which constitute the male gametes of the plants (Witter et al., 2014b).

Canola is a self-supporting plant, and thus, both self-pollination and cross-pollination result in fruits and seeds. Flowers are more likely to be susceptible to self-pollination during the beginning of anthesis, when the anthers are inwardly facing the flower, and the stigma is below them, making it easier for the pollen to fall on the stigma. However, when the stigma exceeds the height of the anthers, it is not possible for the pollen of the flower itself to fall on the stigma. Self-pollinating by a pollinating agent, such as an insect, is still possible, but researchers believe that this is a mechanism that favors cross-pollination (Witter et al., 2014b).

Insect pollination, mainly by bees, increases the production of canola plants because these pollinators deposit more pollen grains on the stigma of the flowers compared to the amount of pollen grains deposited on the stigma by self-pollination (spontaneous) without the action of bees, as shown in the work done by Witter et al. (2015). Previous studies have shown that the presence of pollinating insects, especially *Apis mellifera* Linnaeus (Hymenoptera: Apidae), increases not only seed productivity (Delaplane & Mayer, 2000) but also the quality of the oil and thus the market value of the crop (Witter & Tirelli, 2014). In addition, despite autogamy, mechanisms for the occurrence of allogamy were found in the reproductive strategies used by this species, such as abundant pollen, nectar and aroma. These mechanisms are important for the genetic improvement of the species, favoring the occurrence of more adaptive characteristics (Mussury & Fernandes, 2000).

In an experiment conducted in Canada, Sabbahi, Oliveira, and Marceau (2005) demonstrated a 46% improvement in grain yield in the presence of three beehives per hectare compared to crops with no hives. In a study conducted in Australia, Manning and Boland (2000) observed that the number of pods per plant decreased with increasing distance between apiaries. In Brazil, Mussury and Fernandes (2000) demonstrated an increase of 31.9% in the number of grains per plant under natural pollination conditions when compared to autogamy conditions. For crops in Rio Grande do Sul, Blochtein, Nunes-Silva, Halinski, Lopes, and Witter (2014) verified that free visits of insects increased yield by 17% in the Hyola 420 cultivar and by approximately 30% in the Hyola 61 cultivar.

Considering the importance of canola cultivation and the recent efforts to expand its cultivation into low latitude regions in Brazil, there is a need to improve knowledge about the influence of pollination on canola seed productivity to optimize production. Thus, the objective of this study was to evaluate the influence of pollinating insects on two different commercial hybrids, Hyola 61 and Hyola 433, in the Cerrado biome of Minas Gerais.

Material and method

The studies were conducted at the Água Limpa Experimental Farm at the Federal University of Uberlândia, located at latitude 19°05'48"S, longitude 48°21'05"W, with an altitude of approximately 800 m. The soil of the area is classified as typical

Dystrophic Red Latosol, a moderate, medium texture soil, and the area had tropical Cerrado subdeciduous vegetation and a mild wavy relief (Embrapa, 2006). The Aw climate, according to the Köppen classification, is marked by two well-defined seasons, one rainy and one dry (Rosa, Lima, & Assunção, 1991). The area contains 104 ha of preserved area, formed by complex vegetation that includes restricted Cerrado, dense Cerrado, Vereda and gallery forest and 151.72 ha with fruit and pasture (Neto, 2008).

A completely randomized block experimental design with two hybrids and two pollination treatments (open and closed to insect visitation) and 10 replications was used, totaling 20 plots. Each plot consisted of six planting four meters in length, with 0.20 m spacing between rows and a density of 18 seeds m^{-1}. The plots were separated by 0.5 m, totaling an experimental area of 121 m^2. The experiment was established on April 2, 2016, with both hybrids flowering after 55 days and presenting senescence after 97 days.

The hybrids used in this experiment were Hyola 61 and Hyola 433. The hybrid Hyola 61 has polygenic resistance to *Phoma lingam* Tode ex. Shaw. Desm., known as blackleg, and performs well under water deficient conditions and severe cold. This hybrid has the characteristics of a medium cycle with 123 to 155 days from emergence to maturation and presents great stability of yield when cultivated under varied conditions. Hyola 433 is a short-cycle hybrid suitable for highly fertile soils (Tomm et al., 2009a).

To control insect pests, mainly aphids of the species *Lipaphis pseudobrassicae* Davis (Hemiptera: Aphididae) and *Myzus persicae* Sulzer (Hemiptera: Aphididae), two applications of Decis 25 EC® were performed on June 4 and 15, and one application of Acetox Nortox® was performed on June 26, prior to the beginning of the evaluations.

To evaluate the influence of bee pollination on the production of canola seeds, an experiment was conducted to exclude floral visitors, and canola seed production was compared among the following four treatments: (a) autogamy in Hyola 61, (b) autogamy in Hyola 433 (the branches were covered with a fine mesh fabric during the entire period of flowering, depriving them of insect visitation), (c) open pollination in Hyola 61, with insects having free access, and (d) open pollination in Hyola 433, with insects having free access. For each treatment, 10 replicates were made, with 8 plants in each replicate.

After the maturing of the pod, the marked branches were collected and submitted to a laboratory analysis evaluating the following parameters: (1) pod weight (n = 800 pod/treatment), (2) number of seeds per pod (n = 800 pod treatment^{-1}), and (3) weight of a thousand seeds (per treatment).

The residual normality, homogeneity of variances and additivity of the blocks were verified by Shapiro-Wilk's, Levene's and Tukey's tests, respectively, for each variable. After fulfillment of the assumptions, analysis of variance (ANOVA) was applied, and if a significant difference was detected, then the means were compared by Tukey's test. All tests were performed at 5% significance. The analyses were performed by SPSS and SISVAR® software (Ferreira, 2011)

Result and discussion

When the influence of pollination was evaluated, there was a significant difference between the treatments with open pollination (OP) and autogamy (AT) for both hybrids for the following characteristics: the mean weight of the pod and mean number of seeds/pod. The flowers that were available for pollination presented 25% and 18.8% heavier pods and 22.5% and 20.6% more seeds per pod in Hyola 61 and Hyola 433, respectively, than those that were bagged without cross-pollination (Table 1).

Table 1. Mean weight of the pod (g), mean number of seeds/pod and weight of one thousand seeds (g) of the hybrids Hyola 61 and Hyola 433 with and without cross-pollination.

Treatment	Average weight of pod (g)	Average number of seeds per pod	Weight of one thousand seeds (g)
Hyola 61 - OP	0.076 ± 0.009 a	14.38 ± 0.894 a	2.28 ± 0.32 a
Hyola 61 - AT	0.057 ± 0.006 c	11.15 ± 1.133 b	2.12 ± 0.35 a
Hyola 433 - OP	0.069 ± 0.006 b	14.88 ± 1.235 a	1.94 ± 0.20 b
Hyola 433 - AT	0.056 ± 0.004 c	11.81 ± 0.659 b	1.86 ± 0.15 b

*Means followed by the same letters in the column do not differ as determined by Tukey's test at 5% probability. OP: open pollination; AT: autogamy.

Regarding the mean seed weight, Williams, Martin, and White (1986) and Adegas and Nogueira-Couto (1992) found an increase of 37.47 and 19.24%, respectively, in the productivity of canola plants with visits from *A. mellifera* when compared to plants with insect exclusion. In the present study, there was no difference in the weight of one thousand seeds for the pollination treatments (OP and AT) in both hybrids. However, Hyola 61 presented heavier seeds than Hyola 433 (Table 1)

In Rio Grande do Sul, according to Rosa, Blochtein, and Lima (2011), the induction of pollination in tests with free visits of insects to *B. napus* (Hyola 432) resulted in a 22.99% increase in grain weight in relation to autogamy. In Canada,

Sabahhi et al. (2005) also recorded increases of up to 30% in seed weight with insect visits when compared to areas deprived of insects. Several factors have been discussed in the literature as determinants for full plant development and crop productivity, such as sowing density and spacing (Von Pinho, Gross, Steola, & Mendes, 2008), genotype (Silveira et al., 2010), cultivar (Krüger et al., 2011), time of sowing (Ungaro, Nogueira, & Nagai, 2000), harvest season (Marchiori et al., 2002) and environmental conditions (Mello et al., 2006). The absence of significant differences between the pollination treatments in this study emphasizes the need to study the adaptation and development of the crop in the Cerrado region and the need to perform new evaluations of canola crops in the presence of pollinators under climatic conditions that differ from those already studied.

There was a significant difference between the treatments for hybrids in relation to the mean weight of the pod. Hyola 61 presented a heavier pod, showing that it responded better to pollination, and was more affected by OP than Hyola 433. In the AT treatment, there were no significant differences (Hyola 61: AT = 0.057 ± 0.006 and Hyola 433: AT = 0.056 ± 0.004).

According to Gavloski (2012), the effect of pollination on grain yield depends not only on the density of pollinators in the field but also on the atmospheric conditions during the flowering period and on the canola variety. The hybrid Hyola 61 is characterized by having great yield stability when cultivated under varied environmental conditions. However, Hyola 433 is more affected by environmental conditions, especially highly fertile soils (Tomm et al., 2009a). These different profiles between the hybrids probably caused the different responses to the environmental conditions and resulted in different productivities. In addition, the availability of resources and the environment can determine the growth potential of specific genotypes and the number of flowers produced per plant. Consequently, the total number of seeds produced is susceptible to variations in not only different locations but also the seasons (Wiens, 1984).

The deposition (quantity and quality) and distribution of pollen grains on flower stigmas through wind and self-pollination, despite being sufficient for the promotion of fruit and formation of seed, are often not able to maximize the productive potential of the plant and ensure that all the seeds and fruits that the crop can potentially produce are in fact produced (Freitas & Nunes-Silva, 2012). In addition, it is necessary that at least 28 grains of pollen be deposited in the stigma of each canola flower since the ovary usually contains 28 eggs to be fertilized (Witter & Tirelli, 2014). The more pollen grains that are deposited on the stigma of the flower, the greater the chance that all the eggs will be fertilized and that all the possible canola grains will be formed. The amount of pollen grains deposited on flower stigmas depends on the number of pollinating insects present in canola crops (Marsaro-Junior et al., 2017).

Witter et al. (2014b) confirmed that there is a positive correlation between the density of insects (mostly bees) and the production of canola grains. In this study, both flowers deprived of insect visitation and those with free visitation were passive to anemophilia since the tissue used for the deprivation of floral visitors did not prevent the transfer of pollen grain between flowers of distinct plants, which was considered a common factor for both results.

The species *A. mellifera*, *Paratrigona lineata* Lepeletier, 1836 (Hymenoptera: Apidae) and *Trigona spinipes* Fabricius, 1793 (Hymenoptera: Apidae) were observed visiting flower buds during the canola flowering period, demonstrating the possibility of pollinating this crop in the Cerrado region. After verifying which pollinating bees are effective in a crop, an increase in the frequency of visits of these species to obtain direct benefits for increased productivity is recommended. Through direct management, the pollination services promoted by *A. mellifera* have been used to increase the production of several crops (Kremen, 2008).

In addition to the increase in seed productivity, there are also resources produced by the bees since canola crops offer large amounts of floral resources (pollen and nectar) (Witter, Nunes-Silva, & Lisboa, 2014a). In addition to honey production, which has been very valued by consumers, canola flowering stimulates the development and strengthening of bee colonies (Marsaro-Junior et al., 2017). In addition, beekeeping is a species conservation activity, one of the few agricultural activities that cover the three pillars of sustainability: the economic pillar, because beekeeping generates income for the farmer; the social pillar, because it is still mainly a family business; and the ecological pillar, because no deforestation is necessary to raise bees (Alcoforado-Filho & Gonçalves, 2000). Therefore, it is desirable for beekeepers and canola producers to strengthen their relationships and establish consortia, as both would benefit.

The protection of natural areas close to canola crops also guarantees pollination services from bees, which benefits productivity. Natural habitats are necessary for bees not only to as places to obtain food resources but also as places to nest and breed (Knight et al., 2009). Areas of peripheral vegetation, such as field borders, live fences, road margins, and riparian forests, also provide nesting sites and can provide corridors where bees and other beneficial insects can feed and migrate through the agricultural landscape (Witter et al., 2014a). At present, the relationship between the quantity and/or the distance from natural areas and the presence of bees (richness and abundance) in crop yields has been recognized in several studies worldwide and in Brazil (Schulp, Lautenbach, & Verburg, 2014).

Halinski, Dorneles, and Blochtein (2015), in experiments in Rio Grande do Sul, found that the productivity of canola seeds decreases as farms are located farther away from forest remnants. The average grain yield at 25 m from natural areas ranged from 3,368 to 4,656 kg ha^{-1}, while at 175 m, the variation ranged from 2,044 to 3,956 kg ha^{-1}. At 325 m, the productivity range was from 1,508 to 3,432 kg ha^{-1}. This finding is important to the sector involved in the production of canola (Blochtein, Witter, & Halinski, 2015) since the fields in southern Brazil are surrounded by homogeneous areas dominated by agricultural lands and scarce remnants of natural areas (Witter et al., 2014b). As a result of this landscape, there is a loss of favorable environments for pollinators and, consequently, a decrease in their permanence in the area. Thus, it is important to keep forest remnants close to crops, allowing pollinators to perform their services and increase grain yields and associated economic values.

Conclusion

Canola shows an increase in the number of its seeds when pollinated by native bees and by *A. mellifera* in the Cerrado of Minas Gerais State. Canola flowers with free visitation by pollinators produce heavier pods and more seeds compared to those submitted to only autogamy. In relation to hybrids, when grown in the Cerrado, Hyola 61 presents heavier pods and seeds than Hyola 433 under both pollination treatments but has a smaller number of seeds per pod.

Acknowledgements

The authors thank the Conselho Nacional de Desenvolvimento Científico e Tecnológico (CNPq) for a scholarship to the first author.

References

Adegas, J. E. B., & Nogueira-Couto, R. H. (1992). Entomophilous pollination in rape (*Brassica napus* L. var. *oleifera*) in Brazil. *Apidologie, 23*(3), 203-209. doi: 10.1051/apido:19920302

Alcoforado-Filho, F. G., & Gonçalves, J. C. (2000). Flora apícola e mel orgânico. In S. L. O. Vilela, & F. G. Alcoforado-Filho (Ed.), *Cadeia produtiva do mel no estado do Piauí* (p. 48-59). Teresina, PI: Embrapa Meio-Norte.

Ávila, M. R., Braccini, A. L., Scapim, C. A., Fagliari, J. R., & Santos, J. L. (2007). Influência do estresse hídrico simulado com manitol na germinação de sementes e crescimento de plântulas de canola. *Revista Brasileira de Sementes, 29*(1), 98-106. doi: 10.1590/S0101-31222007000100014

Blochtein, B., Nunes-Silva, P., Halinski, R., Lopes, L. A., & Witter, S. (2014). Comparative study of the floral biology and of the response of productivity to insect visitation in two rapeseed cultivars (Brassica napus L.) in Rio Grande do Sul. *Brazilian Journal of Biology, 74*(4), 784-794. doi: 10.1590/1519-6984.02213

Blochtein, B., Witter, S., & Halinski, R. (2015). *Plano de manejo para polinização da cultura da canola:* conservação e manejo de polinizadores para agricultura sustentável, através de uma abordagem ecossistêmica. Rio de Janeiro, RJ: Funbio.

Companhia Nacional de Abastecimento [CONAB]. (2016). *Conjuntura mensal agosto de 2016: Safra 2015/2016*. Brasília, DF: Ministério da Agricultura Pecuária e Abastecimento.

De Mori, C., Tomm, G. O., & Ferreira, P. E. P. (2014). *Aspectos econômicos e conjunturais da cultura da canola no mundo e no Brasil*. Passo Fundo, RS: Embrapa Trigo.

Delaplane, K. S., & Mayer, D. F. (2000). *Crop pollination by bees*. New York, US: CABI Publishing.

Durán, X. A., Ulloa, R. B., Carrillo, J. A., Contreras, J. L., & Bastidas, M. T. (2010). Evaluation of yield component traits of honeybee-pollinated (*Apis mellifera* L.) rapeseed canola (*Brassica napus* L.). *Chilean Journal of Agricultural Research, 70*(2), 309-314. doi: 10.4067/S0718-58392010000200014

Empresa Brasileira de Pesquisa Agropecuária [EMBRAPA´]. (2006). *Sistema brasileiro de classificação de solos*. Rio de Janeiro, RJ: Centro Nacional de Pesquisa de Solos.

Ferreira, D. F. (2011). Sisvar: a computer statistical analysis system. *Ciência e Agrotecnologia, 35*(6), 1039-1042.

Freitas, B. M., & Nunes-Silva, P. (2012). Polinização agrícola e sua importância no Brasil. In Imperatriz- V. L. Fonseca, D. A. L. Canhos, D. A. Alves, & A. M. Saraiva (Ed.), *Polinizadores no Brasil: contribuição e perspectivas para a biodiversidade, uso sustentável, conservação e serviços ambientais* (p. 103-118). São Paulo, SP: Edusp.

Gavloski, J. (2012). *Bees on canola: what are the benefits?* CA: Manitoba. Retrieved on Feb. 13, 2017 from https://www.gov.mb.ca/agriculture/crops/insects/pubs/beesoncanolafactsheet.pdf

Halinski, R., Dorneles, A. L., & Blochtein, B. (2015). Bee assemblage in habitats associated with *Brassica napus* L. *Revista Brasileira de Entomologia*, *59*(3), 222-228. doi: 10.1016/j.rbe.2015.07.001

Knight, M. E., Osborne, J. L., Sanderson, R. A., Hale, R. J., Martin, A. P., & Goulson, D. (2009). Bumblebee nest density and the scale of available forage in arable landscapes. *Insect Conservation and Diversity*, *2*(2), 116-124. doi: 10.1111/j.1752-4598.2009.00049.x

Kremen, C. (2008). Crop pollination services from wild bees. In R. R. James, & T. Pitts-Singer (Ed.), *Bee pollination in agricultural ecosystems* (p. 10-26). Oxford, UK: Oxford University Press.

Krüger, C. A. M. B., Silva, J. A. G., Medeiros, S. L. P., Dalmago, G. A., Sartori, C. O., & Schiavo, J. (2011). Arranjo de plantas na expressão dos componentes da produtividade de grãos de canola. *Pesquisa Agropecuária Brasileira*, *46*(11), 1448-1453. doi: 10.1590/S0100-204X2011001100005

Luz, G. L., Medeiros, S. L. P., Tomm, G. O., Bialozor, A., Amaral, A. D., & Pivoto, D. (2012). Temperatura base inferior e ciclo de híbridos de canola. *Ciência Rural*, *42*(9), 1549-1555. doi: 10.1590/S0103-84782012000900006

Mader, E., Shepherd, M., Vaughan, M., Black, S. H., & Lebuhn, G. (2011). *Attracting native pollinators:* protecting North America's bees and butterflies. North Adams, MA: Storey Publishing.

Manning, R., & Boland, J. (2000). A preliminary investigation into honey bee (*Apis mellifera*) pollination of canola (*Brassica napus* cv. Karoo) in Western Australia. *Australian Journal of Experimental Agriculture*, *40*(3), 439-442. doi: 10.1071/EA98148

Marchiori Jr., O., Inoue, M. H., Braccini, A. L., Oliveira JR, R. S., Ávila, M. R., Lawder, M., & Constantin, J. (2002). Qualidade e produtividade de sementes de canola (*Brassica napus*) após aplicação de dessecantes em pré-colheita. *Planta Daninha*, *20*(2), 253-261. doi: 10.1590/S0100-83582002000200012

Marsaro-Júnior, A. L., Halinski, R., Blochtein, B., Pereira, P. R. V. S., Tomm, G. O., & Ferreira, P. E. P. (2017). *Diversidade de abelhas na cultura da canola no Rio Grande do Sul*. Passo Fundo, RS: Embrapa Trigo.

Mello, R., Nörnberg, J. L., Restle, J., Neumann, M., Queiroz, A. C., Costa, P. B.; ... David, D. B. de. (2006). Características fenológicas, produtivas e qualitativas de híbridos de girassol em diferentes épocas de semeadura para produção de silagem. *Revista Brasileira de Zootecnia*, *35*(3), 672-682. doi: 10.1590/S1516-35982006000300007

Mussury, R. M., & Fernandes, W. (2000). Studies of the floral biolgy and reproductive system of *Brassica napus* L. (Cruciferae). *Brazilian Archives of Biology and Technology*, *43*(1), 111-117.

Neto, P. L. (2008). *Levantamento planimétrico n° 36.243.* Uberlândia, MG: Prefeitura Municipal de Uberlândia.

Rosa, R., Lima, S. C., & Assunção, W. L. (1991). Abordagem preliminar das condições climáticas de Uberlândia (MG). *Sociedade e Natureza*, *3*(1), 91-108.

Rosa, A. S., Blochtein, B., & Lima, D. K. (2011). Honey bee contribution to canola pollination in Southern Brazil. *Scientia Agricola*, *68*(2), 255–259. doi: 10.1590/S0103-90162011000200018

Sabbahi, R., Oliveira, D., & Marceau, J. (2005). Influence of Honey Bee (Hymenopetra: Apidae) Density on the production of canola (Cruciferae: Brassicaceae). *Journal of Economic Entomology*, *98*(2), 367-372.

Schulp, C. J. E., Lautenbach, S., & Verburg, P.H. (2014). Quantifying and mapping ecosystem services: demandand supply of pollination in the European Union. *Ecological Indicators*, *36*(01), 131-141. doi: 10.1016/j.ecolind.2013.07.014

Silveira, G., Carvalho, F. I. F., Oliveira, A. C., Valério, I. P., Benin, G., Ribeiro, G., ... Silva, J. A. G. (2010). Efeito da densidade de semeadura e potencial de afilhamento sobre a adaptabilidade e estabilidade em trigo. *Bragantia*, *69*(1), 63-70. doi: 10.1590/S0006-87052010000100009

Tomm, G. O. (2005). *Situação em 2005 e perspectivas da cultura de canola no Brasil e em países vizinhos*. Passo Fundo, RS: Embrapa Trigo.

Tomm, G. O., Ferreira, P. E. P., Aguiar, J. L. P., Castro, A. M. G., Lima, S. M. V., & De Mori, C. (2009a). *Panorama atual e indicações para aumento de eficiência da produção de canola no Brasil*. Passo Fundo, RS: Embrapa Trigo. Retrieved on Jan. 5, 2017 from http://www.cnpt.embrapa.br/biblio/do/p_do118.pdf

Tomm, G. O., Wiethölter, S., Dalmago, G. A., & Santos, H. P. (2009b). *Tecnologia para produção de canola no Rio Grande do Sul*. Passo Fundo, RS: Embrapa Trigo.

Ungaro, M. R. G., Nogueira, S. S. S., & Nagai, V. (2000). Parâmetros fisiológicos, produção de aquênios e fitomassa de girassol em diferentes épocas de cultivo. *Bragantia*, *59*(2), 205-211. doi: 10.1590/S0006-87052000000200012

United States Department of Agriculture [USDA]. (2016). *Canola*. Washington, D.C.: USDA.

Von Pinho, R. G., Gross, M. R., Steola, A. G., & Mendes, M. C. (2008). Adubação nitrogenada, densidade e espaçamento de híbridos de milho em sistema de plantio direto na região sudeste do Tocantins. *Bragantia*, *67*(3), 733-739. doi: 10.1590/S0006-87052008000300023

Wiens, D. (1984). Ovule survivorship, brood size, life history, breeding systems, and reproductive success in plants. *Oecologia*, *64*(1), 47-53. doi: 10.1007/BF00377542

Williams, I. H., Martin, A. P., & White, R. P. (1986). The pollination requirements of oil-seed rape (*Brassica napus* L.). *Journal of Agricultural Science*, *106*(1), 27-30. doi: 10.1017/S0021859600061670

Witter, S., & Tirelli, F. (2014). Polinizadores nativos presentes em lavouras de canola no Rio Grande do Sul. In S. Witter, P. Nunes-Silva, & B. Blochtein

(Ed.), *Abelhas na polinização de canola:* benefícios ambientais e econômicos. Porto Alegre, RS: Edipucrs.

Witter, S., Nunes-Silva, P., & Lisboa, B. B. (2014a). Práticas amigáveis aos polinizadores em áreas agrícolas. In S. Witter, P. Nunes-Silva, & B. Blochtein (Ed.), *Abelhas na polinização da canola:* benefícios ambientais e econômicos (p. 49-60). Porto Alegre, RS: Edipucrs.

Witter, S., Blochtein, B., Nunes-Silva, P., Tirelli, F. P., Lisboa, B. B., Bremm, C., & Lanzer, R. (2014b). The bee community and its relationship to canola seed production in homogenous agricultural areas. *Journal of Pollination Ecology*, *12*(3), 15-21.

Witter, S., Nunes-Silva, P., Lisboa, B. B., Tirelli, F. P., Sattler, A., Hilgert-Moreira, S. B., & Blochtein, B. (2015). Stingless bees as alternative pollinators of canola. *Journal of Economic Entomology*, *108*(3), 880-886. doi: 10.1093/jee/tov096

25

Bromatological composition of sorghum, millet plant and midget-guandu at different cut times in intercropping and monoculture

Carlos Sérgio Tiritan[1], Diego Henriques Santos[2*], Cláudia Regina Minutti[1], José Salvador Simoneti Foloni[3] and Juliano Carlos Calonego[1]

*[1]Centro de Ciências Agrárias, Universidade do Oeste Paulista, Presidente Prudente, São Paulo, Brazil. [2]Companhia de Desenvolvimento Agrícola de São Paulo, Centro de Negócios de Presidente Prudente, Rod. Raposo Tavares, km 564, 19053-205, Presidente Prudente, São Paulo, Brazil. [3]Centro Nacional de Pesquisa de Soja, Empresa Brasileira de Pesquisa Agropecuária, Londrina, Paraná, Brazil. *Author for correspondence. E-mail: dhenriques@codasp.sp.gov.br*

ABSTRACT. To determine the chemical composition of intercropping and monoculture cultivars of sorghum, millet and midget-guandu, we determined the production of dry matter (DM), ashes, gross fiber, ethereal extract, gross protein (GP), nutritious digestive total, and extractive without oxygen. The experimental design used was completely random blocks with four repetitions split into 18 treatments: monoculture sorghum, monoculture midget-guandu, monoculture millet, intercropping sorghum and midget-guandu, millet and midget-guandu and millet and sorghum, with cutting times at 30, 60 and 90 days after sowing (DAS). Millet had the highest production of dry biomass at 30 and 60 DAS in monoculture and intercropping. The highest DM was obtained by intercropping (90 DAS). Sorghum intercropping had the highest crude protein at 30 DAS, and the midget-guandu promoted increased crude protein at intercropping.

Keywords: consortium of cultures, productivity, legume, grass.

Introduction

No-tillage (NT) systems have begun to be adopted in various regions in Brazil, but with little knowledge of the cover crops that are able to produce the amount of dry matter (DM) sufficient for the system and thus to maintain or increase the soil fertility and productivity of crops. Therefore, it is necessary to determine the correct way to implement the system, studying the cultivation of grasses and legumes as cover crops.

Low natural soil fertility, a limiting factor in the productivity and sustainability of tropical pastures (TIRITAN et al., 2008), can promote nutrient deficiency, especially of nitrogen (N). An increase in the supply of N can be obtained by increasing the organic matter (OM) content in the soil by gradually releasing N mineralization and introducing microorganisms to the soil (SANTOS et al., 2011), through the application of N fertilizer or the use of legumes in association with grasses or through atmospheric N fixation. N fixed by legumes increases the quality of the diet (PACIULLO et al., 2003) and benefits livestock (EUCLIDES et al., 1998). The contribution is made indirectly by the transfer of fixed N to grass, which increases the carrying capacity of pasture and extends its productive capacity (VIERA VARGAS et al., 1995). Another advantage of legumes is their lower seasonal variation in nutritional value compared with grasses (KLUSMANN, 1988).

The intercropping of legumes and grass species, provided they are compatible in terms of plant development, is of great benefit to production systems through outputs such as increase in biomass production, biological fixation of atmospheric N and quality of forage for animal feed. According to Giacomini et al. (2004), to protect the soil and add N, the consortium between plant species coverage should provide an output of DM whose C/N ratio is intermediate with that of species isolated in cultures, providing a longer period of ground cover and synchronization between the supply and demand of N by crops.

Amado et al. (2000) experimented with cover crops in the autumn / winter cultivation period to identify the best combinations of species and proportions of seeds for the implementation of intercropping. They attested to the efficiency of the cultivation of winter grasses and legumes at supplying N immobilization or less for the main crop coupled with the durability of the straw layer on the surface of the NT.

Cover crops can also be used in animal feed, especially when grown in the fall / winter, coinciding with the shortage of pasture. According to Mello and Nornberg (2004), crop–livestock integration is a practice that has contributed to the economic viability of farms and the NT system. Thus, the cultivation of cover crops with high biomass production and tolerance to water stress is essential to enable the rotation of crops in NT and reform the areas of pasture.

Crop areas support livestock through food production, whether in the form of grain, silage or hay as forage available for grazing; they increase the efficiency of land use; they permit the sale of animals in the off-season; and they provide a better distribution of income during the year. Therefore, species with rapid growth and intense regrowth are favored because they may be used as fodder as a first stage and, after regrowth, in the form of straw for the NT system in the subsequent summer season (ALVARENGA et al., 2003).

According to Heinrichs et al. (2001), the main advantages of intercropping cover crops compared with cultivation alone, are the greater accumulation of nutrients and yield of DM, depletion of the available soil N by the grass in intercropping and the stimulation of N_2 fixation by legumes. Additionally, the presence of grasses and legumes in the mix adds to the soil biomass, with a C/N ratio intermediate to that of isolated cultivated grasses, while providing protection and a supply of soil N to the crops in succession.

Sorghum has been grown especially in regions where droughts occur frequently, limiting the production of fodder. In addition, after harvesting of the original crop, the sorghum plant root system remains alive, allowing regrowth at adequate temperatures and soil moisture (BOTELHO et al., 2010). According to Scheffer-Basso et al. (2004), another summer forage crop that can fill gaps in cultivation is autumn millet, which originated from Africa and is widespread throughout the country. In this context, midget-guandu (*Cajanus cajan*), which is adapted to the tropical environment and able to fix atmospheric N, can be used to cover the soil and for crops for forage or for grain production (SILVEIRA et al., 2005).

Much of the meat and milk production in Brazil is based on the use of pasture, with very low levels of animal productivity as a consequence of the quality of forage species used, low soil fertility and inadequate management of pastures. Thus, there is a need for studies that contribute to reversing this situation. This study aimed to evaluate the effect of both intercropping and of monoculture crops of sorghum, millet and midget-guandu in the production of dry biomass content and the behavior of crude fiber (CF), crude protein (CP), total digestive nutrients (TDNs) and non-nitrogen extract (NNE).

Material and methods

The experiment was conducted in the agricultural area of the Universidade do Oeste Paulista - Unoeste in Presidente Prudente city, São Paulo state, from February to May in the 2006 agriculture year. The climate, according to the Köppen classification, is CWA, having a tropical rainy season and a well-defined hot climate from September to March and dry winter conditions with mild temperatures from April to August.

The experimental area soil was classified as dystrophic (EMBRAPA, 2006), with a wavy soft relief and good drainage. The experimental area had been cultivated for over ten years with conventional tillage, with corn in summer and fallow in winter. Samples were collected for the characterization of chemical attributes (RAIJ et al., 2001) and granulometric analysis (EMBRAPA, 1997) at a depth of 0 to 20 cm with the following results: pH ($CaCl_2$ 1 mol L^{-1}) 5.9; 18 g dm^{-3} OM; 16 mg dm^{-3} P_{resin}; 27 $mmol_c$ dm^{-3} H+Al; 1.2 $mmol_c$ dm^{-3} K; 38 $mmol_c$ dm^{-3} Ca; 12 $mmol_c$ dm^{-3} Mg; 52 $mmol_c$ dm^{-3} sum of bases (SB); 69 $mmol_c$ dm^{-3} of cation exchange capacity; 74% base saturation (V); 740 g kg^{-1} sand; 80 g kg^{-1} silt; and 180 g kg^{-1} clay.

Each plot consisted of eight plant rows that were 5.0 m long and spaced 0.40 m apart. Monoculture cover crop species were sown in eight rows, and intercropped species were sown in alternate rows.

A randomized complete block design was adopted with four replicates in a split-plot with 18 treatments: monoculture sorghum, monoculture midget-guandu, monoculture millet, intercropping sorghum and midget-guandu, millet and midget-guandu and millet and sorghum, with cutting times at 30, 60 and 90 days after sowing (DAS). The plants contained in a meter length in the four sub-samples per experimental plot were cut close to the soil surface. A distance of 0.5 m was left as a boundary between the times of collection and at the ends of the plots.

Tilling was performed in February, with plowing and harrowing, on the spontaneously occurring vegetation of the area. Then, furrows were dug using a mechanized planter, and 200 kg ha^{-1} of 8-28-16 fertilizer was added with the manual sowing of sorghum (*Sorghum bicolor* x *Sorghum sudanensis* cv. BRS-800), millet (*Penissetum glaucum* cv. BN-2) and midget-guandu (*Cajanus cajan* cv. anão).

At 17 days after sowing, the weeds in the rows of the cover crops were removed by hand. At 30 DAS, further manual weeding was undertaken. The plant stands at four randomly chosen points in the useful area of the plots were quantified as follows: sorghum: 761,000 stems ha^{-1} (± 91,320); midget-guandu: 422,000 stems ha^{-1} (± 33,760); and millet: 1,140,000 stems ha^{-1} (± 159,600). During counting, all structures that characterize stem tillers and/or mother plants were considered. The cuts were made at 30, 60 and 90 DAS.

After cutting, the samples were dried in an oven with forced ventilation at 55°C for 72 hours until constant weight. The samples were then milled in a Wiley type mill and passed through sieves with mesh holes a millimeter wide for the evaluation of DM, mineral matter (MM), CP, ether extract (EE) and CF, according to the techniques of the Association of Official Analytical Chemists, described by Hortws (1975).

The micro-Kjeldahl method was used for the determination of N. The N content was multiplied by a factor of 6.25 to give the CP content, which was corrected for DM. The non-nitrogen extract (NNE) was obtained by calculating NNE = 100 - (% moisture +% CP +% CF +% EE +% MM). Because the TDN was obtained by calculating the total carbohydrate (TC), where TC = 100 - (% CP + EE + 5% ⋆ Ashes) and TDN% = CP + CF + 2.25 ⋆ EE + TC. Statistical analysis consisted of analysis of variance and a t-test at a 5% significance level to compare the means of the experimental treatments.

Results and discussion

The findings for plant shoot dry biomass over three seasons are evaluated in Table 1. Significant results were found for the different species evaluated in the three seasons. At 30 and 60 days, millet produced the greatest amount of dry matter in both cropping and intercropping with midget-guandu and sorghum. At 90 days, sorghum showed a large increase in DM yield, but the highest yields were obtained in intercropping. According to Neumann et al. (2008), the DM content rises faster in hybrid sorghum, medium or low (if BRS 800), due to the greater participation of panicles in the total dry matter.

Using intercropping, and considering the total output of DM in the consortium at 60 days, the highest value was obtained with the millet+sorghum. At 90 days, the millet+sorghum ratio was reduced, though the millet remained as the main constituent responsible for total DM production in the three periods evaluated. Likewise, Andreola et al. (2000) and Perin et al. (2004) stated that in grass and legume intercropping the grass contributes to relatively high amounts of biomass, which are characterized by a high C/N ratio and which increase the persistence of land cover over time. However, high biomass may cause problems in subsequent cultivation due to the temporary immobilization of N in the process of the biological decomposition of straw.

The results indicate that with respect to DM production for the use of grazing or silage, the behavior of species and associations are very different. The millet at 30 days shows monoculture spot grazing, a feature that was maintained for up to 60 days. Millet intercropped with midget-guandu showed the same behavior as a monoculture millet crop, although in this case 30 days grazing would most likely cause major problems for midget-guandu productivity due to its low dry matter yield.

The sorghum culture showed the highest DM yield at 90 days, both in the monoculture crop and intercropping with midget-guandu and pearl millet. The millet presented its maximum DM production at 60 days in the monoculture cropping system, while the DM yield was higher at 90 days in intercropping with midget-guandu and sorghum, demonstrating the importance of the addition of culture intercropping DM. Therefore, the higher millet DM for silage was 60 days in cropping and 90 days in intercropping (Table 1).

Table 1. Dry matter of plants shoots of the midget-guandu, millet and sorghum at 30, 60 and 90 days after sowing (DAS), in monoculture and intercropping.

Species Coverage	Species 1			Species 2			Total	
	------------------------ (kg ha^{-1}) --------------------------							
				(30 Days After Sowing - DAS)				
Midget-guandu (monoculture)	-			-			144	b
Millet (monoculture)	-			-			2573	a
Sorghum (monoculture)	-			-			699	b
M.guandu (S1) + Millet (S2)	109	b B	(3%)	3362	a A	(97%)	3471	a
M.guandu (S1) + Sorghum (S2)	305	b A	(35%)	575	b A	(65%)	880	b
Millet (S1) + Sorghum (S2)	3009	a A	(90%)	347	b B	(10%)	3356	a
				(60 Days After Sowing - DAS)				
Midget-guandu (monoculture)	-			-			1397	e
Millet (monoculture)	-			-			6318	c
Sorghum (monoculture)	-			-			4085	d
M.guandu (S1) + Millet (S2)	501	e B	(7%)	7130	bc A	(93%)	7631	ab
M.guandu (S1) + Sorghum (S2)	1003	e B	(21%)	3768	d A	(79%)	4771	d
Millet (S1) + Sorghum (S2)	6894	bc B	(82%)	1538	e C	(18%)	8432	a
				(90 Days After Sowing - DAS)				
Midget-guandu (monoculture)	-			-			3278	d
Millet (monoculture)	-			-			5585	c
Sorghum (monoculture)	-			-			7086	b
M.guandu (S1) + Millet (S2)	1349	e C	(16%)	6902	b B	(84%)	8251	a
M.guandu (S1) + Sorghum (S2)	2785	d C	(30%)	6452	bc B	(70%)	9237	a
Millet (S1) + Sorghum (S2)	6279	bc B	(71%)	2507	d C	(29%)	8786	a
Causes of variation				F calculated values				
Cover crop				83.19**				
Collect time				321.25**				
Culture x Time				9.04**				
Interaction DMS (kg ha^{-1})				1150				
CV (%)				20.98				

S1: Midget-guandu and millet in consortium. S2: millet and sorghum in intercropping. The following averages for the same letter in the line do not differ by the Tukey test at 5% of probability, within each collect time. Lower case letters compare columns and upper letters compare lines. Percentages in parentheses represent the amounts of biomass of each species in relation to total production in intercropping.

The sorghum and midget-guandu intercropping showed low DM production at 30 and 60 days compared with the DM production of millet, so they are not recommended for grazing or cultivation beds (Table 1). Teixeira et al. (2005) evaluated the production of the dry biomass of millet, beans and pork in monoculture cropping and intercropping dwarf pea at 119 days after sowing and observed that the lowest yield was obtained with the midget-guandu monoculture crop. They also noted that for the consortium of millet and beans with pork, the contribution of the species to the production of total dry biomass was very similar, unlike the consortium of millet and midget-guandu where dry millet contributed the most, because millet is a grass with higher DM production potential.

The results for the CP content of the plant shoots in the three seasons are shown in Table 2. Meaningful results were found for the different species evaluated in the three seasons. At 30 days, the sorghum+midget-guandu and sorghum+millet intercropping and the millet and midget-guandu monoculture showed CP contents higher than the millet and midget-guandu intercropping and sorghum monoculture (Table 2). Therefore, at 30 days, millet presenting a high amount of DM and had high levels of CP, so it can be used as cover plants for grazing.

The millet and midget-guandu intercropping had a lower CP content at 30 days than midget-guandu and the millet monoculture, and midget-guandu with sorghum and millet with sorghum, but had higher crude protein content than sorghum monoculture (Table 2). At 60 DAS, the highest CP content was observed in the midget-guandu monoculture, followed by midget-guandu+millet and midget guandu + sorghum intercropping.

A lower protein content was also observed in the monoculture sorghum in comparison with intercropping sorghum and midget-guandu and millet monoculture.

At 90 DAS, the midget-guandu continued to have the highest CP content, followed by millet+midget-guandu and sorghum+midget-guandu intercropping. The midget-guandu and sorghum intercropping had a higher CP content than the sorghum monoculture. A similar situation occurred with the millet. The sorghum monoculture had the lowest CP content.

At 60 and 90 DAS, the midget-guandu monoculture, followed by midget-guandu intercropped with sorghum or millet, always had the highest CP content. This effect, in the presence of midget-guandu (protein source), even resulted in a lower amount of DM, possibly increasing the CP. Thus, midget-guandu crops intercropped with millet and sorghum may be an option for raising CP levels, especially when the object is to produce silage or grazing.

Table 2. Gross protein content measured at 30, 60 and 90 days after sowing of the monoculture and intercropping of the midget-guandu, millet and sorghum.

Species Coverage	Days After Sowing (DAS)						
	30		60		90		Average
	(Gross Protein Content, in % DM)						
Midget-guandu	14.19	ab A	14.67	a A	12.15	a B	13.67
Millet	14.64	a A	9.70	c B	7.48	c C	10.61
Sorghum	11.25	c A	8.89	c B	6.24	d C	8.79
M.-guandu + Millet	13.19	b A	11.95	b B	9.16	b C	11.43
M.guandu + Sorghum	15.08	a A	11.09	b B	8.76	b C	11.64
Millet + Sorghum	14.36	a A	9.53	c B	6.39	cd C	10.09
Average	13.79		10.97		8.36		
	F calculated values						
Causes of variation	Gross Protein Content						
Cover crop	32.54**						
Collect time	176.75**						
Culture x Time	6.38**						
CV (%)	7.42						

The following averages for the same letter in the line do not differ by the Tukey test at 5% of probability. Lower case letters compare columns and upper letters compare lines. *and **significant at the 5% and 1% level of probability, respectively. ns: not significant.

Similarly, the consortium of pasture and legumes improved the CP, neutral detergent fiber and acid detergent fiber according to Schaefer et al. (2010) in experiments with different systems of forage cultures constituted by *Pennisetum americanum*, *Arachis pintoi* and *Stylosanthes guianensis* associated with *Cynodon dactylon* and both isolated species. Vitor et al. (2008), working with legume species intercropped with grasses, also reported that the consortium, in general, presented a greater forage CP concentration compared with signalgrass fertilized with nitrogen.

Millet has been the object of evaluation for use in animal feed (KOLLET et al., 2006; RIBEIRO et al., 2004; ROMAN et al., 2008). In addition to its resistance to drought, high temperatures and acidic soils, millet retains good development in soils with low OM content (GUIMARÃES JÚNIOR et al., 2009) and is promising for many Brazilian soils. The CP content of pearl millet is higher than corn, so there is a reduction in amino acids in the diet. According to Bastos et al. (2002), millet grain has an amino acid content similar to wheat, barley and rice. Data from the NRC (1998) show a CP content of 11.1% for millet and 8.3% for maize. Rostagno et al. (2000) evaluated the chemical composition of pearl millet and maize obtained from experiments performed in Brazil and found higher values of CP (12.08 vs. 8.57%), fat (4.19 vs. 3.46%) and gross energy (3980 vs. 3933 kcal kg^{-1}) for maize. These values are close to those obtained in the experiment and to those obtained by Adeola and Orban (1995) and Lawrence et al. (1995). Gomes et al. (2006), evaluating the CP content of different parts of the sorghum plant at 112 DAS, observed a higher CP content in the panicle (9.72%) compared with the leaves (6.31%) and stems (1.70%).

Table 3 presents the results for the crude fiber (GB) content of the shoots at 60 and 90 days. Significant responses were presented by the different species in the two periods evaluated. At 60 days, midget-guandu, pearl millet and midget-guandu+sorghum had the highest concentrations of crude fiber. At 90 days, the highest GB concentrations were observed in millet+midget-guandu and sorghum+millet. However, the lowest GB concentration was observed in the sorghum monoculture. Midget-guandu and sorghum showed a reduction in GB from 60 to 90 DAS. The results show that the average GB content was similar to that observed 60 and 90 DAS, and the production of monoculture and consortium was satisfactory for the production of forage high in fiber. Similar results were observed for Leonel et al. (2008).

The results of total digestible nutrients (TDN) of the shoots at 60 and 90 DAS are shown in Table 4. Significant responses were found for the different species in the two periods evaluated. At 60 days, midget-guandu, because being a legume, presented the highest TDN concentrations. However, millet was the crop with the lowest TDN concentration. At 90 days, the highest value was also observed for midget-guandu and the lowest for millet + sorghum.

Table 3. Gross fiber (GB) content measured at 60 and 90 days after sowing of the monoculture and intercropping of the midget-guandu, millet and sorghum.

Species Coverage	Days After Sowing (DAS)				
	60 dias		90 dias		Average
	(Gross Fiber Content, in % DM)				
Midget-guandu	28.46	abc A	26.70	c B	27.58
Millet	28.62	ab A	29.88	b A	29.25
Sorghum	27.81	bc A	24.56	d B	26.18
M.guandu + Millet	27.14	c B	31.77	a A	29.45
M.guandu + Sorghum	29.44	a A	26.02	c B	27.75
Millet + Sorghum	27.50	bc B	31.55	a B	29.53
Average	28.16		28.42		
Causes of variation	F calculated values				
	Gross Fiber Content				
Cover crop	16.84**				
Collect time	0.95				
Culture x Time	29.81**				
CV (%)	3.28				

The following averages for the same letter in the line do not differ by the Tukey test at 5% of probability. Lower case letters compare columns and upper letters compare lines. *and **significant at the 5% and 1% level of probability, respectively. ns: not significant.

The results show that from 60 to 90 DAS, midget-guandu also accumulated total digestible nutrients when the plant was at the height of its production, increasing dry matter and nutrient accumulation. For millet, there was a decrease in the TDN concentration from 60 to 90 DAS, indicating the plant was undergoing the process of senescence, decreasing the nutrient ratio to the shoots and thereby decreasing the chemical quality (millet crop intercropping).

Table 4. Total digestive nutrients (TDN) at 60 and 90 days after sowing of monoculture and intercropping of the midget-guandu, millet and sorghum.

Species Coverage	Days After Sowing (DAS)				
	60		90		Average
	(TDN Contents in % DM)				
Midget-guandu	56.05	a B	58.09	a A	57.07
Millet	51.93	c A	50.29	c B	51.11
Sorghum	53.41	bc B	55.51	b A	54.46
M.guandu + Millet	54.45	b A	49.12	c B	51.79
M.guandu + Sorghum	53.82	b B	55.48	b A	54.65
Millet + Sorghum	53.55	b A	47.28	d B	50.42
Average	53.87		52.63		
Causes of variation	F calculated values				
	TDN Contents				
Cover crop	42.82**				
Collect time	15.06**				
Culture x Time	23.68**				
CV (%)	2.08				

The following averages for the same letter in the line do not differ by the Tukey test at 5% of probability. Lower case letters compare columns and upper letters compare lines. *and **significant at the 5% and 1% level of probability, respectively. ns: not significant.

The ethereal extract (EE) results at 60 and 90 DAS are shown in Table 5. There were significant responses in the different species in the two periods evaluated. The midget-guandu and the millet at 60 days in monoculture (no intercropping) had the highest EE percentage. At 90 DAS, the midget-guandu monoculture already presented the highest EE values, which represent higher energy density. With intercropping, the results obtained at 60 DAS with midget-guandu+sorghum were higher than with grass monoculture (sorghum).

Table 5. Ethereal extract (EE) contents at 60 and 90 days after sowing of the monoculture and intercropping of the midget-guandu, millet and sorghum.

Species Coverage	Days After Sowing (DAS)				
	60		90		Average
	(EE Contents in % DM)				
Midget-guandu	2.64	ab B	3.52	a A	3.08
Millet	2.21	ab A	2.20	b A	2.20
Sorghum	1.96	b A	2.09	b A	2.03
M.guandu + Millet	2.39	ab A	2.19	b A	2.29
M.guandu + Sorghum	2.69	a A	2.59	b A	2.64
Millet + Sorghum	2.32	ab A	2.02	b A	2.17
Average	2.37		2.43		
Causes of variation	F calculated values				
	EE Contents				
Cover crop	5.44**				
Collect time	0.24^{ns}				
Culture x Time	1.62^{ns}				
CV (%)	19.68				

The following averages for the same letter in the line do not differ by the Tukey test at 5% of probability. Lower case letters compare columns and upper letters compare lines. *and **significant at the 5% and 1% level of probability, respectively. ns: not significant.

The results of NNE are presented in Table 6. Meaningful results were found for the different species in the two periods evaluated. The sorghum monoculture at 60 DAS did not differ in the production of NNE, but at 90 DAS, the sorghum intercropped with midget-guandu had the highest response in the production of NNE when compared with the consortium or monoculture crops. The MM results at 60 and 90 DAS are shown in Table 7.

Table 6. Non-nitrogen extract contents (NNE) at 60 and 90 days after sowing of the monoculture and intercropping of the midget-guandu, millet and sorghum.

Species Coverage	Days After Sowing (DAS)				
	60		90		Average
	(NNE Content in % DM)				
Midget-guandu	46.19	c B	51.47	cd A	48.83
Millet	51.21	b B	53.81	c A	52.51
Sorghum	54.31	a B	61.06	a A	57.68
M.guandu + Millet	50.06	b A	49.87	e A	49.97
M.guandu + Sorghum	50.14	b B	57.22	b A	53.68
Millet + Sorghum	50.11	b A	52.13	cd A	51.12
Average	50.34		54.26		
Causes of variation	F calculated values				
	NNE Contents				
Cover crop	36.12**				
Collect time	83.74**				
Culture x Time	7.63**				
CV (%)	2.84				

The following averages for the same letter in the line do not differ by the Tukey test at 5% of probability. Lower case letters compare columns and upper letters compare lines. *and **significant at the 5% and 1% level of probability, respectively. ns: not significant.

Table 7. Mineral matter (MM) contents at 60 and 90 days after sowing of the monoculture and intercropping of the midget-guandu, millet and sorghum.

Species Coverage	Days After Sowing (DAS)				
	60		90		Average
	(MM contents in % DM)				
Midget-guandu	8.25	a A	6.22	bc B	7.24
Millet	8.17	a A	6.60	abc B	7.38
Sorghum	6.88	b A	5.90	c B	6.39
M.guandu + Millet	8.76	a A	7.13	ab B	7.94
M.guandu + Sorghum	6.65	b A	5.99	c A	6.32
Millet + Sorghum	8.88	a A	7.36	a B	8.12
Average	7.93		6.53		
Causes of variation	F calculated values				
	MM Contents				
Cover crop	11.23**				
Collect time	57.73**				
Culture x Time	1.19^{ns}				
CV (%)	8.80				

The following averages for the same letter in the line do not differ by the Tukey test at 5% of probability. Lower case letters compare columns and upper letters compare lines. *and **significant at the 5% and 1% level of probability, respectively. ns: not significant.

Significant responses to the different species in the two seasons evaluated were observed. At 60 DAS of monoculture, millet and midget-guandu had a higher response to the production of minerals. At 90 DAS, sorghum + millet and midget-guandu+millet intercropping and millet monoculture showed higher results than midget+sorghum intercropping and sorghum monoculture in MM production. The percentage of ash (DM) did not differ between species and the collection times. Such distinctions may be related to the method of nutrient uptake, both as best in the rooting soil volume used, as in the increased capacity of cultivars to exploit mineral nutrients in this experiment. The way as the phosphate fertilizer was located in the phosphate fertilizer court system may have provided better utilization of this nutrient, which is extremely important for the establishment of roots, favoring the uptake of nutrients by the plant and enriching the forage with nutrients, expressed as ash in DM.

Conclusion

Millet had the highest production of dry biomass at 30 and 60 DAS with monoculture and intercropping. The highest DM were obtained with intercropping (90 DAS). Sorghum intercropping had the highest crude protein at 30 DAS, and midget-guandu promoted increased crude protein with intercropping.

References

ADEOLA, O.; ORBAN, J. I. Chemical composition and nutrient digestibility of pearl millet (*Pennisetum glaucum)* fed to growing pigs. **Journal of Cereal Science**, v. 22, n. 2, p. 177-184, 1995.

ALVARENGA, R. C.; CABEZAS, W. A. L.; CRUZ, J. C.; SANTANA, D. P. Plantas de cobertura de solo para sistema plantio direto. **Informe Agropecuário**, v. 22, n. 208, p. 25-36, 2003.

AMADO, T. J. C.; MIELNICZUK, J.; FERNANDES, S. B. V. Leguminosas e adubação mineral como fontes de nitrogênio para o milho em sistemas de preparo do solo. **Revista Brasileira de Ciência do Solo**, v. 24, n. 1, p. 179-189, 2000.

ANDREOLA, F.; COSTA, L. M.; OLSZEVSKI, N.; JUCKSCH, I. A cobertura vegetal de inverno e a adubação orgânica e, ou mineral influenciando a sucessão feijão / milho. **Revista Brasileira de Ciência do Solo**, v. 24, n. 4, p. 867-874, 2000.

BASTOS, A. O.; LANDELL FILHO, L. C.; PASSIPIERI, M.; BASTOS, J. F. P. Diferentes níveis de grão de milheto (*Pennisetum glaucum* (L.) R. Brown) na alimentação de suínos. **Revista Brasileira de Zootecnia**, v. 31, n. 4, p. 1753-1760, 2002.

BOTELHO, P. R.; PIRES, D. A. A.; SALES, E. C. J.; ROCHA JÚNIOR, V. R.; JAYME, D. G.; REIS, S. T. Avaliação de genótipos de sorgo em primeiro corte e rebrota para produção de silagem. **Revista Brasileira de Milho e Sorgo**, v. 9, n. 3, p. 287-297, 2010.

EMBRAPA-Empresa Brasileira de Pesquisa Agropecuária. Centro Nacional de Pesquisa de Solos. **Manual de métodos de análise de solo**. 2. ed. Rio de Janeiro: Embrapa, 1997.

EMBRAPA-Empresa Brasileira de Pesquisa Agropecuária. Centro Nacional de Pesquisa de Solos. **Sistema brasileiro de classificação de solos**. 2. ed. Rio de Janeiro: Embrapa Solos, 2006.

EUCLIDES, V. P. B.; MACEDO, M. C. M.; OLIVEIRA, M. P. Produção de bovinos em pastagens de *Brachiaria* spp. consorciadas com *Calopogonium mucunoides* nos cerrados. **Revista Brasileira de Zootecnia**, v. 27, n. 2, p. 238-245, 1998.

GIACOMINI, S. J.; AITA, C.; CHIAPINOTTO, I. C.; HÜBNER, A. P.; MARQUES, M. G.; CADORE, F. Consorciação de plantas de cobertura antecedendo o milho em Plantio Direto. II - nitrogênio acumulado pelo milho e produtividade de grãos. **Revista Brasileira de Ciência do Solo**, v. 28, n. 4, p. 751-762, 2004.

GOMES, S. O.; PITOMBEIRA, J. B.; NEIVA, J. N. M.; CÂNDIDO, M. J. D. Comportamento agronômico e composição químico-bromatológico de cultivares de sorgo forrageiro no Estado do Ceará. **Revista de Ciências Agronômicas**, v. 37, n. 2, p. 221-227, 2006.

GUIMARÃES JÚNIOR, R.; GONÇALVES, L. C.; RODRIGUES, J. A. S. **Utilização do milheto para produção de silagem**. Planaltina: Embrapa Cerrados, 2009. (Documentos 259).

HEINRICHS, R.; AITA, C.; AMADO, T. J. C.; FANCELLI, A. L. Cultivo consorciado de aveia e ervilhaca: relação C/N da fitomassa e produtividade do milho em sucessão. **Revista Brasileira de Ciência do Solo**, v. 25, n. 2, p. 331-340, 2001.

HORTWS, W. **Official methods of analyses of association of the official analytical chermist**. 17th ed. Washington, D.C.: AOAC, 1975.

KLUSMANN, C. Trees and shrubs for animal production in tropical and subtropical areas. **Plant Research and Development**, v. 27, n. 1, p. 92-104, 1988.

KOLLET, J. L.; DIOGO, J. M. S.; LEITE, G. G. Rendimento forrageiro e composição bromatológica de variedades de milheto (*Pennisetum glaucum* (L.) R. BR.). **Revista Brasileira de Zootecnia**, v. 35, n. 4, p. 1308-1315, 2006.

LAWRENCE, B. V.; ADEOLA, O.; ROGLER, J. C. Nutrient digestibility and growth performance of pigs fed pearl millet as a replacement for corn. **Journal of Animal Science**, v. 73, n. 7, p. 2026-2032, 1995.

LEONEL, F. P.; PEREIRA, J. C.; COSTA, M. G.; MARCO JÚNIOR, P.; LARA, L. A.; SOUSA, D. P.; SILVA, C. J. Consórcio capim-braquiária e soja, produtividade das culturas e características qualitativas das silagens. **Revista Brasileira de Zootecnia**, v. 37, n. 11, p. 2031-2040, 2008.

MELLO, R.; NORNBERG, J. L. Fracionamento dos carboidratos e proteínas de silagens de milho, sorgo e girassol. **Ciência Rural**, v. 34, n. 5, p. 1537-1542, 2004.

NEUMANN, K.; OST, P. R.; PELLEGRINI, L. G.; DEFAVERI, F. J. Comportamento de híbridos de milho (*Zea mays*) e sorgo (*Sorghum bicolor*) para silagem na região centro-sul do Paraná. **Ambiência**, v. 4, n. 2, p. 237-250, 2008.

NRC-National Research Council. **Nutrient requeriments of swine**. 20th ed. Washington D.C.: National Academy Press, 1998.

PACIULLO, D. S. C.; AROEIRA, L. J. M.; ALVIM, M. J.; CARVALHO, M. M. Características produtivas e qualitativas de pastagem de braquiária em monocultivo e consorciada com estilosantes. **Pesquisa Agropecuária Brasileira**, v. 38, n. 3, p. 421-426, 2003.

PERIN, A.; SANTOS, R. H. S.; URQUIAGA, S.; GUERRA, J. G. M.; CECON, P. R. Produção de fitomassa, acúmulo de nutrientes e fixação biológica de nitrogênio por adubos verdes em cultivo isolado e consorciado. **Pesquisa Agropecuária Brasileira**, v. 39, n. 1, p. 35-40, 2004.

RAIJ, B.; ANDRADE, J. C.; CANTARELLA, H.; QUAGGIO, J. A. **Análise química para avaliação da fertilidade de solos tropicais**. Campinas: IAC, 2001.

RIBEIRO, C. V. D. M.; PIRES, A. V.; SIMAS, J. M. C.; SANTOS, F. A. P.; SUSIN, I.; OLIVEIRA JÚNIOR, R. C. Substituição do grão de milho pelo milheto (*Pennisetum americanum*) na dieta de vacas holandesas em lactação. **Revista Brasileira de Zootecnia**, v. 33, n. 5, p. 1351-1359, 2004.

ROMAN, J.; ROCHA, M. G.; GENRO, T. C. M.; SANTOS, D. T.; FREITAS, F. K.; MONTAGNER, D. B. Características produtivas e estruturais do milheto e sua relação com o ganho de peso de bezerras sob suplementação alimentar. **Revista Brasileira de Zootecnia**, v. 37, n. 2, p. 205-211, 2008.

ROSTAGNO, H. S.; ALBINO, L. F. T.; DONZELE, J. L.; GOMES, P. C.; FERREIRA, A. S.; OLIVEIRA, R. F.; LOPES, D. C. **Composição de alimentos e exigências nutricionais de aves e suínos**: tabelas brasileiras. Viçosa: UFV, 2000.

SANTOS, D. H.; SILVA, M. A.; TIRITAN, C. S.; FOLONI, J. S. S.; ECHER, F. R. Qualidade tecnológica da cana-de-açúcar sob adubação com torta de filtro enriquecida com fosfato solúvel. **Revista Brasileira de Engenharia Agrícola e Ambiental**, v. 15, n. 5, p. 443-449, 2011.

SCHAEFER, P. E.; PIZZANI, R; LOVATO, T.; GOULART, R. Z.; LUDWIG, L. R. Produção de fitomassa da parte aérea e ciclagem de nutrientes de diferentes sistemas forrageiros. **Revista Enciclopédia Biosfera**, v. 6, n. 1, p. 1-8, 2010.

SCHEFFER-BASSO, S. M.; AGRANIONIK, H.; FONTANELI, R. S. Acúmulo de biomassa e composição bromatológica de milhetos das cultivares comum e africano. **Revista Brasileira de Agrociência**, v. 10, n. 4, p. 483-486, 2004.

SILVEIRA, P. M.; BRAZ, A. J. B. P.; KLIEMANN, H. J.; ZIMMERMANN, F. J. P. Acumulação de nutrientes no limbo foliar de guandu e estilosantes. **Revista Agropecuária Tropical**, v. 35, n. 3, p. 133-138, 2005.

TEIXEIRA, C. M.; CARVALHO, G. J.; NETO, A. E. F.; ANDRADE, M. J. B.; MARQUES, E. L. S. Produção de biomassa e teor de macronutrientes do milheto, feijão-de-porco e guandu-anão em cultivo solteiro e consorciado. **Ciência e Agrotecnologia**, v. 29, n. 1, p. 93-99, 2005.

TIRITAN, C. S.; FOLONI, J. S. S.; SANTOS, D. H.; SATO, A. M.; DOMINGUES, W. L. Resposta a calagem da *Brachiaria brizantha* submetida a diferentes doses de adubação nitrogenada. **Colloquium Agrariae**, v. 4, n. 2, p. 18-26, 2008.

VIERA VARGAS, M. S.; SOUTO, C. M.; URQUIAGA, S.; BODDEY, R. M. Quantification of the contribution of N2 fixation to tropical forage legumes and transfer to associated grass. **Soil Biology and Biochemistry**, v. 27, n. 1, p. 1193-1200, 1995.

VITOR, C. M. T.; FONSECA, D. M.; MOREIRA, L. M.; FAGUNDES, J. L.; JÚNIOR, D. N.; JÚNIOR, J. I. R.; PEREIRA, A. L. Rendimento e composição química do capim-braquiária introduzido em pastagem degradada de capim-gordura. **Revista Brasileira de Zootecnia**, v. 37, n. 12, p. 2107-2114, 2008.

26

Technical and economic viability and nutritional quality of mombasa guinea grass silage production

Fernando Shintate Galindo[1*], Tatiane Beloni[2], Salatiér Buzetti[1], Marcelo Carvalho Minhoto Teixeira Filho[1], Elisângela Dupas[1] and Mariana Gaioto Ziolkowski Ludkiewicz[1]

*[1]Departamento de Fitossanidade, Engenharia Rural e Solos, Faculdade de Engenharia, Universidade Estatual Paulista "Julio de Mesquita Filho", Rua Monção, 830, 15385-000, Ilha Solteira, São Paulo, Brazil. [2]Escola Superior de Agricultura "Luiz de Queiroz", Universidade de São Paulo, Piracicaba, São Paulo, Brazil. *Author for correspondence. E-mail: fs.galindo@yahoo.com.br*

ABSTRACT. The objective of this study was to evaluate the viability of pasture with high production capacity and the potential for production of silage using different sources and doses of nitrogen (N) by quantifying the productivity of dry matter (DM), some qualitative characteristics, the total cost of production (TC) and total operating cost (TOC) of silage production. The experimental design was randomized block in a factorial scheme with four replications including two sources of nitrogen (urea and ammonium nitrate) and five N doses (0, 50, 100, 150, and 200 kg ha^{-1} by cutting). The increase in N provided greater DM and crude protein content and reduced the fiber content in neutral detergent. The source of N did not affect the chemical composition, and therefore, it is recommended to use urea due to the ease of acquisition and higher economic return. The production of Mombasa silage can be economically viable due to high DM and crude protein content combined with low production costs. The smallest TC and TOC were obtained in the absence of nitrogen fertilization; however, the proper management of N ensures high productivity in cattle feed with higher total profit when using 100 kg ha^{-1} N by cutting with ammonium nitrate.

Keywords: nitrogen fertilization; chemical analysis; production cost; tropical grass; *Panicum maximum*.

Introduction

Brazil has approximately 180 million hectares of pastures and is one of the largest global commercial producers of cattle, which are dependent on pastures (Silva et al., 2009a). Thus, pastures in Brazil are extremely important for the production of beef, and pastures occupy a broad extension the territory in the country. Pastures are considered the base of cattle feed, since they are a less costly means to produce animal protein for human consumption (Fernandes, Buzetti, Dupas, Teixeira Filho, & Andreotti, 2015). This low cost is possible because of climatic factors favoring the production of fodder in different locations and periods of the year. However, forage plants usually do not receive fertilization and over the years lose their developmental potential, reducing quality and productivity (Bennett, Buzetti, Silva, Bergamaschine, and Fabricio, 2008); pasture degradation then occurs after a few years (Silveira et al., 2010).

Several species are used in pasture formations in Brazil, and among them, Mombasa is widely used in breeding and rearing of cattle. The grass species *Panicum maximum* (syn. *Megathyrsus maximus*) presents one of the greatest known potentials for production of dry matter (DM) in subtropical and tropical environments, and can reach annual dry matter production of approximately 33 t ha^{-1} (Freitas et al., 2007). Proper management of soil fertility and knowledge of the nutritional requirements of this grass are extremely important for the practice of pasture management and result in higher productivity and availability of food for animals (Silva et al., 2009b; Barth Neto et al., 2010; Silveira et al., 2010; Pereira et al., 2012).

The use of fertilizer, particularly nitrogen, can significantly increase the production of fodder, allowing greater capacity and resulting in increased production of milk and meat per unit area. Furthermore, nitrogen fertilization may also improve production efficiency of grasses, which allows alternating the use of fodder and silage, the latter for use in the dry season. Food scarcity during the winter period, especially in the Cerrado region, has led producers to store food with higher nutritional value produced in more favorable conditions (spring and summer) in the form of silage for animal nutrition in the period when food is lacking (autumn and winter) (Costa, Andreotti, Bergamaschine, Lopes, & Lima, 2015). In this context, the preservation of green fodder in the form of silage is a strategy to cope with the pasture shortage problem in the dry season (Oliveira et al., 2014). Another important aspect, as discussed by Penteado et al. (2007), is that production potential for tropical grass silage increases with the intensification of the production system, warranting interest in additional research on the subject.

Currently, increased competitiveness in the livestock market in relation to agricultural land use and in seeking the best returns on the capital invested, complementary production technologies have been sought to provide benefits that can be integrated into the production system as a whole. Thus, the use of stored grass silage in the dry season from the summer production surplus has resulted in efficiency gains in pasture management and has minimized feeding costs for livestock due to a lower cost per ton of DM obtained with these silages (Coan et al., 2008).

Given the above, there is a need for viable alternatives to increase the efficiency and utilization of nitrogen fertilization, thereby allowing reduction of N use in the production system without decreasing production or quality of pasture, and to maximize dry matter production of forage during the year. However, there are few studies in the literature that show the economic effects of nitrogen fertilization using different sources and doses on the production of grass silage. The objective of this work was to evaluate the productivity and chemical composition of Mombasa Guinea grass dry matter in silage production, as well as the economic viability of this preservation technique, according to different sources and doses of N in the Northeast Paulista region.

Material and method

The experiment was conducted without irrigation in the region of Ilha Solteira, São Paulo State, Brazil, on the left bank of the Paraná river (20°21' S and 51°22' W). The study site has an altitude of 326 m in an area previously occupied by *P. maximum* cv. Mombasa under grazing (since 2006). The soil is classified as ultisol with sandy texture (Embrapa, 2013). Total precipitation during the experimental period was 1,320 mm, average air humidity was 70%, and average temperature was 25.3°C with average minimum and maximum temperatures ranging from 19.8 to 32.2°C. The climate type is Aw according to Koppen system, which is characterized as humid tropical with a rainy season in the summer and a dry season in the winter. Figure 1 shows the average weather data for each cut madc.

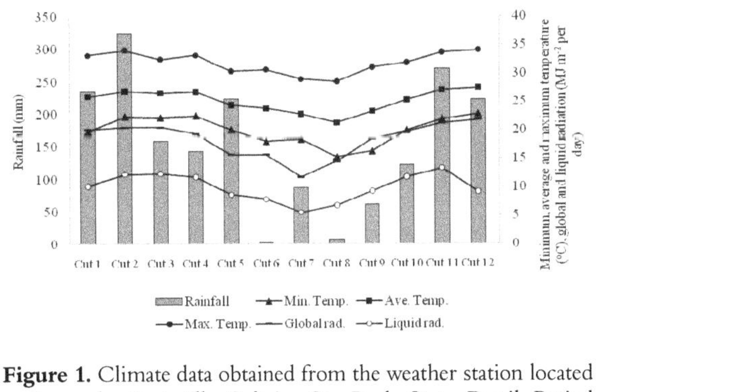

Figure 1. Climate data obtained from the weather station located at the study site in Ilha Solteira, São Paulo State, Brazil. Period December 2012 to January 2014.

Soil samples were collected at a depth of 0-20 cm before the cutting of grass in 2012 to determine the chemical characteristics of the soil, with the following results: P resin = 13 mg dm^{-3}, pH $CaCl_2$ = 5.2; K, Ca,

Mg, Al + H, SB, and cation exchange capacity (CEC) = 2.2, 35.0, 7.0, 16.0, 44.2, and 60.2 $mmol_c\ dm^{-3}$, respectively, organic matter (O.M.) = 25 g dm^{-3} and V = 73%.

We utilized a randomized block experimental design with ten treatments and four replicates and with two N sources: urea (45% of N) as the most commonly used nitrogen fertilizer and the type likely to be volatilized and ammonium nitrate (32% of N) as a comparative source. Five N doses were used for each N source (0, 50, 100, 150, and 200 kg ha^{-1}, totaling 0, 300, 600, 900, and 1,200 kg ha^{-1} for a period) in a 2 x 5 factorial scheme (two sources and five doses). Each plot area was 6.0 m^2 (3 x 2 m) with 2 m distance between plots.

Grass was cut at 25 cm above the soil at initial fertilization, and further growth for the evaluation of cuts was made on July 7, 2012. The other cuts were made from December 2012 to January 2014. At every third cut, 60 kg ha^{-1} of K_2O was applied. Due to the number of cuts (12) and for better interpretation of the data, the sections were divided into periods of the year: the dry season (4/20/2013, 5/25/2013, 6/28/2013, 8/03/2013, 9/26/2013, and 11/1/2013) and the rainy season (12/16/2012, 1/5/2013, 2/16/2013, 3/26/2013, 12/11/2013, and 1/18/2014).

Samples of the aerial portion of Mombasa Guinea grass were collected manually at 25 cm above soil in the center of the plot. Plots of 0.5 m^2 were delimited using a metallic square (1.0 x 0.5 m), and an iron frame was used to cut grass at intervals based on height, varying by approximately 90 to 100 cm. Overall, there was almost one cut per month, except for the first cut due to low rainfall that led to lower plant growth in that period. The harvested forage was packed in paper bags and subsequently dried in a forced-circulation oven at approximately 65°C for 72 hours. The samples were then weighed to quantify the DM (dry matter) produced in the representative area and ground in a Wiley Mill equipped with a screen with a 1 mm sieve for subsequent analysis.

After each cut, the N doses were topdress-applied to each plot. The DM of Mombasa Guinea grass was calculated based on the amount of green matter (kg m^{-2}), and the original amount of dry matter was obtained. Fiber content in neutral detergent (NDF) and acid detergent fiber (ADF) were measured following the methodologies of Casali et al. (2009) and Van Soest (1994). The nitrogen content (NT, %) was determined by the semimicro-Kjeldahl analytical method, and protein content (CP) was obtained by multiplying the content by 6.25 following Silva and Queiroz (2006).

Data were submitted to analysis of variance (F-test). We used a Tukey test for comparisons of means for N sources and regression analysis for N doses. Statistical analyses were performed using the SISVAR program (Ferreira, 2011).

After verification of bromatological and production results (e.g., high PMS and satisfactory bromatological values), an economic analysis was performed using an analytical structure based on the total operating cost (TOC) of production according to the Institute of Agricultural Economics (IEA) and following Matsunaga et al. (1976). It consisted of calculating the sum of operating expenses, including transactions, inputs (fertilizers, seeds, pesticides etc.), labor, machinery and irrigation, called the effective operational cost (EOC). In addition to the TOC, in this work, we considered other expenses and costs of interest (5% of EOC) (Matsunaga et al., 1976), resulting in a total operating cost (TOC) that was extrapolated to one hectare, which increased the compensation to the fixed capital, being considered 5% of the TOC; from the remuneration of the land, we obtained the total cost of production (CT) according to the methodology proposed by Martin, Serra, Oliveira, Ângelo, and Okawa (1998).

The total profit (PT) was calculated according to the total revenue obtained (RT) by subtracting the total cost of production (CT) in R$ ha^{-1}. We considered for basic calculations feed conversion values in kg of DM intake kg^{-1} LWG (weight gain) = 9; i.e., it is necessary to for an animal to eat 9 kg DM to gain 1 kg of live weight in animals showing initial live weight of 13 kg @ and an expected weight gain near 1 kg per day (adapted from Coan et al., 2008 and Difante et al., 2010) and with a price of @ = R$ 133.00, which is the mean value from the previous 3 years (IEA, 2016).

The data presented here were obtained under experimental conditions; however, to determine the economic factors, the values were extrapolated to real field conditions according to the practices typically employed by producers in the region. For Mombasa silage, a simulation of production costs was carried out on the basis of production data and bromatological results for the grass. The costs of mechanized operations were obtained from Agrianual (2015) and Anualpec (2015) and were adjusted to reflect the average rent values practiced in the region including the tractor driver and the cost of fuel and lubricants, as well as depreciation of machinery and equipment. The technical coefficients and unit values used were obtained from the technical and regional producers working with these crops and with similar levels of technology. For other expenses, we considered the average price spent in the study area adjusted to those prevailing in commercial crops in São Paulo State for the agricultural year of 2014/2015.

Result and discussion

In general, the nitrogen sources urea and ammonium nitrate did not differ in the levels of ADF, CP and accumulated DM regardless of the evaluation period, except for NDF content in the dry season with nitrate ammonium, which was lower compared to urea (Table 1). The results agree with those found by Silveira et at. (2015) working with the N sources ammonium nitrate, ammonium sulfate, urea, urea treated with Agrotain, Super N and ammonium sulfonitrate and under different N doses (0, 60, and 120 kg ha^{-1} per year) in bahiagrass (*Paspalum notatum*). Our results also differed from those of Fernandes et al. (2015) in a study with urea, ammonium sulfonitrate, ammonium nitrate, ammonium sulfate and sulfammo at doses of 0, 50, 100, 150, and 200 kg ha^{-1} in Mombasa Guinea grass, in which the authors found no differences between the sources in productivity and chemical composition.

The N doses influenced the NDF and CP levels in the dry and rainy seasons; however, ADF content was not affected by N dose regardless of the period (Table 1).

With regard to the NDF in the dry period, the value was fit to the second-degree equation with a minimum point at approximately 155 kg ha^{-1} (Figure 2A). In the rainy season, it fit decreasing linear function; i.e., N dose increments up to 200 kg ha^{-1} per cut reduced NDF content (Figure 2C). Bennett, Buzetti, Silva, Bergamaschine, and Fabricio (2008) also found that the NDF decreased with increasing N doses, and according to the authors higher doses of N applied at certain times, depending on environmental conditions, can change the NDF content of the forage, since the magnitude of the effect on NDF is related to climate and soil conditions, management of species used and achieved productivity. Further, according to Magalhães et al. (2012) and Martuscello et al. (2015), increased levels of applied N can increase the ratio between leaves and stems and cause competition for higher active photosynthetic rate and nutritional value since the leaves present lower fiber content and higher content of crude protein compared to the stem. Furthermore, according to Pereira et al. (2012), the possible increase in the specific leaf area with N fertilization is related to the lower leaf thickness at high doses of N, where under conditions of greater N inside the plant, leaf lignification tends to be lower. For ADF content, fodder with values approximately 40% or more have low consumption and digestibility. The results obtained in this study show that Mombasa grass as fodder has good composition of FDA, with less than 40% of dry matter irrespective of the dose and source of N.

Table 1. NDF, ADF, CP and accumulated DM of Mombasa guinea grass due to doses and sources of N per season. Ilha Solteira, São Paulo State, Brazil, 2012-2014.

		Dry season		
Doses of N	NDF	ADF	CP	Accumulated DM
---kg ha^{-1}---		---------% D.M.---------		---kg ha^{-1}---
0	68.25*	35.40^{ns}	11.53**	9,358**
50	67.42	34.83	12.70	10,518
100	67.21	35.12	13.35	11,423
150	65.19	33.29	14.74	11,988
200	67.22	35.55	14.14	11,780
Mean	67.06	34.84	13.29	11,013
C.V. (%)	0.98	2.75	7.19	12.69
Sources of N				
Urea	67.48 a	34.67 a	12.91 a	10,938 a
Ammonium nitrate	66.63 b	35.00 a	13.68 a	11,088 a
L.S.D. (5%)	0.66	0.97	0.97	906.5
		Rainy season		
	NDF	ADF	CP	Accumulated DM
Doses of N				
---kg ha^{-1}---		---------% D.M.---------		---kg ha^{-1}---
0	69.73*	39.46^{ns}	9.55**	21,948**
50	69.42	39.78	10.34	23,985
100	69.62	39.77	10.86	25,775
150	67.49	37.80	12.88	28,368
200	68.34	38.51	13.86	27,295
Mean	68.92	39.06	11.50	25,474
C.V. (%)	1.62	3.64	8.90	10.76
Sources of N				
Urea	69.08 a	38.84 a	11.06 a	25,150 a
Ammonium nitrate	68.76 a	39.29 a	11.94 a	25,798 a
L.S.D. (5%)	1.13	1.44	1.04	1,778.2

Means followed by the same letter in the column do not differ by Tukey test at 5% probability. ** Mean $p < 0.01$; Mean * $0.01 < p < 0.05$; ns: not significant.

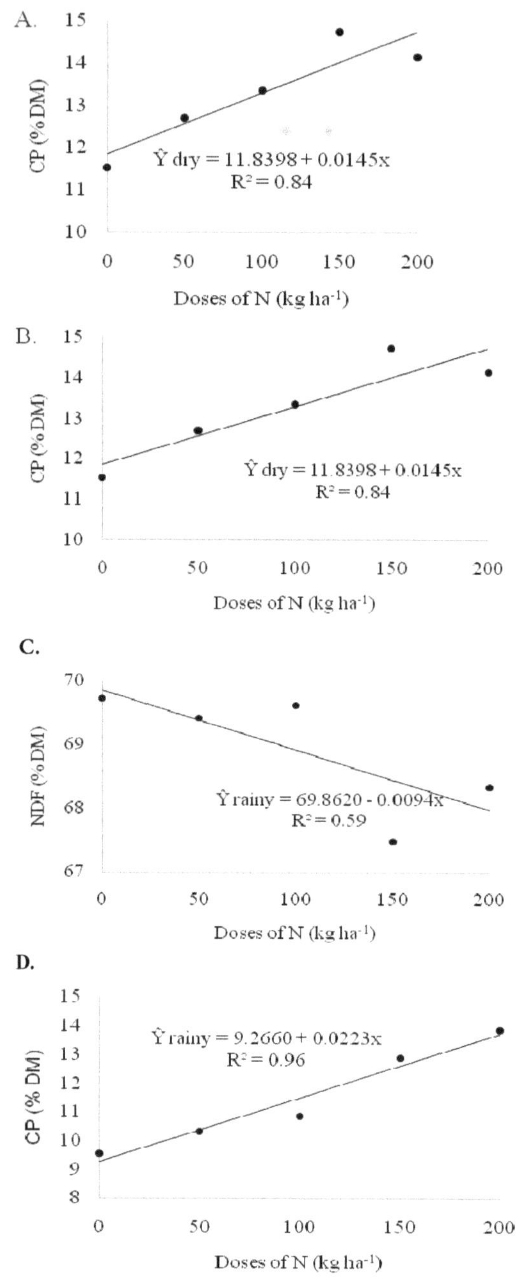

Figure 2. NDF of mombasa guinea grass according to N doses in the dry season (A); CP of mombasa guinea grass according to N doses in the dry season (B); NDF of mombasa guinea grass according to N doses in the rainy season (C); CP of mombasa guinea grass according to N doses in the rainy season (D). Ilha Solteira, São Paulo State, Brazil, 2012-2014.

Regarding the CP, in both the rainy and dry seasons, there was an adjustment to the increasing linear function, i.e., increased N doses led to an increase in CP content (Figure 2B and D). This corroborates the work of Silva et al. (2009b) in a study of N doses (0, 100, 300, and 500 kg ha^{-1} of N) in Mombasa Guinea grass showing that in the dry and rainy seasons, increased N resulted in increased CP content, with values ranging from 6.5 to 10.6%. Nitrogen fertilization increased the concentration of N in Mombasa, which is directly related to the increase in CP content. According to Van Soest (1994), CP content less than 7% in fodder reduces digestibility because of inadequate nitrogen levels for ruminal microorganisms, reducing their populations and consequently reducing the digestibility and intake of dry matter. Thus, higher levels of CP are necessary to meet the protein requirements of animals. Crude protein content in this study was above 7% and thus should not negatively affect rumen microorganism populations, preserving digestibility and dry matter intake.

The interaction between N dose and source was significant for DM accumulated in both the dry season and the rainy season (Tables 2 and 3). It was adjusted to an increasing linear function for dose for both urea and ammonium nitrate (Figure 3A and B).

Table 2. Interaction between doses and sources of N related to the productivity of dry matter accumulation of mombasa guinea grass in the dry season in Ilha Solteira region, São Paulo State, Brazil.

Sources	Doses of N (kg ha^{-1})				
	0	300	600	900	1,200
Urea**	9,820 a	10,540 a	10,745 a	12,025 a	11,560 a
A.N.**	8,895 a	10,495 a	12,100 a	11,950 a	12,000 a
L.S.D. (5%)			2,027		

Means followed by the same letter in the column do not differ by Tukey test at 5% probability. ** Significant $p < 0.01$, * significant $0.01 < p < 0.05$, ns: not significant.

Table 3. Interaction between doses and sources of N related to the productivity of dry matter accumulation of mombasa guinea grass in the rainy season in Ilha Solteira region, São Paulo State, Brazil.

Sources	Doses of N (kg ha^{-1})				
	0	300	600	900	1,200
Urea**	21,835 a	23,585 a	24,215 a	28,075 a	28,040 a
A.N.**	22,060 a	24,385 a	27,335 a	28,660 a	26,550 a
L.S.D.(5%)			3,976		

Means followed by the same letter in the column do not differ by Tukey test at 5% probability. ** Significant $p < 0.01$, * significant $0.01 < p < 0.05$, ns: not significant.

The increase in DM in the control (0 kg ha^{-1} of N) for the higher dose (1,200 kg ha^{-1} of N) in the dry and rainy seasons was 2,422 kg ha^{-1}, 5,347 kg ha^{-1}, and 1,425 kg ha^{-1}, representing 5.88 and 24.36%, respectively. On average, the DM in the rainy season was 131.31% higher than in the dry period, varying from 11,013 to 25,474 kg ha^{-1}, respectively, and bringing to light the problems related to lack of forage for cattle in the dry season, which is the main limiting factor for livestock in Central Brazil.

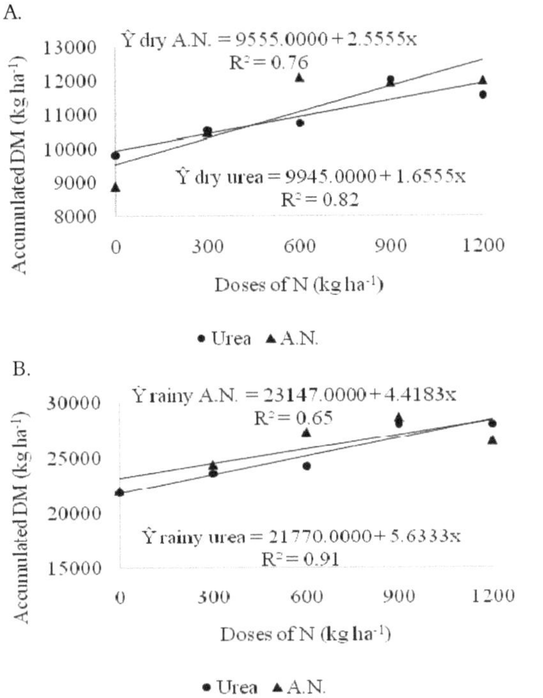

Figure 3. Breakdown of interaction between doses and sources of N in accumulated DM of mombasa guinea grass in the dry period (A); unfolding of the interaction between doses and sources of N in accumulated DM mombasa guinea grass in the rainy season (B).

The results for DM in the rainy season were higher than those found by Paziani et al. (2009) and were obtained by evaluating the DM of corn for silage, with an average yield of 18,690 kg ha^{-1}. These yields can be considered above average compared to the values obtained in practice, thus keeping high productive potential of green matter and silage dry matter per unit area, which is the advantage of using silage grass compared to corn and sorghum silage.

The DM produced due to the nitrogen fertilization was considered satisfactory to ensure stability of the pasture and animal production since the values were above 1,600 kg ha^{-1} by cutting, a value cited by Fernandes et al. (2015) as sufficient to ensure satisfactory forage consumption. However, this amount may vary according to the stocking dose used and period or season, which explains the magnitude of responses to nitrogen fertilization among different cuts. According to Paziani et al. (2009), the most desirable features in silage crop are high production of green and dry matter in both the silage point and in physiological maturity, and high concentrations of crude protein and energy factors that ensure greater digestibility of forage. Thus, the high yields obtained in this study are due to the technology adopted in these production systems, with proper fertilization to meet the requirements of Mombasa.

Considering the technical results (e.g., high DM, good NDF, ADF, and CP) combined with the durability, low risk of loss and greater flexibility in harvesting as well as the possibility of exclusive supply or as bulk for cattle complementary feed during seasonal food shortages, it is justifiable to analyze the economic feasibility of Mombasa Guinea grass silage production, mainly due to the need for livestock farming to become more competitive by reducing costs and increasing productivity.

Initial investment in terms of soil preparation, liming and forage deployment were not considered in this study, since these practices were not carried out in the undergrazed area with Mombasa already deployed. This contributed to the reduction of initial cost of deployment for silage. To evaluate the economic performance of silage production, the total and actual operating costs recorded in the experiment were extrapolated to one hectare.

Investments with additives to assist in lowering the pH and fermentation of the silage were not considered in this study, as these techniques are not adopted by producers of grass silage in the Ilha Solteira, São Paulo State region, mainly due to the increase in production costs and controversial results in relation to efficiency in improving the quality of silage grass surplus. However, it is noteworthy that this technique, when used rationally on specific areas such as those intended only for silage, may be a viable alternative compared to acquisition of foreign silage or the adoption of risk deferral. Use of the technique may also influence the fermentation process by favoring conservation and improving the nutritional value, reducing surface losses and the exposed layer of silage, increasing the useful lifetime, increasing the value energy, and improving fiber and dry matter digestibility, collectively resulting in improved animal performance (Ferreira Júnior, Paulino, Possenti, & Lucenas, 2009).

Table 4 shows that spending was highest on mechanized operations, followed by fertilizers, corresponding to 68.46 and 19.51% of the TOC, respectively. This result is in agreement with Costa et al. (2015) in a study of the cost of silage production in the Cerrado region in the 2010/2011 and 2011/2012 seasons, in which the authors applied urea at dose of 200 kg ha^{-1}, and also found higher detachment with mechanized operations, corresponding to more than 57% of the TOC. It is noteworthy that with increasing nitrogen doses and change in N source, there is a tendency toward increase in the percentage of spending in relation to the TOC of fertilizers.

Table 4. Total operating costs structure model of Mombasa guinea grass silage for the treatment control (zero kg ha^{-1} N in top-dressing, as urea form), per hectare. Ilha Solteira, São Paulo State, Brazil, 2012 until 2014.

Description	Specification[1]	Times	Coefficient	Unitary Value (R$)	Total Value (R$)
		A. Operations			
Hoeing	HM*	6.00	1.00	65.00	390.00
Top dressing	HM	6.00	0.30	65.00	117.00
Potassium fertilization	HM	2.00	0.30	65.00	39.00
Transport	HM	6.00	0.80	70.00	336.00
Compaction	HM	6.00	1.00	60.00	360.00
Subtotal A					1,242.00
		B. Agricultural input			
Potassium Chloride	T	2.00	0.10	1,770.00	354.00
Urea	T	6.00	0.00	1,780.00	0.00
Plastic canvas	m^2	1.00	0.50	160.00	80.00
Subtotal B					434.00
Effective operating costs (EOC)					1,676.00
Other expenses					83.80
Interest cost					54.47
Total operating cost (TOC)					1,814.27
Remuneration of fixed capital					90.71
Remuneration of land					850
Total cost (TC)					2,754.98

* Area with mombasa guineagrass already deployed. * HM = hour machine. 2014/2015 average exchange rate: R$ 2.97 = U$ 1.00.

Garcia et al. (2012), working with the corn crop submitted to different consortia with forage for silage in Selvíria region, MS, verified higher costs related to inputs with corn seeds (46.46%), followed by fertilization costs (39.01%), and higher operating costs of the center pivot sprinkler irrigation (45.28%) with yields ranging between 6,300 and 8,220 kg ha^{-1} of corn using urea as a source of nitrogen fertilizer at a dose of 100 kg ha^{-1}. This demonstrates the main technical advantage of the production of grass silages, with the reduction in spending on inputs and high production per area, making grasses a strategically interesting food as nutritional reserves due to lower cost compared to other additional bulky options. Additionally, Jobim, Sarti, Santos, Branco, and Cecato (2006), in a study of resulting animal performance and economic viability of elephant grass silage as a replacement for corn silage for dairy cows, concluded that elephant grass silages showed potential for production and chemical composition of milk equivalent to corn silage when given to cows in the intermediate period of lactation. The deficiencies of elephant grass silage compared to corn silage can be offset by the lower cost of production, showing that this culture has great potential for exploitation.

The costs of nitrogen fertilizer per period, season and cut with increasing doses of N ranged from 38.29 to 68.27% of the TOC for urea and 41.74 to 70.88% for ammonium nitrate. Regarding the COT and CT (Table 5), the highest values refer to the treatment using ammonium nitrate at a dose of 1200 kg ha^{-1} of N per period, 600 kg ha^{-1} per station or 200 kg ha^{-1} per cut. The lowest value for the TOC and CT corresponds to the treatments without nitrogen fertilization (0 kg ha^{-1}). However, it is worth noting that soil N reserves will be depleted, as occurs in the extraction of the nutrient, if this is not reset to the ground, jeopardizing the productivity of pastures over time. The TOC of treatments without nitrogen fertilization (dose 0 kg ha^{-1}) had the same values regardless of the source. This occurred because in the absence of nitrogen fertilizer, the relative cost is equal to zero for both sources.

The possibility of dry matter production and storage to feed, whether in the form of silage or hay, should consider possible improvements to chemical, physical and biological soil caused by fertilizer in addition to the possible improvement of nutritional value in the formed pasture for feeding of ruminants. Additionally, based on the technical results, consideration of the effect of nitrogen fertilization is essential to guarantee high DM with sufficient levels of crude protein and fiber, improving the quality and quantity of forage production in Mombasa.

Another important aspect relates to conserved forage acquisition costs, which can be higher compared to the production of silage grass remnants. Moreover, techniques used by farmers for deferral can be viable alternatives in this area, because rains and rotting will eventually occur, and there may also be a risk of fires. Typically, deferred pastures have high quantities of fodder but with low quality, and during the deferral period, much of the vegetative tillers (without inflorescences) develop on reproductive tillers (with inflorescences), and these are placed in the category of dead tillers. At this time, there is also a reduction in the amount of green leaf and an increase in the mass of dry leaves, dried stems and green stalks in the pasture (Santos, Fonseca, Gomes, Balbino, & Magalhães, 2010), in addition to problems with layering, which directly affect forage quality.

Table 5. Total operating costs and total cost of mombasa guinea grass silage production according to sources and doses of N. Ilha Solteira, São Paulo State, Brazil, 2012-2014.

	Period			
	Urea		A.N.#	
Doses	TOC	TC	TOC	TC
kg ha^{-1}	(R$)			
0	1,814.27	2,754.98	1,814.27	2,754.98
300	3,098.84	4,103.78	3,309.94	4,325.44
600	4,383.40	5,452.57	4,805.61	5,895.89
900	5,667.97	6,801.37	6,301.28	7,466.34
1,200	6,952.54	8,150.16	7,796.94	9,036.79
	Season			
	Urea		A.N.#	
Doses	TOC	TC	TOC	TC
kg ha^{-1}	(R$)			
0	907.14	1,802.49	907.14	1,802.49
150	1,549.43	2,476.89	1,654.97	2,587.72
300	2,191.70	3,151.29	2,402.80	3,372.94
450	2,833.99	3,825.68	3,150.64	4,158.17
600	3,476.27	4,500.08	3,898.47	4,943.40
	Cut			
	Urea		A.N.#	
Doses	TOC	TC	TOC	TC
kg ha^{-1}	(R$)			
0	302.38	1,167.50	302.38	1,167.50
50	516.47	1,392.30	551.66	1,429.24
100	730.57	1,617.10	800.93	1,690.98
150	944.66	1,841.90	1,050.21	1,952.72
200	1,158.76	2,066.69	1,299.49	2,214.47

#Ammonium nitrate 2014/2015 average exchange rate: R$ 2.97 = U$ 1.00.

Table 6. Total profit production mombasa guinea grass silage according to sources and doses of N. Ilha Solteira, São Paulo State, Brazil, 2012-2014.

	Total profit				
	R$ ha^{-1}				
	Doses of N (kg ha^{-1}) per period				
Sources	0	300	600	900	1,200
Urea	18,756.54	19,131.81	18,403.69	20,857.70	19,474.43
A.N.#	18,978.21	19,698.30	21,034.15	20,769.07	17,119.88
	Doses of N (kg ha^{-1}) per season				
Sources	0	150	300	450	600
Ureia	9,378.27	9,565.91	9,201.85	10,428.85	9,737.22
A.N.#	9,489.10	9,849.15	10,517.07	10,384.53	8,559.94
	Doses of N (kg ha^{-1}) per cut				
Sources	0	50	100	150	200
Ureia	3,126.09	3,188.64	3,067.28	3,476.28	3,245.74
A.N.#	3,163.00	3,283.05	3,505.69	3,461.51	2,853.31

#Ammonium nitrate.

Even with lower production costs in the absence of nitrogen fertilizer, management with nitrogen provided higher total profit from the production of silage. Although there were no significant differences between N sources for the technical parameters assessed, the use of ammonium nitrate at a dose of 600 kg ha^{-1} per period, 300 kg ha^{-1} per season, and 100 kg ha^{-1} per cut increased profitability compared to urea by R$ 176.45 (0.84%) and R$ 2,055.94 (9.77%) compared to the highest profit from urea (dose of 900, 450 and 150 kg ha^{-1} for a period, season and cut, respectively) and the control (without N) (Table 6, Figure 4).

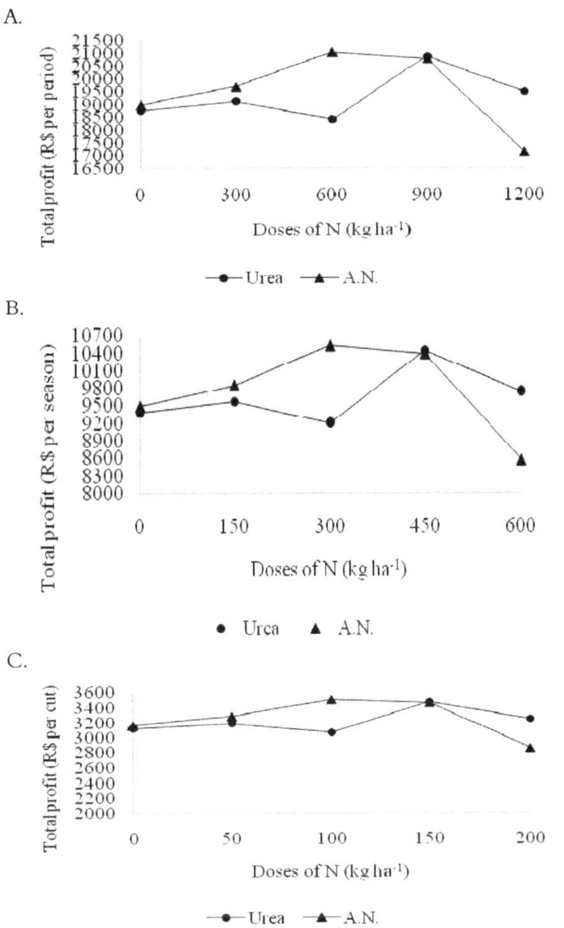

Figure 4. Total profit obtained on the basis of production mombasa guinea grass silage per period (A); Total profit on the production of mombasa silage per station (B), and total profit due to the production mombasa guinea grass silage cutting (C) 2014/2015 average exchange rate: R$ 2.97 = U$ 1.00.

According to Ferreira et al. (2009) and Gonçalves et al. (2010), the use of grass silage made with surplus production from summer in confinement systems in the dry season has allowed gains in efficiency in pasture management and minimization of feeding costs in confinement. This is due the lower cost per ton of DM obtained with these silages, corroborating the economic results obtained in the present work. In view of the greater competitiveness of livestock farming in relation to agriculture through land use, and aiming at better remuneration over invested capital, we have sought complementary production technologies that provide integrated benefits to the production system as a whole (Monteiro, Abreu, Cabral, Ribeiro, & Reis, 2011). Thus, as suggested by Vieira et al. (2010), the agronomic management of crop to obtain satisfactory nutritive value and with high-yielding pastures would justify the use of this forage as a preserved food, allied to the process of intensification of meat and milk production in pastures.

Looking at economies scaled for profit per area (R$ ha^{-1}), especially in places where the cost of land is high, silage becomes an interesting alternative due to the direct impact on productivity per hectare, where bulky foods with high capacity have the tendency to dilute costs of implementation, fertilizers and management in general.

Thus, although corn and sorghum crops are the most widely used for the production of silage, their production costs have been the limiting factor for use by farmers. This highlights the importance of studies concerning the feasibility, costs, and advantages of the production of silage grasses. In addition, the commonly used forage types are subject to seasonal production, resulting in considerable quantitative and qualitative deficiency of forage in the dry period in the Cerrado. Silage produced in the summer can thus serve as a high nutritional value food in the winter period.

Conclusion

Nitrogen doses positively influenced crude protein and dry matter yield of Mombasa Guinea grass and negatively influenced the fiber content in neutral detergent in both seasons up to a dose of 200 kg ha^{-1} by cutting. The nitrogen sources did not differ in technical evaluations.

The production of Mombasa silage can be economically viable due to the high productivity of dry matter and crude protein and low production costs in an established pasture. The lowest total cost and total operating cost were achieved in the absence of nitrogen fertilization; however, the proper management of nitrogen ensures high productivity and yield in cattle feed.

The use of ammonium nitrate source at a dose of 100 kg ha^{-1} by cutting provided higher total profit from silage Mombasa production.

References

Agrianual. (2015). *Anuário da Agricultura Brasileira*. São Paulo, SP: AgraFNP.

Anualpec. (2015). *Anuário da Pecuária Brasileira*. São Paulo, SP: Instituto FNP.

Barth Neto, A., Boleta, V. S., Pancera Júnior, E. J., Almeida, G. M., Canto, M. W., Gasparino, E., & Baltazar, L. F. (2011). Nitrogênio e época de colheita nos componentes da produtividade de forragem e sementes de capim-mombaça. *Pesquisa Agropecuária Brasileira, 45*(11), 1312-1320. doi: 10.1590/S0100-204X2010001100012

Benett, C. C. S., Buzetti, S., Silva, K. S., Bergamaschine, A. F., & Fabricio, J. A. (2008). Produtividade e composição bromatológica do capim-marandu a fontes e doses de nitrogênio. *Ciencia e Agrotecnologia, 32*(5), 1629-1636. doi: 10.1590/S1413-70542008000500041

Casali, A. O., Detmann, E., Valadares Filho, S. C., Pereira, J. C., Cunha, M., Detmann, K. S. C., & Paulino, M. F. (2009). Estimação de teores de componentes fibrosos em alimentos para ruminantes em sacos de diferentes tecidos. *Revista Brasileira de Zootecnia, 38*(1), 130-138. doi: 10.1590/S1516-35982009000100017

Coan, R. M., Reis, R. A., Resende, F. D. D., Sampaio, R. L., Schocken-Iturrino, R. P., Garcia, G. R., & Berchielli, T. T. (2008). Viabilidade econômica, desempenho e características de carcaça de garrotes em confinamento alimentados com dietas contendo silagem de capins tanzânia ou marandu ou silagem de milho. *Revista Brasileira de Zootecnia, 37*(2), 311-318. doi: 10.1590/S1516-35982008000200018

Costa, N. R., Andreotti, M., Bergamaschine, A. F., Lopes, K. S. M., & Lima, A. E. L. (2015). Custo da produção de silagens em sistemas de integração lavoura-pecuária sob plantio direto. *Revista Ceres, 62*(1), 9-19. doi: 10.1590/0034-737X201562010002

Difante, G. D. S., Euclides, V. P. B., Nascimento Júnior, D. D., Silva, S. C. D., Barbosa, R. A., & Torres Júnior, R. A. D. A. (2010). Desempenho e conversão alimentar de novilhos de corte em capim-tanzânia submetido a duas intensidades de pastejo sob lotação rotativa. *Revista Brasileira de Zootecnia, 39*(1), 33-41. doi: 10.1590/S1516-35982010000100005

Empresa Brasileira de Pesquisa Agropecuária [Embrapa]. (2013). *Sistema Brasileiro de Classificação de Solos* (3a ed.). Brasília, DF: Embrapa-CNPS.

Fernandes, J. C., Buzetti, S., Dupas, E., Teixeira Filho, M. C. M., & Andreotti, M. (2015). Sources and rates of nitrogen fertilizer used in Mombasa guineagrass in the Brazilian Cerrado region. *African Journal of Agricultural Research, 10*(19), 2076-2082. doi: 10.5897/AJAR2014.9276

Ferreira, A. C. H., Neiva, J. N. M., Rodriguez, N. M., Campos, W. E., & Borges, I. (2009). Avaliação nutricional do subproduto da agroindústria de abacaxi como aditivo de silagem de capim-elefante. *Revista Brasileira de Zootecnia, 38*(2), 223-229. doi: 10.1590/S1516-35982009000200002

Ferreira, D. F. (2011). Sisvar: a computer statistical analysis system. *Ciência e Agrotecnologia, 35*(6), 1039-1042. doi: 10.1590/S1413-70542011000600001

Ferreira Júnior, E., Paulino, V. T., Possenti, R. A., & Lucenas, T. L. (2009). Aditivos em silagem de capim elefante paraíso (*Pennisetum hybridum* cv. paraíso). *Archivos de Zootecnia, 58*(222), 185-194.

Freitas, K. R., Rosa, B., Ruggiero, J. A., Nascimento, J. L., Heineman, A. B., Macedo, R. F., ... Oliveira, I. P. (2007). Avaliação da composição químico - bromatológica do capim Mombaça (*Panicum maximum* Jacq.) submetido a diferentes doses de nitrogênio. *Bioscience Journal, 23*(3), 1-10.

Garcia, C. M. P., Andreotti, M., Tarsitano, M. A. A., Teixeira Filho, M. C. M., Lima, A. E. S., & Buzetti, S. (2012). Análise econômica da produtividade de grãos de milho consorciado com forrageiras dos gêneros *Brachiaria* e *Panicum* em sistema plantio direto. *Revista Ceres, 59*(2), 157-163. doi: 10.1590/S0034-737X2012000200002

Gonçalves, J. R. S., Pires, A. V., Susin, I., Lima, L. G. D., Mendes, C. Q., & Ferreira, E. M. (2010). Substituição do grão de milho pelo grão de milheto em dietas contendo silagem de milho ou silagem de capim-elefante na alimentação de bovinos de corte. *Revista Brasileira de Zootecnia*, *39*(9), 2032-2039. doi: 10.1590/S1516-35982010000900023

Instituto de Economia Agrícola [IEA]. (2016). *Banco de dados*. Retrieved on July 28, 2016 from http://www.iea.sp.gov.br/out/bancodedados.html

Jobim, C. C., Sarti, L. L., Santos, G. T., Branco, A. F., & Cecato, U. (2006). Desempenho animal e viabilidade econômica do uso da silagem de capim-elefante em substituição a silagem de milho para vacas em lactação. *Acta Scientiarum. Animal Sciences, 28*(2), 137-144. doi: 10.4025/actascianimsci.v28i2.849

Magalhães, J. A., Carneiro, M. S. S., Andrade, A. C., Pereira, E. S., Souto, J. S., Pinto, M. S. C., ... Mochel Filho, W. J. E. (2012). Eficiência do nitrogênio, produtividade e composição do capim-andropogon sob irrigação e adubação. *Archivos de Zootecnia*, *61*(236), 577-588.

Martin, N. B., Serra, R., Oliveira, M. D. M., Ângelo, J. A., & Okawa, H. (1998). Sistema integrado de custos agropecuários - "CUSTAGRI". *Informações Econômicas*, *28*(1), 7-28.

Martuscello, J. A., Silva, L. P., Cunha, D. N. F. V., Batista, A. C. S., Braz, T. G. S., & Ferreira, P. S. (2015). Adubação nitrogenada em capim-massai: morfogênese e produção. *Ciência Animal Brasileira*, *16*(1), 1-13. doi: 10.1590/1089-68916i118730

Matsunaga, M., Bemelmans, P. F., Toledo, P. N. E., Dulley, R. D., Okawa, H., & Pedroso, I. A. (1976). Metodologia de custo de produção utilizada pelo IEA. *Agricultura em São Paulo*, *23*(1), 123-139.

Monteiro, I. J. G., Abreu, J. G., Cabral, L. D. S., Ribeiro, M. D., & Reis, R. H. P. (2011). Silagem de capim-elefante aditivada com produtos alternativos. *Acta Scientiarum. Animal Sciences*, *33*(4), 347-352. doi: 10.4025/actascianimsci.v33i4.12629

Oliveira, E. R., Monção, F. P., Moura, L. V., Gabriel, A. M. A., Tonissi, R. H., Góes, B., ... Nascimento, F. A. (2014). Valor nutricional de silagem de capim-mombasa com aditivos agroindustriais. *Semina: Ciências Agrárias*, *35*(3), 1543-1556. doi: 10.5433/1679-0359.2014v35n3p1543

Paziani, S. F., Duarte, A. P., Nussio, L. G., Gallo, P. B., Bittar, C. M. M., Zopollato, M., & Reco, P. C. (2009). Características agronômicas e bromatológicas de híbridos de milho para produção de silagem. *Revista Brasileira de Zootecnia*, *38*(3), 411-417. doi: 10.1590/S1516-35982009000300002

Penteado, D. C. S., Santos, E. M., Carvalho, G. G. P., Oliveira J. S., Zanine, A. M., Pereira, O. G., & Ferreira, C. L. L. F. (2007). Inoculação com *Lactobacillus plantarum* da microbiota em silagem de capim-mombaça. *Archivos de Zootecnia*, *56*(214), 191-202.

Pereira, O. G., Rovetta, R., Ribeiro, K. G., Santos, M. E. R., Fonseca, D. M., & Cecon, P. R. (2012). Crescimento do capim-tifton 85 sob doses de nitrogênio e alturas de corte. *Revista Brasileira de Zootecnia*, *41*(1), 30-35. doi: 10.1590/S1516-35982012000100005

Pereira, V. V., Fonseca, D. M., Martuscello, J. A., Cecon, P. R., Santos, M. V., & Braz, T. G. S. (2012). Biomass accumulation in mombasa guineagrass plants under different levels of nitrogen supply and plant densities. *Revista Brasileira de Zootecnia*, *41*(5), 1118-1126. doi: 10.1590/S1516-35982012000500006

Santos, M. E. R., Fonseca, D. M. F., Gomes, V. M., Balbino, E. M., & Magalhães, M. A. (2010). Estrutura do capim-braquiária durante o diferimento da pastagem. *Acta Scientiarum. Animal Sciences*, *32*(2), 193-145. doi: 10.4025/actascianimsci.v32i2.7922

Silva, D. J., & Queiroz, A. C. (2006). *Análise de alimentos*: métodos químicos e biológicos (3a ed.). Viçosa, MG: UFV.

Silva, S. C. D., Bueno, A. A. D. O., Carnevalli, R. A., Uebele, M. C., Bueno, F. O., Hodgson, J., ... Morais, J. P. G. (2009a). Sward structural characteristics and herbage accumulation of *Panicum maximum* cv. Mombasa subjected to rotational stocking managements. *Scientia Agricola*, *66*(1), 8-19. doi: 10.1590/S0103-90162009000100002

Silva, A. G., França, A. F. S., Miyagi, E. S., Mello, S. Q. S., Ferreira, J. L., & Carvalho, E. R. (2009b). Frações proteicas do capim-mombasa submetido a doses de nitrogênio em duas alturas de corte. *Arquivo Brasileiro de Medicina Veterinária e Zootecnia*, *61*(5), 1148-1155. doi: 10.1590/S0102-09352009000500018

Silveira, M. C. T., Nascimento Júnior, D., Cunha, B. A. L., Difante, G. S., Pena, K. S., Silva, S. C., & Sbrissia, A. F. (2010). Effect of cutting interval and cutting height on morphogenesis and forage accumulation of guinea grass (*Panicum maximum*). *Tropical Grasslands*, *44*(2), 103-108.

Silveira, M. L., Vendramini, J. M. B., Sellers, B., Monteiro, F. A., Artur, A. G., & Dupas, E. (2015). Bahiagrass response and N loss from selected N fertilized sources. *Grass Forage Science*, *70*(1), 154-160. doi: 10.1111/gfs.12078

Van Soest, P. J. (1994). *Nutritional ecology of the ruminant*. Ithaca, US: Cornell Universtity.

Vieira, B. R., Obeid, J. A., Pereira, O. G., Valadares Filho, S. C., Carvalho, I. P. C., & Azevedo, J. A. G. (2010). Consumo, digestibilidade dos nutrientes e parâmetros ruminais em bovinos alimentados com silagem de capim-mombaça. *Arquivo Brasileiro de Medicina Veterinária e Zootecnia*, *62*(5), 1148-1157. doi: 10.1590/S0102-09352010000500018

27

Gherkin cultivation in saline medium using seeds treated with a biostimulant

Maria Lilia de Souza Neta, Francisco de Assis de Oliveira, Salvador Barros Torres*, Antonia Adailha Torres Souza, Dennis Darkyan Almeida da Silva and Sandy Thomaz dos Santos

*Centro de Ciências Agrárias, Universidade Federal Rural do Semi-Árido, Av. Francisco Mota, 572, 59625-900, Mossoró, Rio Grande do Norte, Brazil. *Author for correspondence. E-mail: sbtorres@ufersa.edu.br*

ABSTRACT. The objective of this study was to evaluate the effects of a biostimulant on gherkin seeds, cultivar Liso de Calcutá, under conditions of salt stress. A split-plot experimental design with five replications was used; the main plot was represented by two levels of irrigation water salinity (0.5 and 3.5 dS m^{-1}), and the subplot was composed of five biostimulant doses applied as a seed treatment (0, 5, 10, 15, and 20 mL kg^{-1}). The following variables were evaluated: the number of leaves; number of branches; length of the largest branch; leaf, stem, fruit and total dry matter; number of fruits; average fruit weight; and fruit production. The values of all analyzed variables decreased with an increase in irrigation water salinity regardless of the biostimulant dose, but in the absence of salt stress, biostimulant use led to an increase in the number of leaves and branches, the length of the largest branch and biomass accumulation. Biostimulation was not effective in reducing the effects of salinity on gherkin production, but it increased fruit production regardless of the salinity level.

Keywords: *Cucumis anguria*, horticulture, salt stress, bioregulator.

Introduction

The gherkin (*Cucumis anguria* L.) belongs to the Cucurbitaceae family and produces fruits that are highly appreciated in the cuisine of Brazil, especially in the Northeast Region. It is an annual and monoecious plant with an indeterminate and prostrate growth habit, lobed leaves and fruits that are highly variable in shape and in the presence or absence of spicules (Modolo & Costa, 2003).

However, the gherkin is still considered a secondary culture (Oliveira et al., 2014a), and according to Oliveira et al. (2008), it is common to find gherkin plants growing subspontaneously among other crops whose production meets the demands for domestic consumption and international markets.

One of the main technologies that is applied to ensure satisfactory results in the cultivation of agricultural products is irrigation. However, in addition to the volume of water made available to plants, another crucial factor is water quality, especially in terms of the concentration of dissolved salts (Oliveira et al., 2014b).

The salts that are present in water and soils reduce the availability of water for plants, and certain ions have specific effects. Excess ions that enter the flow of water due to transpiration injure leaves, thus reducing growth or negatively influencing the absorption of essential elements, and the result is a decrease in crop yields (Munns, 2005).

Several studies have been conducted to evaluate the effects of salinity on plant growth and yield, particularly in species of agronomic interest. In most cases, the results have shown deleterious effects of salinity on the growth and yield of plants, e.g., cucurbits such as pumpkin (Carmo et al., 2011) and zucchini (Strassburger, Peil, Fonseca, Aumonde, & Mauch, 2011).

With respect to gherkin, there have been few studies on the stages of germination and early seedling development. Oliveira et al. (2014b) found that an increase in the salinity of irrigation water caused a significant decrease in plant growth and yield.

In these studies, the results have indicated that gherkin is sensitive to salinity, highlighting the need to adopt technologies that mitigate the problem of salinity to promote better plant growth.

The use of synthetic biostimulants may be an alternative because such substances increase crop growth and development by stimulating cell division and increasing the absorption of water and nutrients by plants (Vieira & Castro, 2002).

Stimulate®, a product considered to be a plant growth promoter, is partially composed of phytohormones that act to mediate physiological processes (Garcia, Gazola, Merlin, Villas Boas, & Crusciol, 2009). In the literature, the use of this biostimulant has been reported to have several positive results on the growth and yield of crops, but most of the research has focused on grains and cereals (Garcia et al., 2009; Oliveira et al., 2015).

The use of Stimulate® in vegetable production, especially vegetable-fruit crops, has been poorly studied, but examples include research conducted by Palangana, Silva, Goto, and Ono (2012) on chili and cucumber, respectively. The results showed beneficial effects from the application of this biostimulant on the number of fruits, the amount of fresh matter and productivity.

Thus, the present study aimed to evaluate the effects of applying this biostimulant to gherkin seeds, cultivar Liso de Calcutá, and growing them under conditions of salt stress.

Material and methods

The experiment was conducted from October 2014 to January 2015 at the experimental area of the Department of Environmental Sciences and Technology of the Federal Rural University of the Semi-Árido (UFERSA), Mossoró, Rio Grande do Norte State, Brazil (5°12'02" S, 37°19'37" W; altitude 18 m).

The climate, according to the Köppen classification, is BSwh, hot and dry; the average annual rainfall is 673.9 mm, which is distributed from February to June with a very low probability of rainfall between August and December. The temperature and the average relative humidity are 27°C and 68.9%, respectively (Carmo Filho & Oliveira, 1995).

Climatic data during the experimental period are shown in Figure 1 and. Although the experiment was developed in the open, rainfall during the study had little effect on the results since even the highest rainfall volume (8 mm) represented only a small fraction of the rainfall that may have accumulated in each pot.

Material from a soil classified as eutrophic Red-Yellow Argisol with cattle manure (10% base volume) was used as the substrate. The soil characteristics are shown in Table 1.

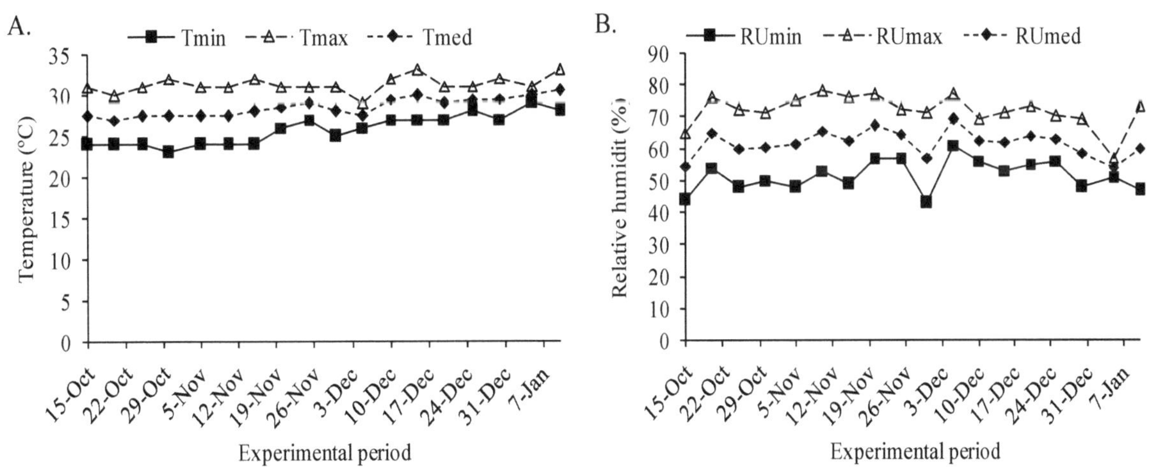

Figure 1. Maximum, average and minimum temperatures (A) and maximum, average and minimum humidity values (B) during the experimental period.

Table 1. Chemical characteristics of the soil used in the experiment.

pH	EC	P	K^+	Na^+	Ca^{2+}	Mg^{2+}	Al^{3+}	(H+Al)	SB	T	CTC	V	M	PST
	dS m^{-1}	----mg dm^{-3} ----			-----------------------			$cmol_c$ dm^{-3}	-----------------------				------ % -------	
6.18	0.06	5.23	44.52	8.61	1.62	1.88	0.0	1.32	3.65	3.65	4.97	73	0.0	1.0

*pH, in water, ratio 1:2.5, N = total nitrogen obtained by the sum of N-NO_3^- and N-NH_4^+ contents, OM = organic matter, P, K and Na extracted using Mehlich-1, N, Ca, Mg, Al and M extracted using KCl 1N.

A split-plot experiment design was used in which the main plot was represented by two levels of irrigation water salinity (0.5 and 3.5 dS m^{-1}), and the subplot included five doses of biostimulant applied via seed treatment (0, 5, 10, 15, and 20 mL kg^{-1} of seeds). The experiment was replicated five times with each replicate represented by two pots, each containing one plant.

The lowest water salinity level (0.5 dS m^{-1}) corresponded to that of the water supplied by UFERSA, and the level of 3.5 dS m^{-1} was obtained by the dissolution of sodium chloride (NaCl). Salinity was adjusted using a Tech-4MP conductivity bench meter (Tecnal®).

The water used to prepare the fertigation solutions was collected from the water supply system of the central campus of UFERSA. It had the following characteristics: pH = 8.30, EC = 0.50 dS m^{-1}, Ca^{2+} = 3.10, Mg^{2+} = 1.10, K^+ = 0.30, Na^+ = 2.30, Cl^- = 1.80, HCO_3 = 3.00, and CO_3^{2-} = 0.20 ($mmol_c$ L^{-1}).

Using the doses established for each treatment, gherkin seeds, cultivar Liso de Calcutá, were treated with a biostimulant (Stimulate®), which is composed of 0.009% kinetin, 0.005% gibberellic acid, 0.005% indolbutiric acid, and 99.981% inert ingredients (Stoller do Brasil, 1998).

The seeds were sown in pots with a capacity of 8 L, and five seeds were sown per pot. Later, plants were thinned, leaving the most vigorous individual in each pot.

The plants were grown in the open at a spacing of 1.50 x 0.75 m, resulting in a population density of 8,889 plants per hectare, and they were cultivated by staking without pruning the main branch. The culture was performed vertically using polythene strips tied to vertical trellises installed along the planting line.

Each water treatment used a separate irrigation system that comprised a PVC reservoir (300 L); a Metalcorte/Eberle EBD250076 auto-ventilated, electric circulation pump (driven by a single-phase, 210-V, 60-Hz motor, similar to that used in washing machines); 12-mm side lines; and emitters with microtubules with an average flow rate of 2.5 L h^{-1}. The irrigations were performed so that one fertigation corresponded to one irrigation event using the nutrient solution recommended by Castellane and Araújo (1994) for hydroponic melon crops (Oliveira et al., 2014a).

The irrigation blade was selected to raise the moisture in the soil to its maximum water retention capacity plus a leaching fraction of 10%. Fertigations were performed at a frequency of five events daily at two-hour intervals (8:00, 10:00, 12:00, 14:00, and 16:00 h) and were controlled by means of a model TE-2 digital timer (Decorlux®). The duration of each irrigation was adjusted throughout the experiment to meet the water requirements of the plants. The applied water depth was defined as that necessary for minimum drainage from the pots, at which time watering was stopped.

The harvests, of which there was a total of five, began 35 days after anthesis with the picking of unripened fruits with an intense green color (Medeiros, Grangeiro, Torres, & Freitas, 2010). After each harvest, the fruits were counted and then weighed to determine the production per plant.

At the end of the experiment (110 days after sowing), plants were cut close to the soil and collected, and their development (number of leaves; number of secondary branches; length of the largest branch; and leaf, stem, fruit and total dry matter) and yield (number of fruits, average fruit weight and fruit production) were evaluated.

These variables were analyzed as follows:

a) number of leaves: determined immediately after collection and starting from the basal leaves to the last open leaf of each replicate considering only green leaves while disregarding yellow and/or dried leaves;

b) number of secondary branches: obtained by counting the number of secondary branches in each replicate;

c) length of the largest branch: plants were placed on the bench, and the length of the longest branch was determined by measuring the branch from the cutting point (close to the soil surface) to the apical meristem of the plant using a tape graduated in cm;

d) leaf, stem, fruit and total dry matter: plants were separated into three parts (leaves, stems and fruits), packaged in paper bags, dried in a forced-air ventilation oven at 65°C to a constant weight and then weighed on a precision scale (0.01 g). The total dry matter was determined as the sum of the dry leaf, stem and fruit matter;

e) number of fruits: obtained by counting the number of fruits in all samples and expressed as fruits per plant;

f) average fruit mass: calculated as the sum of the fruit fresh mass divided by the number of fruits and expressed in grams (g); and

g) fruit production: calculated as the fresh mass of the commercial-quality fruits of each plant and expressed as yield per plant (g $plant^{-1}$).

The data were subjected to analyses of variance using the F-test. The mean values of the variables for each salinity treatment were compared using Tukey's test at the 5% level of probability. Means related to the effects of the biostimulant doses were analyzed by regression analysis, and the equations were fitted to polynomial models. The analyses were performed using Sisvar software (Ferreira, 2011).

Results and discussion

There was a significant effect of the interaction between salinity and the biostimulant, as well as salinity alone, for all variables at the 1% probability level, except for the length of the largest branch ($p > 0.05$) and fruit dry matter ($p < 0.05$). There was also a significant effect of the biostimulant, at 1% probability, on the other growth variables (Table 2).

The absence of salt stress resulted in higher values of all variables regardless of biostimulant use. With salinity, the greatest losses were observed in the number of leaves, stem dry matter, fruit dry matter and total dry matter with decreases of 29.9, 47.3, 71.7, and 49.6%, respectively (Table 2).

One of the most common effects of salinity on plants is growth limitation due to increased osmotic pressure from the medium, as well as a decrease in the volume of water available to plants, which affects cell division and elongation. The decrease in the leaf area of plants grown under salinity conditions represents an important adaptive mechanism since, under such conditions, transpiration decreases, which is accompanied by a decrease in the transport of Na^{+} and Cl^{-} ions in the xylem that maintains a high water potential (Taiz & Zeiger, 2009).

A decrease in gherkin growth under salt stress was also reported by Oliveira et al. (2014b), and decreases have been observed in other species of agronomic importance in the same botanical family, such as melons (Freitas, Figueirêdo, Porto Filho, Costa, & Cunha, 2014) and zucchini (Strassburger et al., 2011).

With respect to the effects of the biostimulant, there was a quadratic response in the number of leaves and the number of secondary branches regardless of the salinity level (Figure 2A and B).

The highest number of leaves occurred with the biostimulant doses of 14.3 and 6.8 mL kg^{-1} of seeds, resulting in 365.9 and 253.2 leaves in plants subjected to the 0.5 and 3.5 dS m^{-1} salinity levels, respectively (Figure 2A). The 9.0 mL kg^{-1} of seeds biostimulant dose resulted in the highest number of secondary branches (12 branches) in the absence of salt stress. In plants irrigated with saline water, the highest number of secondary branches (10 branches) occurred with the dose of 14.6 mL biostimulant kg^{-1} of seeds, evidencing the positive effects of the biostimulant on this variable (Figure 2B).

Table 2. Summary of the results of the analyses of variance and the mean values for the number of leaves (NL); number of secondary branches (NSB); largest branch length (LBL); leaf (LDM), stem (SDM), and fruit dry matter (FDM); and the total dry matter (TDM) of gherkin, cultivar Liso de Calcutá, using seeds treated with different doses of biostimulant and grown under conditions of salt stress.

Sources of variation	GL	Average squares						
		NL	NSB	LBL	LDM	SDM	FDM	TDM
Salinity (S)	1	121228.9**	38.72**	0.32**	21.41**	807.30**	6535.67**	12967.58**
Resíduc – 1	8	653.74	0.85	0.03	1.73	0.89	15.24	10.65
Biostimulant (B)	4	9995.0**	12.73**	0.02^{ns}	33.38**	11.40**	18.06*	123.35**
S X B	4	14581.9**	11.47**	0.23**	10.48**	11.20**	36.06**	71.39**
Resíduc – 2	32	658.74	1.85	0.02	1.97	2.69	5.06	9.36
CV 1(%)		9.16	9.85	11.24	8.71	7.25	18.83	6.68
CV 2 (%)		9.19	14.53	9.56	9.29	12.61	10.86	6.26
Doses of Biostimulant (mL kg^{-1} of seeds)	Salinity dS m^{-1}	Test of averages						
		NL	NSB	LBL (m)	LDM (g)	SDM (g)	FDM (g)	TDM (g)
0	0.5	269.4 a#	9.8 a	1.27 b	12.35 a	14.78 a	29.65 a	56.77 a
	3.5	235.8 b	6.0 b	1.51 a	11.61 a	8.56 b	8.72 b	28.88 b
5	0.5	419.2 a	10.4 a	1.51 a	16.37 a	16.17 a	31.21 a	63.75 a
	3.5	247.6 b	8.8 a	1.42 b	16.92 a	10.05 b	9.82 b	36.79 b
10	0.5	264.6 a	11.6 a	1.49 a	15.92 a	20.05 a	31.13 a	67.10 a
	3.5	259.4 a	9.2 b	1.42 a	16.36 a	9.13 b	10.60 b	36.10 b
15	0.5	332.2 a	11.8 a	1.60 a	17.46 a	17.85 a	31.19 a	66.49 a
	3.5	218.2 b	9.0 b	1.32 b	13.30 b	8.32 b	9.31 b	30.94 b
20	0.5	356.8 a	7.6 b	1.80 a	16.71 a	16.21 a	37.66 a	70.56 a
	3.5	188.8 b	9.4 a	1.21 b	14.12 b	8.82 b	8.06 b	30.91 b

^{ns}, ** and * non significant, significant at 5 and 1% probability, by T test. # Averages followed by the same letter in columns, per salinity, did not differ by Tukey test ($p < 0.05$).

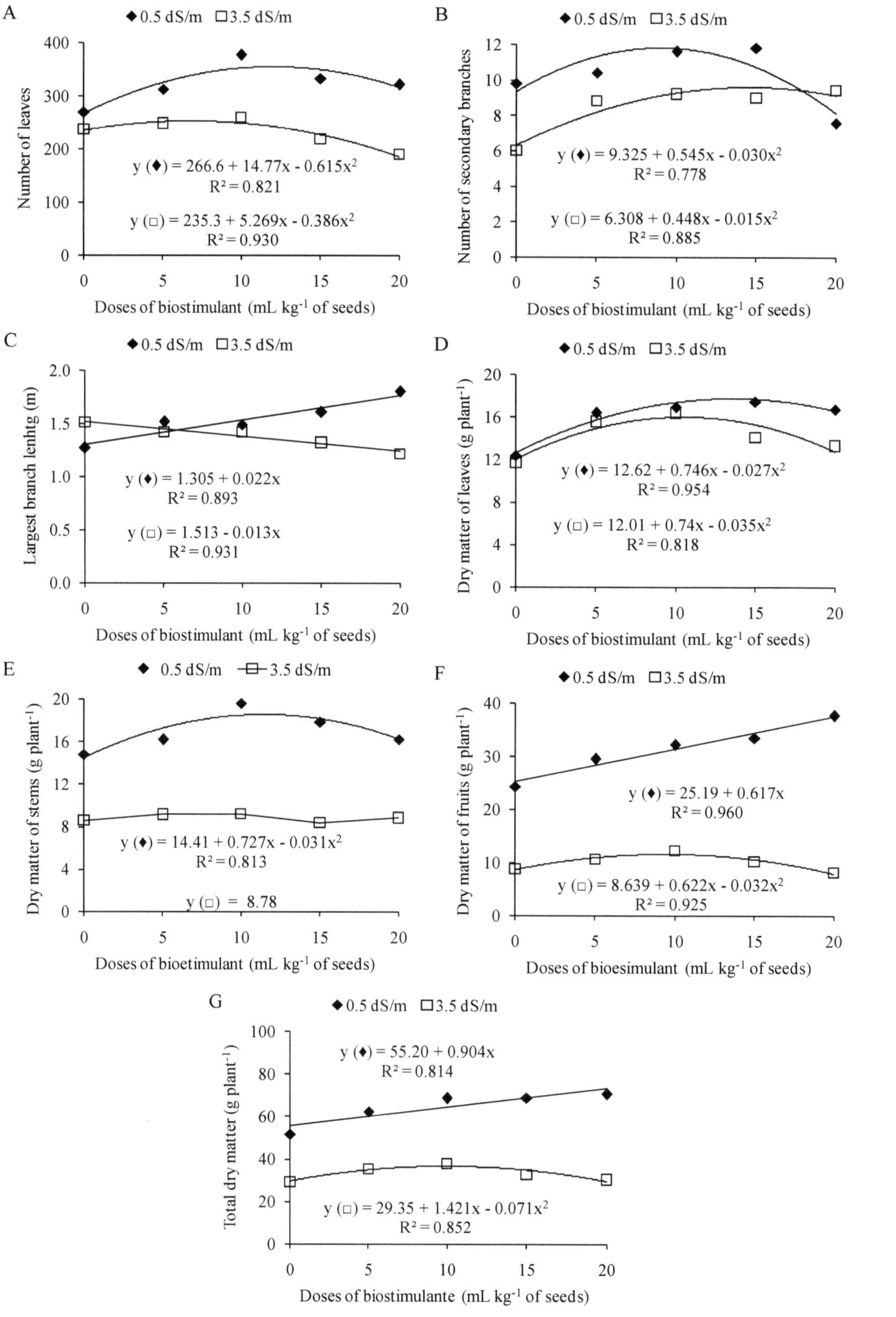

Figure 2. Number of leaves (A); number of secondary branches (B); length of the largest branch (C); and the leaf (D), stem (E), fruit (F), and total dry matter (G) of gherkin plants, cultivar Liso de Calcutá, grown from seeds treated with different doses of biostimulant and cultivated under conditions of salt stress.

Contrary to the results of this study, different concentrations of Stimulate® were not found to promote significant differences in the number of leaves in watermelons when plants were sprayed with doses of 0, 0.25, 0.5, 0.75, and 1% thirteen days after sowing in a previous study (Silva, Bolfarini, Rodrigues, Ono, & Rodrigues, 2014a). This result may be related to the way the treatment was applied, i.e., based on the phenological stage of the crop. According to Taiz and Zeiger (2009), cytokinins present in the biostimulant may induce the formation of new branches.

The increase in the number of branches contributes to gherkin fruit production. Therefore, the highest concentration of female flowers and, consequently, fruits occurs on the secondary and tertiary branches (Modolo & Costa, 2003).

The length of the largest branch responded linearly to the different biostimulant doses under both levels of salinity. Thus, in plants grown without salt stress, the length increased. No biostimulant (0 mL kg^{-1}) resulted in the shortest branch length (1.27 m), but as the dosage increased, this variable increased to a maximum length of 1.80 m with the highest dose (20 mL kg^{-1}). Divergent results were found for plants cultivated under salt stress; the lowest dose of the biostimulant increased the length of the largest branch (1.51 m). In contrast, the application of the biostimulant at the highest dose reduced this variable by 19.6% (1.21 m) (Figure 2C).

The accumulation of biomass in response to the biostimulant varied according to the part of the plant. For leaf dry matter, a quadratic response to the two salinity conditions was observed so that the highest values, 17.2 and 15.5 g $plant^{-1}$, were estimated for the doses of 14.5 and 10.4 mL kg^{-1} of seeds for plants grown under 0.5 and 3.5 dS m^{-1} salinity, respectively (Figure 2D).

These results differ from those obtained by Izidório, Lima, Vendrusculo, Ávila, and Alvarez (2015), who evaluated the effects of a foliar application of Stimulate® while transplanting lettuce. These authors observed a decrease in leaf dry matter with an increase in the concentration of the biostimulant; the highest dose of the product reduced leaf dry matter by 7.0 g $plant^{-1}$, representing a 51.1% decrease relative to the control. Therefore, transplantation does not seem to be the ideal time to apply a bioregulator.

No significant effects of the biostimulant were found on the stem dry matter (Figure 2E) of plants irrigated with saline water (3.5 dS m^{-1}); the average value was 8.98 g $plant^{-1}$. However, in plants irrigated with low-salinity water, there was a quadratic response. The maximum biomass accumulation in stems, 18.8 g $plant^{-1}$, was obtained at a biostimulant dose of 11.5 mL kg^{-1} of seeds (Figure 2E).

Regarding the accumulation of fruit dry matter (Figure 2F), significant differences were found between the different biostimulant doses and salinity levels. For plants irrigated with low-salinity water, there was a quadratic response; the increase in dosage caused an increase in this variable. The highest value occurred at the dose of 20 mL kg^{-1} of seeds (35.36 g $plant^{-1}$), which was 18.1% higher than that without biostimulant (0 mL kg^{-1} of seeds). There was also a quadratic response under salt stress conditions with an increase in fruit dry matter until 9.1 mL biostimulant kg^{-1} of seeds (10.3 g $plant^{-1}$), after which biomass decreased.

Analyzing the accumulation of total dry matter by the plants (Figure 2G), different responses were found with the evaluated doses of biostimulant at both salinity levels (absence and presence). For plants irrigated with low-salinity water, the data better fitted the increasing linear equation, in which the lowest value (56.77 g $plant^{-1}$) was obtained without the biostimulant (0 mL kg^{-1} of seeds), while the highest value (70.56 g $plant^{-1}$) was obtained with the highest dose (20 mL kg^{-1} of seeds), corresponding to a 19.54% increase in the production of total dry matter.

For plants grown under conditions of salt stress, the data were better represented by the quadratic equation. Initially, there was an increase in the total dry matter, and the best results were observed at the biostimulant dose of 9.7 mL kg^{-1} of seeds (35.63 g $plant^{-1}$), after which the increase in biostimulant caused a decrease in the total dry matter (Figure 2G).

The beneficial effects of treating seeds with biostimulants were due to the effects on the plant components. Such substances increase plant growth and plant development and stimulate cell division, and they may also increase the absorption of water and nutrients (Vieira & Castro, 2002).

In terms of the effects of the treatments on the performance variables, there were no significant effects of the interaction between salinity and biostimulant treatment on production. However, there was a significant effect on the number of fruits at the 5% level of probability. The interaction among factors and the factors in isolat significantly affected the average fruit weight at the 1% probability level (Table 3).

Table 3. Summary of the results of the analyses of variance and the average number of fruits per plant (NF), average fruit weight (AFW) and production (PROD) of gherkin, cultivar Liso de Calcutá, grown from seeds treated with different doses of biostimulant under conditions of salt stress.

Sources of variation	GL	Average squares		
		NF	AFW	PROD
Salinity (S)	1	1607.45**	875.38**	1648590.24**
Resídue – 1	8	1.41	1.66	645.94
Biostimulant (B)	4	7.57**	12.49**	6668.01**
S X B	4	2.01*	14.65**	622.45$^{n.s}$
Resídue – 2	32	0.64	1.69	841.24
CV 1(%)		8.08	6.11	7.60
CV 2 (%)		5.44	6.16	8.67
Doses of Biostimulant (mL kg^{-1} of seeds)	**Salinity dS m^{-1}**	**Test of averages**		
		NF	AFW (g fruit^{-1})	PROD (g plant^{-1})
0	0.5	19.20 a#	25.66 a	492.42 a
	3.5	9.40 b	14.39 b	132.79 b
5	0.5	20.00 a	26.46 a	523.77 a
	3.5	8.10 b	18.66 b	151.33 b
10	0.5	21.80 a	24.77 a	540.32 a
	3.5	10.50 b	19.24 b	201.65 b
15	0.5	20.80 a	23.88 a	533.88 a
	3.5	9.00 b	17.07 b	153.53 b
20	0.5	20.10 a	25.60 a	489.25 a
	3.5	8.20 b	15.17 b	124.53 b

^{ns},** and * non significant, significant at 5 and 1% probability, by T test. # Averages followed by the same letter in columns, per salinity, did not differ by Tukey test ($p < 0.05$).

When the effects of the factors were disentangled, it was also observed that the absence of salt stress resulted in higher values for the plant yield variables compared to growth under salt stress conditions, as shown in Table 3. The observed decreases were 55.6, 33.1, and 70.4% for the number of fruits, average fruit mass and fruit production, respectively, when plants were grown under salt stress regardless of the biostimulant dose.

Salinity decreased the number of gherkin fruits per plant regardless of the biostimulant dose, which is consistent with the results obtained by Oliveira et al. (2014b) for the Do Norte gherkin cultivar.

A decrease in the number of fruits in response to salt stress has been reported by several authors for different species in the same botanical family as gherkin, such as muskmelon (Medeiros, Medeiros, Dias, Barbosa, & Lima, 2011; Terceiro Neto, Gheyi, Medeiros, Dias, & Campos, 2013) and pumpkin (Silva, Lima, Silva, Oliveira, & Medeiros, 2014b). According to Amor, Martinez, and Cerdá (1999), the decrease in the number of fruits in response to salinity is due to a high rate of abortions.

It is likely that the decrease in the number of fruits per plant with an increase in salinity is due to a change in osmotic potential, that is, a decrease in water consumption by plants and a consequent decrease in nutrient uptake, which may cause a decrease in the fruit fixation index. Similar results were obtained for other vegetables such as melon (Melo et al., 2011) and eggplant (Marques et al., 2012).

The decrease in production in response to salinity may be attributed to lower fixation by the fruits as well as a decrease in the translocation of assimilates, which was also observed by Oliveira et al. (2014b) for gherkin, cv. Do Norte.

With respect to the effects of the biostimulant on production variables, the number of fruits varied depending on the salinity of the irrigation water. For plants grown without salt stress, the data best fit the quadratic polynomial equation. The highest number of fruits per plant (21.24 fruits) was promoted by the biostimulant dose of 11.5 mL kg^{-1} of seeds, beyond which there was a decrease in fruit number (Figure 3A).

In plants irrigated with saline water (3.5 dS m^{-1}), biostimulation had no effect on the number of fruits; the mean value was 9.04 fruits per plant (Figure 3A). These results demonstrate that the effects of the biostimulant may be modified or inhibited according to the environmental conditions, which was verified by Oliveira et al. (2015).

The significant effect of the interaction between the studied factors on average fruit mass indicates that the effects of the biostimulant depended on salt stress. For this variable, there were no differences in the effects of the applied biostimulant doses when the plants were grown in the absence of salt stress; the mean value was 25.48 g fruit^{-1}. For plants grown under stress, the data best fit the polynomial equation, initially presenting an increase in average fruit weight up to the dose of 10.01 mL biostimulant kg^{-1} of seeds (19.06 g. fruit^{-1}) and then a decrease to a 20% loss in plants whose seeds were treated with a biostimulant dose of 20 mL kg^{-1} of seeds (Figure 3B).

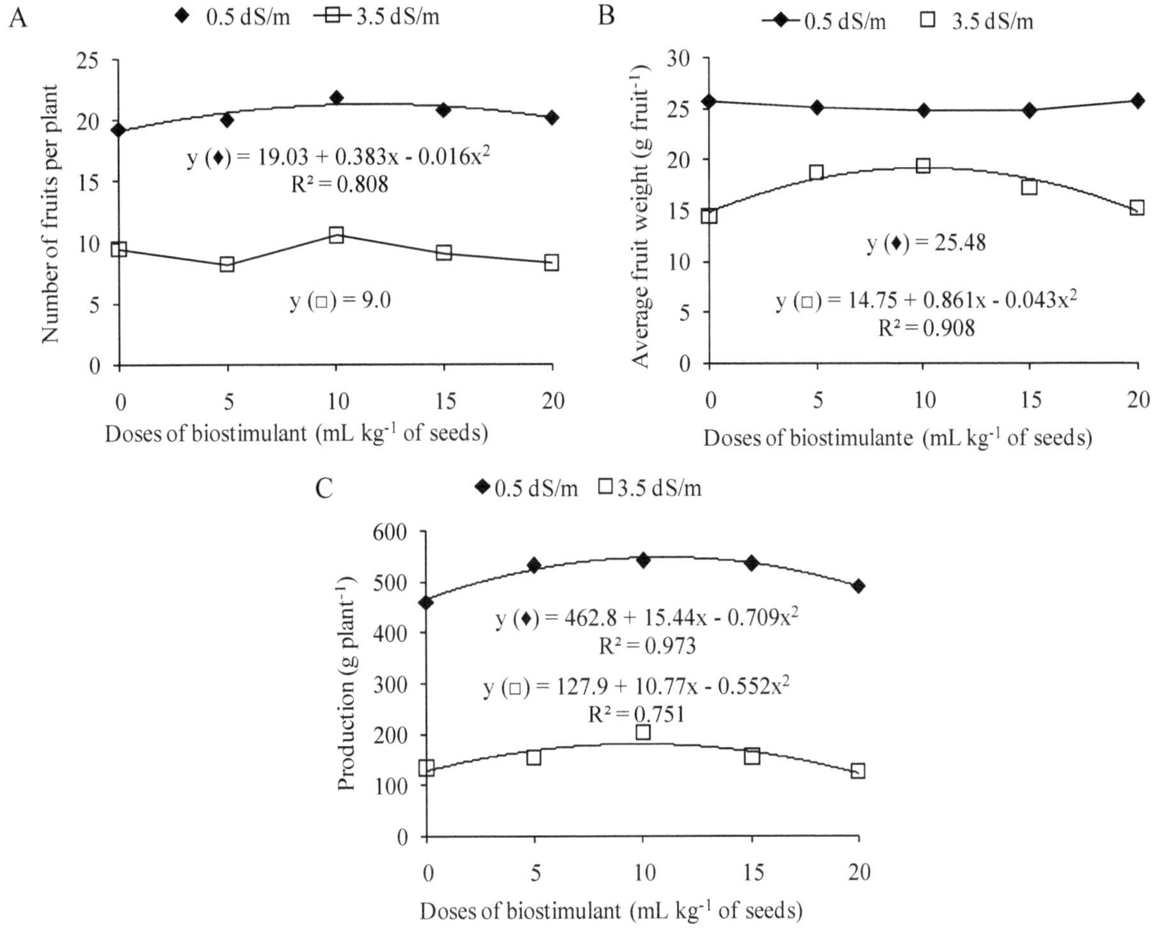

Figure 3. Number of fruits per plant (A), average fruit weight (B), and fruit production (C) in gherkin, cultivar Liso de Calcutá, grown from seeds treated with different doses of biostimulant under conditions of salt stress.

At both salinity concentrations, the biostimulant doses appeared to affect fruit production, and in both cases, the quadratic equation best fit the values. Initially, there was an increase in production in response to increasing doses of biostimulant. The highest yields occurred at the doses of 9.74 and 10.87 mL kg^{-1} of seeds for plants grown in the absence and presence of salt stress, respectively, obtaining maximum values of 546.84 and 180.45 g $plant^{-1}$ (Figure 3C).

In summary, the results of this study show that treating seeds with biostimulant has variable effects on the development and production of gherkin depending on the presence or absence of salt stress. These results partially refute those for soy found by Mortele, Santos, Braccini, Scapim, and Barbosa (2008), who assumed that the effectiveness of the bioregulator would be more pronounced and produce the best results under stress conditions.

On the other hand, the results corroborate reports by several authors for different cultures, such as cowpea (Oliveira et al., 2015) and cotton (Ávila, Barizão, Gomes, Fedri, & Albrecht, 2010). These authors found that the effects of the biostimulant on plants change under stress conditions, which indicates that a set of eco-physiological factors, together or separately, may interfere with the performance of bioregulators (Albrecht, Braccini, Scapim, Ávila, & Albrecht, 2012).

However, the results of this study only potentially and not definitely confirm the hypothesis of this research that treating seeds with a biostimulant could inhibit or mitigate the effects of salinity, so further tests are needed to confirm the actual effectiveness of the use of this product under conditions of abiotic stress (Albrecht et al., 2011).

Studies are also necessary to identify possible product concentrations and other methods of applying the biostimulant that could increase gherkin crop productivity. Finally, further research is needed to verify whether this biostimulant promotes an increase in productivity and to assess the economic feasibility of its use in gherkin production.

Conclusion

Saline irrigation water decreases the number of leaves, the number of secondary branches and the length of the largest branch as well as the accumulation of biomass and gherkin production variables regardless of the biostimulant dose.

Treating seeds with the biostimulant in the absence of salt stress increases the number of leaves and branches as well as the length of the largest branch and biomass accumulation.

Applying biostimulant to seeds is not effective at reducing the effects of salinity on gherkin production, but it promotes an increase in fruit production regardless of salinity level.

References

Albrecht, L. P., Braccini, A. L., Scapim, C. A., Ávila, M, R., & Albrecht, A. J. P. (2012). Biorregulador na composição química e na produtividade de grãos de soja. *Revista Ciência Agronômica, 43*(4), 774-782.

Albrecht, L. P., Braccini, A. L., Scapim, C. A., Ávila, M. R., Albrecht, A. J. P., & Ricci, T. T. (2011). Manejo de biorregulador nos componentes de produção e desempenho das plantas de soja. *Bioscience Journal, 27*(6), 865-876.

Amor, F. M. D., Martinez, V., & Cerdá, A. (1999). Salinity duration and concentration affect fruit yield and quality, and growth and mineral composition of melon plants grown in perlite. *HortScience, 34*(7), 1234-1237.

Ávila, M. R., Barizão, D. A. O., Gomes, E. P., Fedri, G., & Albrecht, L. P. (2010). Cultivo de feijoeiro no outono/inverno associado à aplicação de bioestimulante e adubo foliar na presença e ausência de irrigação. *Scientia Agraria, 11*(3), 221-230.

Carmo Filho, F., & Oliveira, O. F. (1995). *Mossoró*: um município do semiárido nordestino, caracterização climática e aspecto florístico. Mossoró, RN: ESAM.

Carmo, G. A., Oliveira, F. R. A., Medeiros, J. F., Oliveira, F. A., Campos, M. S., & Freitas, D. C. (2011). Teores foliares, acúmulo e partição de macronutrientes na cultura da abóbora irrigada com água salina. *Revista Brasileira de Engenharia Agrícola e Ambiental, 15*(5), 512-518.

Castellane, P. D., & Araújo, J. A. C. (1994). *Cultivo sem solo*: hidroponia. Jaboticabal, SP: Funep.

Ferreira, D. F. (2011). Sisvar: a computer statistical analysis system. *Ciência e Agrotecnologia, 35*(6), 1039-1042.

Freitas, L. D. A., Figueirêdo, V. B., Porto Filho, F. Q., Costa, J. C., & Cunha, E. M. (2014). Crescimento e produção do meloeiro cultivado sob diferentes níveis de salinidade e nitrogênio. *Revista Brasileira de Engenharia Agrícola e Ambiental, 18*(supl.), S20-S26.

Garcia, R. A., Gazola, E., Merlin, A., Villas Boas, R. L., & Crusciol, C. A. C. (2009). Crescimento aéreo e radicular de arroz de terras altas em função da adubação fosfatada e bioestimulante. *Bioscience Journal, 25*(4), 65-72.

Izidório, T. H. C., Lima, S. F., Vendrusculo, E. P., Ávila, J., & Alvarez, R. C. F. (2015). Bioestimulante via foliar em alface após o transplantio das mudas. *Revista de Agricultura Neotropical, 2*(2), 49-56.

Marques, D. J., Broetto, F., Lobato, A. K. S., Silva, E. C., Carvalho, J. G., Ávila, F. W., & Andrade, I. P. (2012). Photosynthetic pigments, nitrogen status, and flower behavior in eggplant exposed to different sources and levels of potassium, *Scientific Research and Essays, 8*(2), 67-74.

Medeiros, D. C., Medeiros, J. F., Dias, P. M. S., Barbosa, M. A. G., & Lima, K. S. (2011). Produção de melão Pele de Sapo híbrido 'Medellín' cultivado com diferentes salinidades da água. *Revista Ciência Agronômica, 42*(3), 628-634.

Medeiros, M. A., Grangeiro, L. C., Torres, S. B., & Freitas, A. V. L. (2010). Maturação fisiológica de sementes de maxixe (*Cucumis anguria* L.). *Revista Brasileira de Sementes, 32*(3), 17-24.

Melo, T. K., Medeiros, J. F., Espínola Sobrinho, E., Figueiredo, V. B., Pereira, V. C., & Campos, M. S. (2011). Evapotranspiração e produção de melão Gália irrigado com água de diferentes salinidades e adubação nitrogenada. *Revista Brasileira de Engenharia Agrícola e Ambiental, 15*(12), 1235-1242.

Modolo, V. A., & Costa, C. P. (2003). Avaliação de linhagens de maxixe paulista cultivadas em canteiros com cobertura de polietileno. *Horticultura Brasileira, 21*(3), 534-538.

Mortele, L. M., Santos, R. F., Braccini, A. L., Scapim, C. A., & Barbosa, M. C. (2008). Efeito da aplicação de biorregulador no desempenho agronômico e produtividade da soja. *Acta Scientiarum. Agronomy, 30*(5), 701-709.

Munns, R. (2005). Genes and salt tolerance: bringing them together. *New Phytologist, 167*(3), 645-663.

Oliveira, A. P., Oliveira, A. N. P., Alves, E. U., Alves, A. U., Leonardo, F. A. P., & Santos, R. R. (2008). Rendimento de maxixe em função de doses de P_2O_5 em solo arenoso. *Ciência e Agrotecnologia, 32*(4), 1203-1208.

Oliveira, F. A., Oliveira, M. K. T., Medeiros, J. F., Silva, O. M. P., Paiva, E. P., & Maia, P. M. E. (2014a). Produtividade do maxixeiro cultivado em substrato e fertirrigado com soluções nutritivas. *Horticultura Brasileira, 32*(4), 464-467.

Oliveira, F. A., Pinto, K. S. O., Bezerra, F. M. S., Lima, L. A., Cavalcante, A. L. G., Oliveira, M. K. T., & Medeiros, J. F. (2014b). Tolerância do maxixeiro, cultivado em vasos, à salinidade da água de irrigação. *Revista Ceres, 61*(1), 147-154.

Oliveira, F. A., Medeiros, J. F., Alves, R. C., Lima, L. A., Santos, S. T., & Régis, L. R. L. (2015). Produção de feijão caupi em função da salinidade e regulador de crescimento. *Revista Brasileira de Engenharia Agrícola e Ambiental, 19*(11), 1049-1056.

Palangana, F. C., Silva, E. S., Goto, R., & Ono, E. O. (2012). Ação conjunta de citocinina, giberelina e auxina em pimentão enxertado e não enxertado sob cultivo protegido. *Horticultura Brasileira, 30*(4), 751-755.

Silva, M. J. R., Bolfarini, A. C. B., Rodrigues, L. F. O. S., Ono, E. O., & Rodrigues, J. D. (2014a). Formação de mudas de melancia em função de diferentes concentrações e formas de aplicação de mistura de reguladores vegetais. *Revista Scientia Plena, 10*(10), 1-9.

Silva, M. V. T., Lima, R. M. S., Silva, N. K. C., Oliveira, F. L., & Medeiros, J. F. (2014b). Produção de abobora sob diferentes níveis de água salina e doses de nitrogênio. *Revista Verde de Agroecologia e Desenvolvimento Sustentável, 9*(1), 287-294.

Stoller do Brasil. (1998). *Stimulate® Mo em hortaliças*: informativo técnico. Cosmópolis, SP: Stoller do Brasil. Divisão Arbore.

Strassburger, A. S., Peil, R. M. N., Fonseca, L. A., Aumonde, T. Z., & Mauch, C. R. (2011). Dinâmica do crescimento da abobrinha italiana em duas estações de cultivo. *Acta Scientiarum. Agronomy, 33*(2), 283-289.

Taiz, L., & Zeiger, E. (2009). *Fisiologia vegetal.* Porto Alegre, RS: Artmed.

Terceiro Neto, C. P. C., Gheyi, H. R., Medeiros, J. F., Dias, N. S., & Campos, M. S. (2013). Produtividade e qualidade de melão sob manejo com água de salinidade crescente. *Pesquisa Agropecuária Tropical, 43*(4), 354-362.

Vieira, E. L., & Castro, P. R. C. (2002). *Ação de stimulante no desenvolvimento inicial de plantas de algodoeiro (Gossypium hirsutum* L.). Piracicaba, SP: USP/Departamento de Ciências Biológicas.

28

Sustainable production of 'Comum' tannia in the hilling and function of seedling types in three crop seasons

Néstor Antonio Heredia Zárate[1*], Maria do Carmo Vieira[1], Fabiane Cargnin Faccin[1], Danilo Augusto Heredia Vieira[2], Rosimeire Pereira Gassi[1] and Anna Luiza Farias dos Santos[1]

[1]*Universidade Federal da Grande Dourados, Faculdade de Ciências Agrárias, Rod. Dourados a Ithaum, km 12, 79804-970, Dourados, Mato Grosso do Sul, Brazil.* [2]*Programa de Pós-graduação em Administração, Universidade Estadual de Maringá, Maringá, Paraná, Brazil.* [*]*Author for correspondence: E-mail: nestorzarate@ufgd.edu.br*

ABSTRACT. This study aims to discover the agro-economic yield of the 'Comum' tannia, which was propagated with five seedling types (Extra, Large, Medium, Small and Tiny) and grown both with and without a hilling operation during three crop seasons. In each agricultural year, the treatments were arranged as a 5 x 2 factorial in a randomized complete block design with four replications. The harvests were conducted 238 (2007-08), 231 (2008-09) and 237 (2009-10) days after planting. The data were submitted to an analysis of variance each year. The plants produced from large seedlings and RM-corms had a higher fresh leaf weight. Additionally, the dry weight of the RFC-commercial cormels and RFNC-non-commercial cormels was higher with extra seedlings. The plants grown during 2008-2009 had a higher fresh weight of RM and of RFNC than the plants grown during 2009-10. The interaction between the seedling types and crop seasons significantly influence the yield and dry weight of the leaves of RM and RFC, the leaf length of RM and RFC and the leaf diameter of RFC. According to the economic analysis, a net income of R$ 2,234.08 was obtained when the tannia were cultivated without a hilling operation using extra propagated seedlings.

Keywords: *Xanthosoma mafaffa*, propagation materials, cultural treatments, yield, income.

Introduction

The increase of agriculturally produced income, especially in familiar agriculture, is dependent on the implementation of technician-management implantation practices that allow for the possibility of better organization and the exploitation of productive activity. In addition to preserving and improving the productive environment, these practices must promote the reduction of costs, the addition of values to production and/or yield increases (DOMIT et al., 2008). In vegetable production, concerns about sustainability are derived from reflections on the relationship between humankind and the environment. Thus, the challenge of sustainability is not solely restricted to the generation of environmentally appropriate solutions; sustainability must also be lucrative and socially desirable (TERRA et al., 2006).

The Araceae family includes approximately 104 genders and 3700 species. Although this family occurs naturally on all continents (except in Antarctica), it is predominantly tropical (FOGAÇA et al., 2007). Tannia (*Xanthosoma mafaffa* Schott), also known as yautia, is consumed by the 'indigenous populations', and it is highly appreciated because of its rhizomes and leaves. All of the available information indicates that Tannia originated in Northeastern South America and then expanded to other regions like Africa, which is currently the world's leading Tannia producer. Tannia is also found in Mexico, Venezuela, Colombia, Panama, Costa Rica, Peru and Brazil, where it was cultivated by the early colonizers and is found in the states of Goiás, Rio de Janeiro, Mato Grosso do Sul, Minas Gerais and São Paulo (SILVA et al., 2011). Because of the peculiar culinary characteristics of its rhizomes, tannia is seasonally commercialized close to production areas. The flavor of its young leaves is comparable to spinach (COSTA et al., 2008).

Technological progress that leads to an increase in the production and commercial quality of the tannia rhizomes will allow it to become a popular product in the national market of farm products (HEREDIA ZÁRATE et al., 2006). However, it is necessary to establish a rational basis for its cultivation, and there exist previous studies that refer to tannia's fertilization, plantation depth, spacing and the type and size of its rhizomes-seed (LEITE et al., 2007).

The lack of plantation material is one of the factors that limits the expansion of tannia cultivation in several vegetatively propagated species. Thus, the efficient exploitation of seedlings is recommended. For these species, the type and quality of the plantation material determines the differences in rooting speed, growth and, consequently, in the production and extension of the vegetative cycle (HEREDIA ZÁRATE; VIEIRA, 2003). Some aspects of the seedlings have not been studied or, when they were studied, the results were not conclusive. As cited by Monteiro and Peressin (1997), Vasconcelos (1972) found that the primary type of the tannia rhizomes-seed (corms), which had a fresh weight varying from 13 to 40 g, were more productive (by 1.5 to 4.5 g) than secondary rhizomes-seed (cormels). The same author also indicated that, beyond their health problems, the primary rhizomes obtained in a previous culture were hardly sufficient, in numerical terms, for the implantation of a new culture. Heredia Zárate et al. (2006) evaluated the productive capacity of 'Comum' tannia plants by examining the secondary rhizomes-seed of different sizes grown under three or four rows of plants in the plot. When considering the total and commercial production, these researchers concluded that it was better to cultivate 'Comum' tannia under three rows of plants in a plot and by planting rhizomes with 3.0 g of fresh weight.

Hilling is a cultural practice used by the producers of some vegetables. During hilling, land is moved to cover part of a stem base and/or a plant root (HEREDIA ZÁRATE et al., 2010b). The practice is highly valued and considered to be an essential operation to optimize technologies and to obtain a high yield. The true necessity and accomplishment of hilling has been questioned, and there have been several previous experiments related to the results found in this study. In the consulted literature, there were no recommendations based on the experimental studies that indicate the execution time and the hilling height. However, these studies do indicate that the results depend on the species, stage of plant growth and execution form, whether manual or mechanized (HEREDIA ZÁRATE et al., 2010a; TERRA et al., 2006). Puiatti (2002) suggests that to lower the production cost of taro (*Colocasia esculenta* (L.) Schott), which is in the same botanical family as tannia, slight hillings during each weeding should be made. One hilling should occur at either 45 to 60 days or only at 120 days after plantation.

As in any economic activity (but especially for the agriculturist), the accompaniment of costs is essential; in addition to the knowledge of the total operational cost, it is necessary to know the participation of the item in the effective operational cost because this interaction reflects the variable costs or the expenses that are effectively carried. It is equally important to know the structure of the fixed costs (the indirect great expenses), which are represented by the costs and administrative incumbencies in a form that details the remuneration attributed to other important production factors, without which the profitability calculation is incorrect (MELO et al., 2009). In the consulted literature, there was no study that examined the effect of hilling on tannia.

The aim of this work was to determine the 'Comum' tannia agro-economic yield, which was propagated with five seedling types and cultivated both with and without hilling during three crop seasons.

Material and methods

The experiment took place during three crop seasons between September 2007 and April 2010, and it was conducted at the Medicinal Plants Garden of the Faculty of Agrarian Science-FCA, of the Federal University of the Grande Dourados, in Dourados, Mato Grosso Sul State, Brazil, which is located at 22°11'44"S of latitude, 54°56'07"W of longitude and 452 m of altitude. According to Köppen, the climate of Dourados is mesothermal humid, Cwa type, with an average temperature and annual rainfall ranging from 20 to 24°C and 1250-1500 mm, respectively.

The soil of the area was classified as a dystrophic red oxisol of clayey texture, and it contained the following chemical characteristics: 5.9 of pH in H_2O; 28.9 g dm^{-3} of organic matter; 38.0 mg dm^{-3} of P; 0.0; 3.5; 46.0; 22.0; 53.0; 71.5 and 124.5 $mmol_c$ dm^{-3} of Al^{+3}, K, Ca, Mg, H+Al, SB and CTC, respectively, and 57.0% of base saturation. Considering the physical properties, the sieve method indicated a soil composed by 80 g kg^{-1} of sand, 130 g kg^{-1} of fine sand, 160 g kg^{-1} of silt and 630 g kg^{-1} of clay.

The factors in this study were the seedling types (Extra, Large, Medium, Small and Tiny) and the number of hilling operations (zero and one, at 47 days after planting). The treatments were arranged in a 5 x 2 factorial fixed in a completely randomized block with four replications. The plots had a total area of 4.5 m^2 (3.0 m in long and 1.50 m in width). The plots had a width of 1.0 m and contained four rows spaced 25.0 cm apart. There was a spacing of 17.5 cm between the plants, which established a population of 150,744 plants ha^{-1}.

Two weeks prior to planting, the soil of the experimental area was prepared by plowing, harrowing and then elevating the plots with a bedshaper rotary cultivator offset. Lime was not used for the soil amendment. Additionally, there was no fertilization technique used during the crop cycle because the results of the soil analysis indicated that this necessity did not exist.

Rhizomes-seed were used as propagating material in the planting, and they were harvested in the area of the Medicinal Plants Garden of the UFGD. The rhizomes-seed were detached from the plant-mother and visually chosen for types. The fresh weight was then determined (Table 1). On the planting day, four furrows for planting were dug in the plot; each furrow was approximately 0.05 m wide x 0.05 m deep. A direct form of planting was conducted; the rhizomes were embedded manually with the top as the apex (HEREDIA ZÁRATE et al., 2005).

Table 1. Average fresh weight (g) of five seedling types used in the propagation of the 'Comum' tannia, in three crop seasons.

Seedling types	Crop seasons			
	2007-08	2008-09	2009-10	Average
Extra	5.51	7.84	3.82	5.72
Large	3.41	5.83	3.14	4.13
Medium	2.44	4.98	2.67	3.36
Small	1.80	4.55	2.40	2.92
Tiny	1.67	3.97	2.10	2.58
Average	2.97	5.43	2.83	3.74

During the crop cycle, the cultural treatments consisted of irrigation using a sprinkler system. The irrigations were applied to each plot for two to three days and aimed to maintain the ground at 70% of the field capacity (SANGALLI et al., 2011). In the two final months, the irrigations were applied two times per week. When the weeds indicated ± 5.0 cm of height, spontaneous vegetation was controlled by weeding between the plots and by manual removal within the plots. There was no damage from the attack of pests or diseases.

The harvests were performed at 238 (2007-08), 231 (2008-09) and 237 (2009-10) days after planting, when the plants showed more than 50% of the dry leaves in the plants had symptoms of senescence, and the fresh and dry weight of the leaves, corms-RM, commercial cormels-RFC and non-commercial cormels-RFNC was evaluated. The cormels that weighed more than 5 g were considered to be commercial. The diameter and length of RM, RFC and RFNC were also determined. The data were submitted to an analysis of variance using the totals of the treatments in each year and when significance was detected by the F test. The averages were compared using Tukey's test at 5% probability.

To determine sustainability, an agro-economic analysis was performed according to the method of Terra et al. (2006) for sweet corn (*Zea mays* L.) and of Rezende et al. (2009) for vegetables grown in a monocrop and intercropping system. This analysis related the factors studied (productive) with the likely returns (economics), i.e., the determination of variable costs (manpower and rental of machinery) and fixed costs (rent of land and improvements), and the contingency reserve, the spending on administration and the payment of interest on capital. For the calculation of gross income, the value of R$1.20 kg^{-1} (the average of the price paid to producers in Dourados, Mato Grosso do Sul State, Brazil, by kg of commercial cormels) was used. Subsequently, the conversions were made per hectare using the obtained yields in each treatment. The net income was calculated by subtracting the production costs relative to the gross income.

Results and discussion

Agronomic analysis

Based on the analysis of variance, it is clear that the interaction between the seedling types and crop seasons significantly influenced the yield of the fresh and dry weight of leaves, corms-RM, commercial cormels-RFC and non-commercial cormels-RFNC. Additionally, this interaction influenced the length and the diameter of RM,

RFC e RFNC of 'Comum' tannia. Significant differences were also observed in the yield of the crop from one season to another. The seedling type induced significant differences in the yield, and hilling did not significantly influence any of the evaluated characteristics. These responses might have occurred because the plants were vegetatively propagated. Moreover, these results are not derived from propagating material obtained from a plant, and they only demonstrate the phenotypic behavior of heterogeneous populations. Moreover, these results are consistent with the results of Larcher (2006), who suggested that ecological systems are capable of self-control because of the balance of the interference relations, and that they have the capacity to adapt to the individual organism and to populations.

In relation to the leaf fresh weight (Table 2), the plants produced from Large type seedling had the highest yield. The yield of the Large type seedlings was similar to the yields of the Extra, Small and Tiny types but different from the plants produced from Medium type seedlings. The Medium type was less productive; in comparison with the highest yield, it demonstrated the smallest value (0.50 ton ha^{-1}). The differences in the leaf fresh weight production occurred because plants of heterogeneous populations can present variable growth rates and morphology characteristics, with modifications at the end of growing season because of environmental factors. However, the response pattern is dependent on the genetic component (HEREDIA ZÁRATE et al., 2003b). Puiatti et al. (2004) had suggested that the propagation material of taro with a higher reserve (seedlings of type corms and large cormels) favored more vigorous plants, with higher values of height (data not shown) and an index of the foliar area.

Regarding the crop season, the plants cultivated in 2008-2009, which consisted of 0.76 ton ha^{-1} (+ 61.79%), statistically surpassed the lower yield obtained in 2009-2010. In trying to determine the reason for the productive differences of the 'Chines' taro at different times of cultivation, Puiatti et al. (2004) indicated that the plants installed in a field during the initial period of temperature increase (such as the winter) grew better than the plants installed during the period with less propitious temperatures.

The fresh weight of RM, RFC and RFNC and the dry weight of RFNC (Table 2) were higher in plants produced from the seedlings of the Extra type. Additionally, these values were similar to the fresh and dry weight of RM and RFNC in plants produced from Large, Small and Tiny types and to the weights of RFC in plants produced from the Large and Small types (Table 2). Compared with plants produced from the less productive Medium type seedlings, the productive differences in the fresh weight of plants produced from Extra type seedlings was 0.73 (+ 37.63%), 1.17 (+ 29.77%) and 1.42 (+ 51.08%) ton ha^{-1} and the difference in the dry weight was 0.31 ton ha^{-1} (+ 46.97 ton ha^{-1}). The evidence suggests that the type of seedling exerts a direct influence on the growth and vigor of the plants, and that the amount of rhizome reserves should not be the only factor considered when attempting to produce vigorous plants (PUIATTI et al., 2003).

Table 2. Fresh weight of leaves, corms, commercial cormels, non-commercial cormels; non-commercial cormels dry weight, corms diameter and diameter and length of non-commercial cormels proceeding from plants propagated with five seedling types, with and without hilling, in three crop seasons.

Factors in study	Fresh weight (ton ha^{-1}) Leaves	RM	RFC	RFNC	Dry weight RFNC (ton ha^{-1})	Diam. RM (mm)	Diam. RFNC (mm)	Length RFNC (mm)
				Seedling types				
Extra	1.63 ab	2.67 a	5.10 a	4.20 a	0.97 a	29.90 a	11.21 a	18.29 a
Large	1.94 a	2.34 ab	4.34 ab	3.54 ab	0.82 ab	28.61 a	10.78 a	16.57 a
Medium	1.44 b	1.94 b	3.93 b	2.78 b	0.66 b	26.97 a	11.65 a	18.68 a
Small	1.55 ab	2.42 ab	4.67 ab	3.46 ab	0.77 ab	29.15 a	10.84 a	18.25 a
Tiny	1.72 ab	2.16 ab	4.06 b	3.47 ab	0.78 ab	27.71 a	10.83 a	17.49 a
				Hilling				
Without	1.63 a	2.41 a	4.34 a	3.63 a	0.84 a	28.93 a	10.99 a	17.77 a
With	1.67 a	2.20 a	4.50 a	3.35 a	0.76 a	27.96 a	11.13 a	17.94 a
				Agricultural year				
2007-08	1.75 a	1.86 b	2.69 c	3.71 a	0.99 a	31.36 a	11.00 a	15.12 b
2008-09	1.99 a	3.22 a	6.55 a	4.68 a	0.87 a	25.17 c	11.37 a	21.77 a
2009-10	1.23 b	1.84 b	4.02 b	2.07 b	0.54 b	28.80 b	10.82 a	16.68 b
C.V. (%)	45.83	16.27	20.35	18.95	20.62	7.81	7.23	9.18
Average	1.65	2.30	4.42	3.49	0.80	28.44	11.06	17.86

RM = corm; RFC = commercial cormel; RFNC = non-commercial cormel; Diam. = Diameter. Averages followed by the same letters in the columns, inside from type of seedling and of crop seasons, do not differ for the Tukey test, and for hilling, by F test, at 5% of probability.

Moreover, the results are consistent with the findings of Heredia Zárate et al. (2004), who indicated that plants with exuberant growth are not good producers because they lose many photo-assimilated compounds in the maintenance of the shoot. This loss of compounds delays the maturity and the beginning of the senescence process of the oldest leaves because of the delay in the translocation of the photo-assimilated compounds of the rhizomes.

Plants grown in 2008-09 produced more fresh weight of RM (+ 1.38 ton ha^{-1}) and of RFNC (+ 2.61 ton ha^{-1}) than those plants grown in 2009-10. Additionally, the plants grown in 2008-0 had a higher fresh weight of RFC (3.86 ton ha^{-1}) than the plants grown in 2007-08. In terms of compensation, the highest dry weight of RFNC was obtained in 2007-08. This value surpassed the dry weight in 2008-09 and 2009-10 by 17.86 and 83.33%, respectively.

The diameter of RM and RFNC and the length of RFNC were neither significantly influenced by the types of seedlings nor by the hilling operations, but were instead influenced by the time of planting during the crop season. The highest values for diameter of RM and RFNC and the length of RFNC occurred in 2007-2008 and 2008-09. These productive behaviors of the 'Comum' tannia plants, which were produced by different types of seedlings and from plants grown in different years, are similar to the behaviors in taro. The plants of Taro can differ with regard to the time it takes to reach maturity, to the amount of photo-assimilated compounds stored in the leaves (limbs and petioles). The compounds stored in the leaves can be translocated for the corms and cormels when the leaves initiate senescence (HEREDIA ZÁRATE et al., 2003a).

The lack of significant differences induced by hilling suggests that the vegetal systems have self-control mechanisms. The self-control mechanism is based on either the capacity of the individual organism and the total populations to adapt or to the balance of the interference relations, as competition for nutrients, water and other elements (LARCHER, 2006).

The average yields of leaves in RM, RFC and RFNC, which represented 13.91, 19.39, 37.27 and 29.43% of the total yield of the plant, indicates that these plants already had reached their maturity and maximum growth. Additionally, these plants experienced a probable increase of the translocation of the photo-assimilated compounds of the shoot to the RM and of RM directly to the cormels, or that a balance of translocation occurred between the shoot and rhizomes (HEREDIA ZÁRATE et al., 2006). According to several authors cited by Heredia Zárate et al. (2009), the sugars synthesized in the limb of the taro are translocated in the rhizomes, passing to a 'temporary storage' in the petiole. Thus, the storage of reserves in rhizomes is highly dependent on the integrity of the aerial structures (limb and petiole). Any morph-physiologic alteration of these structures will affect the synthesis, amount and speed of translocation of the assimilated compounds, which impacts the growth and yield of the rhizomes.

The average total yield of the rhizomes obtained in this experiment (10.21 ton ha^{-1}) and of plants produced from seedlings with an average weight of 3.74 g was inferior to the 16.24 ton ha^{-1} obtained by Silva et al. (2011), who used corms with an average weight of 55 g. However, the average total yield of the rhizomes was similar to the 10.34 ton ha^{-1} obtained with seedlings formed from type 1 cormels (average weight of 10 g), and the rhizomes had a superior yield to the 6.46 ton ha^{-1} of seedlings formed from the type 2 cormels (average weight of 5.0 g).

The yield of the dry weight of leaves of RM and RFC, the length of RM and RFC and the diameter of RFC were significantly influenced by the interaction between seedlings types and crop seasons (Table 3). The highest values for leaf dry weight were in plants produced from all types of seedlings and grown in the 2007-2008-crop season. The lowest values were of the plants grown in 2009-2010 and propagated with Extra, Large, Small and Tiny seedlings. In relation to the effect of the seedling, the results indicated that the sprout capacity of the seedlings is an intrinsic character of the clone, and that it probably had modifying responses that adapted the plants to the environmental conditions during the sprout and growth of the shoot (HEREDIA ZÁRATE et al., 2002).

Variable responses in relation to the interactions between seedling type and agricultural year were observed in the dry weight of RM. The highest values were observed in 2009-2010 in plants propagated with Extra seedlings, and in the Large and Medium seedling during the 2008-2009 crop season. Excluding the plants produced from the Extra seedlings in 2009-2010, the highest yields of RFC dry weight were from the plants grown from all types of seedlings in the 2008-2009 crop season.

Among the reasons for the variable yields, the environmental variations can be cited. Environmental variations normally occur during the cultivation times, and they influence the capacity of the phenotypic expression of the plants in response to the genetic differences inside of a population and in response to the effect of the seedling type used in the propagation.

Table 3. Dry weight of leaves, corms (RM) and commercial cormels (RFC), length of RM and diameter and length of RFC proceeding from plants propagated with five seedling types, in three crop seasons.

Factors in study		Dry weight (ton ha^{-1})			Length	Diameter	Length
Seedlings	Year	Leaves	RM	RFC	RM (mm)	RFC (mm)	RFC (mm)
	2007-08	1.92 a	0.43 c	1.19 b	32.16 c	20.49 a	32.63 b
Extra	2008-09	0.21 b	0.62 b	1.29 b	54.89 a	18.92 a	39.23 a
	2009-10	0.06 b	0.91 a	1.65 a	40.45 b	20.39 a	35.53 ab
	2007-08	1.04 a	0.37 b	0.54 c	32.76 b	20.60 ab	34.10 a
Large	2008-09	0.23 b	0.73 a	1.22 a	53.23 a	21.55 a	34.30 a
	2009-10	0.12 b	0.40 b	0.90 b	38.70 b	18.90 b	30.63 a
	2007-08	1.04 a	0.46 a	0.61 b	34.10 b	20.14 a	31.14 ab
Medium	2008-09	0.19 b	0.61 a	1.23 a	45.83 a	21.33 a	35.63 a
	2009-10	0.26 b	0.24 b	0.56 b	33.46 b	18.15 b	28.92 b
	2007-08	1.00 a	0.44 a	0.70 b	36.23 b	21.84 a	32.52 ab
Small	2008-09	0.19 b	0.56 a	1.25 a	50.62 a	18.72 b	36.98 a
	2009-10	0.10 b	0.45 a	0.90 b	36.74 b	17.89 b	30.88 b
	2007-08	1.16 a	0.43 a	0.91 ab	34.67 b	18.16 ab	27.34 b
Tiny	2008-09	0.18 b	0.45 a	1.09 a	47.69 a	16.85 b	33.02 a
	2009-10	0.17 b	0.33 a	0.76 b	37.11 b	19.16 a	28.85 b
C.V. (%)		15.19	15.81	29.10	6.44	3.73	6.19
Average		0.51	0.49	0.99	40.59	19.54	32.74

RM = corm; RFC = commercial cormel. Averages followed by the same letters in the columns, inside from type of seedling and of crop seasons, do not differ for the Tukey test at 5% of probability.

This fact coincides with the non-sequential variability observed in the length of RM and RFC and in the diameter of RFC, all of which are dependent on the translocation of photo-assimilated compounds of the leaves to the rhizomes.

In studying the effect of two sizes of rhizomes-seed (large and small, 5 and 2 g, respectively) and of three times of planting of tannia (beginning, middle and end of October 1985), Monteiro and Peressin (1997) observed that, in general, the large type rhizomes-seed surpassed the small type rhizomes-seed in production. Moreover, there was interaction between the size of the rhizomes-seed and the plantation time, and the effect of the size of the seedling was always more evident during the first planting time.

Economic analysis

The production costs per hectare varied between the highest cultivation costs of R$ 892.31 (+ 18.27%) and the lowest cost of R$ 592.32 (+ 10.83%). The highest cultivation costs were associated with the use of Extra seedlings, and the lower costs were associated with the use of Tiny seedlings for cultivation without (Table 4) and with (Table 5) a hilling operation. These results indicate that a higher average weight of the cormels used as seedling correlates with the higher weight of this component in the production cost. The results also indicate that the increment in the average weight of seedlings (Table 1) does not always lead to the proportional increment in the yield of the rhizomes' fresh weight (Table 2).

In relation to the total cost, in the cultivation without hilling, the variable costs represented 73.35 and 71.73% between the use of Extra and Tiny seedlings, respectively. In the cultivation with the hilling operation, the variable costs represented between 77.83 and 72.85%. The costs of the seedlings indicated a variation of R$ 733.67 between the cost of Extra seedlings, which weighed the most (average of 5.72 g), and the cost of the Tiny seedlings (average of 2.58 g), which weighed the least. In relation to the total cost, the cost of the Large seedlings represented between 22.06 and 23.14%, while the cost of the Tiny seedlings represented between 11.03 and 12.34% in relation to the cultivation with and without hilling, respectively. The increase of the total costs of production in response to the increase of the weight of the seedling type used in the tannia propagation cultivated with and without hilling confirms that as the average weight of the cormel used as the seedling increases, then the participation of this component in the production cost will be higher (PUIATTI et al., 2004).

The cost of the manpower in the cultivation without hilling represented 36.36% when seedlings of Extra type were used and 43.00% when the Tiny seedlings were used. With the same types of seedlings, the percentages varied between 42.58 and 47.19% in the cultivation with hilling. In relation to the total cost, the high percentages of the production costs related with manpower used in tannia cultivation highlight the importance of this practice as a source that generates agricultural employment.

Using the productive averages of commercial cormels from the 2007-2008, 2008-2009 and 2009-2010 crop seasons, the economic analysis for the tannia cultivation that was propagated using five types of seedlings indicated that it was better to cultivate tannia without hilling and to propagate with seedlings of Extra type, which netted an income of R$ 2,234.08 (Table 6).

Table 4. Yield cost of one hectare of 'Comum' tannia, propagated with five seedling types, without hilling operation. Average of 2007-08, 2008-09 and 2009-10 crop seasons.

Cost components	Cost in function of the seedling type (R$)				
	Extra	Large	Médium	Small	Tiny
		1. Variable costs			
Inputs: [1]Seedlings	1,293.39	933.86	759.75	660.26	583.38
[2]Transport	43.11	31.13	25.33	22.01	19.45
	1,336.50	964.99	785.08	682.27	602.83
		[3]Manpower			
Planting	300.00	300.00	300.00	300.00	300.00
Irrigation	360.00	360.00	360.00	360.00	360.00
Weeding	540.00	540.00	540.00	540.00	540.00
Harvest	900.00	900.00	900.00	900.00	900.00
	2,100.00	2,100.00	2,100.00	2,100.00	2,100.00
		Machinery			
Irrigation pump	560.00	560.00	560.00	560.00	560.00
Tractor	240.00	240.00	240.00	240.00	240.00
	800.00	800.00	800.00	800.00	800.00
Subtotal 1 (R$)	4,236.50	3,864.99	3,685.08	3,582.27	3,502.83
		2. Fixed costs			
Boon	352.50	352.50	352.50	352.50	352.50
Remuneration of land	160.00	160.00	160.00	160.00	160.00
Subtotal 2 (R$)	512.50	512.50	512.50	512.50	512.50
		3. Other costs			
Contingency (10% ST1 + ST2)	474.90	437.75	419.76	409.48	401.53
Administration (5% ST1+ ST2)	237.45	218.88	209.88	204.74	200.77
Subtotal 3	712.35	656.68	629.64	614.22	602.30
Total	5,461.35	5,034.12	4,827.22	4,708.99	4,617.63
Quarterly interest (0,72%)	314.57	289.97	278.05	271.24	265.98
Total Geral	5,775.92	5,324.09	5,105.27	4,980.23	4,883.61

Adapted of Terra et al. (2006) and Rezende et al. (2009). ST1 = Subtotal 1; ST2 = Subtotal 2. [1]Cost of 1.0 kg of seedling = R$ 1.50. [2]Cost of transport of 1.0 t of seedling = R$ 50.00. [3]Manpower = R$ 30.00 day man^{-1} (D H^{-1}). Harvest = 30 D H^{-1}. Hour of tractor = R$ 60.00. Hour of irrigation pump = R$ 10.00. Land rent year^{-1} = 240.00.

Table 5. Yield cost of one hectare of 'Comum' tannia, propagated with five seedling types, with hilling operation. Average of 2007-08, 2008-09 and 2009-10 crop seasons.

Cost components	Cost in function of the seedling type (R$)				
	Extra	Large	Médium	Small	Tiny
		1. Variable costs			
Inputs: [1]Seedlings	1,293.39	933.86	759.75	660.26	583.38
[2]Transport	43.11	31.13	25.33	22.01	19.45
	1,336.50	964.99	785.08	682.27	602.83
		[3]Manpower			
Planting	300.00	300.00	300.00	300.00	300.00
Irrigation	480.00	480.00	480.00	480.00	480.00
Weeding	360.00	360.00	360.00	360.00	360.00
Harvest	540.00	540.00	540.00	540.00	540.00
Planting	900.00	900.00	900.00	900.00	900.00
	2,580.00	2,580.00	2,580.00	2,580,00	2,580.00
		Machinery			
Irrigation pump	560.00	560.00	560.00	560.00	560.00
Tractor	240.00	240.00	240.00	240.00	240.00
	800.00	800.00	800.00	800.00	800.00
Subtotal 1 (R$)	4,716.50	4,344.94	4,165.08	4,062.27	3,982.83
		2. Fixed costs			
Boon	352.50	352.50	352.50	352.50	352.50
Remuneration of land	160.00	160.00	160.00	160.00	160.00
Subtotal 2 (R$)	512.50	512.50	512.50	512.50	512.50
		3. Other costs			
Contingency (10% ST1 + ST2)	522.90	485.75	467.76	457.48	449.53
Administration (5% ST1+ ST2)	261.45	242.88	233.88	228.74	224.77
Subtotal 3	784.35	728.63	701.64	686.22	674.30
Total	6,013.35	5,586.12	5,379.22	5,260.99	5,169.63
Quarterly interest (0,72%)	346.37	321.76	309.84	303.03	297.77
Total Geral	6,059.72	5,907.88	5,689.06	5,564.02	5,467.40

Adapted of Terra et al. (2006) and Rezende et al. (2009). ST1 = Subtotal 1; ST2 = Subtotal 2. [1]Cost of 1.0 of kg seedling = R$ 1.50. [2]Cost of transport of 1.0 t of seedling = R$ 50.00. [3]Manpower = R$ 30.00 day man^{-1} (D H^{-1}). Hilling = 16 D H^{-1}. Harvest = 30 D H^{-1}. Hour of tractor = R$ 60.00. Hour of irrigation pump = R$ 10.00. Land rent year $^{-1}$ = 240.00.

Table 6. Economic analysis of two taro clones propagated with six types of seedlings. Average of 2007-2008, 2008-2009 and 2009-2010 crop seasons.

Seedling types	Hilling	Commercial production (ton ha^{-1})	Gross income (R$ ha^{-1})	Total cost (R$ ha^{-1})	Net income (R$ ha^{-1})
Extra	Without	5.34	8,010.00	5,775.92	2,234.08
	With	4.86	7,290.00	6,059.72	1,230.28
Large	Without	3.69	5,535.00	5,324.09	210.93
	With	4.99	7,485.00	5,907.88	1,577.12
Medium	Without	4.79	7,185.00	5,105.27	2,079.73
	With	3.08	4,620.00	5,689.06	-1,069.06
Small	Without	4.28	6,420.00	4,980.23	1,439.77
	With	5.06	7,590.00	5,564.02	2,025.98
Tiny	Without	3.59	5,385,00	4,883.61	501.39
	With	4.52	6,780.00	5,467.40	1,312.60

This higher net income surpassed the lower income by R$ 3,303.14, and the lowest income used the Medium type seedlings and hilling operation as a cultural treatment.

These results confirm that the determination of some indices of the economic result should be made known in more detail to structure productive activity and to make the necessary changes to increase efficiency (PEREZ JÚNIOR et al., 2006). In general, profitability is determined by the comparison of proceeds with the production cost. The activity will only be productive if it provides a return that exceeds the alternative cost or opportunity (VILELA; MACEDO, 2000).

Conclusion

Under the conditions in which the experiments were conducted during the 2007-2008, 2008-2009 e 2009-2010 crop seasons, and in considering the higher average yield of commercial cormel fresh weight and the higher net income, it is recommended that the 'Comum' tannia should be produced from seedlings of the Extra type and without a hilling operation.

Acknowledgements

Thanks are given to CNPq and Fundect-MS for their scholarships and financial support.

References

COSTA, C. A.; RAMOS, S. J.; ALVES, D. S.; FERNANDES, L. A.; SAMPAIO, R. A.; MARTINS, E. R. Nutrição mineral do mangarito num Latossolo Vermelho Amarelo. **Horticultura Brasileira**, v. 26, n. 1, p. 102-106, 2008.

DOMIT, L. A.; DALBOSCO, M.; SANTOS, R. M. S.; GUIMARÃES, M. F. Transferência de tecnologia para a cultura da soja – a experiência da COPACOL 61-63. **Semina: Ciências Agrárias**, v. 29, n. 2, p. 255-264, 2008.

FOGAÇA, C. M.; CORDEIRO, D. C.; CORREIA, T. D.; LAURIANO, M. P.; SANT'ANNA-SANTOS, B. F.; FINGER, F. L.; OTONI, W. C. Microtuberização de *Colocasia esculenta* L. Schott (Araceae) *in vitro*. **Revista Brasileira de Biociências**, v. 5, supl. 1, p. 123-125, 2007.

HEREDIA ZÁRATE, N. A.; VIEIRA, M. C. Produção de clones de taro em função dos tipos de mudas. **Horticultura Brasileira**, v. 21, n. 4, p. 646-648, 2003.

HEREDIA ZÁRATE, N. A.; VIEIRA, M. C.; MARTINS, F. M. Produção de massa fresca dos inhames 'Cem/Um' e 'Macaquinho', em três densidades de plantas. **Horticultura Brasileira**, v. 21, n. 1, p. 119-122, 2003a.

HEREDIA ZÁRATE, N. A.; VIEIRA, M. C.; MINUZZI, A. Brotação de seis tipos de mudas dos clones de inhame Roxo e Mimoso. **Ciência e Agrotecnologia**, v. 26, n. 4, p. 699-704, 2002.

HEREDIA ZÁRATE, N. A.; VIEIRA, M. C.; HELMICH, M.; MARIA, M. A. Tamanho de rizoma semente e fileiras de plantas no canteiro na produção do mangarito 'Comum'. **Ciência e Agrotecologia**, v. 30, n. 5, p. 907-913, 2006.

HEREDIA ZÁRATE, N. A.; VIEIRA, M. C.; OLIVEIRA, A. C. P.; LIMA, A. A. Produção e renda bruta de dois cultivares de taro, em cultivo solteiro e consorciado com alface. **Semina: Ciências Agrárias**, v. 26, n. 3, p. 283-290, 2005.

HEREDIA ZÁRATE, N. A.; VIEIRA, M. C.; ROSA JÚNIOR, E. J.; ALVES, J. C. Cama-de-frango de corte semidecomposta na produção dos taros Chinês e Macaquinho. **Acta Scientiarum. Agronomy**, v. 25, n. 1, p. 177-181, 2003b.

HEREDIA ZÁRATE, N. A.; VIEIRA, M. C.; ROSA JÚNIOR, E. J.; SILVA, C. G. Desempenho de taro em função de doses de cama de aviário, sob sistema orgânico de produção. **Pesquisa Agropecuária Tropical**, v. 34, n. 2, p. 111-117, 2004.

HEREDIA ZÁRATE, N. A.; MATTE, L. C.; VIEIRA, M. C.; GRACIANO, J. D.; HEID, D. M.; HELMICH, M. Amontoas e cobertura do solo com cama-de-frango na produção de cebolinha, com duas colheitas. **Acta Scientiarum. Agronomy**, v. 32, n. 3, p. 449-454, 2010a.

HEREDIA ZÁRATE, N. A.; SANGALLI, C. M. S.; VIEIRA, M. C.; GRACIANO, J. D.; MUNARIN, E. E. O.; PAULA, F. S. Cobertura do solo com cama-de-frango, com e sem amontoa, na produção de beterraba. **Ciência e Agrotecnologia**, v. 34, p. 1598-1603, 2010b. (Edição Especial).

HEREDIA ZÁRATE, N. A.; VIEIRA, M. C.; GRACIANO, J. D.; FIGUEIREDO, P. G.; BLANS, N. B.; CURIONI, B. M. Produtividade de mandioquinha-salsa sob diferentes densidades de plantio e tamanho das mudas. **Ciência e Agrotecnologia**, v. 33, n. 1, p. 139-143, 2009.

LARCHER, W. **Ecofisiologia vegetal**. São Carlos: Rima-Artes e Textos, 2006.

LEITE, G. L. D.; SILVA, F. W. S.; JESUS, F. M.; COSTA, C. A.; GUANABENS, R. E. M.; GUAMÃO, C. A. G. Efeito da adubação orgânica, espaçamento e tamanho de rizomas-semente sobre artrópodes em mangarito *Xanthosoma mafaffa* Schott. **Arquivos do Instituto Biológico**, v. 74, n. 4, p. 343-348, 2007.

MELO, A. S.; COSTA, B. C; BRITO, M. E. B.; AGUIAR NETTO, A. O.; VIÉGAS, P. R. A. Custo e rentabilidade na produção de batata doce nos perímetros irrigados de Itabaiana. **Pesquisa Agropecuária Tropical**, v. 39, n. 2, p. 119-123, 2009.

MONTEIRO, D. A.; PERESSIN, V. A. Efeito do tamanho do rizoma-semente, da época e do local de plantio, na produção de rizomas de mangará. **Bragantia**, v. 56, n. 1, p. 155-161, 1997.

PEREZ JUNIOR, J. H.; OLIVEIRA, L. M.; COSTA, R. G. **Gestão estratégica de custos**. 5. ed. São Paulo: Atlas, 2006.

PUIATTI, M. Handling of the taro culture. In: CARMO, C. A. S. (Ed.). **Inhame e taro**: sistemas de produção familiar. Vitória: Incaper, 2002. p. 203-240.

PUIATTI, M.; PEREIRA, F. H. F.; AQUINO, L. A. Crescimento e produção de taro 'Chinês' influenciados por tipos de mudas e camadas de bagaço de cana-de-açúcar. **Horticultura Brasileira**, v. 22, n. 4, p. 722-728, 2004.

PUIATTI, M.; KATSUMOTO, R.; PEREIRA, F. H. F.; BARRELLA, T. P. Crescimento de plantas e produção de rizomas de taro 'Chinês' em função do tipo de muda. **Horticultura Brasileira**, v. 21, n. 1, p. 110-115, 2003.

REZENDE, L. A.; BARROS JÚNIOR, A. P.; CECÍLIO FILHO, A. B.; PÔRTO, D. R. Q.; MARTINS, M. I. E. G. Custo de produção e rentabilidade das culturas de alface, rabanete, rúcula e repolho em cultivo solteiro e consorciadas com pimentão. **Ciência e Agrotecnologia**, v. 33, n. 1, p. 305-312, 2009.

SANGALLI, A.; VIEIRA, M. C.; HEREDIA ZÁRATE, N. A.; SCALON, S. P. Q. Desenvolvimento e produção da carobinha (*Jacaranda decurrens* Cham. subsp.*symmetrifoliolata* Farias & Proença) cultivada sob dois arranjos de plantas, com ou sem cobertura de cama-de-frango no solo. **Revista Brasileira de Plantas Medicinais**, v. 13, n. 4, p. 439-446, 2011.

SILVA, A. C.; CALDEIRA JÚNIOR, C. F.; COSTA, C. A. Produção do mangarito, em função do tamanho do rizoma-semente. **Bioscience Journal**, v. 27, n. 5, p. 706-709, 2011.

TERRA, E. R.; HEREDIA ZÁRATE, N. A.; VIEIRA, M. C.; MENDONÇA, P. S. M. Proposta de cálculo e forma de adubação, com e sem amontoa, para a produção e renda do milho Superdoce 'Aruba'. **Acta Scientiarum. Agronomy**, v. 28, n. 1, p. 75-82, 2006.

VILELA, N. J.; MACEDO, M. M. C. Fluxo de poder no agronegócio: o caso das hortaliças. **Horticultura Brasileira**, v. 18, n. 2, p. 88-94, 2000.

29

Sugarcane cultivation with source potassium low water-soluble

Gaspar Henrique Korndorfer, Hamilton Seron Pereira*, Ivaniele Nahas Duarte, Laerte Rocha Neves Pinto and Luciano Ferreira da Fonseca

Universidade Federal de Uberlândia, Av. Amazonas, s/n., Bloco 2E, S. 122, Cx. Postal 593, 38400-902, Uberlândia, Minas Gerais, Brazil.
**Author for correspondence. E-mail: hspereira@iciag.ufu.br*

ABSTRACT. Thermopotash is a fertilizer with low solubility but has a release level compatible with plant demands. This study aimed to compare the effect of Thermopotash to that of KCl in sugarcane. Field work was conducted during two crop seasons, in a randomised block design. Treatments were applied to the bases of the furrows before planting the sugarcane, arranged in a 2 x 5 factorial design to evaluate two K sources (TK47 - Thermopotash, KCl - potassium chloride) at five levels of K (0, 50, 100, 150, and 200 kg ha^{-1} K_2O). The variables analysed included leaf and soil K, Ca, Mg and Si, soil pH and productivity. The effects of the TK47 fertilizer on sugarcane stalk production were not initially superior to that of the soluble source, but over time, differences between the soluble and insoluble sources increased. The maximum expected sugarcane productivity for both sources was 133 Mg ha^{-1}, which was achieved at an application level of 121 kg ha^{-1} K_2O. TK47 fertilizer provided K, Si, Ca and Mg to the sugarcane in both the first and second harvests. Thus, TK47 may be a viable source of K for sugarcane.

Keywords: thermopotash; potassium chloride; soil; *Saccharum* spp.

Introduction

Potassium (K) deposits containing soluble K minerals occur predominantly in the northern hemisphere, mainly in Canada and Russia. Production tonnage and prices do not take into account the needs of farmers; therefore, as needs continually evolve, alternatives to soluble salts have been explored to produce K fertilizers (Ciceri, Manning, & Allanore, 2015). The investigation of K silicate minerals and their host rocks as alternative sources of K is especially relevant in systems with highly weathered soils that lack significant cation exchange capacity (Manning, 2010).

Verdete is a sedimentary rock that stands out among potential potassium sources and is found in Serra da Saudade (Alto Paranaíba Region of Minas Gerais State, Brazil), which lies atop the San Francisco craton. This rock has low water-soluble K content, indicating a certain limitation in its use as a K fertilizer in its natural state, especially for short cycle crops. Nevertheless, the total K content of Verdete indicates the potential of this rock as a raw material for the production of K fertilizer (Santos et al., 2015). One means to increase the K availability of this rock is through chemical and thermal processing and mixing with $CaCO_3$ at 1,100°C. This process results in a fertilizer called Thermopotash that has lower solubility then soluble K minerals, such as sylvite (KCl) and arcanite (K_2SO_4), but has a release level compatible with plant demands and a possible residual effect. This is a desirable feature given that the effects of the fertilizer last longer and losses from leaching are minimised compared to soluble fertilizers (Nunes, Kautzmann, & Oliveira, 2014).

Assessing the dynamics of nutrient release from sources with low water solubility requires long-term field studies. Fertilization in sugarcane increases plant height and plant dry matter (Thengavelu, 2009). Sugarcane is a good candidate for this type of study given its long crop cycle, extremely high K requirements, and economic significance in Brazil.

Therefore, this study aimed to compare the effect of Thermopotash (TK47 fertilizer) to that of KCl on sugarcane productivity, nutrient uptake and soil nutrient availability.

Material and method

The experiment was conducted with sugarcane cultivar SP 832847 grown in an Ustoxic soil and was set up in June, 2011 in Chaveslandia, Minas Gerais State, Brazil (19°04'07.49"S and 50°22'19.70"W). The 0-20 cm layer of the soil had the following characteristics: pH in water = 4.4, P (Mehlich 1) = 2.0 mg dm^{-3}, K (Mehlich 1) = 0.04 $cmol_c\,dm^{-3}$, Ca^{2+} (KCl) = 0.6 $cmol_c\,dm^{-3}$, Mg^{2+} (KCl) = 0.3 $cmol_c\,dm^{-3}$, Si ($CaCl_2$ 0.01 mol L^{-1}) = 3.0 mg dm^{-3}, H+Al = 2.44 $cmol_c\,dm^{-3}$, CEC = 3,34 $cmol_c\,dm^{-3}$, sand = 860 g kg^{-1}, silt = 10 g kg^{-1}, and clay = 130 g kg^{-1}.

Accumulated rainfall totalled 1,323 mm before the first sugarcane harvest and 1,223 mm before the second harvest. Soil preparation consisted of an application of dolomitic limestone (1 Mg ha^{-1}) over the entire experimental area followed by harrowing and levelling. Before planting the sugarcane, 250 kg ha^{-1} monoammonium phosphate (MAP) was applied to the base of the furrow using a plough. Next, the K sources (KCl and TK47 fertilizer) at the various application levels were distributed manually in the sugarcane furrow. After the first harvest, gypsum (1 t ha^{-1}) and dolomitic limestone (1 t ha^{-1}) were superficially applied throughout the entire experimental area.

The experiment was carried out at a sugarcane plant and set up in a randomised block design with five replication blocks for a total of 50 plots. Each experimental plot covered 150 m^2 and consisted of five 20-m-long rows of sugarcane spaced 1.5 m apart. Treatments were arranged in a 2 x 5 factorial design, evaluating two K sources (TK47 fertilizer - Thermopotash and KCl) at five levels of K (0, 50, 100, 150, and 200 kg ha^{-1} K_2O).

TK47 is a mineral fertilizer developed by Verde Fertilizantes and is obtained by mixing limestone, Verdete and special reagents at a ratio of approximately 1 ($CaCO_3$): 5 (Verdete). After mixing, the particle size is greatly reduced, and the mixture is then heat treated in a thermal reactor at approximately 1,000°C.

The K sources were characterised physically and chemically (MAPA, 2017). TK47 is a grained fertilizer (granulometry is between 2.83 and 4.80 mm) and is characterised chemically by total K_2O = 7%, soluble K_2O in 2% citric acid = 6%, total Si = 28%, total Ca = 22%, and total Mg = 1%. It has a salt index of 0.5, and the neutralising power was determined as 55% $EqCaCO_3$. KCl has K_2O = 60%, Cl = 47%, and a salt index of 116. Thus, these sources are capable of providing other nutrients in addition to K. Therefore, to provide 200 kg of K_2O, 2,857 kg ha^{-1} TK47 and 333 kg ha^{-1} KCl were applied (Table 1).

After application of the treatments, sugarcane sets of cultivar SP 832847 were manually planted in the furrows at a density of 17 buds per metre. Earthing-up and an application of ammonium nitrate (80 kg ha^{-1}) were carried out 90 days after planting.

To evaluate the residual effect of the K applications on the ratoon cane, all plots received applications of nitrogen (150 kg ha^{-1}) as ammonium nitrate (33% N) and 70 kg ha^{-1} P_2O_5 as triple superphosphate (46% P_2O_5). There were no reapplications of the K sources.

Table 1. Quantities of Ca, Mg, Si and K supplied by KCl or TK47 K fertilizers.

Application levels[1]							
TK47					KCl		
Source	K_2O	Ca	Mg	Si	Source	K_2O	Cl
kg ha^{-1}					kg ha^{-1}		
0	0	0	0	0	0	0	0
714	50	157	7	200	83	50	39
1,429	100	314	14	400	167	100	78
2,143	150	471	21	600	250	150	118
2,857	200	629	29	800	333	200	157

[1]Application levels based on the total nutrient levels found in the sources TK47 = Thermopotash (K_2O total = 7%; Si total = 28%, Ca total = 22%, Mg total = 1%) and KCl = Potassium chloride (K_2O total = 60%; Cl total = 47%).

Leaves were collected in February 2012 (cane plant) and in February 2013 (ratoon cane), and an analysis of K, Ca, Mg (Raij, Andrade, Cantarella, & Quaggio, 2001) and Si (Datnoff, Snyder, & Korndörfer, 2001) content was undertaken. Soil samples were taken immediately after sugarcane harvest (cane plant and ratoon) at depths of 0-20 cm and 20-40 cm in rows. Measurements were taken of K, Ca, Mg, pH (Raij et al., 2001) and Si (Datnoff et al., 2001). After 12 months of growth, the sugarcane was harvested manually and was not burned. The stalks were weighed with the aid of a tensile load cell.

The results were submitted for analysis of variance using the SISVAR program (Ferreira, 2011). Averages were compared at the 5% significance level. Quantitative treatments were evaluated using regression analysis, and the qualitative results were evaluated using the Tukey test.

Result and discussion

In the first harvest, the TK47 fertilizer produced the same stalk yield (128.6 Mg ha^{-1}) as that of KCl (123.7 t ha^{-1}), considering the average of the K levels applied (Table 2). In this harvest, according to a quadratic model ($y = 0.0012X^2 + 0.2905X + 115.03$), sugarcane productivity was 115.03 t ha^{-1} for the control treatment (without K application) but reached 132.61 t ha^{-1} with application of 121.04 kg ha^{-1} K_2O, regardless of source. Sugarcane stalk productivity decreased at higher application levels (Figure 1A, harvest 1).

Averaging the different K levels, sugarcane productivity in the second harvest was 5.44% higher using TK47 fertilizer application versus the use of KCl (Table 2). Figure 1A shows a linear response of production to dose increase.

Table 2. Effects of K levels and sources on stalk productivity for sugarcane variety SP832847, 2011/12 and 2012/13 crops (Usina Vale do São Simão, Chaveslandia, Minas Gerais State, Brazil).

K_2O levels	1st harvest			2nd harvest		
	TK47	KCl	Mean	TK47	KCl	Mean
kg ha^{-1}	-----------------------Stalk Productivity, Mg ha^{-1} ---					
0	113.0	113.6	113.3	108.2	107.8	108.0
50	135.7	122.8	129.3	119.5	111.9	115.7
100	136.1	132.5	134.3	124.3	115.6	119.9
150	123.4	128.8	126.1	125.2	117.2	121.2
200	134.8	120.8	127.8	126.9	118.7	122.8
Mean	128.6 A	123.7 A		120.8 A	114.3 B	
	CV= 10.57%; LSDsource= 7.65			CV= 7.79%; LSDsource= 5.25		

Means followed by distinct letters within a row differ according to the Tukey test at the 0.05 significance level. TK47 = Thermopotash; KCl = Potassium chloride.

The accumulated result (sum of the 1st and 2nd harvests) clearly shows that, regardless of K level, sugarcane stalk productivity was higher using the TK47 fertilizer versus using KCl (Figure 2). Furthermore, a quadratic fit shows that without K applications, productivity was 224.3 Mg ha^{-1}, but with K applications, regardless of source, productivity increased up to a maximum of 253.3 Mg ha^{-1} with an application of 135 kg ha^{-1} K_2O (Figure 1B). These data demonstrate that for sugarcane production, the residual effect of TK47 is more effective than that of KCl. Duarte, Sousa, Korndorfer, and Pereira (2015) studied different K sources and reported that millet absorbed K from Alfa 02, a K source produced by mixing limestone and Verdete with subsequent heat treatment, in both the first and second crops, thereby demonstrating a residual effect from this source.

The TK47 fertilizer increased Si availability in the soil (Tables 4 and 5) (Figures 4 and 5) and Si leaf concentrations (Table 3 and Figure 3). The supply of Si may also have contributed to increases in sugarcane productivity. Datnoff et al. (2001) observed that silicate applications were associated with increases of up to 17% in cane plant productivity and up to 20% in ratoon productivity.

An evaluation of cane plant nutrition showed that, in the first harvest, the only significant interaction between source and application level was related to leaf K and Ca. KCl was associated with higher leaf K levels compared to the TK47 fertilizer at application levels of 100, 150, and 200 kg ha^{-1} K_2O in the cane plant and at all application levels in the ratoon cane (Table 3). However, average leaf K in crops for both K sources ranged between 10 and 20 g kg^{-1}, which is adequate for sugarcane (Fageria, Baligar, & Jones, 2011).

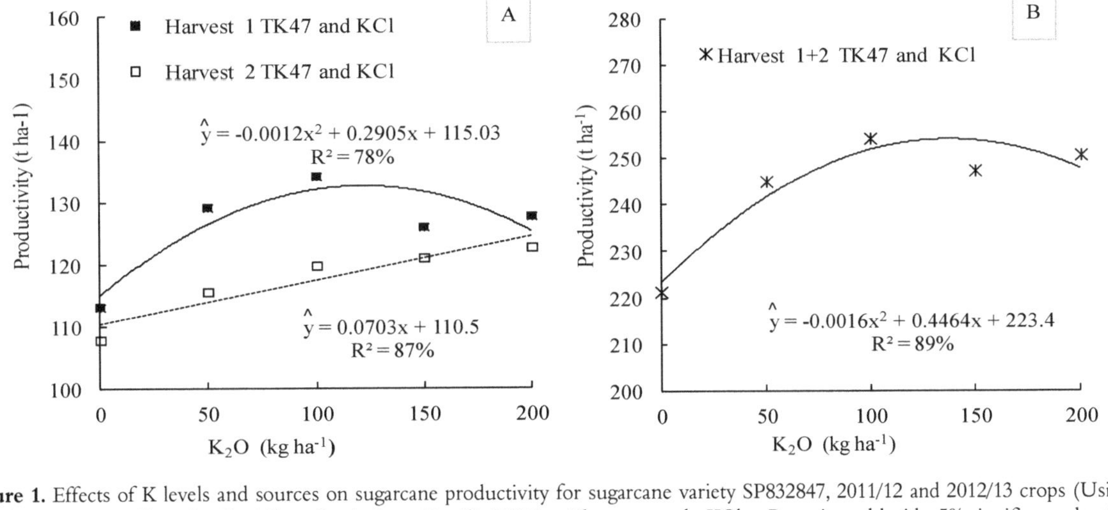

Figure 1. Effects of K levels and sources on sugarcane productivity for sugarcane variety SP832847, 2011/12 and 2012/13 crops (Usina Vale do São Simão, Chaveslandia, Minas Gerais State, Brazil). TK47 = Thermopotash; KCl = Potassium chloride; 5% significance level.

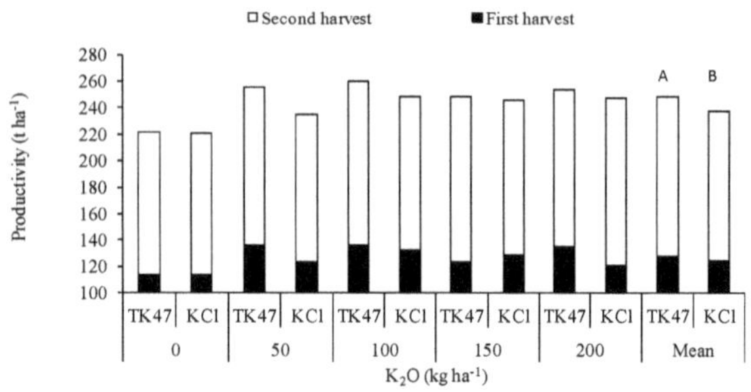

Figure 2. Total productivity (1st + 2nd harvests) of the sugarcane (Usina Vale do São Simão, Chaveslandia, Minas Gerais State, Brazil). Means followed by distinct lower case letters differ according to the Tukey test at 5% significance regarding K sources (LSD = 10.95). TK47 = Thermopotash; KCl = Potassium chloride.

Although sugarcane absorbs more K when it is fertilised using KCl than when it is fertilised using TK47, this difference does not translate into increased productivity (Table 2), indicating that these plants had K levels that are more than sufficient. According to Setiyono, Walters, Cassman, Witt, and Dobermann (2010), plants have the ability to absorb more K than is required for full development, a situation that is commonly referred to as luxury consumption.

Leaf Ca concentration was higher when using TK47 fertilizer than when using KCl at levels of 150 and 200 kg ha^{-1} K_2O in the plant cane and for all levels in the ratoon cane (Table 3) and appeared to be within the optimum range (Raij, 2011) in both harvests. This occurred because the limestone used to correct the soil acidity also supplies Ca, whereas the TK47 fertilizer contains CaO from limestone. The K sources did not affect leaf Mg and Si in the plant cane crop; however, in the second harvest, the levels of leaf Mg were higher for the TK47 treatment, regardless of application level (Table 3). However, leaf Si was higher when using the TK47 fertilizer than when using KCl at 150 and 200 kg ha^{-1} K_2O (Table 3), but these concentrations were still not high enough to be considered ideal for sugarcane cultivation (6 g kg^{-1}) (McCray et al., 2010).

A linear fit or direct relationship was found between higher K application levels and increases in leaf nutrient levels. This relationship was observed for KCl in the first sugarcane harvest and for KCl and the TK47 fertilizer in the second harvest (Figure 3A). This direct relationship existed for leaf Ca the TK47 fertilizer treatment in the first harvest (Figure 3B) and for leaf Si in the TK47 fertilizer treatment after the second harvest (Figure 3C).

Table 3. Effects of K levels and sources on leaf concentrations of K, Ca, Mg, and Si for sugarcane variety SP832847, 2011/12 and 2012/13 crops (Usina Vale do São Simão, Chaveslandia, Minas Gerais State, Brazil).

K_2O levels	1st harvest					2nd harvest				
	TK47		KCl		Mean	TK47		KCl		Mean
kg ha^{-1}	K g kg^{-1}									
0	14.9	A	14.7	A	14.8	8.4		9.3		8.8
50	15.1	A	15.2	A	15.2	8.8		9.2		9.0
100	14.0	B	15.5	A	14.7	9.1		10.0		9.5
150	14.0	B	16.6	A	15.3	9.0		10.5		9.7
200	14.7	B	17.0	A	15.8	9.1		11.1		10.1
Mean	14.5		15.8			8.9	B	10.0	A	
	CV = 5.15%; LSDsource = 1,0					CV = 9.54%; LSDsource = 0.52				
	Ca g kg^{-1}									
0	3.02	A	3.19	A	3.10	2.48		1.87		2.18
50	3.26	A	3.96	A	3.61	3.41		2.39		2.90
100	4.29	A	3.98	A	4.14	2.80		2.15		2.48
150	4.96	A	3.93	B	4.44	3.06		2.09		2.57
200	5.04	A	3.81	B	4.43	2.65		2.04		2.34
Mean	4.11		3.77			2.88	A	2.10	B	
	CV = 15.77%; LSDsource = 0.80					CV = 22.59%; LSDsource = 0.32				
	Mg g kg^{-1}									
0	1.91		1.78		1.84	1.76		1.54		1.65
50	2.05		1.97		2.01	1.91		1.88		1.89
100	2.12		1.97		2.04	1.90		1.73		1.82
150	2.11		1.88		2.00	1.85		1.67		1.76
200	2.01		1.94		1.98	1.96		1.69		1.83
Mean	2.04	A	1.91	A		1.88	A	1.70	B	
	CV = 12.61%; LSDsource = 0.14					CV = 12.48%; LSDsource = 0.13				
	Si g kg^{-1}									
0	2.52		2.86		2.69	2.22	A	2.27	A	2.25
50	3.86		2.98		3.42	2.66	A	2.44	A	2.55
100	2.93		3.14		3.03	2.73	A	2.45	A	2.59
150	3.42		3.17		3.30	3.12	A	2.37	B	2.75
200	3.01		3.20		3.11	3.18	A	2.43	B	2.80
Mean	3.15	A	3.07	A		2.79		2.39		
	CV = 24.64%; LSDsource = 0.44					CV = 13.40%; LSDsource = 0.44				

Means followed by distinct, uppercase letters within a row differ according to the Tukey test at the 0.05 significance level. TK47 = Thermopotash; KCl = Potassium chloride.

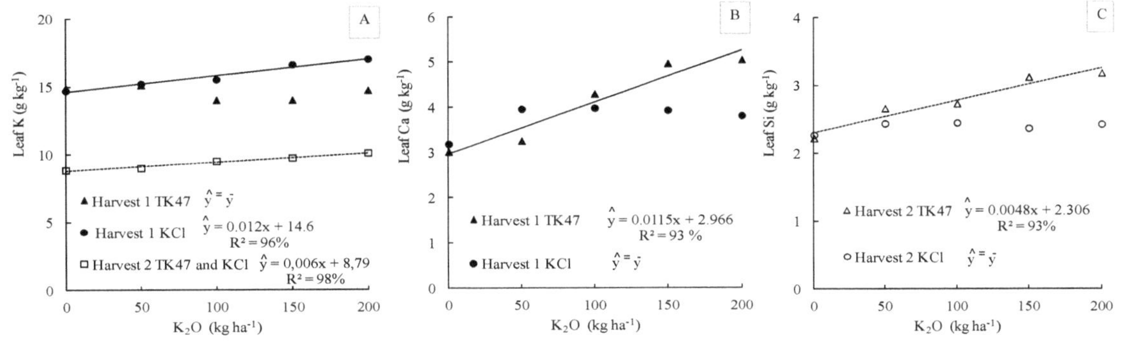

Figure 3. Effects of K levels and sources on leaf concentrations of K, Ca, and Si for sugarcane variety SP832847, 2011/12 and 2012/13 crops (Usina Vale do São Simão, Chaveslandia, Minas Gerais State, Brazil). TK47 = Thermopotash; KCl = Potassium chloride; 5% significance level (*).

Table 4. Effects of K levels and sources on soil availability of K, Ca, Mg, and Si in the 0-20 cm soil layer after the sugarcane harvest for variety SP832847, 2011/12 and 2012/13 crops (Usina Vale do São Simão, Chaveslandia, Minas Gerais State, Brazil).

K_2O Levels	1st harvest					2nd harvest				
Cane Plant	TK47		KCl		Mean	TK47		KCl		Mean
Depth 0-20 cm										
kg ha^{-1}	K Mehlich ($cmol_c dm^{-3}$)									
0	0.05	A	0.05	A	0.05	0.06		0.05		0.05
50	0.07	A	0.08	A	0.08	0.07		0.05		0.06
100	0.18	A	0.07	B	0.13	0.12		0.06		0.09
150	0.25	A	0.06	B	0.16	0.14		0.06		0.10
200	0.28	A	0.07	B	0.18	0.13		0.07		0.10
Mean	0.17		0.07			0.11	A	0.06	B	
	CV = 15.77%; LSDsource = 0.80					CV = 44.96%; LSDsource = 0.02				
	Ca ($cmol_c dm^{-3}$)									
0	0.94	A	0.74	A	0.84	1.10	A	1.06	A	1.08
50	0.98	A	0.69	B	0.84	1.35	A	0.96	B	1.15
100	1.26	A	0.73	B	0.99	1.73	A	1.24	B	1.49
150	1.38	A	0.65	B	1.03	1.37	A	1.00	B	1.19
200	1.40	A	0.68	B	1.04	1.54	A	1.10	B	1.32
Mean	1.19		0.70			1.42		1.07		
	CV = 21.47%; LSDsource = 0.26					CV = 21.78%; LSDsource = 0.34				
	Mg ($cmol_c dm^{-3}$)									
0	0.26		0.26		0.26	0.45	A	0.42	A	0.43
50	0.27		0.26		0.27	0.45	A	0.51	A	0.48
100	0.31		0.24		0.28	0.52	B	0.87	A	0.69
150	0.32		0.24		0.28	0.41	B	0.79	A	0.60
200	0.30		0.26		0.28	0.49	A	0.40	A	0.44
Mean	0.30	A	0.25	A		0.47		0.60		
	CV = 31.41%; LSDsource = 0.05					CV = 17.93%; LSDsource = 0.12				
	Si ($mg dm^{-3}$)									
0	1.40	A	1.44	A	1.42	1.63	A	1.58	A	1.61
50	1.65	A	1.32	A	1.48	2.06	A	1.43	A	1.74
100	1.96	A	1.20	B	1.58	3.07	A	1.97	B	2.52
150	2.02	A	1.11	B	1.56	2.83	A	1.80	B	2.31
200	2.14	A	1.10	B	1.62	2.89	A	2.20	B	2.50
Mean	1.84		1.23			2.49		1.79		
	CV = 19.50%; LSDsource = 0.38					CV = 23.67%; LSDsource = 0.66				

Means followed by distinct uppercase letters within the same row differ by the Tukey test at the 0.05 significance level. TK47 = Thermopotash; KCl = Potassium chloride.

Leaf K levels in the first harvest were higher than those in the second harvest. These differences indicate lower K availability in the soil due to the absence of subsequent K applications. The TK47 fertilizer produced higher soil K availability compared to KCl in the 0-20 cm layer at application levels of 100, 150, and 200 kg ha^{-1} K_2O after the first harvest and at all levels of K_2O after the second harvest (Table 4). In the 20-40 cm layer, the TK47 fertilizer produced higher soil K availability compared to KCl at application levels of 150 and 200 kg ha^{-1} K_2O (after the first and second harvests) (Table 5).

Soil Si was higher using the TK47 fertilizer than when using KCl (Tables 4 and 5). This occurred in the 0-20 cm layer after the first and second harvests at levels of 100, 150, and 200 kg ha^{-1} K_2O (Table 4). The same was found in the 20-40 cm layer at levels of 150 and 200 kg ha^{-1} K_2O after the first harvest and at 50, 100, 150, and 200 kg ha^{-1} K_2O after the second harvest (Table 5). Si availability in the soil was higher under the TK47 treatment compared to KCl given that Thermopotash contains Si (Table 1) and that this Si is available to the soil (Figures 4D and 5D) and plant (Figure 3C).

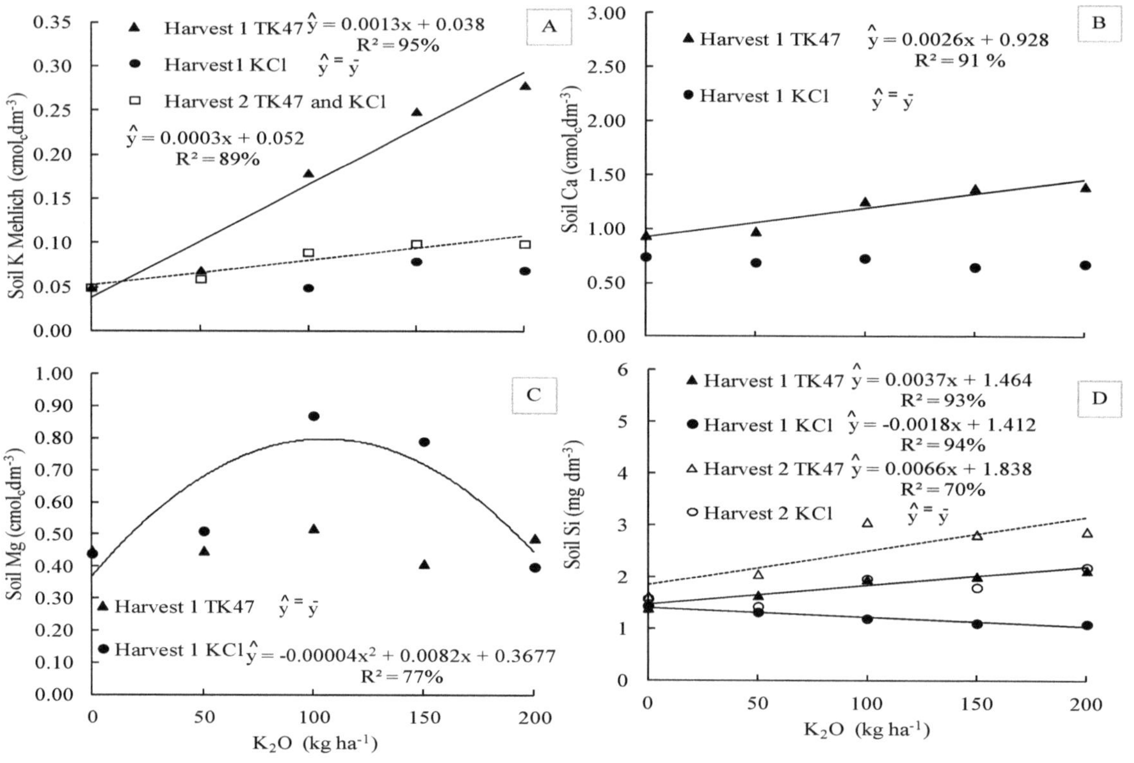

Figure 4. Effects of K levels and sources on soil availability of K, Ca, Mg, and Si in the 0-20 cm soil layer after the sugarcane harvest for variety SP832847, 2011/12 and 2012/13 crops (Usina Vale do São Simão, Chaveslandia, Minas Gerais State, Brazil).

Table 5. Effects of K levels and sources on soil availability of K, Ca, Mg, and Si in the 20-40 cm soil layer after the sugarcane harvest for variety SP832847, 2011/12 and 2012/13 crops (Usina Vale do São Simão, Chaveslandia, Minas Gerais State, Brazil).

K_2O Levels Cane Plant	1st harvest					2nd harvest				
	TK47		KCl		Mean	TK47		KCl		Mean
	Depth 20-40 cm									
kg ha^{-1}	K Mehlich ($cmol_c dm^{-3}$)									
0	0.03	A	0.05	A	0.03	0.05	A	0.03	A	0.04
50	0.05	A	0.04	A	0.05	0.10	A	0.04	A	0.07
100	0.03	A	0.04	A	0.04	0.11	A	0.05	A	0.08
150	0.22	A	0.04	B	0.13	0.17	A	0.06	B	0.12
200	0.30	A	0.05	B	0.18	0.30	A	0.06	B	0.18
Mean	0.13		0.04			0.15		0.05		
	CV = 47.70%; LSDsource = 0.07					CV = 51.97%; LSDsource = 0.09				
	Ca ($cmol_c dm^{-3}$)									
0	0.31	A	0.34	A	0.32	0.68	A	0.70	A	0.69
50	0.33	A	0.37	A	0.35	1.34	A	0.69	B	1.01
100	0.44	A	0.28	A	0.36	1.50	A	0.98	B	1.24
150	0.80	A	0.16	B	0.48	1.58	A	0.68	B	1.40
200	0.78	A	0.17	B	0.48	1.87	A	0.89	B	1.38
Mean	0.53		0.26			1.39		0.79		
	CV = 47.49%; LSDsource = 0.24					CV = 32.17%; LSDsource = 0.45				
	Mg ($cmol_c dm^{-3}$)									
0	0.19	A	0.20	A	0.19	0.15		0.15		0.15
50	0.16	A	0.22	A	0.19	0.19		0.15		0.17
100	0.19	A	0.18	A	0.19	0.22		0.23		0.22
150	0.23	A	0.14	B	0.18	0.17		0.17		0.17
200	0.24	A	0.14	B	0.18	0.24		0.22		0.23
Mean	0.20		0.18			0.19	A	0.18	A	
	CV = 28.43%; LSDsource = 0.06					CV = 35.35 %; LSDsource =0.03				
	Si (mg dm^{-3})									
0	1.11	A	1.20	A	1.15	1.23	A	1.12	A	1.17
50	1.26	A	1.24	A	1.25	1.94	A	1.09	B	1.52
100	1.28	A	1.26	A	1.26	2.45	A	1.12	B	1.78
150	1.51	A	1.23	B	1.37	2.94	A	1.17	B	2.06
200	2.02	A	1.28	B	1.65	2.96	A	1.46	B	2.21
Mean	1.43		1.24			2.30		1.19		
	CV = 14.52%; LSDsource = 0.25					CV = 27.37%; LSDsource = 0.61				

Means followed by distinct uppercase letters within a row differ by the Tukey test at the 0.05 significance level. TK47 = Thermopotash; KCl = Potassium chloride.

Sugarcane has a high demand for potassium. The response of sugarcane to potassium fertilization is dependent on the available K of the soil (Kumawat, Rajawat, & Rathod, 2016). Available K at both depths and after the first and second sugarcane harvests is considered low and very low (as classified by Raij, 2011) for the KCl treatment, respectively, while they are considered adequate for TK47, thereby demonstrating the residual effect of the Thermopotash. According to Chiba, Natale, Cruz, Teixeira, and Centurion (2008), the amount of K absorbed by the plants was often greater than the variation of exchangeable K content, suggesting that non ex changeable forms are being taken up.

The TK47 fertilizer yielded higher soil Ca after both harvests in the 0-20 cm layer (Table 4) and in the 20-40 cm layer (Table 5), except at 0 kg ha^{-1} K_2O for both layers and in the 20-40 cm layer at 50 and 100 kg ha^{-1} K_2O (after the first and second harvests). This soil Ca availability is greater than 0.7 $cmol_c\ dm^{-3}$ and is therefore considered high (Raij et al., 2001). The higher levels of available soil Ca in the TK47 fertilizer treatments are a result (Table 1) of the limestone used in the production of this fertilizer. The application of KCl and the TK47 fertilizers did not affect soil Mg availability in the 0-20 cm (Table 4) and 20-40 cm (Table 5) layers after the first and second harvests. According to Raij (2011), soil Mg availability is considered high when it is greater than 0.8 $cmol_c\ dm^{-3}$, a level that was not achieved in the present study.

Polynomial regressions showed that there was a linear fit only for Mehlich K (Figure 4), Ca (Figure 4B), and Si (Figure 4D) in the 0-20 cm soil layer after the first harvest. The K and Si availability increased, whereas the soil Ca decreased with higher TK47 fertilizer application levels. Conversely, soil Si decreased with higher KCl application levels. After the second harvest, soil K availability (Figure 4A) increased with higher K application levels for both sources. However, Si availability only increased with higher TK47 fertilizer levels (Figure 4D). A quadratic fit was found between soil Mg (0-20 cm layer) and KCl availability (Figure 4C). Soil Mg increased until a maximum of 0.71 $cmol_c\ dm^{-3}$ with an application of 85 kg ha^{-1} K_2O (KCl), and at higher application levels, soil Mg tended to decrease. In the 20-40 cm layer, after the first and second harvests, higher TK47 fertilizer applications increased soil K (Figure 5A), Ca (Figure 5B) and Si (Figure 5D). Mg (Figure 5C) and soil pH (Figure 5E) also increased with greater TK47 fertilizer applications, but only after the first harvest.

The aim of this experiment was not to evaluate the corrective capacity of this fertilizer; therefore, lime was applied at recommended levels throughout. Nevertheless, the TK47 fertilizer did neutralise soil acidity because of the Ca and Mg oxides in this product.

Sandy soils with low levels of organic matter have low cation-exchange capacity (CEC) and therefore retain small amounts of cations and are more susceptible to nutrient leaching (Raij, 2011). Thus, the higher soil pH caused by the TK47 fertilizer probably contributed to a higher soil CEC that in turn led to the greater availability of K, Ca, and Mg in the soil and consequently the sugarcane. This occurs because the soil used in the present study has a low CEC (Raij, 2011) and contains clay minerals with predominantly pH-dependent charges. Additionally, it was found that in the 20-40 cm layer, after the first harvest, KCl reduced Ca (Figure 5B) and Mg (Figure 5C) availability as soil K availability increased. This was probably due to the chlorine released from the KCl.

According to Raij et al. (2001), most soils have a predominance of negative charges, and anions such as NO^{3-}, Cl^- and SO^{4-} are not adsorbed and can thus be leached. These anions in the soil solution obey the electroneutrality principle and thus carry equivalent quantities of cations such as Ca^{2+}, Mg^{2+}, and K^+.

Therefore, many of the Cl^- anions stay in soil solution and can take up cations. This may have occurred with Ca and Mg, which decreased with higher levels of K from KCl.

In the current experiment, the highest sugarcane productivity was achieved with the application of 121 kg ha^{-1} K_2O. The application level to be used for the TK47 fertilizer is a viable 1.72 t ha^{-1}. Even though the K level in TK47 (7% K_2O) is low, its effect is residual, avoiding the need for repeated applications. Furthermore, TK47 can also supply other nutrients (Ca, Mg, and Si) and increase soil pH.

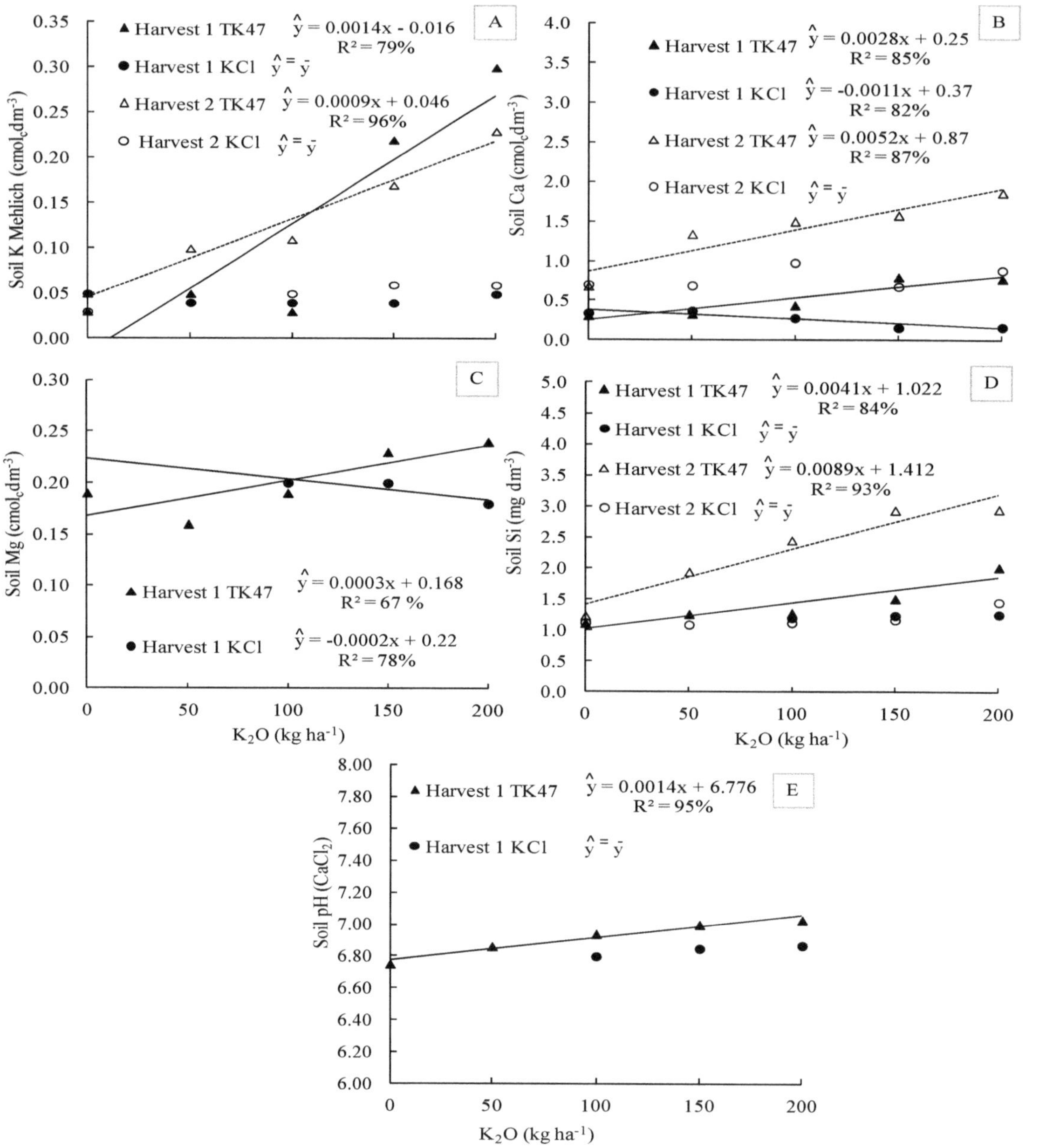

Figure 5. Effects of K levels and sources on soil availability of K, Ca, Mg, and Si and pH in the 20-40 cm soil layer after the sugarcane harvest for variety SP832847, 2011/12 and 2012/13 crops (Usina Vale do São Simão, Chaveslandia, Minas Gerais State, Brazil). TK47 = Thermopotash; KCl = Potassium chloride; 5% significance level (*).

Conclusion

The TK47 fertilizer has a residual effect and can be used as a K source for sugarcane. This source of poorly water-soluble K was not initially advantageous compared to soluble sources, but over time, this source was found to be superior to soluble sources.

The maximum expected productivity of sugarcane variety SP832847 using TK47 fertilizer is 132.61 t ha^{-1} at an application level of 121 kg ha^{-1} K_2O.

The TK47 fertilizer applied to the sugarcane plant provided K, Si, Ca, and Mg and raised the soil pH in both the cane crop and the first ratoon crop.

Acknowledgements

We are grateful to the Brazilian organisations Capes, Fapemig and CNPq. We are also grateful to Verde Fertilizante for supplying Thermopotash and to the owners and employees of the Usina Vale do São Simão.

References

Chiba, M. K., Natale,W., Cruz, M. C. P., Teixeira, L. A. J., & Centurion, J. F. (2008). Potássio nas frações texturais de um Latossolo. *Acta Scientiarum. Agronomy*, *30*(4), 581-587. doi: 10.4025/actasciagron.v30i4.5321.

Ciceri, D., Manning, D. A. C., & Allanore, A. (2015). Historical and technical developments of potassium resources. *Science of The Total Environment*, *502*(2), 590-601. doi: 10.1016/j.scitotenv.2014.09.013.

Datnoff, L. E., Snyder, G. H., & Korndörfer, G. H. (2001). *Silicon in agriculture*. Amsterdam, NE: Elsevier.

Duarte, I. N., Sousa, R. T. X., Korndorfer, G. H., & Pereira, H. S. (2015). Alternative potash fertilizer source for millet crop. *Journal of Plant Nutrition*, *38*(12), 1961–1972. doi: 10.1080/01904167.2015.1009097.

Fageria, N. K., Baligar, V. C., & Jones, C. A. (2011). *Growth and mineral nutrition of field crops* (3rd ed.). London, UK: CRC Press.

Ferreira, D. F. (2011). SISVAR: Um sistema computacional de análise estatística. *Ciência e Agrotecnologia, 35*(6), 1039-1042. doi: 10.1590/S1413-70542011000600001.

Kumawat, P. D., Rajawat, B. S., & Rathod, A. D. (2016). Role of potassium nutrition in improving growth, yield and quality of sugarcane. *Asian Journal of Science and Technology*, *7*(9), 3609-3613.

Manning, D. A. C. (2010). Mineral sources of potassium for plant nutrition. *Agronomy for Sustainable Development*, *30*(2), 281-294. doi: 10.1051/agro/2009023.

McCray, J. M., Ji, S., Powell, G., Montes, G., Perdomo, R., & Luo, Y. (2010). Boundary lines used to determine sugarcane production limits at leaf nutrient concentrations less than optimum. *Communications in Soil Science and Plant Analysis*, *41*(5), 606-622. doi: 10.1080/00103620903531185.

Ministério da agricultura, Pecuária e Abastecimento. [MAPA] (2017). *Manual de métodos analíticos oficiais para fertilizantes e corretivos*. Brasília, DF: MAPA.

Nunes, J. M .G., Kautzmann, R. M., & Oliveira, C. (2014). Evaluation of the natural fertilizing potential of basalt dust waste from the mining district of Nova Prata (Brazil). *Journal of Cleaner Production*, *84*(1), 649-656. doi: 10.1016/j.jclepro.2014.04.032.

Raij, B. van, Andrade, J. C., Cantarella, H., & Quaggio, J. A. (2001). *Análise química para avaliação da fertilidade de solos tropicais*. Campinas, SP: Editora do Instituto Agronômico.

Raij, B. van (2011). *Fertilidade do solo e manejo dos nutrientes*. Piracicaba, SP: Editora International Plant Nutrition Institute.

Setiyono, T. D., Walters, D. T., Cassman, K. G., Witt, C., & Dobermann, A. (2010). Estimating maize nutriente uptake requirements. *Field Crops Research, 118*(2), 158–168. doi: 10.1016/j.fcr.2010.05.006.

Santos, W. O., Mattielo, E. M., Costa, L. M., Abrahão, W. A. P., Novais, R. F., & Cantarutti, R. B. (2015). Thermal and chemical solubilization of verdete for use as potassium fertilizer. *International Journal of Mineral Processing*, *140*(10), 72-78. doi: 10.1016/j.minpro.2015.05.003

Thangavelu, S. (2009). *Mineral nutrition in sugarcane crop production and improvement*. Texas, US: Stadium Press LLC.

30

Improvement of production and fruit quality of pitayas with potassium fertilization

Denison Ramalho Fernandes[1], Rodrigo Amato Moreira[2*], Maria do Céu Monteiro da Cruz[1], Josimara Mendes Rabelo[1] and Jéssica de Oliveira[1]

[1]*Instituto Federal do Norte de Minas, Fazenda São Geraldo, s/n, Estrada Januária, km 06, 39480-000, Januária, Minas Gerais, Brazil.*
[2]*Universidade Federal dos Vales do Jequitinhonha e Mucuri, Diamantina, Minas Gerais, Brazil. *Author for correspondence. E-mail: amatomoreira@yahoo.com.br*

ABSTRACT. Pitaya is a fruit that is becoming promising for cultivation around the world due to the quality of its fruits in terms of taste and their peculiar external shape. Because it is a recenttly cultivated crop, information about fertilization is essential to enable its production. In this sense, this study was performed aiming to evaluate different doses of K_2O for pitaya cultivation. *Hylocereus undatus* and *Hylocereus polyrhizus* were studied under field conditions. Pitayas were cultivated at a 3 × 3m spacing, and four doses of K_2O were tested: 0, 50, 100, and 200 g plant^{-1}, with a 2 × 4 factorial arrangement distributed in a randomized block design. The production per plant (kg and number of fruits) was evaluated in the production cycles of 2013/2014, 2014/2015, and 2015/2016. The fruit quality was also evaluated: longitudinal diameter, transverse diameter, mass, soluble solids content, titratable acidity and soluble solids/titratable acidity ratio. Potassium fertilization increased production and provided a better quality of pitaya fruits.K_2O fertilization should be from 106.0 to 133.3 g K_2O in the first year and 200 g K_2O in the second and third years after planting.

Keywords: chemical fertilizer; exotic fruit; *Hylocereus undatus*; *Hylocereus polyrhizus*.

Introduction

The consumption of exotic fruits has significantly increased over recent years, promoting the interest of producers and traders. Among the exotic fruit species with great potential for marketing is pitaya (*Hylocereus* sp.), a vine cacti native to the tropical forests of Mexico and Central and South America (Moreira, Ramos, Araújo, & Marques, 2011).

In Brazil, pitaya cultivation is in the expansion phase, with the largest producing area located in the state of São Paulo, especially the region of Catanduva (Bastos et al., 2006). Although it is a small fruit compared to other commercialized fruits, it is becoming more popular in marketing centers due to its exotic appearance and organoleptic properties (Marques, Moreira, Ramos, Araújo, & Silva, 2011; Moreira et al., 2011).

To expand the area of pitaya cultivation in this country, knowledge about crop management is necessary, especially in regard to fertilization.

Although it is considered to be a hardy species due to its adaptation to different climatic and soil conditions, pitaya needs balanced fertilization to supply its nutritional demand, satisfactorily enabling its growth and production (Corrêa, Almeida, Marques, Vale, & Aquino, 2014).

In Brazil, fertilization is performed as needed sporadically or based on doses recommended in other countries with different climatic and soil conditions (Cavalcante et al., 2011) because there is no information regarding plant nutrition for growth and satisfactory production.

For accurate nutrient recommendations, knowledge about the nutritional state of the species is necessary, which takes into account the physiological ecology and the plant growth rate. Thus, the extrapolation of results that have been obtained in other cropping systems is not recommended (Souza, Pio, Coelho, Rodas, & Silva, 2015).

Among the essential nutrients for culture, potassium is highly required, participates in many biochemical processes, and, when properly applied, favors the quality of pitaya fruit (Then, 2013).

Studies about potassium fertilization for pitaya in Brazil are still in the initial stages. In this sense, this study was performed with the aim to identify suitable doses of potassium in the production and fruit quality of two pitaya species.

Material and methods

The experiment was conducted in an orchard located 726 meters above sea level, 18° 04' 15"S latitude and 43° 28' 15"W longitude.

The climate according to Koppen and Gleiger is AW, classified as tropical with a dry winter season, with an average temperature of 24°C in the hottest month and 18.4°C in the coldest month and an average annual temperature of 21.5°C. The average annual rainfall is 1,246 mm, with well-defined dry and rainy seasons.

During the research period, the variation in temperature was recorded using a digital MT 240 MINIPA thermo hygrometer, and rainfall data were collected with the use of a rain gauge, both of which were installed in the experimental area (Figure1).

Two species of pitaya, *Hylocereus undatus* (Haw.) Britton & Rose and *Hylocereus polyrhizus* (Web.) Britton & Rose, which were ten months old and established from 1 m long cuttings, were planted in January 2013. Before planting the pitayas, soil analyses of the experimental area at the depths of 0-20 cm and 20-40 cm were performed (Table 1).

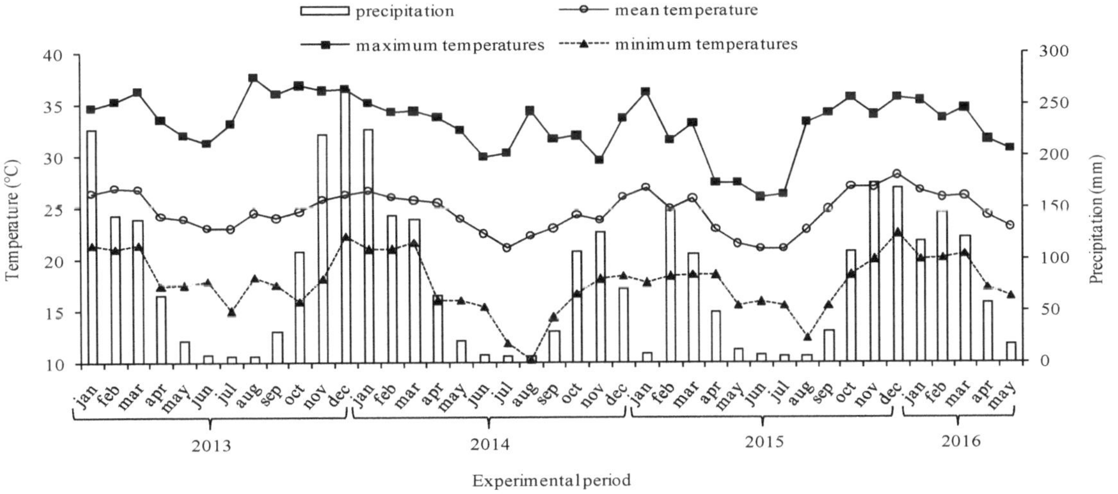

Figure 1. Monthly average temperatures (°C), maximum temperatures (°C), minimum temperatures (°C) and precipitation (mm) from January 2013 to May 2016.

Table 1. Chemical analysis of the soil in the experimental area before plantingat the depths of 0-20 and 20-40 cm.

Depth (cm)	pH	P	K	Ca	Mg	Al^{3+}	H + Al
	H_2O	----mg dm^{-3}----			-----------------$cmol_c$ dm^{-3}-----------------		
0-20	5.1	1.8	40.7	1.0	0.3	0.5	3.7
20-40	4.9	5.09	40.7	0.6	0.3	0.7	4.2

Depth (cm)	SB	T	T	V	M	O.M
	----------$cmol_c$ dm^{-3}----------			-------%-------		dag dm^{-3}
0-20	1.4	1.9	5.1	28	26	0.4
20-40	1.0	1.7	5.2	19	40	0.1

pH(H_2O): soil-water relation 1:2.5; P and K: Mehlich-1 extractor; Ca, Mg and Al: KCl 1 mol L^{-1} extractor; T: cation exchange capacity at pH 7.0; m: aluminum saturation; V: base saturation; OM: organic matter, determined by multiplying the result of organic carbon by 1.724 based on the Walkey-Black method; sand, silt and clay: pipette method.

The holes in which the pitayas were transplanted were 0.5 × 0.5 × 0.5 m, prepared with 200 g of dolomite lime (PRNT = 87%), 20 L of manure, and 300 g of single superphosphate (P_2O_5 = 18.0%, Ca = 18.0%, and S = 20.0%).

The pitayas were attached to eucalyptus fence posts at 1.80 m in height at a spacing of 3 × 3 m. A 1 m long beam was positioned at the end of each fence post, forming a "T"-shaped structure. The plants were attached by a single cladode to reach the desired height; they were later pruned to form the productive shoots.

The experimental design was randomized blocks, with a 2 × 4 factorial design. The factors were two pitaya species and four doses of K_2O: 0, 50, 100, and 200 g per plant, with four blocks and three plants per plot.

The source of K used was potassium chloride (58% of K_2O). The tested doses were split in three months (November, January, and April of each year). In these months, 100 g of N per plant was also applied using ammonium sulfate (20% of N and 22% of S).

One hundred and twenty grams of Yoorin Master® was applied, split between January and April of 2014 and between January and April of 2015, and 5 L of manure was applied in June of 2014 and in June of 2015. Yoorin Master® contains 17.5% P_2O_5, 18.0% Ca, 7.0% Mg, 0.1% B, 0.08% Cu, 0.3% Mn, 10% Si, and 0.55% Zn.

Irrigation was performed weekly, and weed control was conducted by hand up to 40 cm from the plant. The rest of the area was mowed.

The production per plant (kg and number of fruits) was evaluated in the production cycles 2013/2014, 2014/2015, and 2015/2016.

Fruit quality was analyzed using representative samples collected on each fertilization treatment. Ten fruits per plot were collected when they presented a peel with red coloration. The following parameters were evaluated: longitudinal diameter (mm), transverse diameter (mm), mass (g), soluble solids content (°Bx), titratable acidity (% of malic acid) and soluble solids/titratable acidity ratio.

In June 2014 and June 2015 soil samples were collected after harvesting the fruits for the analysis of the nutrient content in the soil. Cladode samples were collected in September 2014 and September 2015.

For the evaluation of soil fertility, samples were taken with a Dutch auger at a distance of 0.2 m from the main cladode and at a depth of 0 - 0.2 m for each plant, forming a sample composed of each experimental plot. The cladodes that were removed to compose the sample were one year old, with an average size of 0.4 m and a healthy appearance. Each sample was composed of five cladodes per plot.

The data were submitted to analysis of variance, and the K_2O doses were submitted to polynomial regression analysis ($p < 0.05$). The species of pitaya were compared by the F test ($p < 0.05$).

Results

In the 2013/2014 productive cycle, differences were observed among the K_2O doses in terms of the number of fruits and production per plant (Figure 2A and B). Increases of 20.7% in the number of fruits per plant of *H. undatus* occurred with an estimated dose of 106g of K_2O per plant. *H. polyrhizus* showed an increase of 24.4% in the plants that received 200 g of K_2O when compared to the control, the plants without potassium fertilization (Figure 2A).

The production per plant (kg) was similar to the number of fruits produced in the 2013/2014 cycle, with an increase of 45.6% in *H. undatus* with the dose of 120 g of K_2O per plant and 25.0% in *H. polyrhizus* with the application of 200 g of K_2O per plant (Figure 2B).

In the 2014/2015 production cycle, *H. undatus* showed increases of 70.6% in the number of fruits per plant for the estimated dose of 135 g of K_2O per plant and 64.6% in production per plant with 200 g of K_2O compared to the plants without potassium fertilization. On the other hand, *H. polyrhizus* had similar production across all K fertilization doses (Figure 2C and D).

In 2015/2016, differences only occurred among K_2O doses in terms of the production per plant. The species produced an average of 36 fruits per plant (Figure 2E), and the productionin kg per plant showed an increase of 54.2% in *H. undatus* with 200 g of K_2O (Figure 2F).

Potassium fertilization also influenced fruit quality. In the 2013/2014 cycle, differences in the mass and diameters of the fruits were observed among theK_2O doses and between the two pitaya species. The largest fruit sizes produced by *H. undatus* were observed in plants fertilized with 124.3 and 133.3 g of K_2O, with increases of 79.8% in mass, 21.2% in transverse diameter and 22.6% in longitudinal diameter. In *H. polyrhizus*, fruits of larger sizes were observed in the plants fertilized with 127.4 and 138.3 g of K_2O, with increases of 36.1% in mass and of 12.9% and 6.9% in transverse and longitudinal diameters, respectively (Figure 3A and B).

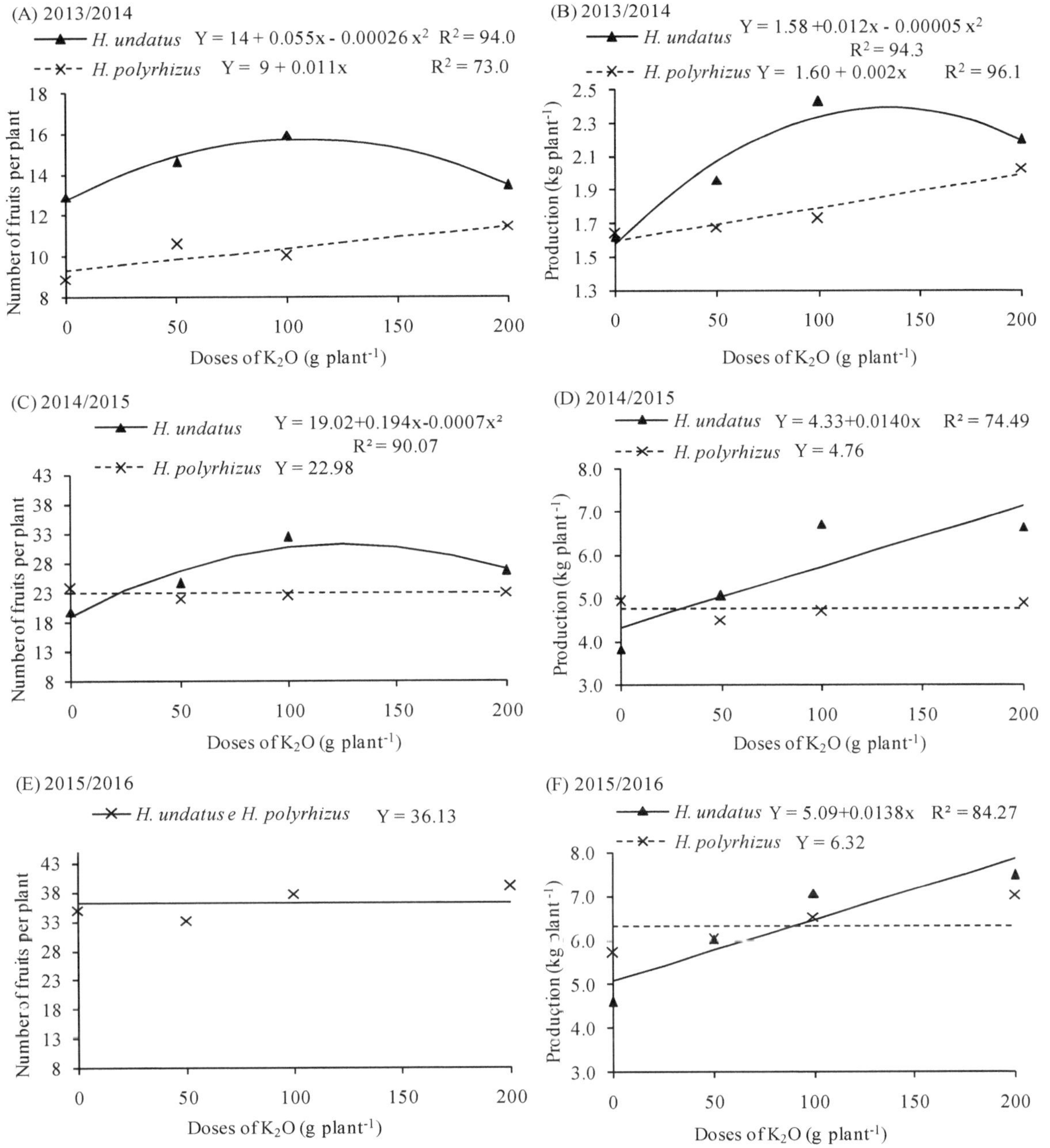

Figure 2. Number of fruits and production per plant of *H. undatus* and *H. polyrhizus* as a function of the dose of K_2O applied to the soil for the production cycles of 2013/2014, 2014/2015 and 2015/2016.

In the following two cycles, differences in the mass and diametera of the fruits were observed only in *H. undatus*. In the 2014/2015 cycle, fruits of *H. undatus* increased 25.3% in mass, 25.2% in transverse diameter and 25% in longitudinal diameter with 200 g of K_2O per plant (Figure 3C and D), while in the 2015/2016 cycle, the fruit mass increased 21.9%, the transverse diameter 22.2%, and the longitudinal diameter 21.9% in the plants also fertilized with 200 g of K_2O (Figure 3E and F).

Regarding the soluble solids, acidity and soluble solids/acidity ratio, differences were observed only between the species of pitaya. *H. polyrhizus* produced fruits with the highest soluble solids, whereas *H. undatus* produced the most acidic fruits. These differences between the species contributed to an increase in the soluble solids/acidity ratio in the fruits of *H. polyrhizus* (Table 2).

The levels of K in cladodes increased linearly as a function of the doses of K_2O applied to the soil (Figure 4). In June 2014, a 91% higher level was observed in plants fertilized with 200 g of K_2O when compared with the control (Figure 4A). In June 2015, the plants fertilized with 200 g of K_2O had a 117.5% increase in the K levels in cladodes (Figure 4B).

The levels of K in the soil also increased with the application of K_2O. In June 2014, the K concentration in the soil associated with plants without potassium fertilization was 87.56 mg dm^{-3}, while it was788.3 mg dm^{-3} in the soil associated with plants fertilized with 200 g K_2O (Figure 5A). In June 2015, the K levels increased from 103.9 mg dm^{-3} in the soil without potassium fertilization to 516.2 mg dm^{-3} in the soil fertilized with 200 g of K_2O (Figure 5B).

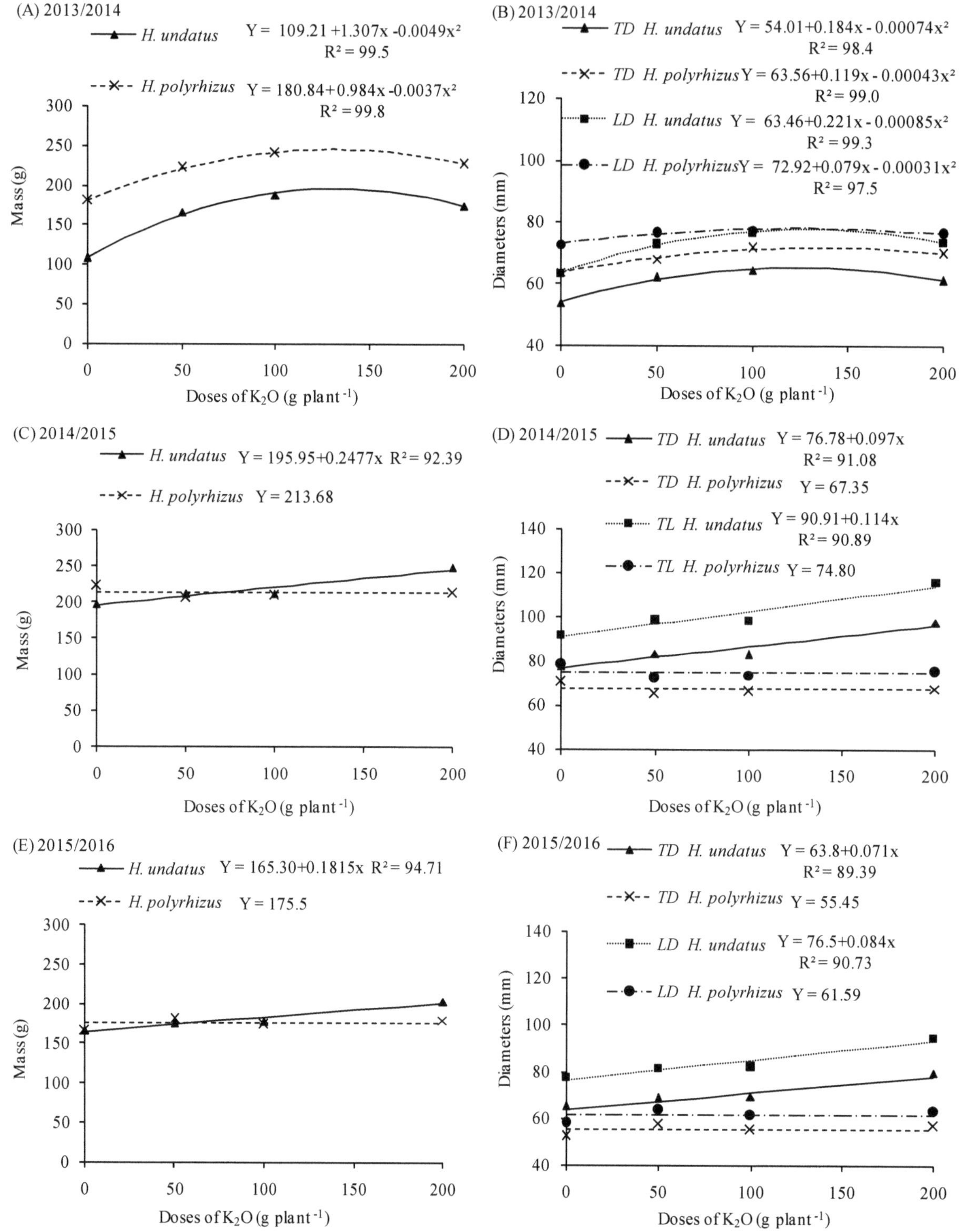

Figure 3. Mass, transverse diameter (TD) and longitudinal diameter (LD) of *H. undatus* and *H. polyrhizus* fruits as a function of the dose of K_2O applied to the soil for the production cycles of 2013/2014, 2014/2015, and 2015/2016.

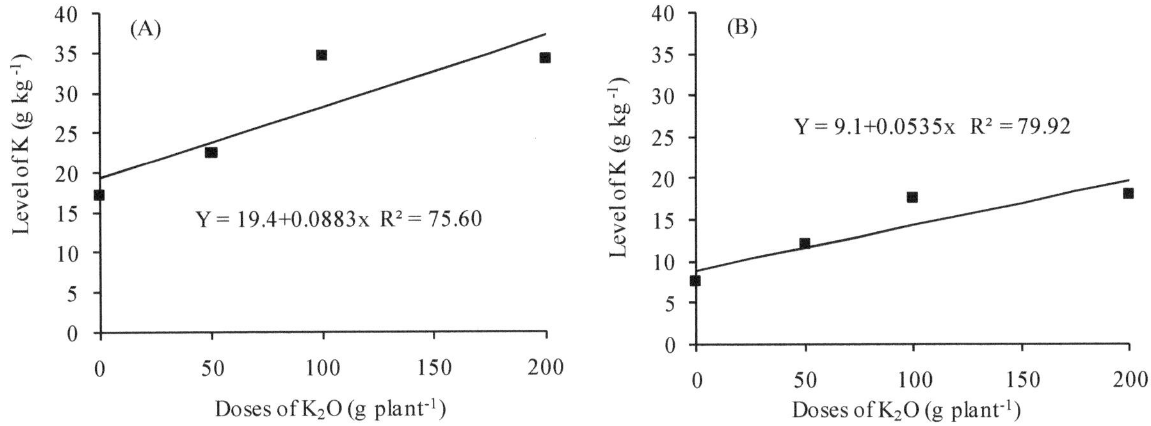

Figure 4. Levels of K in cladodes of *H. undatus* and *H. polyrhizus* as a function of the dose of K_2O applied to the soil in September 2014 (A) and September 2015 (B).

Table 2. Soluble solids (SS), titratable acidity (TA) and soluble solids/acidity ratio (RAT) of *H. undatus* and *H. polyrhizus* fruits as a function of the dose of K_2O applied to the soil in the production cycles of 2013/2014, 2014/2015 and 2015/2016.

K_2O doses (g plant^{-1})	SS (°Bx)		TA (%)		RAT	
	H. undatus	*H. polyrhizus*	*H. undatus*	*H. polyrhizus*	*H. undatus*	*H. polyrhizus*
			2013/2014			
0	15.54	16.87	0.45	0.27	35.23	61.76
50	15.17	16.41	0.46	0.24	34.88	68.65
100	16.58	17.67	0.45	0.32	37.46	56.46
200	15.42	16.34	0.47	0.24	35.70	68.13
Average	15.67 b	16.82 a	0.46 a	0.27 b	35.8 b	63.7 a
CV (%)	5.43		8.43		8.62	
			2014/2015			
0	16.62	20.25	0.29	0.08	65.24	266.72
50	16.87	19.62	0.25	0.10	73.29	191.63
100	17.98	19.62	0.26	0.09	71.74	234.40
200	16.62	21.50	0.27	0.12	66.61	176.46
Average	17.02 b	20.24 a	0.27 a	0.10 b	69.2 b	217.3 a
CV (%)	6.40		38.71		35.63	
			2015/2016			
0	16.65	19.75	0.51	0.21	33.14	98.97
50	17.50	21.37	0.51	0.24	34.51	90.55
100	17.50	22.00	0.49	0.25	35.87	88.93
200	17.75	21.00	0.53	0.22	36.36	76.66
Average	17.35 b	21.03 a	0.51 a	0.23 b	34.9 b	88.7 a
CV (%)	5.69		13.92		38.48	

Means followed by the same lowercase letter within a row do not differ from one another based on the F test ($p < 0.05$).

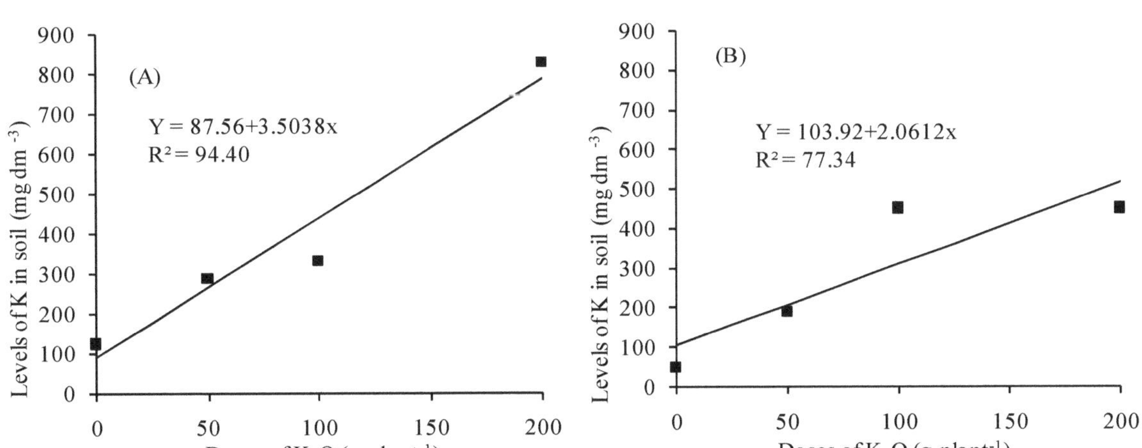

Figure 5. Levels of K at 0-20 cm in soil samples cultivated with *H. undatus* and *H. polyrhizus* as a function of the dose of K_2O applied to the soil in June 2014 (A) and June 2015 (B).

Discussion

H. undatus was more productive than *H. polyrhizus* under different K_2O applications because they are different species (Figure 2) with distinct needs in relation to potassium fertilization.

The response of the plants to potassium fertilization occurred because the soil had low K availability according to the initial analysis performed at the beginning of the orchard installation in this study (Table 1). Thus, the fertilization allowed K to become available to the plants, performing essential functions such as assimilation and storage of assimilates (Meurer, 2006).

An increase in pitaya production has been reported under different growing conditions and fertilizer combinations. In Bangladesh, *H. polyrhizus* yield increased from 12.88 t ha^{-1} to 31.64 t ha^{-1} with 250 g of K_2O, 540 g of N, and 310 g of P per plant (Chakma, Harunor Rashid, Roy, & Islam, 2014).

The response of plants to potassium fertilization depends on the species and the soil fertility conditions in which they are cultivated. In the present study, an increase in *H. polyrhizus* production occurred only in the 2013/2014 cycle, the first year after orchard establishment (Figure 2A and B). In the two harvests evaluated later, the K levels in the soil increased, even in the plants without K_2O application (Figure 2C, D, E, and F). This consequently contributed to the increased production. This observation may be explained by the addition of 5 L of bovine manure that was performed every year for all plants, contributing to the elevation in K levels in the soil, even for those plants without K_2O application.

The yields observed in the third year (8.7 t ha^{-1} for *H. undatus* and 7 t ha^{-1} for *H. polyrhizus*) are promising because in a *H. undatus* orchard aged 3 years old with bovine and chicken manure fertilization, the productivity was only 2.6 t ha^{-1} (Marques, Ramos, Araújo, & Moreira, 2012). In the present experiment, chemical fertilization contributed to causing the plants to begin production in the first year after planting, and in the second cycle, pitaya productivity was approximately 8 t ha^{-1} for *H. undatus* and 5.3 t ha^{-1} for *H. polyrhizus* (Figure 2D). These results make the importance of adequate fertilization to obtain satisfactory pitaya yields evident.

Pitaya orchards reach production stability between the 5th and 6th year of cultivation and, depending on the management employed, can producefor 15 to 20 years (Nerd, Tel-Zur, & Mizrahi, 2002). The average yield is 10 to 30 t ha^{-1}, varying according to the soil and climatic conditions, orchard age and cultivation techniques (Le Bellec, Vaillant, & Imbert, 2006).

The highest fruit size observed with the application of K_2O to the soil (Figure 3) indicates the contribution of fertilization to the improvement in fruit quality because K promotes the transport of photoassimilates to storage organs such as fruits and participates in the formation and translocation of sugars (Senapati & Santra, 2009).

An increase in *H. polyrhizus* fruit size with potassium had already been verified by Then (2013). This author verified that potassium application promoted a higher number of fruits classified in category A, with a mass of approximately 450 g per fruit, similar to that observed in the present study, in which fruits of greater mass were observed in plants fertilized with 200 g of K_2O. Size is an important characteristic in the market because the largest fruits garner the best prices.

Plants well supplied with K experience an increase in K in their tissues and consequently a reduction in water potential, favoring a greater accumulation of water and an increase in fruit mass (Montoya, Spinoia, Garcia, & Paredes, 2002).

Although the results observed in the last two cycles show linear responses, the species does not respond indefinitely to an increase in potassium fertilization. The responses are dependent on an internal balance of nutrients, which also involves physiological processes of incorporation (Marodin et al., 2010).

Regarding the chemical characteristics of the fruits, the observed differences indicate that each species shows intrinsic properties and the ability to synthesize different organic compounds since the pitaya species were submitted to the same management practices and presented different results in relation to the soluble solids and titratable acidity (Table 2).

These differences observed in soluble solids and acidity between the fruits of *H. polyrhizus* and *H. undatus* may be attributed to their genetic characteristics, which may be influenced by fertilization as well as the soil temperature and humidity in irrigated orchards or rainfall during fruit harvest (Figure 1).

Although there was no difference in soluble solids and titratable acidity with potassium fertilization, it should be noted that there was an increase in production and fruit size (Figures 2 and 3). Higher yields and fruit size require a greater amount of photoassimilates, and they are therefore distributed between the sink and the dilution of sugars in the fruit pulp. Nevertheless, the results observed were superior to *H. undatus* fertilized with different organic sources, which showed fruits with soluble solids from 11.87 to 12.95°Bx (Moreira et al., 2011), and to fruits harvested in commercial orchards in Nicaragua, with soluble solids from 7 to 11°Bx (Vaillant, Perez, Davila, Dornier, & Reynes, 2005).

It can be noticed that high levels of soluble solids improve fruit flavor, making them more attractive to consumers, who consider pitaya fruit as mild-tasting due to the low concentration of sugars and acidity (Cordeiro, Silva, Mizobutsi, Mizobutsi, & Mota, 2015).

Potassium fertilization did not directly influence the chemical characteristics of fruits, possibly due to the adequate nutrient content in the plants at all fertilization levels. Therefore, it is important to apply fertilization with this element in a balanced way because this nutrient can be lost by leaching. Moreover, excessive application may inhibit absorption due to competition in soil with other nutrients, such as calcium and magnesium (Andreotti, Rodrigues, Crusciol, Souza, & Büll, 2001). Another important point to be taken into consideration is that excess application contributes to higher production costs, thus reducing profit.

As for the increasing of K levels observed in cladodes due to the application of K_2O (Figure 4), these can be explained by the so-called "luxury consumption" element, which occurs when a nutrient that is available in abundance is often absorbed by plants in quantities that are higher than necessary.

The linear increase in K levels available in the soil was expected because this element was used in increasing quantities with the fertilizations (Figure 5), similar to the results obtained by Almeida, Corrêa, Crisostomo, Araújo, and Silva (2014).

These results demonstrate the need to carry out potassium fertilization, which favors an increase in K levels in the dry matter of plants and also promotes higher production and an improvement in fruit quality.

Conclusion

Potassium fertilization increased production and provided a better quality of pitaya fruits.

K_2O fertilization should be from 106.0 to 133.3 g K_2O in the first year and 200 g K_2O in the second and third years after planting.

References

Almeida, E. I. B., Corrêa, M. C. M., Crisostomo, L. A., Araújo, N. A., & Silva, J. C. V. (2014). Nitrogênio e potássio no crescimento de mudas de pitaia [*Hylocereus undatus* (Haw.) Britton & Rose]. *Revista Brasileira de Fruticultura, 36*(4), 1018-1027. doi: 10.1590/0100-2945-296/13.

Andreotti, M., Rodrigues, J. D., Crusciol, C. A. C., Souza, E. C. A., & Büll, L. T. (2001). Crescimento do milho em função da saturação por bases e da adubação potássica. *Scientia Agricola, 58*(1), 145-150. doi: 10.1590/S0103-90162001000100022

Bastos, D. C., Pio, R., Scarpare Filho, J. A., Libardi, M. N., Almeida, L. F. P., Galuchi, T. P. D., & Bakker, S. T. (2006). Propagação de pitaia vermelha por estaquia. *Ciência e Agrotecnologia, 30*(6), 1106-1109. doi: 10.1590/S1413-70542006000600009

Cavalcante, I. H. L., Martins, A. B. G., Silva Júnior, G. B., Rocha, L. F., Falcão Neto, R., & Cavalcante, L. F. (2011). Adubação orgânica e intensidade luminosa no crescimento e desenvolvimento inicial da pitaya em Bom Jesus-PI. *Revista Brasileira de Fruticultura, 33*(3), 970-983. doi: 10.1590/S0100-29452011005000086

Chakma, S. P., Harunor Rashid, A. S. M., Roy, S., & Islam, M. (2014). Effect of NPK doses on the yield of dragon fruit (*Hylocereus costaricensis* [F. A. C. Weber] Britton & Rose) in Chittagong Hill Tracts. *American-Eurasian Journal of Agricultural & Environmental Sciences, 14*(6), 521-526. doi: 10.5829/idosi.aejaes.2014.14.06. 12346

Cordeiro, M. H. M., Silva, J. M., Mizobutsi, G. P., Mizobutsi, E. H., & Mota, W. F. (2015).Caracterização física, química e nutricional da pitaia-rosa de polpa vermelha. *Revista Brasileira de Fruticultura, 37*(1), 20-26. doi:1 0.1590/0100-2945-046/14

Corrêa, M. C. M., Almeida, E. I. B., Marques, V. B., Vale, J. C., & Aquino, B. F. (2014). Crescimento inicial de pitaia em função de combinações de doses de fósforo-zinco. *Revista Brasileira de Fruticultura, 36*(1), 23-38. doi: 10.1590/0100-2945-297/13

Le Bellec, F., Vaillant, F., & Imbert, E. (2006). Pitahaya (*Hylocereus* spp.): a new fruit crop, a market with a future. *Fruits, 61*(4), 237-250. doi: 10.1051/fruits: 2006021

Marodin, J. C., Resende, J. T. V., Morales, R. G. F., Camargo, C. K., Camargo, L. K. P., & Pavinato, P. S. (2010). Qualidade físico-química de frutos de morangueiro em função da adubação potássica. *Scientia Agraria Paranaensis, 9*(3), 50-57.

Marques, V. B., Moreira, R. A., Ramos, J. D., Araújo, N. A., & Silva, F. O. R. (2011). Fenologia reprodutiva de pitaia vermelha no município de Lavras, MG. *Ciência Rural, 41*(6), 984-987. doi: 10.1590/S0103-84782011005000071

Marques, V. B., Ramos, J. D., Araújo, N. A., & Moreira, R. A. (2012). Custo de produção e rentabilidade na cultura da pitaia sob o efeito de adubação orgânica. *Científica, 40*(2), 138-149. doi: 10.15361/1984-5529.2012v40n2p138+-+149

Meurer, E. J. (2006). *Nutrição Mineral das Plantas*. Viçosa, MG: Sociedade Brasileira de Ciência do Solo.

Montoya, R. B., Spinoia, A. G., Garcia, P. S., & Paredes, D. G. (2002). Demanda de potasio del tomate tipo sadette. *Terra Latino americana, 20*(4), 391-399.

Moreira, R. A., Ramos, J. D., Araújo, N. A., & Marques, V. B. (2011). Produção e qualidade de frutos de pitaia-vermelha com adubação orgânica e granulado bioclástico. *Revista Brasileira de Fruticultura, 33*(1), 762-766. doi: 10.1590/S0100-29452011000500106

Nerd, A., Tel-Zur, N., & Mizrahi, Y. (2002). Fruits of vine and columnar cacti. In P. S. Nobel (Ed.), *Cacti: biology and uses* (p. 185- 197). Berkeley, US: University of California Press. doi: 10.1525/california/9780520231573.003.0011

Senapati, H. K., & Santra, G. H. (2009). *Potassium management in vegetables, spices and fruit crops. Optimizing crop production through judicious use of K.* Bhubaneswar, Orissa, India: Bijupattnaik Hall, OUAT

Souza, F. B. M., Pio, R., Coelho, V. A. T., Rodas, C. L., & Silva, I. P. (2015). Sintomas visuais de deficiência de macronutrientes, boro e ferro e composição mineral de amoreira-preta. *Pesquisa Agropecuária Tropical, 45*(2), 241-248.

Then, T. H. (2013). The effects of foliar fertilizers on the red pitaya (*Hylocerus polyrhizus*) fruit weight. *Acta Horticulturae, 984*(25), 227-230. doi: 10.17660/ActaHortic.2013.984.25

Vaillant, F., Perez, A., Davila, I., Dornier, M., & Reynes, M. (2005). Colorant and antioxidant properties of red pitahaya (*Hylocereus* sp.). *Fruits, 60*(1), 1-7. doi: 10.1051/ fruits:2005007

Permissions

All chapters in this book were first published in ASA, by Editora da Universidade Estadual de Maringá; hereby published with permission under the Creative Commons Attribution License or equivalent. Every chapter published in this book has been scrutinized by our experts. Their significance has been extensively debated. The topics covered herein carry significant findings which will fuel the growth of the discipline. They may even be implemented as practical applications or may be referred to as a beginning point for another development.

The contributors of this book come from diverse backgrounds, making this book a truly international effort. This book will bring forth new frontiers with its revolutionizing research information and detailed analysis of the nascent developments around the world.

We would like to thank all the contributing authors for lending their expertise to make the book truly unique. They have played a crucial role in the development of this book. Without their invaluable contributions this book wouldn't have been possible. They have made vital efforts to compile up to date information on the varied aspects of this subject to make this book a valuable addition to the collection of many professionals and students.

This book was conceptualized with the vision of imparting up-to-date information and advanced data in this field. To ensure the same, a matchless editorial board was set up. Every individual on the board went through rigorous rounds of assessment to prove their worth. After which they invested a large part of their time researching and compiling the most relevant data for our readers.

The editorial board has been involved in producing this book since its inception. They have spent rigorous hours researching and exploring the diverse topics which have resulted in the successful publishing of this book. They have passed on their knowledge of decades through this book. To expedite this challenging task, the publisher supported the team at every step. A small team of assistant editors was also appointed to further simplify the editing procedure and attain best results for the readers.

Apart from the editorial board, the designing team has also invested a significant amount of their time in understanding the subject and creating the most relevant covers. They scrutinized every image to scout for the most suitable representation of the subject and create an appropriate cover for the book.

The publishing team has been an ardent support to the editorial, designing and production team. Their endless efforts to recruit the best for this project, has resulted in the accomplishment of this book. They are a veteran in the field of academics and their pool of knowledge is as vast as their experience in printing. Their expertise and guidance has proved useful at every step. Their uncompromising quality standards have made this book an exceptional effort. Their encouragement from time to time has been an inspiration for everyone.

The publisher and the editorial board hope that this book will prove to be a valuable piece of knowledge for researchers, students, practitioners and scholars across the globe.

List of Contributors

Ana Paula Cecatto, Eunice Oliveira Calvete, Alexandre Augusto Nienow, Rosiani Castoldi da Costa, Heloísa Ferro Constâncio Mendonça and Aislam Celso Pazzinato
Faculdade de Agronomia e Medicina Veterinária, Universidade de Passo Fundo, Passo Fundo, Rio Grande do Sul, Brazil

Luiziene Soares Alves
Programa de Pós-graduação em Agronomia, Ciência do Solo, Universidade Federal Rural do Rio de Janeiro, Seropédica, Rio de Janeiro, Brazil

Elvia Mariam Lis Martinez Stark, Everaldo Zonta and Manlio Silvestre Fernandes
Departamento de Solos, Universidade Federal Rural do Rio de Janeiro, Seropédica, Rio de Janeiro, Brazil

André Marques dos Santos and Sonia Regina de Souza
Departamento de Química, Laboratório de Bioquímica de Plantas, Universidade Federal Rural do Rio de Janeiro, BR-465, km 7, 23897-000, Seropédica, Rio de Janeiro, Brazil

Vilson Benz, Alessandro Dal'Col Lúcio and Sidinei José Lopes
Programa de Pós-graduação em Agronomia, Departamento de Fitotecnia, Universidade Federal de Santa Maria, Av. Roraima, 1000, 97105-900, Camobi, Santa Maria, Rio Grande do Sul, Brazil

Daniel Fernandes da Silva
Programa de Pós-graduação em Botânica Aplicada, Departamento de Biologia, Universidade Federal de Lavras, 37200-000, Lavras, Minas Gerais, Brazil

Rafael Pio and Joyce Dória Rodrigues Soares
Departamento de Agricultura, Universidade Federal de Lavras, Lavras, Minas Gerais, Brazil

Paulyene Vieira Nogueira and Pedro Maranha Peche
Programa de Pós-graduação em Fitotecnia, Departamento de Agricultura, Universidade Federal de Lavras, Lavras, Minas Gerais, Brazil

Fabíola Villa
Centro de Ciências Agrárias, Universidade Estadual do Oeste do Paraná, Marechal Cândido Rondon, Paraná, Brazil

Thiago Palhares Farias, André Trochmann and Fatima Maria Souza Moreira
Departamento de Ciências do Solo, Universidade Federal de Lavras, Lavras, Minas Gerais, Brazil

Bruno Lima Soares
Departamento de Agricultura, Universidade Federal de Lavras, Lavras, Minas Gerais, Brazil

Genivaldo David de Souza-Schlick
Programa de Pós-graduação em Agronomia-Agricultura, Faculdade de Ciências Agronômicas, Universidade Estadual Paulista, Botucatu, São Paulo, Brazil

Rogério Peres Soratto and Mauricio Dutra Zanotto
Departamento de Produção e Melhoramento Vegetal, Faculdade de Ciências Agronômicas, Universidade Estadual Paulista, Rua José Barbosa de Barros, 1780, 18610-307, Botucatu, São Paulo, Brazil

Daniela Pinto Gomes, Daniel Fonseca de Carvalho, Marinaldo Ferreira Pinto and David da Cunha Valença
Departamento de Engenharia, Universidade Federal Rural do Rio de Janeiro, Br-465, km 7, 23897-000, Seropédica, Rio de Janeiro, Brazil,

Leonardo Oliveira Medici
Departamento de Ciências Fisiológicas, Universidade Federal Rural do Rio de Janeiro, Seropédica, Rio de Janeiro, Brazil

Gustavo Henrique Freiria, Josemeyre Bonifácio da Silva and Cássio Egídio Cavenaghi Prete
Programa de Pós-graduação em Agronomia, Centro de Ciências Agrárias, Universidade Estadual de Londrina, Rod. Celso Garcia Cid, Pr-445, Km 380, Campus Universitário, Londrina, Paraná, Brazil

Wilmar Ferreira Lima
Instituto Agronômico do Paraná, Londrina, Paraná, Brazil

Rodrigo Santos Leite and José Marcos Gontijo Mandarino
Empresa Brasileira de Pesquisa Agropecuária, Centro Nacional de Pesquisa de Soja, Embrapa/Soja, Londrina, Paraná, Brazil

Renan Soares de Souza, Roberto Rezende, Tiago Luan Hachmann, Cláudia Salim Lozano, André Felipe Barion Alves Andrian and Paulo Sérgio Lourenço de Freitas
Departamento de Agronomia, Universidade Estadual de Maringá, Av. Colombo, 5790, 87020-900, Maringá, Paraná, Brazil

Camila Karen Reis Barbosa, Fernando Luiz Finger and Vicente Wagner Dias Casali
Departamento de Fitotecnia, Universidade Federal de Viçosa, Av. Peter Henry Rolfs, s/n, 36570-900, Viçosa, Minas Gerais, Brazil

Núbia Maria Correia
Departamento de Fitossanidade, Universidade Estadual Paulista "Julio de Mesquita Filho", Via de Acesso Prof. Paulo Donato Castellane, s/n., 14884-900, Jaboticabal, São Paulo, Brazil

Bruno Daniel and Marcela Basile Leite
Curso de Agronomia, Universidade Estadual Paulista "Julio de Mesquita Filho", Jaboticabal, São Paulo, Brazil

Edson Shigueaki Nomura, Erval Rafael Damatto Junior and Eduardo Jun Fuzitani
Agência Paulista de Tecnologia dos Agronegócios, Polo Regional Vale do Ribeira, Registro, São Paulo, Brazil

Francine Lorena Cuquel
Universidade Federal do Paraná, Curitiba, Paraná, Brazil

Ana Lúcia Borges
Empresa Brasileira de Pesquisa Agropecuária, Embrapa Mandioca e Fruticultura, Cruz das Almas, Bahia, Brazil

Francisca Ligia de Castro Machado
Instituto Federal de Educação, Ciência e Tecnologia, Campus Tianguá, Rod. CE-075, s/n, 62320-000, Aeroporto, Tianguá, Ceará, Brazil

Janaína de Paula da Costa
Departamento de Engenharia de Pesca, Universidade Federal do Ceará, Fortaleza, Ceará, Brazil

Adunias dos Santos Teixeira
Departamento de Engenharia Agrícola, Universidade Federal do Ceará, Fortaleza, Ceará, Brazil

José Maria Correia da Costa
Departamento de Tecnologia de Alimentos, Universidade Federal do Ceará, Fortaleza, Ceará, Brazil

Welington Secundino
Instituto Capixaba de Pesquisa, Assistência Técnica e Extensão Rural, São Mateus, Espírito Santo, Brazil

Rodrigo Sobreira Alexandre, Edilson Romais Schmildt, Omar Schmildt and Gizele Cristina Magevski
Centro Universitário Norte do Espírito Santo, Universidade Federal do Espírito Santo, Rod. BR-101 Norte, km 60, 29932-540, São Mateus, Espírito Santo, Brazil

João Paulo Rodrigues Martins
Departamento de Biologia, Universidade Federal de Lavras, Lavras, Minas Gerais, Brazil

Mayara Neves Santos Guedes
Departamento de Agricultura, Universidade Federal de São João del-Rei, Sete Lagoas, Minas Gerais, Brazil

Rafael Pio
Departamento de Agricultura, Universidade Federal de Lavras, Lavras, Minas Gerais, Brazil

Luana Aparecida Castilho Maro
Empresa de Pesquisa Agropecuária e Extensão Rural de Santa Catarina, Itajaí, Santa Catarina, Brazil

Fabíola Fonseca Lage, Celeste Maria Patto de Abreu and Adelir Aparecida Saczk
Departamento de Química, Universidade Federal de Lavras, Lavras, Minas Gerais, Brazil

Tatiana Pires Barrella, Mário Puiatti and Ricardo Henrique Silva Santos
Departamento de Fitotecnia, Centro de Ciências Agrárias, Universidade Federal de Viçosa, Av. Peter Henry Rolfs, s/n, 36570-000, Viçosa, Minas Gerais, Brazil

Paulo Roberto Cecon
Departamento de Informática, Universidade Federal de Viçosa, Viçosa, Minas Gerais, Brazil

Gabriel Alves Botelho de Mello, Daniel Fonseca de Carvalho, Aldir Carlos Silva, Daniela Pinto Gomes and Marinaldo Ferreira Pinto
Departamento de Engenharia, Universidade Federal do Rio de Janeiro, BR-465, km 7, 23897-000, Seropédica, Rio de Janeiro, Brazil

Leonardo Oliveira Medici
Departamento de Ciências Fisiológicas, Universidade Federal do Rio de Janeiro, Seropédica, Rio de Janeiro, Brazil

Kerlley Cristina de Assis
Empresa Brasileira de Pesquisa Agropecuária, Embrapa Clima Temperado, Pelotas, Rio Grande do Sul, Brazil

Flávia Dionísio Pereira
Cooperativa Agroindustrial dos Produtores Rurais do Sudoeste Goiano, Rio Verde, Goiás, Brazil

Juliana Silva Rodrigues Cabral and Fabiano Guimarães Silva
Instituto Federal de Educação, Ciência e Tecnologia Goiano, Rod. Sul Goiana, km 1, Rio Verde, Goiás, Brazil

José Waldemar Silva
Universidade Federal de Uberlândia, Uberlândia, Minas Gerais, Brazil

Silvia Correa dos Santos
Universidade Federal de Goiás, Jataí, Goiás, Brazil

Lusiene Barbosa de Sousa, Newton Pereira Stamford, Wagner da Silva Oliveira, Emmanuella Vila Nova da Silva, Marllon dos Santos Martins and Carolina Etienne de Rosália e Silva Santos
Departamento de Agronomia, Universidade Federal Rural de Pernambuco, Rua Dom Manuel de Medeiros, s/n, Dois Irmãos, 52171-900, Recife, Pernambuco, Brasil

Edmilson José Ambrosano, Patrícia Patri and Celina Maria Henrique
Agência Paulista de Tecnologia dos Agronegócios, Pólo Regional Centro Sul, Rodovia SP-127, km 30, 13400-970, Piracicaba, São Paulo, Brazil

Gabriela Cristina Salgado and Paulo César Tavares de Melo
Departamento Produção Vegetal, Escola Superior de Agricultura "Luiz de Queiroz", Universidade de São Paulo, Piracicaba, São Paulo, Brazil

Ivani Pozar Otsuk
Agência Paulista de Tecnologia dos Agronegócios, Instituto de Zootecnia, Nova Odessa, São Paulo, Brazil

Marines Batalha Moreno Kirinus, Caroline Farias Barreto, Marcelo Barbosa Malgarim and José Carlos Fachinello
Departamento de Fitotecnia, Faculdade de Agronomia Eliseu Maciel, Universidade Federal de Pelotas, Centro Agropecuário, Avenida Eliseu Maciel, s/n, Capão do Leão, Rio Grande do Sul, Brazil

Pricila Santos da Silva
Departamento de Fitotecnia, Faculdade de Agronomia, Universidade do Estado de Santa Catarina, Centro Agroveterinário, Lages, Santa Catarina, Brazil

Roberto Pedroso de Oliveira
Empresa Brasileira de Pesquisa Agropecuária, Embrapa Clima Temperado, Pelotas, Rio Grande do Sul, Brazil

Guiliano Rangel Alves, Itamar Rosa Teixeira, Francisco Ramos Melo and Raniele Tadeu Guimarães Souza
Departamento de Engenharia Agrícola, Universidade Estadual de Goiás, Campus Henrique Santillo, BR-153, Fazenda Barreiro do Meio, 75132-400, Anápolis, Goiás, Brazil

Alessandro Guerra Silva
Departamento de Agronomia, Universidade de Rio Verde, Rio Verde, Goiás, Brazil

Marcelo Angelo Campagnolo
Universidade Estadual do Oeste do Paraná, Marechal Cândido Rondon, Paraná, Brazil.

Rafael Pio
Departamento de Agricultura, Universidade Federal de Lavras, Lavras, Minas Gerais, Brazil

Leandro Fuzaro, Nayara Luchini Xavier and Flávia Andrea Nery Silva
Instituto de Ciências Agrárias, Universidade Federal de Uberlândia, Avenida Amazonas, s/n°, Bloco 2E, Sala 2E 152, Campus Umuarama, Uberlândia, Minas Gerais, Brazil

Fábio Janoni Carvalho and Vanessa Andaló
Instituto de Ciências Agrárias, Universidade Federal de Uberlândia, Campus Monte Carmelo, Monte Carmelo, Minas Gerais, Brazil

Stephan Malfitano Carvalho
Departamento de Entomologia, Universidade Federal de Lavras, Campus Universitário, Lavras, Minas Gerais, Brazil

Carlos Sérgio Tiritan, Cláudia Regina Minutti and Juliano Carlos Calonego
Centro de Ciências Agrárias, Universidade do Oeste Paulista, Presidente Prudente, São Paulo, Brazil

Diego Henriques Santos
Companhia de Desenvolvimento Agrícola de São Paulo, Centro de Negócios de Presidente Prudente, Rod. Raposo Tavares, km 564, 19053-205, Presidente Prudente, São Paulo, Brazil

José Salvador Simoneti Foloni
Centro Nacional de Pesquisa de Soja, Empresa Brasileira de Pesquisa Agropecuária, Londrina, Paraná, Brazil

Fernando Shintate Galindo, Salatiér Buzetti, Marcelo Carvalho Minhoto Teixeira Filho, Elisângela Dupas and Mariana Gaioto Ziolkowski Ludkiewicz
Departamento de Fitossanidade, Engenharia Rural e Solos, Faculdade de Engenharia, Universidade Estatual Paulista "Julio de Mesquita Filho", Rua Monção, 830, 15385 000, Ilha Solteira, São Paulo, Brazil

Tatiane Beloni
Escola Superior de Agricultura "Luiz de Queiroz", Universidade de São Paulo, Piracicaba, São Paulo, Brazil

Maria Lilia de Souza Neta, Francisco de Assis de Oliveira, Salvador Barros Torres, Antonia Adailha Torres Souza, Dennis Darkyan Almeida da Silva and Sandy Thomaz dos Santos
Centro de Ciências Agrárias, Universidade Federal Rural do Semi-Árido, Av. Francisco Mota, 572, 59625-900, Mossoró, Rio Grande do Norte, Brazil

Néstor Antonio Heredia Zárate, Maria do Carmo Vieira, Fabiane Cargnin Faccin, Rosimeire Pereira Gassi and Anna Luiza Farias dos Santos
Universidade Federal da Grande Dourados, Faculdade de Ciências Agrárias, Rod. Dourados a Ithaum, km 12, 79804-970, Dourados, Mato Grosso do Sul, Brazil

Danilo Augusto Heredia Vieira
Programa de Pós-graduação em Administração, Universidade Estadual de Maringá, Maringá, Paraná, Brazil

Gaspar Henrique Korndorfer, Hamilton Seron Pereira, Ivaniele Nahas Duarte, Laerte Rocha Neves Pinto and Luciano Ferreira da Fonseca
Universidade Federal de Uberlândia, Av. Amazonas, s/n., Bloco 2E, S. 122, Uberlândia, Minas Gerais, Brazil

Denison Ramalho Fernandes, Maria do Céu Monteiro da Cruz, Josimara Mendes Rabelo and Jéssica de Oliveira
Instituto Federal do Norte de Minas, Fazenda São Geraldo, s/n, Estrada Januária, km 06, 39480-000, Januária, Minas Gerais, Brazil.

Rodrigo Amato Moreira
Universidade Federal dos Vales do Jequitinhonha e Mucuri, Diamantina, Minas Gerais, Brazil

Index

CPSIA information can be obtained
at www.ICGtesting.com
Printed in the USA
LVHW060855030122
707669LV00002B/162

9 781641 165259